HISTORY OF THE EARTH

A SERIES OF BOOKS IN GEOLOGY

EDITORS: James Gilluly and A. O. Woodford

Strata	Thick-ness	Springs	Fossils, Petrifactions, etc. etc.	Descriptive Characters & Situations.
1. Chalk	300	Intermitting on the Downs	Echinites, Pyrites, Mytilites, Dentalia, Funnel-shaped Corals, & Madrepores, Nautilites, Strombites, Cochliae, Ostreae, Serpulae	Strata of Silex, imbedded.
2. Sand	70			The fertile vales intersecting Salisbury Plain & the Downs.
3. Clay	30	Between the Black Dog and Berkley		
4. Sand & Stone	30	Hinton, Norton, Woolverton, Bradford Leigh		Imbedded is a thin stratum of calcareous Grit. The stones flat, smooth and rounded at the edges.
5. Clay	15			
6. Forest Marble	10		A Mass of Anomiae and High-waved Cockles with calcareous Cement	The Cover of the upper Bed of Freestone or Oolyte.
7. Freestone	60		Scarcely any fossils besides the Coral	Oolyte resting on a thin bed of Coral. – Prior Park, Southstoke, Twinny, Winsley, Farley Castle, Westwood, Berfield, Conkwell, Monkton Farley, Colehorn, Marshfield, Coldashton.
8. Blue Clay	6	} Above Bath		
9. Yellow Clay	8			
10. Fuller's Earth	6			Visible at a distance by the slips on the declivities of the Hills round Bath.
11. Bastard Ditto and Sundries	80		Striated Cardia, Mytilites, Anomiae, Pundibs and Duck-muscles.	
12. Freestone	30		Top-covering Anomiae with calcareous Cement, Strombites, Ammonites, Nautilites, Cochliae Hippocephaloides, fibrous Shell resembling Amianth, Cardia, prickly Cockle Mytilites, lower Stratum of Coral, large Scollop, Nidus of the Muscle with its Cables	Lincombe, Devonshire Buildings, Englishcombe, Englishbatch, Wilmerton, Dunkerton, Coomhay, Monkton Combe, Wellow, Mitford, Stoke, Freshford, Claverton, Bathford, Batheaston, and Hampton, Charlcombe, Swainswick, Tadwick, Langridge.
13. Sand	30	Round Bath	Ammonites, Belemnites	Sand Burs
14. Marle Blue	40		Pectenites, Belemnites, Gryphites, High-waved Cockles	Ocre Balls. – Mineral Springs of Lincombe, Middle Hill, Cheltenham.
15. Lias Blue	25		Same as the Marl with Nautilites, Ammonites, Dentalia, and fragments of the Enchrinni.	The fertile Marl lands of Somersetshire. Twerton, Newton, Preston, Clutton, Stanton Prior, Timsbury, Paulton Marksbury, Farmborough, Corston, Hunstreat, Burnet, Keynsham, Whitchurch, Salford, Kelston, Weston, Pucklechurch, Quencharlton, Norton-malreward Knowle, Charlton, Kilmersdon, Babington.
16. Ditto White	15			
17. Marl stone, Indigo and black Marle	15		Pyrites and Ochre	A rich Manure
18. Red-ground	180		No fossil known	Pits of Ruddle. Beneath this bed no fossil, shells, or animal remains are found; above it no vegetable Impressions. The Waters of this Stratum petrify in the trunks in which they are conveyed, so as to fill them, in about fifteen years, with red Watride, which takes a fine polish. – Highlittleton.
19. Milstone				
20. Pennant Stone			Impressions of unknown plants resembling Equisetum	Fragments of Coal and Iron Nodules. – Hanham, Brislington, Mangotsfield, Downend, Winterbourn, Forest of Dean, Pensford, Pablow, Chelwood, Cumptondando, Hallowsrow Near Stratford on Avon, Stonebench on the Severn, four Miles from Gloucester.
21. Grays				
22. Cliff			Impressions of ferns, olive, Stellate plants, Threnax parviflora or Dwarf fan Palm of Jamaica	Stourbridge or Fire-clay.
23. Coal				

William Smith's Table of Strata near Bath, England, dictated in 1799 to the Reverend Joseph Townsend.

SECOND EDITION

HISTORY OF
THE EARTH

AN INTRODUCTION TO HISTORICAL GEOLOGY

Bernhard Kummel
HARVARD UNIVERSITY

W. H. FREEMAN AND COMPANY
San Francisco

Printed in the United States of America
Library of Congress Catalog Card Number: 73–114579
International Standard Book Number: 0–7167–0249–5

9 8 7 6 5 4 3 2 1

to my parents

PREFACE
to the Second Edition

In the preface to the first edition of this book I stressed the complexity and vastness of the data relevant to the study of the earth's history, and pointed out that there was much difference of opinion about how the subject should be presented. At the same time, I expressed my belief that no matter what the emphasis—physical, biological, or geochemical—a broad understanding of the major features of the entire earth was indispensible. Events of the past decade have confirmed this conviction. It is clearly evident from even a cursory survey of the recent literature of geology that to adopt anything short of a worldwide perspective is bound to result in a distorted view. Continental drift and zoogeography, to name only two of the many active areas of current interest, cannot be understood if the study of geological history is confined to a limited part of the earth's surface. Thus, the goal of this book is to present, in the briefest form consistent with accuracy, a "classical" treatment of the geological history of each of the continents—to provide the basic background for any student who wishes to understand what geological history is and why it is studied. In any course the specialized interest of the instructor will determine the details that he brings out in his lectures, and it is his ability, knowledge, and enthusiasm that make the course a success. But unless his listeners have been given an opportunity to acquire enough knowledge and information to provide a framework within which to evaluate his discussion, he will be fighting a losing battle. It was my intention, in writing this book, to provide such knowledge and information, and the reception of the first edition has indicated that I have been at least partly successful.

I have had two distinct readers in mind while writing: the student in his first year of geology, and the advanced student. Each of these will of course read the book differently. The first-year student's reading should be primarily passive—that is, he should read all the text but should not get bogged down in details; he must be aware of them, but only to the degree that they contribute to the general picture. It is the overall tectonic evolution of the continents, the patterns of paleogeography, and the broad features of the evolution of life that are important to him. The advanced student, on the other hand, may well find among the details much that is new to him and that will stimulate him to further study; he will tackle the book as a whole or in pieces, depending on his interest. The stratigraphic charts of Appendix II and the bibliographies are intended primarily for him.

The past decade has been extremely significant for historical geology. Before 1960 geologists had a fairly comprehensive understanding of the Phanerozoic history of North America and Europe, but the data for most other parts of the world was so incomplete that synthesis was at best difficult, and understanding all but impossible. This is no longer true. Extensive programs of geological exploration sponsored by federal and state governmental agencies and private organizations have added significantly to our geological knowledge of such parts of the world as Antarctica, South America, the Middle East, the Himalayas, China, Japan, Australia, Africa, and Siberia. There is still much to be discovered, but at least we are at a stage where meaningful synthesis is possible. One manifestation of this is the renewed interest in the theory of continental drift, which many geologists believe to be substantiated by recent discoveries.

Much, too, has been learned about the Precambrian. Perfection of radiometric dating techniques and their wide application have added greatly to our understanding of the large Precambrian terranes, and the continuing study of the paleontology of the Precambrian has given us a completely new insight into the history of life.

In this edition I have tried to take into account as many of the new developments as possible. Innumerable minor changes have been made in the text, many new illustrations and maps have been added, and many sections are new or have been completely rewritten. In Chapter 1 the section on the radiometric time scale is new, and in Chapter 2 the section on Precambrian life has been greatly expanded. Almost all of Chapter 4, on the Precambrian Era, has been rewritten, as have the parts of Chapters 5, 8, and 12 that deal with California and Nevada. Only those sections of Chapters 6, 9, and 13 that concern Europe are essentially unchanged; those on the rest of the world are almost entirely new. Chapter 11, on Gondwana, has been extensively revised, and the section of Chapter 14 on fossil primates has been rewritten. The results of recent research on fossil man and the Pleistocene climate have been incorporated into Chapter 15. Many of the former gaps in the correlation charts (Appendix II) have been filled, all charts have been brought up to date, and a separate index to them is included.

This edition contains approximately one hundred new illustrations. The source of each illustration (old and new) is given in its accompanying legend, but many persons have been so helpful that I must add their names to those in the preface to the first edition; their kindness is greatly appreciated: E. S. Barghoorn, P. C. Bateman, W. B. N. Berry, R. S. Dietz, A. G. Fischer, C. S. Fleming, MacLain Forman, John Haller, John Holden, Ralph Imlay, E. G. Kauffman, A. R. Palmer, P. E. Playford, Pamela Robinson, J. M. Schopf, J. W. Schopf, E. L. Simons, N. F. Sohl, J. P. Sprinkle, F. G. Stehli, W. Stürmer, R. Trumpy, and F. J. Vine.

Of the colleagues whose criticisms and discussions facilitated the preparation of the first edition—to whom I am continuingly grateful—M. P. Billings, B. Patterson, A. S. Romer, and R. Siever gave me the benefit of their special knowledge when I prepared the revisions and additions to this one, as did John Haller, J. F. Hays, and J. B. Thompson. Among the many students at Harvard who have stimulated me to continue my own study of earth history, W. J. Koch and J. P. Sprinkle have been decidedly devilish devil's advocates. Finally I wish to acknowledge the assistance of Miss Victoria Kohler, who patiently helped in every way throughout the preparation of this edition. The long and tedious chore of typing was admirably done by Mrs. Agnes Pilot.

June 1970   *Bernhard Kummel*

PREFACE
to the First Edition

Few sciences appeal more to the beginning student than that which deals with the earth and its history. The subject is so vast, however, dealing as it does with physical, chemical, and biological data, that there is much difference of opinion as to how it should be presented.

In few other college science courses do the specialty and personality of the lecturer and the geographic location of the college have so great an influence. The paleontologist, the physical stratigrapher, the geochemist—each tends to spend much time on the phase that he knows and likes best; and a college near good exposures of rock may base its course in historical geology on the local geology. It is excellent thus to utilize the special knowledge and ingenuity of the instructor and the local geology; yet one should guard against a provincial frame-

work of thought. Historical geology, dealing as it does with the history of the whole earth, is anything but a provincial subject. A cosmopolitan point of view is required if the problems of the earth's history are to be seen in their proper perspective.

If the lectures of a course in historical geology reflect—quite properly—not only the personality and special interest of the instructor but also the local geologic setting, a textbook that provides a broad view becomes an essential part of the course. Ideally, of course, a text should complement the lectures. This text is designed to give the student what I consider the proper background reading for nearly any course of lectures in historical geology. The introductory chapters outline the nature of the record, the methods of analysis, and the problems involved in the interpretation of the earth's history. The major theme of the remaining chapters is the interplay between mobile and immobile belts in the evolution of continents. The geologic history of each continent is presented, for each era, in terms of the changing spatial distribution of rocks, and that changing distribution is related to the changing distribution and nature of ancient seas, to mountain-making movements, etc. Not every little corner of the earth, obviously, can be mentioned in a text of this size, but I have attempted to present enough wide-ranging material to give a balanced picture of the earth's history, at least since the Precambrian. Complete documentation of all the statements is impossible, but I have discussed a number of areas for each era in detail to illustrate the nature of the available data and the possible interpretations. Additional documentation and discussion are left to the discretion of the instructor. The paleontological record of each era is discussed in terms of the evolution and distribution of faunas and floras and in relation to the physical history of the earth.

Even though the introductory chapters are more extensive than those in most texts on historical geology, I regard them as providing the bare minimum of information that the student must have in order to understand how geologists interpret the earth's history. These chapters presuppose a thorough reading of a text on the principles of geology. Any additional treatment of the subjects of the introductory chapters should be given by the lecturer, and the local geologic setting provides an excellent framework for this.

Even though formational names are generally avoided, it may be felt that the text contains too much for a student to absorb in one semester. My own experience, in lecturing according to the organization and content of this book, tells me that the student can handle the material; the better students, in fact, ask for more data. The extent to which a student should be expected to retain the information in this book must vary with the objectives of the course. The major theme of the physical and biological history of the earth is, I believe, sufficiently emphasized, and mastery of that theme can be the minimum requirement. The detailed data and discussion that substantiate this theme become a body of passive knowledge. The individual instructor can determine for himself the extent to which the student should master the data in terms of an active vocabulary.

Students who have no familiarity with the major phyla of animals and plants will find the summary of them in Appendix I. Advanced students who wish to learn some of the formational names that are avoided in the text will find them in Appendix II. This consists of twenty-three correlation charts, each of which consists of fifteen stratigraphic columns in which the major kinds of rock and the thickness of many formations are indicated. There are two charts for each geologic system, one for North America, the other for the remaining continents, and one chart is devoted to the Gondwana formations of the southern hemisphere.

In the illustrations I have attempted to achieve a balanced geographic distribution. Maps, cross-sections, and sketches are emphasized because they generally tell more than photographs. The photographs that are used have been selected to illustrate the relations or the terrane of certain sites within the major continental areas.

Credit for illustrative material is given in the legends of the figures, but many institutions and individuals have been so helpful that I wish to acknowledge their kindness here. Among the institutions are the U. S. Geological Survey, the American Museum of Natural History, the Chicago Museum of Natural History, the Denver Museum of Natural History, the U. S. National Museum, the Geological Survey of Canada, the Royal Canadian Air Force, the Geological Survey of Great Britain, the

British Museum (Natural History), the Geological Survey of India, and the Geological Survey of the Union of South Africa. Among the individuals who deserve special acknowledgment for illustrative material are J. V. Harrison, R. V. Melville, J. Augusta, A. Sriramadas, Brian Skinner, and Lester King. Photographs used in the text were also received from R. H. Dott, Jr., J. T. Robinson, K. E. Caster, G. A. Cooper, J. H. Wellington, Hans Frebold, E. T. Tozer, H. Futagami, H. K. Erben, C. A. Fleming, Marshall Kay, E. A. Rudd, and S. Sakagami.

Special thanks are due to McLain Forman, David Raup, and John J. Wilson, who have read the whole manuscript. Among my colleagues, R. Siever, R. M. Garrels, M. P. Billings, B. Patterson, and A. S. Romer have been particularly helpful in discussion of a wide range of subjects.

October 1960 *Bernhard Kummel*

CONTENTS

HISTORY OF THE EARTH

And some rin up hill and down dale,
knapping the chunky stanes to pieces wi' hammers,
like sae many road makers run daft.
They say it is to see how the world was made.

— Sir Walter Scott, *St. Ronan's Well* (1824)

1

INTRODUCTION

Historical geology deals with the history of the earth. Chiefly, of course, it concerns itself with the changes in the earth's crust and in the life that has existed upon that crust. As history, it seeks to discover the temporal relations of all that has happened to and on that crust. Examining all the evidence collected by many other sciences, it strives to present an orderly chronological arrangement of the accumulated data that bear upon its subject. It embraces nearly all aspects of geology. In combination with biology, and to a lesser extent with chemistry and physics, it gives us a three-dimensional framework for a history of the earth. With the aid of the other sciences we discover not only *what* has happened but *how* and *why;* as historians we are also concerned with *when.*

Dealing, as it does, with the origin and evolution of the earth and of the life upon the earth's surface, historical geology has often run head on into cultural beliefs, superstitions, and dogmas that have impeded its development and engendered bitter controversy. The conflict between scientific interpretation and theological dogma created a most difficult environment for the growth of a geological science. For this reason few sciences have had a more interesting development than the one that deals with the history of the earth and of life upon the earth.

Until the latter half of the eighteenth century, many, perhaps most, investigators approached geological data with the desire to prove certain theological doctrines. Once the true meaning of fossils was widely accepted, however, it became possible to establish historical geology as a distinct science. The acceptance of a few simple but fundamental concepts took the study of the earth's history away from the theoretical speculators and stimulated the phenomenally rapid development that is still going on today. Let us investigate briefly the history that led to our modern science of historical geology.

1-1 EARLY VIEWS OF THE EARTH'S HISTORY

The ages before the Christian era and the first eighteen centuries of that era contributed little more to the history of the earth than a knowledge of the organic nature of fossils (from the Latin *fossilis*, "dug up"). That statement does not mean that people were not curious about the history of the planet upon which they were living. On the contrary, nearly all primitive cultures, as they elaborated their religious doctrines, sought to explain the origin of the earth and of the life upon it. Their explanations were usually fanciful and were not hampered by any investigation of the evidence. Not until the Greeks do we find any use of observed phenomena as a basis for a history of the earth.

The Hellenic philosophers gave much time to speculation about the origin of the earth. They were aware that the surface of the earth was not static but was changing continuously. Xenophanes of Colophon (ca. 570–ca. 480 B.C.), Xanthus of Sardis (ca. 500 B.C.), and Herodotus of Halicarnassus (484?–425? B.C.), observing fossils, inferred that they were the remains of organisms and that at one time the sea had covered the site of their occurrence. Aristotle (384–322 B.C.) summarized these views as follows: "The distribution of land and sea in particular regions does not endure throughout all time, but it becomes sea in those parts where it was land, and again

it becomes land where it was sea. . . . As time never fails, and the universe is eternal, neither the Tanais nor the Nile can have flowed forever. . . . So also of all other rivers; they spring up, and they perish; and the sea also continually deserts some land and invades others. The same tracts, therefore, of the earth are not, some always sea and others always continents, but everything changes in the course of time."

Aristotle, who is often considered the father of zoology, not only introduced a new element into scientific thought—the value of observing natural phenomena in the field—but developed some of the first scientific ideas on the evolution of life. He believed in a complete, linear, gradual ascent among the animals, from the most imperfect to the perfect. On his scale the sponges and sea anemones were at the bottom and man at the top. His views on evolution, as we shall see later, retained their authority until the seventeenth century.

Three writers of the Roman world—Strabo, geographer (ca. 63 B.C.), Seneca, philosopher (ca. 3 B.C.–A.D. 65), and Pliny the Elder, historian and naturalist (A.D. 23–79)—investigated, during their extensive travels in the Mediterranean area, volcanoes, the rise and fall of the sea, and fossils. Strabo, especially, is noteworthy for his views on the volcanic origin of certain oceanic islands.

None of the writers of the pre-Christian era, however, interesting though their observations and speculations may be, ever understood the agencies that were changing the earth or correlated the changes they observed with the changes of earlier eras. They failed to establish a relation between modern and ancient forms of life, and they never comprehended the antiquity of the organic world. Nor did their immediate successors, the early Christian scholars and churchmen, go beyond them; well schooled in Hellenic philosophy, the early Christians accepted Aristotle as the final authority in natural history.

The downfall of the western Roman Empire threw Europe into turmoil and was followed by nearly a thousand years of wars, mass migrations, and intellectual sterility. Scholarship retreated to the monasteries, where the Schoolmen pondered over the writings of past ages and erected logical systems of theologically derived ideas but contributed nothing new in the way of empirical knowledge. There is no evidence that anyone in Europe, during this time, observed or recorded any aspect of the earth's surface upon which man lived. Only among the Arabs were the remnants of Greek science kept alive. Arabian scholars translated and embellished Greek manuscripts and later passed them on to European scholars. Avicenna (980–1037), for example, an Arabian translator and commentator on Aristotle, explained fossils as the results of unsuccessful attempts of the *vis plastica,* or creative force of nature, to produce the organic out of the inorganic; and Albertus Magnus of Cologne (1193–1280), who depended on Latin translations from the Arabic for his knowledge of Aristotle, explained most fossils as originating from a *virtus formativa* in the earth.

In the fifteenth century, the period in which the great exploratory voyages by Italian, Spanish, and Portuguese adventurers began, the world in which man lived took on new interest for him. The nature of fossils became the subject of active discussion, especially in Italy, and elicited comments from many of the leading literary and scientific men of the times.

Leonardo da Vinci (1452–1519), a great scientist and an even greater painter, approaching the problem of fossils by direct observation and rational interpretation, maintained the organic origin of fossils. The following paragraph, translated from one of his manuscripts, illustrates his genius and emphasizes the absurdity of many of the other explanations that were offered in the following three hundred years.

When the floods from rivers turbid with fine mud deposited this mud over the creatures which live under the water near sea shores, these animals remained pressed into the mud, and being under a great mass of this mud, had to die for lack of the animals on which they used to feed. And in the course of time, the sea sank, and the mud, being drained of the salt water, was eventually turned into stone. And the valves of such molluscs, their soft parts having already been consumed, were filled with the mud; and as the surrounding mud became petrified, the mud which was within the shells, in contact with the former through their apertures, also became turned into stone. And so all the tests [shells] of such molluscs remained between the two stones, that is, the stone in which they were and that which covered them; and these are still found in many places. And nearly all the molluscs petrified in the stones of the mountains still have their natural shell, especially those which were old enough, which would be preserved by their hardness; and the young ones, already calcined, were penetrated in great part by the viscous and petrifying liquid.*

What more could be said in explanation of fossils and their occurrence? Yet, despite Da Vinci's clear exposition, the debate was just beginning, for the problem was complicated by the great revolution that was taking place in the intellectual life of Europe.

Knowledge derived from observation was increasing—in both quantity and prestige. This increase was, in a sense, a revolt against the authority of Aristotle. At the same time there was a great change in the attitude of the Church toward natural history. Early leaders of the Church, such as St. Augustine, and those of the thirteenth century, such as St. Albertus Magnus and St. Thomas Aquinas, had not interpreted the Bible in a strictly literal sense. Now, however, *Genesis* was to be understood literally. The Bible, instead of Aristotle,

Guide to an Exhibition Illustrating the Early History of Paleontology (British Museum, London, 1931), p. 24.

FIGURE 1-1

Pseudofossils, or "Lügensteine," of Beringer. (From the British Museum, Natural History.)

took an authoritative place in the discussion. The new churches established by the Protestant Reformation were sometimes even more fanatical in regarding the Bible, literally translated, as the guide to all thinking on natural history. As we briefly review some of the opinions held about fossils at this period, we must not forget that in the sixteenth century one was likely to lose one's life if one disputed the authority of the Bible.

The *vis plastica* of Avicenna and the *virtus formativa* of Albertus Magnus were still considered worthy of discussion; and to such unnaturalistic explanations of fossils was added the *aura seminalis*, or germ-laden air, which had fallen into crevices and developed inside the rocks. Some thought of fossils as "sports of nature," some as mere mineral concretions, others as formed of fatty matter fermented by heat or by a "lapidific juice."

The idea that fossils were merely sports of nature was finally killed by ridicule in the early part of the eighteenth century. Johann Beringer, a professor at the University of Würzburg, enthusiastically argued against the organic nature of fossils.

In 1726 he published a paleontological work entitled *Lithographia Wirceburgensis*, which included drawings of many true fossils but also of objects that represented the sun, the moon, stars, and Hebraic letters. It was not till later, when Beringer found a "fossil" with his own name on it, that he realized that his students, tired of his teachings, had planted these "fossils" and carefully led him to discover them for himself. Some of Beringer's pseudofossils (*Lügensteine*) are illustrated in Figure 1-1.

If fossils were merely "sports of nature," there was no conflict with *Genesis;* as the organic nature of fossils became generally recognized, however, an adjustment between paleontology and *Genesis* became necessary. The Flood of Noah was, of course, the obvious explanation; yet, though this explanation had been suggested as early as the *De Pallio* of Tertullian (ca. 160–230 A.D.), it was only near the end of the seventeenth century that fossils were generally thought to be relics of the Flood. More than a hundred years after Da Vinci were spent in senseless arguments about the true nature of fossils, for the entrance of Noah's Flood into the argument brought along countless theologians to support it, and it took nearly two centuries to explode this hypothesis.

To illustrate the essence of the Diluvialist teachings (those attributing fossils to the Flood), we quote a passage from *An Essay Toward a Natural History of the Earth*, published in 1695 by John Woodward, a professor of medicine:

> That during the time of the Deluge, whilst the water was out upon, and covered the Terrestrial Globe, All the Stone and Marble of the Antediluvian Earth: all the metalls of it: all Mineral Concretions: and in a word, all Fossils whatever that had before obtained any Solidity, were totally dissolved, and their constituent corpuscles all disjoined, their cohesion perfectly ceasing. That the said corpuscles of those which were not before solid, such as Sand, Earth, and the like: as also all Animal Bodies, and parts of Animals, Bones, Teeth, Shells: Vegetables, and parts of Vegetables, Trees, Shrubs, Herbs; and, to be short all Bodies whatsoever that were either upon the Earth, or that constituted the Mass of it if not quite down to the Abyss, yet at least to the greatest depth we ever dig: I say all these were assumed up promiscuously into the Water, and sustained in it, in such manner that the Water, and Bodies in it, together made one common confused Mass.
>
> That at length all the Mass that was thus borne up in the Water, was again precipitated and subsided towards the bottom. That this subsidence happened generally, and as near as possibly could be expected in so great a confusion, according to the Laws of Gravity: that Matter, Body, or Bodies, which had the greatest quantity or degree of Gravity, subsiding first in order, and falling lowest: that which had the next, or a still lesser degree of gravity, subsiding next after, and settling upon the precedent: and so on in their several Courses.

What a striking contrast to Da Vinci's writings of nearly two hundred years before!

Another Diluvialist that is still remembered was a Swiss, Johann Scheuchzer (1672–1733), an enthusiastic follower of Woodward. Scheuchzer published descriptions and illustrations of what he thought to be "the bony skeleton of one of those infamous men whose sins brought upon the world the dire misfortune of the deluge" (Fig. 1-2). This fossil, which he named *Homo diluvii testis*, was later shown to be nothing but the skeleton of a salamander. In 1709 Scheuchzer published a treatise (*Herbarium Diluvianum*) on the fossil plants that figure conspicuously in the Flood discussions of the time. From the "tender, young, vernal" state of some seed cones Scheuchzer concluded that the Flood took place in May. A contemporary disagreed; the number of "ripe" fruits proved, he said, that the Flood took place in the autumn. All this discussion of fossils did little to advance the knowledge of the earth.

FIGURE 1-2

Homo diluvii testis ("man a witness of the deluge"): the skeleton of a giant salamander, *Andrias scheuchzeri*, from the upper Miocene, Oeningen, Baden, Germany. (From the British Museum, Natural History.)

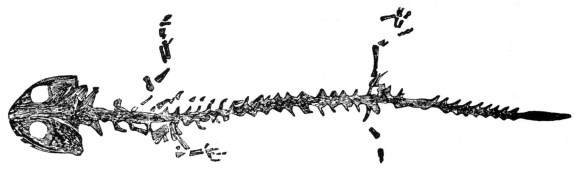

To Nicolaus Steno—Latinized name of Niels Stensen (1638–1687), a Dane who spent most of his life in Italy—we owe one of the early fundamental concepts of historical geology, the superposition of strata. Steno's conclusions as to the formation of a stratigraphic succession can be condensed as follows: (1) a definite layer of deposit can form only upon a solid base; (2) the lower stratum must therefore be consolidated before a fresh deposit is precipitated upon it; (3) any one stratum must either cover the whole earth or be limited laterally by other solid deposits; (4) since, while a deposit is accumulating, only the water from which it is precipitated is above it, the lower layers in a series of strata must be older than the upper. The simplicity of Steno's observations is obvious; their importance will be developed later in this chapter.

Robert Hooke (1635–1703), a contemporary of Steno's, was not only a chemist, physicist, and mathematician, but one of the outstanding geologists of the seventeenth century. His posthumous book entitled *Discourse of Earthquakes* (1705) contains a wealth of pregnant and original suggestions, among them the possibility of using fossils as a chronological index, the extinction of species, variation and progression due to changed conditions, and climatic changes inferred from fossils. Hooke's suggestion on the use of fossils as a chronologic index is such an important idea that we quote it here:

> I do therefore humbly conceive (tho' some possibly may think there is too much notice taken of such a trivial thing as a rotten shell, yet) that men do generally too much slight and pass over without regard these records of antiquity which nature have [*sic*] left as monuments and hieroglyphic characters of preceding transactions in the like duration or transaction of the body of the earth, which are infinitely more evident and certain tokens than anything of antiquity that can be fetched out of coins or medals, or any other way yet known, since the best of those ways may be counterfeited or made by art and design . . . but those characters are not to be counterfeited by all the craft in the world, nor can they be doubted to be, what they appear, by anyone that will impartially examine the true appearances of them: And tho' it must be granted, that it is very difficult to read them, and to raise a Chronology out of them, and to state the intervals of the time wherein such, or much catastrophies [*sic*] and mutations have happened; yet 'tis not impossible, but that, by the help of those joined to other means and assistance of information, much may be done even in the part of information also.

1-2 THE BIRTH OF MODERN HISTORICAL GEOLOGY

In the eighteenth century the long controversy over the organic nature of fossils was coming to an end. Although fossils continued to be a focal point for study, a few pioneers began to observe the variety of rocks exposed on the earth's surface and to speculate on certain features of their relations to one another. We have already recognized Steno's important contribution, the law of superposition of strata. Two eighteenth-century pioneers who deserve special mention are Johann Gottlieb Lehmann of Germany and Giovanni Arduino of Italy. These two men published independently the first observations of the sequence of strata in the earth's crust.

Working in the region of the Harz Mountains of Germany, Lehmann recognized three classes of mountains: (1) the most ancient, with structurally complex, hard rocks; (2) the *Flötzgebirge* (horizontal mountains), composed of successive flat deposits of water-laid sediments, containing fossil animals and plants; (3) the mountains formed from time to time by local accidents. Lehmann was able to work out and describe the succession of strata of his Flötz class in many localities.

Arduino, after studying the Venetian province of northeastern Italy, classified the rocks of the region into Primary, Secondary, and Tertiary divisions. The basis of his classification was entirely lithological. The Primary division included schist, gneiss, and highly folded rocks with quartz veins. The Secondary division comprised limestones, marls, and clays—all containing fossils. The Tertiary division included the youngest strata, which consisted of highly fossiliferous limestone, sand, marl, clay, etc. Arduino observed also that the material of many of the Tertiary strata could have been derived from Secondary strata. The Primary and Secondary divisions have long since been discarded in English-speaking countries; but, as we shall see shortly, the Tertiary division, at least in name, is still in wide use.

Lehmann's and Arduino's contributions to a history of the earth were not widely recognized in their time, but their concepts were adopted by a man who became—during the last quarter of the eighteen century—one of the most influential and controversial figures in the history of geology. This man was Abraham Gottlob Werner (1749–1817), professor of mineralogy at the Mining Academy of Freiberg, Germany. Werner published little, but by his gift of impressing his opinions on others and of arousing the enthusiasm of his students he transformed the Freiberg Academy from a provincial school into one of the most important schools of mineralogy and mining in Europe. Although inclined to dogma and intolerant of dissenting opinions, he took little part in the controversies that arose over his ideas and left the defense of those ideas to his students. The geological thought of Europe was divided, for a time, between pro-Werner and anti-Werner schools.

Early in his career Werner developed a theory of the succession of rock strata, which he altered little during his lifetime. One of his fundamental concepts was that of universal formations. He taught that the earth, at first, was completely enveloped by the ocean, and that from this ocean were precipitated the rocks of the earth's crust. The first-formed rocks of Werner's scheme, which he called Primitive, included granite, gneiss, slate, and basalt; all were chemical precipitates, and none included fossils. They were followed by a succession of graywackes and limestones, also of chemical origin but including some detrital materials and the first organic remains, and called by Werner the "Transition class." This class was overlain by sandstone, limestone, gypsum, rock salt, coal, basalt, obsidian, porphyry, etc., and for it Werner adopted Lehmann's name, Flötz. The youngest of Werner's classes, the Alluvial included the superficial sands, gravels, etc. Although Werner had developed his succession of strata without ever leaving Saxony, he boldly claimed his classes to be universal, and his dogma was dutifully applied and defended by scores of his students all over Europe. At the height of his influence, however, an active opposition developed, and his opponents began to accumulate data that eventually led to the refutation of his doctrines. Both Werner's admirers and his opponents agreed on one point: that he was to be credited with bringing into prominence the doctrine of a geological succession; it was the interpretation of the succession that was the source of disagreement.

Of the many other distinguished investigators of the period we shall mention only two, James Hutton (1726–1797) and William Smith (1769–1839), both of whom contributed ideas of fundamental importance to the science of historical geology.

Hutton, lawyer, physician, and farmer as well as geologist, was an ardent student of the Scottish countryside. He was one of the leading advocates of the idea that basalt originated from a molten state, not (as Werner maintained) as a marine precipitate, but his most significant contribution to historical geology was his demonstration that the processes of transportation, erosion, weathering, etc., which we can observe going on today, have been going on since very ancient times. In other words, he observed a close parallel between the dynamic geologic processes of today and those of ancient times. The logical conclusion is that a study of present processes and their results is an important means of understanding ancient rocks. This idea—that the present is the key to the past—was later established as the principle of uniformitarianism. From his studies of geology Hutton came to the conclusion that "we find no vestige of a beginning—no prospect of an end."

Although Hutton's discovery was a great step along the way toward an understanding of the earth's history, it is another investigator whom we look upon today as the father of historical geology. William Smith was a surveyor who, early in life, had developed an interest in rocks and fossils. Although he had little formal education, it was he—not any of the many well-educated and even brilliant men who had previously taken an interest in the physical nature of the earth—who produced the fundamental concept of historical geology. His professional duties made him familiar with the rock strata and fossils of a large part of England; and, not being burdened with the theoretical dogma of the day, he was able to make fresh and uninhibited observations in the field. While working on a canal project, he first observed that the different strata were characterized by unique assemblages of fossils, and that, even though the physical character of the succession of strata might alter, the succession of fossil assemblages remained constant; thus particular strata could be identified by their fossil content.

Encouraged by a friend, who did the actual writing, Smith, in 1799, dictated the results of his extensive observations: he identified a succession of the strata in central and southeastern England, specifying their thicknesses, their characteristic fossils, and their lithology (see the frontispiece). In the next dozen years he ener-

getically pursued his geological observations, and in 1815 he was able to publish a geologic map of England and Wales, with a list of the strata from oldest to youngest. This was the first geologic map of a large area. Although fossils had been known to man for more than 2,000 years before Smith's observation, the very simple idea that one could use fossils to identify rock strata was new, and it opened a completely new vista to the science. The principles laid down by Smith inspired all students of the history of the earth, for they quickly realized that those principles laid the basis for a chronology, a calendar, of that history.

1-3 THE GEOLOGIC COLUMN

Forty years after Smith dictated to his friend the succession of strata in England and demonstrated the chronological value of fossils, a time scale of universal application had been introduced. This time scale is the unifying concept of historical geology.

Let us review briefly the state of knowledge that prevailed at the time Smith was doing his pioneer work so that we can appreciate the method by which the time scale was pieced together. Before Smith the succession of strata established by Lehmann, as modified by Werner, was in vogue. In this scheme the youngest deposits comprised the sands, gravels, and clays of alluvial origin. Next in antiquity came the Flötz class, or stratified rocks—limestones, sandstones, shales, coals, gypsum deposits, etc. Beneath this class was Werner's Transition class, comprising dark sandstones and limestones. The oldest division, called the Primitive by Werner, included granite, gneiss, schist, and slates. In Great Britain and central Europe strata of the Flötz class are widely distributed, have abundant fossils, and are generally horizontal or dip only gently. In contrast to these, strata of the Transition and Primitive classes generally lack fossils and have complex structures. Under these circumstances it is understandable that much more effort was concentrated on the Flötz strata than on the older rocks. By the time Smith published his geologic map, considerable work had already been done on local successions of Flötz strata in England and Europe.

To illustrate this point, let us examine the geologic map and succession of strata for England and Wales published by William Phillips in 1821 (Fig. 1-3). The text to accompany this map appeared in 1822 under the authorship of W. D. Conybeare and Phillips. This map is used as an example rather than Smith's original map and stratal succession of 1815 because it contains a little more information and also the earliest formal name of a geologic division. A modern geologic map of England and Wales is not very different from this early map. Most of the names given to the various map units are still being used. The formational names used in the following two paragraphs are those of the legend to Conybeare and Phillips' map. In this succession of strata the Flötz class includes the strata from the Old Red Sandstone up through the Upper Marine beds. Most of these mapped units are lithologically very distinctive. Some—such as the Old Red and New Red sandstones—are characterized by their color, others—such as the Chalk and the London clay—by their distinctive lithologies. Note that between the Old and New Red sandstones the authors introduced the name Carboniferous for a single limestone unit (in their text, however, they referred to a Carboniferous series). This formal name—Carboniferous—was soon adopted for all the strata associated with the coal deposits of England and continental Europe. Thus began the custom of giving names to sequences of strata that had similar lithological, faunal, and structural characters.

The strata from the Lias through the Purbeck beds, which are very fossiliferous over much of Europe, had been named the Oolitic group or series by Smith, but the name was never widely adopted outside England, and later these strata came to be known as Jurassic from the fine exposures in the Jura Mountains of France.

Of the strata above the Purbeck beds and below the Plastic clay, the most conspicuous unit is the Chalk, which is widespread over southeastern England and north-central Europe. From the Latin word for chalk, creta, comes the name Cretaceous, given to these strata.

The strata above the chalk had long been a focus of study because of their rich and abundant fossil faunas—faunas that had a distinctly modern aspect. For these strata Arduino's old term Tertiary was not used by Conybeare and Phillips even though the term was widely used at the time. The leading student of these strata was Charles Lyell, who, in 1833, proposed a threefold subdivision of the Tertiary Period (Eocene, Miocene, and Pliocene) according to the percentage of Recent species present in the fossil faunas: a stratum was Eocene if 1–5 percent of its fossils consisted of Recent species, Miocene if 20–40 percent were Recent, and Pliocene if 50–100 percent were Recent. For the Pliocene, however, Lyell recognized two divisions, an Older Pliocene (50–90 percent Recent) and a Newer Pliocene (90–100 percent Recent), and later he introduced the term Pleistocene for the Newer Pliocene. The incompleteness of Lyell's divisions was soon recognized, and two more divisions were subsequently introduced: the name Oligocene was adopted for the strata previously included in the upper Eocene and lower Miocene, and the name Paleocene for the strata Lyell had placed in the lower Eocene.

The interval between the Carboniferous and the Jurassic in England and northern Europe consists mostly of red terrestrial sandstones and shales, with extensive beds of gypsum and rock salt. In England these strata were long known as the New Red Sandstone. In Germany, however, the upper half of this interval was occupied by three

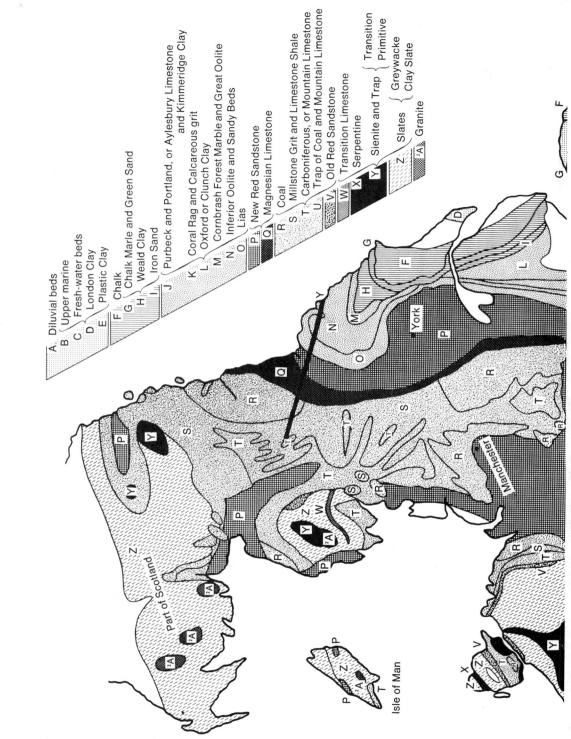

A. Diluvial beds
B. Upper marine
C. Fresh-water beds
D. London Clay
E. Plastic Clay
F. Chalk
G. Chalk Marle and Green Sand
H. Weald Clay
I. Iron Sand
J. Purbeck and Portland, or Aylesbury Limestone and Kimmeridge Clay
K. Coral Rag and Calcareous grit
L. Oxford or Clunch Clay
M. Cornbrash Forest Marble and Great Oolite
N. Inferior Oolite and Sandy Beds
O. Lias
P. New Red Sandstone
Q. Magnesian Limestone
R. Coal
S. Millstone Grit and Limestone Shale
T. Carboniferous, or Mountain Limestone
U. Trap of Coal and Mountain Limestone
V. Old Red Sandstone
W. Transition Limestone
X. Serpentine
Y. Sienite and Trap } Transition
Z. Slates { Greywacke
     Clay Slate
²A. Granite } Primitive

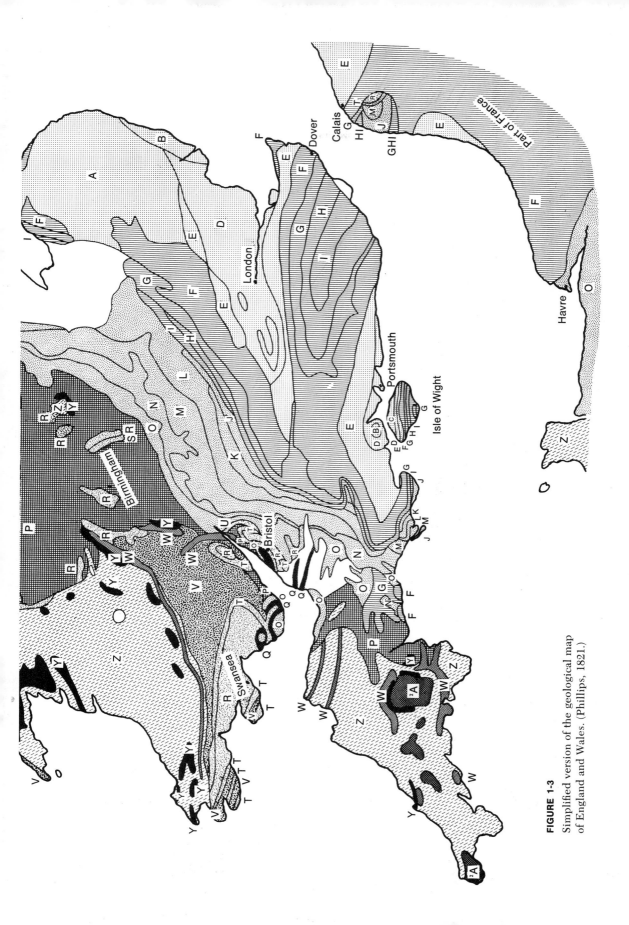

FIGURE 1-3

Simplified version of the geological map of England and Wales. (Phillips, 1821.)

series of strata consisting of red sandstone and shale (the lowest), marine limestone (the middle), and red shale and sandstone (the uppermost). To these three series of strata the name Triassic was applied.

The geologic divisions defined so far have involved mostly fossiliferous strata of little structural complication. While the work on these divisions was going on, many people were also studying Werner's Transition class, but the sparseness of fossils and the complex structure of the Transition rocks made their study difficult. Two pioneers who attacked the problem of these older strata were Adam Sedgwick (1785–1873) and Roderick Murchison (1792–1871). It was widely known that great thicknesses of Transition graywackes cropped out in Wales; and both men, attracted by the difficulty of the problem, began field studies in Wales in the summer of 1831.

Murchison concentrated his studies on southern Wales, where he was able to use the Old Red Sandstone as a starting point in his mapping. He gradually established the existence of a sequence of strata, characterized by a peculiar assemblage of fossils, below the Old Red Sandstone. For this sequence he introduced, in 1835, the name Silurian (from the name of the tribe, the Silures, that inhabited this area at the time of the Roman conquest). He divided his Silurian System into a Lower and an Upper division.

Sedgwick concentrated his studies on northern Wales, a region of high, rugged terrain and highly complicated geology. Whereas Murchison started his work at a known stratigraphic horizon, the base of the Old Red Sandstone, Sedgwick found himself in a stratigraphic no-man's-land. After several years of brilliant field work, however, he was able to establish the sequence of strata for which he proposed the name Cambrian System. (Cambria is a Latin form of the native Welsh name for Wales.) The Cambrian System, as defined by Sedgwick in northern Wales, was divided into a Lower, a Middle, and an Upper division.

At the time these two geologic systems were proposed, neither of the authors had a clear idea of how they joined. It was discovered eventually that the Lower Silurian of Murchison had the same fossils as the Upper Cambrian of Sedgwick. This discovery led Murchison to conclude that all of Sedgwick's Upper Cambrian was merely a part of the Silurian—a conclusion that Sedgwick strongly opposed. The argument turned a warm friendship into an enmity that lasted the lifetime of the two men. Murchison was supported by the Geological Society of London and the Geological Survey of Great Britain. Feelings were so intense that little could be done to resolve the issue while either of the original authors was alive. Finally, in 1879, Charles Lapworth proposed the name Ordovician (after the tribe that had once occupied northern Wales, the Ordovices) to include the Upper Cambrian of Sedgwick and the Lower Silurian of Murchison, and his proposal was generally accepted.

Before their controversy became too intense, Murchison and Sedgwick had spent some time in Cornwall and Devonshire investigating a sequence of graywackes and limestones that were highly folded and faulted. On lithologic grounds the two distinguished geologists first concluded that the Devonshire rocks belonged in the Middle and Upper Cambrian. But William Lonsdale, who studied the fossils of these rocks, came to the startling conclusion that they were intermediate between the Silurian and the Carboniferous. This discovery implied an equivalence of the Devonshire rocks with the Old Red Sandstone, a conclusion that Sedgwick and Murchison were reluctant to accept. Lonsdale finally won his point; and Sedgwick and Murchison, in 1839, established the Devonian System, which, lying between the Silurian and the Carboniferous, included the Old Red Sandstone. Soon afterward these two authors were able to confirm their views in studies of Devonian rocks in the Rhine valley, which, as we now know, is more nearly complete and has fewer structural complications than the type area of Devonshire.

Very soon after the Devonian question had been settled, Murchison accepted an invitation from the Russian Czar to study the rock strata of European Russia. On this extended tour he discovered thick deposits of limestone that contained fossils younger than those of the Carboniferous and older than those of the Triassic. This limestone graded upward into deposits of red sandstone and shale that contained beds of gypsum and anhydrite. Believing the limestone, red sandstone, and shale to be of the same age as the lower New Red Sandstone and the Magnesian Limestone of England and similar deposits in Germany, and finding them to have distinctive faunas and stratigraphic positions (above the Carboniferous and below the Triassic), Murchison named them the Permian System after Perm, the Russian province in which he had studied them.

Throughout this short period in which the geologic time scale was being put together, the study of fossils was not advancing as fast as stratigraphic geology. Even so, enough data were accumulating to give a broad picture of the whole of life from the Cambrian to the Recent. Since the older terms of the early writers — Primary, Transition, Secondary, and Tertiary — no longer had any real meaning, John Phillips, in 1840, following out the ideas of William Smith, grouped the fossiliferous strata from the Cambrian to the Recent into three eras, the Paleozoic, the Mesozoic, and the Cenozoic — ancient life, middle life, and recent life. The Paleozoic includes the Cambrian and Permian Systems and those between; the Mesozoic includes the Triassic, Jurassic, and Cretaceous Systems; and the Cenozoic is the Tertiary of earlier authors. The geological eras — Paleozoic, Mesozoic, Cenozoic — are referred to collectively as the Phanerozoic.

The standard geologic column and time scale are presented in Table 1-1.

1-4 MODERN CONCEPTS OF THE GEOLOGIC TIME SCALE

During the second quarter of the twentieth century, as stratigraphical data were accumulating rapidly, it became necessary to adopt a uniform nomenclature for the various divisions of time and for the rocks deposited during those divisions. The nomenclature now in general use is based on the dual classification shown in Table 1-2.

Time units are simply subdivisions of geologic time. Time-rock units consist of the strata deposited during the time units. The fundamental time and time-rock units are, respectively, the period and the system. The Cambrian Period was the time during which the rocks of the Cambrian System were deposited.

Subdivisions of periods are called epochs, and subdivisions of systems are called series. For the Mesozoic and Cenozoic Eras these subdivisions have intercontinental status. For the Paleozoic Era, however, the terminology is much less uniform; each continent, for instance, commonly has its own nomenclature. It is customary to apply geographic adjectives, formed with the ending "ian" or "an," to epochs and series; where geographic names have not come into common use, an epoch is simply the Early, Middle, or Late part of its period (for example, Late Cretaceous), and a series is the Lower, Middle, or Upper part of its system (for example, Lower Devonian).

The epochs and series are divided, respectively, into ages and stages. These units also have geographic names; the Campanian, for example, is a stage of the Upper Cretaceous. The system, series, and stage names used in North America and on the other continents are included in Appendix II.

TABLE 1-1

Geologic column and time scale

Era	System or Period (rocks) (time)	Series or Epoch (rocks) (time)	Approximate age in millions of years (beginning of unit)
Cenozoic (*recent life*)	Quaternary (an addition to the old tripartite 18th-century classification)	Holocene	.01
		Pleistocene (*most recent*)	2.0 to 3.0
	Tertiary (Third, from the 18th-century classification	Pliocene (*very recent*)	7
		Miocene (*moderately recent*)	25
		Oligocene (*slightly recent*)	40
		Eocene (*dawn of the recent*)	60
		Paleocene (*early dawn of the recent*)	68 to 70
Mesozoic (*intermediate life*)	Cretaceous (*chalk*)		135
	Jurassic (Jura Mountains, France)		180
	Triassic (from three-fold division in Germany)		225
Paleozoic (*ancient life*)	Permian (Perm, a Russian province)		270
	Carboniferous (from abundance of coal)		
	Pennsylvanian°		325
	Mississippian°		350
	Devonian (Devonshire, England)		400
	Silurian (an ancient British tribe, the Silures)		440
	Ordovician (an ancient British tribe, Ordovices)		500
	Cambrian (Cambria, a Latin form of the native Welsh name for Wales)		550 to 600
Precambrian	Many local systems and series are recognized, but no well- established worldwide classification has yet been delineated.		3500 or more

Sources: Approximate ages from Holmes, 1964; Evernden, Savage, Curtis, and James, 1964; and The Phanerozoic Time Scale of the Geological Society of London, 1964.

Notes: Definitions in italics are from the Greek.

Many provincial series and epochs have been recognized in various parts of the world for Mesozoic and older strata. Most of the systems have been divided into Lower, Middle, and Upper Series, to which correspond Early, Middle, and Late Epochs, as the times during which the respective series were deposited.

° Pennsylvanian and Mississippian Systems, named for States of the U.S.A., are not generally recognized outside of North America; elsewhere the Carboniferous System is regarded as a single system.

The smallest time-rock unit recognizable over a wide area is the zone, which is a group of strata characterized by a distinct assemblage of fossil species. Zones are generally named after one of the species present—for example, the *gracilitatis* zone. No equivalent time unit is in general use.

Rock units are distinct assemblages of strata that can be mapped. They are ob-jective units that can be recognized by any geologist and do not necessarily coincide with time units. The basic rock unit is the formation, which is defined merely as a distinct mappable unit.

Formational names are binomial, usually consisting of a geographic noun and a descriptive lithologic term (Mancos Shale), both names being capitalized. When a for-

TABLE 1-2
Classification of stratigraphic units

Time units	Time-rock units
Era	———
Period	System
Epoch	Series
Age	Stage
———	Zone

mation cannot be designated by any single appropriate lithologic term, the word "formation" is used (Thaynes Formation).

When, for the sake of clarity in mapping, or to reveal structure, or for economic reasons, we wish to map and recognize a part of a formation, such a part is called a member if it has considerable geographic extent, a lentil if it is of only local distribution, a tongue if, as an extension of a larger body of similar sediments, it wedges out in one direction between sediments of different lithology.

If we consider the random manner in which the geologic time scale was put together, it is amazing that it has undergone only minor changes in the past hundred years. Most attempts to change the framework of geologic systems established in the first half of the nineteenth century have been unsuccessful. One exception is the Carboniferous System. Two divisions of the Carboniferous were recognized early; in Europe they were named simply the Lower and Upper Carboniferous, but in the United States they received formal systematic names. The Mississippian System, named from exposures in the Mississippi valley, was proposed in 1868 and is approximately equivalent to the Lower Carboniferous of other countries. The Pennsylvanian, named from the state in which splendid developments of the coal measures crop out, was proposed in 1871 and is approximately equivalent to the Upper Carboniferous of other countries. These names are used only in North America.

At the time the systematic units were being proposed, it became customary to put the boundaries between adjoining systems at significant breaks in the stratigraphic succession. It was almost axiomatic that systems should be separated by striking unconformities, series by smaller unconformities, and eras by the greatest unconformities. These unconformities, in the regions where our geologic time scale was first established, as well as the distinctness of the fauna of each system, led to the idea that great catastrophes had occurred between the systems. Some prominent authors of the period even believed that all life was wiped out after each stratal unit and that the earth was completely repopulated, each time, with new forms. The wide distribution of many unconformities led many geologists to believe that here was "the ultimate basis of correlation." Members of this school regarded the earth's history as a series of episodes separated by intervals of universal deformation during which mountain ranges were formed and biologic changes initiated.

This thesis no longer has wide support. At the time it was formulated, paleontology was in its infancy, and the practicing paleontologist did not believe in evolution. It was not until the idea of evolution became established that the nature of the differences between successive faunas was really appreciated. As geologic study was extended to all parts of the world, it was realized that the striking physical boundaries that are found between some systems in Europe are not found in other areas and that the succession of faunas nevertheless remains the same. The geologic systems are of universal application because of their fossil content and not by lithologic or structural criteria.

The revolution in geologic thought that William Smith began is beautifully illustrated in the preface to Hugh Miller's classic book *The Old Red Sandstone*, in which he states: "Such is the state of progression in geological science, that the geologist who stands still but a very little, must be content to find himself left behind. Nay, so rapid is the progress, that scarce a geological work passes through the press in which some of the statements of the earlier pages have not to be modified, restricted, or extended in the concluding ones." This statement was written in 1841.

1-5   A RADIOMETRIC TIME SCALE

Just how old is the earth? The Silurian is above, and therefore younger than, the Ordovician, but this tells us nothing about the duration of either of these periods, nor does it tell us how ancient these periods actually are. If we had to depend on the scale established in the early years of the nineteenth century, we still could think only of relative age.

Knowing what we do about the early development of historical geology, we can well imagine what a fertile field for speculation the problem of age offered to theologians, philosophers, and cosmologists. James Ussher, Archbishop of Armagh (Ireland), calculated in 1654, from studies of the Scriptures, that the Creation had taken place on the twenty-sixth of October, in the year 4004 B.C., at nine o'clock in the morning. This date, inserted as a marginal reference, by some unknown authority, into the later editions of the King James version of the Bible, became implanted in religious thinking, and for long afterward it was an act of heresy among English-speaking Protestants to deny its validity. Other estimates came close to Ussher's, and the genealogies of the Bible do, in fact, require approximately such dating.

As historical geology was firmly established in the nineteenth century, it became obvious to many that the archbishop's date could not possibly be right. Hutton gave proper perspective to the problem of age with his doctrine of uniformitarianism. To Hutton and his followers the physical and chemical processes then shaping the earth made it plain that the age of the earth must be very great to allow for the development of all the depositional and structural features observable at the earth's surface.

Several methods of estimating the earth's age have been used. At the end of the nineteenth century one investigator considered the total sodium salts in the oceans and the amount added each year from the erosion of igneous rocks. A simple bit of arithmetic gave 100,000,000 years as the approximate age of the earth. This estimate was widely accepted for a time, but it was eventually seen to include too many intangibles and not to leave enough time for all of the earth's history and for organic evolution. Another method was to consider the total thickness of the deposits in the whole stratigraphic column and to divide this by the rate of deposition of the main lithologic types. Although various adjustments for differences in the rates of deposition of limestone, shale, and sandstone were generally introduced, this method also was eventually seen to involve many unknowns and by itself to yield results that were not even approximately acceptable.* The processes on which these methods were based — weathering and addition of salt to the oceans, deposition of limestone, sandstone, and shales — depend on variables too complex for measurement, but we can show that the rates of those processes have not been uniform over the whole span of geologic time.

Variation in annual rainfall, for instance, affects the amount of salt carried to the sea and of sediment carried to oceans, lakes, and other depositional basins. We know that sunspot cycles affect the weather; they therefore affect the erosion cycle. Great

---

*Although the method proved unsatisfactory as a guide to geologic age, it did give us the maximum known thickness of each geologic system (Table 1-3). Since the data of Table 1-3 were first assembled, however, many new measurements have shown that the maximum thicknesses of several systems are actually greater than the table shows. The table is presented in its original form because its data were used, along with radiometric measurements of age, in estimates of the duration of the different geologic periods.

TABLE 1-3

Maximum known thicknesses (in feet)
of the geologic systems (and series)
(cumulative thicknesses in italics)

	North America		World	
Pleistocene	1,600	*1,600*	4,000	*4,000*
Pliocene	10,000	*11,600*	18,000	*22,000*
Miocene	20,800	*32,400*	21,000	*43,000*
Oligocene	15,000	*47,400*	15,000	*58,000*
Eocene (including Paleocene)	14,000	*61,400*	23,000	*81,000*
Cretaceous	64,200	*125,000*	64,000	*145,000*
Jurassic	22,200	*147,800*	22,000	*167,000*
Triassic	15,000	*162,800*	25,000	*192,000*
Permian	9,000	*171,800*	18,000	*210,000*
Carboniferous	23,800	*195,600*	40,000	*250,000*
Devonian	12,700	*208,300*	37,000	*287,000*
Silurian	6,400	*214,700*	20,000	*307,000*
Ordovician	23,100	*237,800*	40,000	*347,000*
Cambrian	33,700	*271,600*	40,000	*387,000*

differences between highlands and adjacent lowlands result in rapid erosion in the former and rapid deposition in the latter; but, as the highlands are cut down and the lowlands fill up, erosion and deposition become slower. These are only a few of the many factors that, by varying the rate of geologic processes, undermine the accuracy of purely geologic methods of determining age. What historical geologists needed was some method of dating that was completely independent of all these variables and unknowns. They have found one in the radiometric dating process.

Chemists and physicists who worked with radioactive elements discovered that such elements lost their radioactivity, or decayed, each at a constant but different rate. The half-life—that is, the time required for loss of half the radioactivity—of each radioactive element has been accurately determined. It has been shown that neither the rate of decay nor the properties of the products of the decay are affected by any conditions attainable in the laboratory. We assume, therefore, that radioactive decay is independent of geologic processes, and that rates of decay determined in the laboratory are applicable to the problem of dating the rocks of the earth.

There are several methods of radiometric dating. We shall merely list them here, without attempting to describe in detail how they are applied: (1) the lead-uranium-thorium method, (2) the potassium-argon method, (3) the helium method, (4) the rubidium-strontium method. All the methods are based on the fact that, as a radioactive element decays, some of its atoms lose electrons and particles of their nuclei. This loss transforms some of the original substance into a new element or elements. The ratio of the amount of any of these decay products to the original element remaining at any time is a measure of the age of the sample. In all of the methods there are sources of possible error: some of the decay products are gases, which are quite likely to escape; other decay products are soluble and may have been leached out of the sample; some of the substances thought to be products of decay may have been present in the sample when it crystallized.

A Phanerozoic time scale based on radiometric age determinations has been gradually evolving. One of the first to attempt to construct such a time scale was Arthur Holmes (a leading student of this subject) in 1913. At that time the decay constants were only roughly known, and the possibility that rates of radioactive decay were affected by the geological environment was still open. For this Phanerozoic time scale Holmes had four ages: (1) a helium date of 30 million years (m.y.) for a sample of early Eocene age, (2) a U-Pb age of 340 m.y. for uraninite from the Portland Granite of Connecticut (dated geologically as post-Lower Carboniferous and pre-Triassic), (3) a U-Pb age of 430 m.y. for uraninite from the Brancheville Granite, Connecticut (dated geologically as probably of Ordovician age), and (4) a U-Pb date of 370 m.y. for a granite, believed to be of

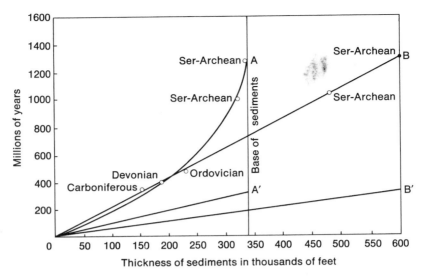

FIGURE 1-4

Radiometric time scale proposed by Arthur Holmes in 1913 based on three Phanerozoic and two Precambrian ages plotted against maximum thickness of sediment. Line *A* is plotted on the assumption that rates of sedimentation were slower in Precambrian time, and line *B* on the assumption that Precambrian rates of sedimentation were the same as those for Phanerozoic time. Lines *A'* and *B'* are ages derived from strictly geological data, which point to an age of 300 m.y. (After Holmes, 1913.)

Devonian age, from the Oslo region. In addition, Holmes had two Precambrian ages. A plot of these age determinations against the maximum thickness of sediments is shown in Figure 1-4. A remarkable feature of this graph is that if Holmes' figure of 250,000 feet is accepted as the total maximum amount of sediment deposited since the beginning of the Cambrian, the date for the base of the Cambrian, 600 m.y., agrees with the accepted figure today! By 1937 Holmes had 12 U-Pb and several U-He age determinations and published another age scheme for the Phanerozoic. This time he estimated the length of the Phanerozoic as 500 m.y. In the next decade a great many new age determinations became available, and in 1947 Holmes presented a new scheme for the Phanerozoic (Fig. 1-5). In spite of having a fairly large number of age determinations, Holmes selected only five, based on U-Pb and Th-Pb methods, as

being the most reliable. The graph in Figure 1-5 shows, in part, two curves, as the stratigraphic position of three samples were in doubt, though the probabilities favored alternative B. Unfortunately, it was subsequently shown that two of these points carried incorrect stratigraphic assignments and that two others yielded incorrect ages because of geochemical alteration. It appears to be fortuitous that these early estimates of the time scale were fair approximations of the scale generally accepted today.

The years immediately following the publication of Holmes' 1947 paper on the geological time scale saw a remarkable outburst of activity in geochronometry. Within a decade the K-Ar and Rb-Sr methods of age determination, little more than academic curiosites in 1947, had been firmly established. A much wider range of rocks and minerals could thus be dated by

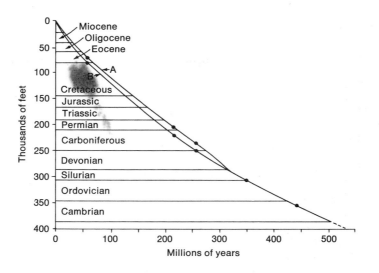

FIGURE 1-5

Geological time scale proposed by Arthur Holmes in
1947. The curves are based on the most probable ages
of radioactive minerals and on the maximum world
thickness of the geological systems. The *A* and *B*
curves correspond to the alternative assignments
shown in Figure 1-6. (After Holmes, 1947.)

methods that were relatively simple and
rapid.

The accumulation of new radiometric
data resulted in three more efforts to establish a time scale, summarized with the older
attempts, in Figure 1-6. The first of these
was by Holmes in 1959 and is close to his
B-curve of 1947 but is more extended, particularly in the Paleozoic. The second contribution was by J. Lawrence Kulp of
Columbia University in 1961. In his com-
pilation Kulp had many data not available
when the Holmes 1959 scale was put together. Finally, the Geological Society of
London published a symposium volume in
1964 entitled "The Phanerozoic Time-
scale" dedicated to Professor Arthur
Holmes. The critical dates arrived at in this
symposium are summarized on Figure 1-6.
The three latest attempts to compile a time
scale are in substantial agreement. Each
successive effort has benefited from new

data not available to the previous author.
Holmes reflected this view when he published his 1959 scale: "The revised time
scale," he wrote, "while appropriate to the
information available in 1959, will also require revision in its turn, since each year
the dated control points become more numerous, more precisely fixed, and less unevenly distributed through the geological
column. Meanwhile, however, it is not
unreasonable to hope that subsequent
modifications will be far less drastic than
those now called for."

At the present time research on radiometric age determinations is extremely
active, and great progress is being made in
both techniques and methods. As dates on
Precambrian rocks accumulate (and they
are accumulating rapidly), the question
"How old are the oldest rocks on earth?"
is often asked. The oldest rocks studied to
date seem to be slightly more than three and

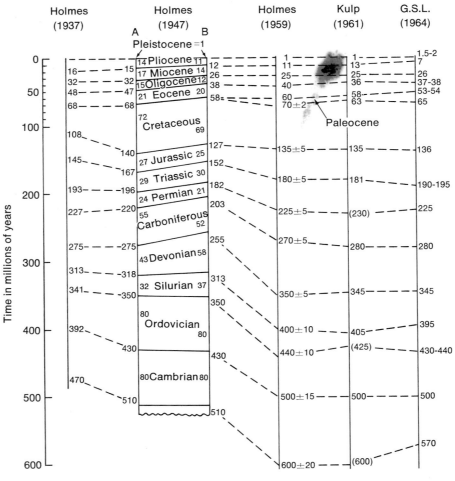

FIGURE 1-6

Comparison of Phanerozoic time scales discussed in the text. G. S. L. refers to Geological Society of London.

one-half billion years old. Another question of great interest is "When did the first living thing appear on the earth?" Minute bacterium-like organisms have been recorded from a black chert formation in South Africa that has been established as older than 3.1 billion years. We shall discuss these fossils fully in Chapter 3.

SUGGESTED READINGS

Adams, F. D., *The Birth and Development of the Geological Sciences* (Williams & Wilkins, Baltimore, 1938; Dover, New York, 1954).

Berry, W. B. N., *Growth of a Prehistoric Time Scale* (W. H. Freeman and Company, San Francisco, 1968).

Dalrymple, G. B., and M. A. Lanphere, *Potassium-argon Dating* (W. H. Freeman and Company, San Francisco, 1969).

Geikie, Sir Archibald, *The Founders of Geology* (Johns Hopkins Press, Baltimore, 1901).

Gillispie, C. C., *Genesis and Geology* (Harvard Univ. Press, Cambridge, 1951).

Haber, F. C., *The Age of the World—Moses to Darwin* (Johns Hopkins Press, Baltimore, 1959).

Harland, W. B., *et al.*, eds., "The Phanerozoic Time Scale, "*Geol. Soc. London Quart. J.*, **120s** (1964).

Holmes, Arthur, *The Age of the Earth* (Harper & Brothers, London, 1913).

———, "The Construction of a Geological Time Scale," *Trans. Geol. Soc. Glasgow*, **21** (1947).

———, "A Revised Geological Time Scale," *Trans. Edinburgh Geol. Soc.*, **17** (1959).

Hurley, P. M., *How Old Is the Earth* (Doubleday, New York, 1959).

Knopf, Adolph, "Measuring Geologic Time," *Sci. Monthly*, **85** (5) (1957).

Kulp, J. L., "Geologic Time Scale," *Science*, **133** (1961).

Wilmarth, M. G., *The Geologic Time Classification of the United States Geological Survey Compared with Other Classifications, Accompanied by the Original Definitions of Era, Period, and Epoch Terms* (U.S. Geological Survey, Bull. 769, 1925).

Zittel, K. A. von, *History of Geology and Paleontology* (Scribner's, New York, 1901).

2

THE FOSSIL RECORD

In Chapter 1 we examined briefly the long controversy that preceded acceptance of the organic nature of fossils. William Smith's demonstration of the value of fossils in the identification and correlation of rock strata, and the adoption of a geochronology based on fossils, clearly established paleontology, the study of ancient life, as an essential part of historical geology. Geology, indeed, could hardly be considered a science until the stratigraphic significance of fossils was recognized.

Let us examine some aspects of ancient animals and plants (and also of modern animals and plants) that are of special importance to the historical geologist. If you are not familiar with the basic morphological features and divisions of the main phyla of plants and animals, be sure to read Appendix 1 before going on with this chapter.

2-1 THE DIVERSITY OF LIFE

The first realization of the abundance and diversity of the life on earth is an amazing experience for anyone. No matter where you live or travel, you are in the midst of teeming life. On a walk through the woods, along the seashore, in the desert, almost anywhere you may go, in fact, you will find, if you are observant, growing plants and animals occupying every environmental niche. The more painstaking you are in your search, the greater the variety of plants and animals you will discover. Just how many kinds of plants and animals there are today no one knows for sure. One estimate of the different species of animals alone puts the number at a million; what the number would be if we added all the kinds that have lived in the past we do not know, even approximately.

For such staggering numbers of kinds of animals some system of classification is obviously necessary. Rudimentary systems were in use even before Aristotle's time, but it was not until the middle of the eighteenth century that a system proposed by the Swedish naturalist Carolus Linnaeus (Carl von Linné 1707–1778) was universally adopted and brought some order out of the chaos. The **Linnaean system of classification** set up a hierarchy of categories, the lowest (the least inclusive) being the distinct animal or plant type called the species. Each category (except, of course, the highest) comprises closely related members of the next higher category. The system and categories are as follows:

Category	Example
Kingdom	Animalia: all animals
Phylum	Chordata: animals (mostly) with backbones
Class	Mammalia: mammals
Order	Primates: monkeys, apes, man, etc.
Family	Hominidae: human and nearly human beings
Genus	*Homo:* man, human being
Species	*Homo sapiens:* the one living species of mankind

The names of some categories (the genus *Homo,* for example) have been taken from classical Latin; others, coined by students of the organisms, may be derived from any source but are always Latinized. All names used in a technical sense are capitalized except the specific epithet (the *sapiens* of *Homo sapiens,* for example), which is not capitalized and is never used without the generic name.

There is no universal agreement on the number of animal phyla, but thirty is the most commonly accepted number at the present time. Of the thirty recognized phyla only twelve are commonly found in the fossil record, but all twelve have had very long geologic histories, extending back to the Ordovician or even the Cambrian. Figure 2-1 diagrammatically compares the approximate number of living species of animals and the number of fossil species. More than three-fourths of the total species belong to the phylum Arthropoda, but most of the arthropod species are insects, which are rare in the fossil records. The remaining phyla (Mollusca, Echinodermata, etc.) comprise only 18 percent of the total. This diagram brings into sharp focus the sparsity of paleontological information on many groups of animals whose geologic record is either dim or nonexistent. The failure to leave a fossil record is usually due to the lack of solid skeletons that are capable of withstanding the normal processes of disintegration and decay.

The fossil record, except for certain reversals, clearly demonstrates a continuous increase in the number of kinds of organisms (Fig. 2-2). This increasing diversity of life reflects the adoption of new ways of life and the occupation of new environments. We shall have occasion to discuss this fact of life several times in the later chapters.

2-2 FOSSILIZATION

In ancient times the term "fossil" meant anything dug up or found buried. A **fossil** is now defined as any recognizable organic structure, or impression of such a structure, preserved from prehistoric times. When we consider the ratio of known fossil species to living species of animals (Fig. 2-1), we realize that the preservation of animals as fossils must require very special circumstances.

As a general rule, only the hard parts of animals and plants are preserved. Preservation requires rapid burial in a substance that protects the organism from destruction by scavengers, bacteria, or weathering. This requirement is oftener met in the sea than on the dry land. Bottom-living marine organisms stand an especially good chance of being rapidly buried after death. If burial is not rapid, scavengers of various types can soon destroy all trace of the organism. Land-living animals have a poorer chance of fossilization, for suitable environments are more restricted on land. In lakes and swamps and on the flood plains of rivers there is at least a chance for an organism to be buried beyond the reach of destructive agents. Falls of volcanic ash also create preservative conditions. Some of the most beautifully preserved insects, plants, and mammals of the Cenozoic come from such formations of ash.

Fossil faunas may represent actual death assemblages or the accumulation of shells or bones transported by currents. A death assemblage that has not been greatly dis-

turbed is shown in Figure 2-3. Here we have a hap-hazard group of limb bones and skulls of an Upper Triassic amphibian. The deposit in New Mexico from which this slab was excavated has yielded about fifty well-preserved skulls and a large number of limbs and other bones. Just what caused the death of these animals it is hard to say, but it may have been a severe drought in which the ponds and streams where they lived gradually dried up. Whatever the cause, the bodies were not moved much, if at all, after death except as the flesh decayed and the skeleton became disarticulated. Such finds as this are by no means common; before this one, in fact, only a few rather poorly preserved examples of this particular amphibian were known.

Among the most spectacular fossils discovered during the past century have been the numerous frozen carcasses of the woolly

FIGURE 2-1

Relative numbers of known species, living and fossil, of various animal phyla. (After Muller and Campbell, 1954.)

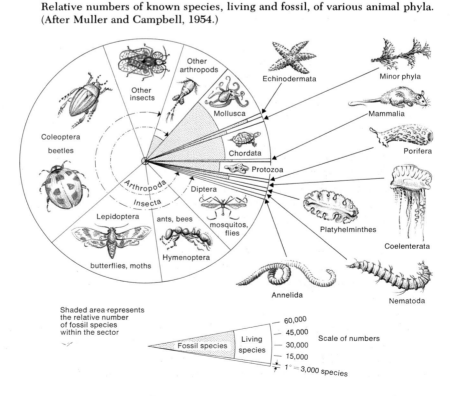

Cenozoic

Cretaceous

Jurassic

Triassic

Permian

Pennsylvanian

Mississippian

Devonian

Silurian

Ordovician

Cambrian

FIGURE 2-2

The expansion of life. The width of the figure is approximately in proportion to the known diversity of organisms at various times in the past. (After Simpson, 1953.)

mammoth found in Siberia. Several specimens with portions of the skin, flesh, and internal organs almost perfectly preserved have been collected. The first report of the mammoth is found in Chinese ceremonial books of the fourth century B.C. Fossil ivory was, for a long time, an important material in the economy of Siberia. Statistics on the trade give a fairly good idea of the number of mammoths discovered. The first fossil ivory reached western Europe in 1611, and from then until World War I a brisk trade in the material was carried on. In 1913, for example (the last year for which statistics are available), 57,600 pounds of fossil ivory were sold at Yakutsk. It has been estimated that 50,000 fossil mammoths have been discovered so far, and some Siberian localities are considered to be as "inexhaustible as a coalfield, and in future, perhaps, the only

FIGURE 2-3

A fossilized graveyard of a Triassic amphibian (*Buettneria perfecta* Case) of New Mexico.

source of animal ivory."[*] Preservation of whole animals—that is, of both the skeleton and the soft parts—is, however, unusual. Another example is the preservation of a rhinoceros embedded, with a mammoth, in a Pleistocene asphalt deposit of eastern Galicia (Fig. 2-4). A most remarkable example, among the invertebrates, of the preservation of soft parts comes from Middle Cambrian shale formations of Alberta (Burgess shale). The fine-grained shale contains, at several horizons, flattened impressions of a large assortment of "worms" and arthropods, which, though they had no hard parts, nevertheless left the outlines of their bodies, their delicate appendages, and even their internal structures (Fig. 2-5).

Most fossils are remains of only the skeletal parts of the animal. There are many ways of preservation. Since most invertebrate animals have hard parts of calcium carbonate, silica, or phosphatic materials, we commonly find such shells completely unaltered. One of the commonest means of preservation of shells, bones, and plants is for the porous structures to become filled with mineral substances carried by ground water. The added mineral substance may be of the same composition as the hard parts of the organism or may be quite different. In either case, in this type of preservation, called **permineralization,** the fossil is always much heavier than the original bone or shell. Figure 2-6 shows brachiopods attached, as in life, to a branching colonial bryozoan. The shell material of both the bryozoan and the brachiopods was originally, and still is, calcium carbonate, and the porous structures have also been filled by calcium carbonate during fossilization. The same type of preservation is illustrated by an Eocene fish from Italy (Fig. 2-7). Note how all the delicate structures of the skeleton are beautifully preserved. Fossil eggs are among the rarest of fossils; the specimen illustrated in Figure 2-8 is from red-

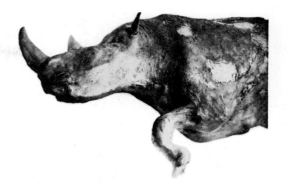

FIGURE 2-4

An extinct fossil rhinoceros embedded (along with a mammoth) in Pleistocene asphalt deposits of eastern Galicia. (From Lomnicki, 1914.)

beds of Pennsylvanian age in Texas and is the oldest fossil egg known.

Fossils that preserve the most delicate structures faithfully are frequently found to have a chemical composition quite different from that of the original hard parts. Solution of the hard parts was accompanied by deposition of some other mineral substance. Fossil shells that were composed during life of calcium carbonate are frequently found to be preserved entirely as silica, pyrite, or hematite, to name a few examples. The commonest replacing substance is silica (Fig. 2-9). An X-ray photograph of a slab of Lower Devonian shale from Germany reveals a nearly complete trilobite with appendages, gill structures, etc. (Fig. 2-10). The specimen is covered by a thin film of pyrite, thus permitting the X-ray image. Mechanical preparation of such a specimen is nearly impossible.

Carbonized films are a common result of fossilization. The volatile components of the animals or plants were released, leaving a thin film of carbon. Plant leaves (Fig. 2-11), graptolites, and arthropods (Fig. 2-12) were often, and fishes and marine reptiles were occasionally, preserved in this way. The remarkable Middle Cambrian fossils illustrated in Figure 2-5 were also preserved as carbonized films.

---

[*]I. P. Tolmachoff, Trans. Amer. Phil. Soc. (n.s.), Vol. 23, Part 1, p. 14 (1929).

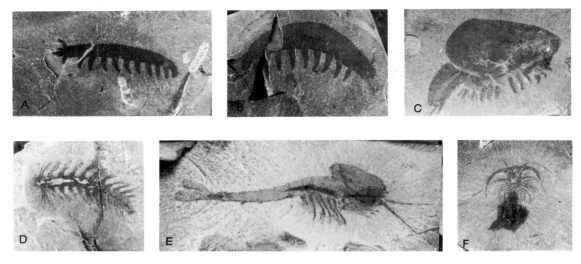

FIGURE 2-5

Carbonized films of soft-bodied worms and arthropods from the Burgess Shale (Middle Cambrian) of British Columbia. (Photographs by H. B. Whittington of specimens in the U.S. National Museum.)

A. *Aysheaia pedunculata*, the only fossil onychopore —
   a minor group of arthropods.
B. *Hymenocaris perfecta*, another extinct group whose
   affinities are not known.
C. *Aysheaia pedunculata*.
D. *Canadia setigera*, a fossil annelid (worm).
E. *Waptia fieldensis*, a primitive shrimp-like arthropod.
F. *Marrella splendens*, a trilobite-like arthropod.

Very often, instead of the structure of the animal itself, we find a mold or cast that represents the impression made by the shell or skeleton on the material of a rock. If the impression is of the outside of the hard parts, it is called an **external mold**; if it is of the inner surface, it is called an **internal mold**. The filling of a mold by a foreign substance forms a **cast** (Fig. 2-13).

Tracks, trails, and borings of animals also form a significant fossil record. Footprints of dinosaurs and various arthropods in fine muds are fairly common. Figure 2-14 shows the track and trail of the walking appendages and dragging tail of a horseshoe crab (*Limulus*). It is an extremely rare specimen in that the track, the trail, and the animal

that made them are all preserved. Worm borings are also rather common fossils. They are illustrated in Figure 2-15.

Finally, pieces of fossilized excrement, known as coprolites, are fairly common fossils. Fecal pellets of marine organisms often make up a considerable part of certain sedimentary formations.

2-3 THE SIGNIFICANCE OF THE FOSSIL RECORD

The study of the relations between organisms and their environments is called **ecology**; when it deals with ancient living

FIGURE 2-6

Several brachiopods attached, as in life, to a fragment of a bryozoan.

FIGURE 2-7

A small fish (*Amphistium paradoxium*) preserved in fine-grained limestone of Eocene age in Italy, showing nearly all of the skeleton in perfect preservation.

FIGURE 2-8

The world's oldest egg. This fossil, obtained from the Carboniferous redbeds of Texas, was the egg of one of the primitive reptiles whose bones have been found in the same deposit but which cannot be definitely associated with a particular species. (Photograph from A. S. Romer.)

FIGURE 2-9

In some of the finest preserved fossils, as in these brachiopods (*Prorichtofenia*) from Permian rocks of western Texas, the calcareous shell has been replaced by silica, a mode of preservation called silicification. (Photograph from the Smithsonian Institution.)

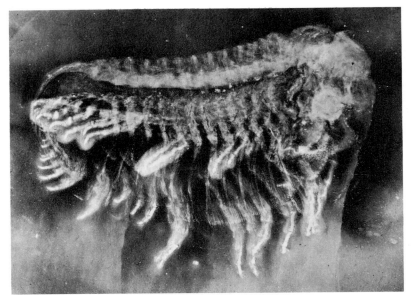

FIGURE 2-10

An X-ray photograph of a slab of Lower Devonian shale from Germany,
revealing a spectacularly complete specimen of a trilobite (*Phacops*).
(X-ray taken by W. Stürmer, Erlangen, West Germany.)

FIGURE 2-11

Carbonized plant leaves (*Sphenopteris gracilis*) from
Pennsylvanian formations of Illinois.

FIGURE 2-12

A slightly carbonized butterfly, showing color markings. (Photograph by F. M. Carpenter.)

communities, it is called **paleoecology.** To realize the full significance of fossils in the study of the earth's history, we must keep in mind that fossils were once animals or plants living in a friendly environment. And, since mere chance has limited our sample of what was once a living community, we must keep in mind not only the incompleteness of the sample but also—and especially—the reason for that incompleteness, for we get information not only from what we find but also from what we do not find. From that information we derive much of our understanding of ancient environments, ancient geography, evolution, and chronology.

FIGURE 2-13

Fossil molds and casts.

A. Natural cast of the interior of a pelecypod, showing the scars of the anterior and posterior muscles, the pallial line, and something of the structure of the hinge plate and teeth.
B. Sandstone cast of the interior of a siliceous sponge that must have looked much like the modern deep-sea sponge known as Venus'-flower-basket.
C. Natural molds of ammonites.
D. Part of a cast of the molds shown in C.

Fossils as Documents
of Ancient Environments

We know that there is a great variety of physical environments for all the kinds of life that we find on land and in the sea, and we recognize that the environmental conditions existing today had their parallels in all geologic eras. We are justified, therefore, in applying—cautiously—the ecology of the present to paleoecology. Since, as we have seen, marine life is a far more abundant source of fossils than land life, let us begin with the ecology of the sea.

The sea has fairly well-defined depth zones (Fig. 2-16), within each of which life is adapted to all the physical factors of the environment. The forms of life found in the littoral zone (between high and low tides) must be able to withstand periodic exposure (as the tides move in and out), the pounding of waves, and the shifting of sand. The neritic zone (from the low-tide line to a depth of 600 feet) is probably the most heavily populated of the marine environments; the animals living there need at least a little light. The bathyal and abyssal zones are less heavily populated than the neritic zone and are characterized by absence of light, high pressure, and extreme cold. In all the zones we find animals adapted to living on the bottom (benthonic forms), free swimmers (nektonic forms), and free floaters on or near the surface (planktonic forms). From the organization and structure of the animals we can tell to which of these ways of life they are adapted.

The relation between marine organisms and their environment is well illustrated by the distribution of faunas at the bottom of the Kattegat, a strait between Denmark and Sweden. Danish marine biologists have recognized six distinct animal communities in this area and have named each after one of its characteristic species. Figure 2-17 shows the distribution of the communities and sketches of what one-quarter of a square meter of the sea bottom within each com-

FIGURE 2-14

Rare cast of a *Limulus* and its trail. One rarely finds both fossilized tracks or trails and fossils of the animals that made them. The photograph is of a plaster cast in the Museum of Comparative Zoology at Harvard University; the original specimen is in the Paleontological Institute of Munich.

munity looks like. It is not necessary to be familiar with the Latin names of the organisms to see that there are differences between the communities. The factors that control their distribution are extremely complex; they include the character of the bottom, the food supply, and the salinity and temperature of the water.

As comparison of the two maps in Figure 2-17 shows, some of the communities are definitely associated with particular types of bottom sediment; the *Brissopsis* community, for example, lives on blue-clay bottoms. The *Macoma* community, however, seems to be independent of the nature of the bottom. Associated with the nature of the bottom as a factor in the distribution of a community are the types of animals that make up the community; for an animal can live only where it can find food without being completely at the mercy of predators. Ophiuroids (brittlestars), for example, feed on organic debris and the larvae of other animals; where brittlestars are abundant, such forms as pelecypods and gastropods, which have free-swimming larvae, are often scarce. The predatory activites of the brittle-stars prevent a large pelecypod or gastropod population from getting established.

Nearly all types of marine life are limited in distribution by depth and temperature. The modern reef-building corals of the South Pacific are a good example. Reef corals thrive best in waters with a depth of less than 150 feet but can grow at 300 feet; they thrive best at temperatures of 75–85°F but survive between 65° and 95°F. The distribution of reef-building corals is much more restricted today (Fig. 2-18) than it has been at many times in the past. Modern corals are rather closely related to corals as old as the Jurassic, and we assume that Jurassic corals had about the same limits of depth and temperature as modern corals. (Corals and reefs older than the Mesozoic are very different, as we shall see in § 7-1.) Many species of sea urchins (echinoids), gastropods, and pelecypods also have definite environmental restrictions that greatly aid the ecologist and paleoecologist in their studies of modern and ancient environments.

The climatic zones of the land are far more sharply defined than those of the sea. Besides the straightforward latitudinal differentiation into tropic, temperate, and arctic zones, with their various subdivisions, there is a vertical stratification, as anyone can note when going up a high mountain. Most modern land plants and animals are associated with restricted ecological niches.

The paleoecological conclusions drawn from the study of fossils are based largely

A

B

FIGURE 2-15

A. Worm burrows in a Cambrian sandstone.
B. These marks were made by worms, eating their way through a sandy mud and digesting such organic matter as they found in it.

FIGURE 2-16

Diagram illustrating marine bathymetric zones and distribution of environmental categories.

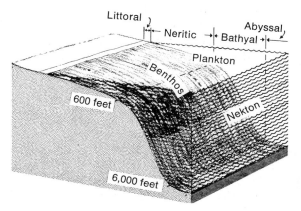

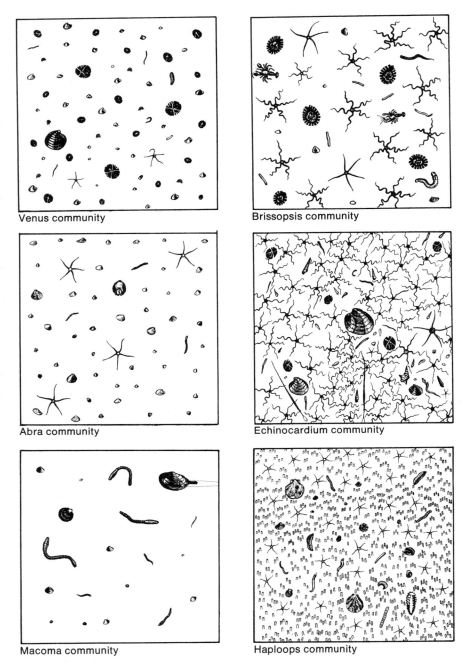

Venus community

Brissopsis community

Abra community

Echinocardium community

Macoma community

Haploops community

FIGURE 2-17

Animal communities and bottom sediments of the Kattegat, between Denmark and Sweden. The diagrams above show the appearance of approximately 0.25 square meter of the bottom within the animal communities indicated in the left-hand map on the opposite page. The right-hand map shows the types of bottom sediment. Note the close correlation between the distribution of some of these communities and the types of bottom sediment. (After Peterson, 1913, 1918.)

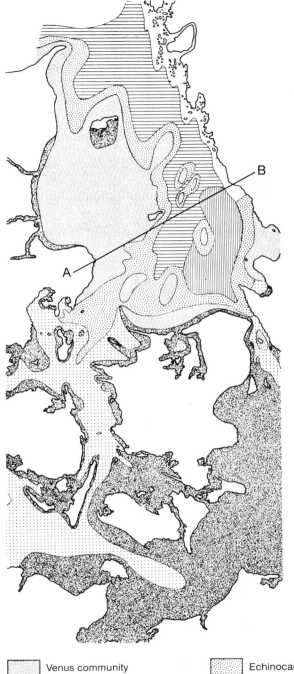

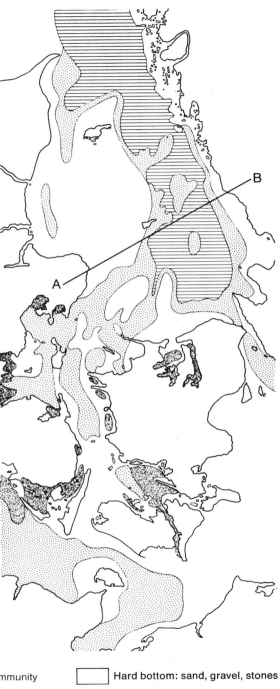

 Venus community	
Abra community	
Macoma community	
Brissopsis community	

Echinocardium community
Haploops community

Hard bottom: sand, gravel, stones
Mixed hard and soft bottom
Blue clay ⎫ Soft bottom
Black mud ⎭

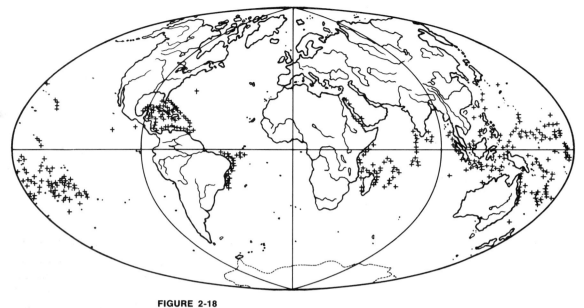

FIGURE 2-18
Distribution of modern reef corals. (After Ekman, 1953.)

on the form and structure of the fossils found in the same or a similar environment, the association of fossil forms, and the character of the entombing rocks. The paleoecologist projects to ancient forms the data derived from his study of modern environments and organisms. Needless to say, the further back in time he tries to carry analogies, the more tenuous they may become.

Sedimentary deposits representing all or most Phanerozoic marine and terrestrial environments are very unevenly distributed in the record. Table 2-1 is a crude attempt to delineate the major environments represented for major segments of Phanerozoic time. Note that for the Paleozoic shallow water marine deposits are dominant and only a small part of the record is represented by terrestrial deposits. There is a gradual change in the percentages of each major environmental type with time. It is not until the Cenozoic Era that we find a fine representation of diverse terrestrial environments. The significance of the data on this table will become more evident as we trace the physical and biological history of the earth for the Phanerozoic.

Fossils as Documents of Ancient Geography

The geographic distribution of modern plants and animals is closely controlled, as we have seen, by environmental limitations. An animal or plant species generally has a definite climatic or environmental range, within which it can live and reproduce, and is not found outside that range. As we examine the animal communities that live on the surface of the earth, we recognize a number of distinct **zoogeographic provinces.** Though environmental barriers are less sharply defined and methods of dispersion generally easier in the sea than on land, it is possible to recognize such provinces even in the sea. The fauna and flora of a province are determined, to a large extent, by its environment and its geologic history.

TABLE 2-1

Estimates (in percent) of the major sedimentary environments represented in the rock record for the major divisions of the Phanerozoic (f means fraction of 1 percent).

	Early Paleozoic	Late Paleozoic	Early Mesozoic	Late Mesozoic	Early Cenozoic	Late Cenozoic
Deep seas	0	0	0	0	0	f
Shallow seas	95	90	85	75	40	20
Lowland fresh-water (coastal plain, delta)	5	10	10	10	15	20
Continental fresh-water	0	0	5	15	40	45
Upland fresh-water	0	0	0	0	5	15

Source: Data from E. C. Olson, *The Evolution of Life*, Mentor, 1965.

The neritic fauna of our modern oceans is divided among a number of zoogeographic provinces (Fig. 2-19). None of these provinces are sharply defined; they have, rather, broad transitional boundaries. The temperature of the water, continental land barriers, and expanses of deep water are decisive factors in the distribution of animals among the provinces. During the present distribution of continents, for one example, members of the tropical Indo–West-Pacific fauna cannot migrate into the Atlantic, for none of them can survive the cold at the southern tip of Africa. The great width and depth of the East Pacific Oceanic Barrier, as a second example, prevent the transport of neritic larvae from the Polynesian island region to the American Pacific coast. This barrier is, in fact, the most pronounced break in the distribution of tropical neritic animals. Of about eighty genera of echinoderms found at the sides of this barrier, only 14 percent are common to both sides. At the Panamanian isthmus, on the other hand, of about eighty genera of strictly warm-water echinoderms, 37 percent are common to the Atlantic and Pacific sides; and in the tropical provinces of the Atlantic, of 60–65 genera, about 45 percent are common to the two sides.

The zoogeographic provinces of land animals were bounded in the late years of the nineteenth century. Figure 2-20 shows the major provinces of land animals (mainly mammals). The dispersion of land animals is even more complex than that of marine animals. Broad land connections generally allow a wide dispersion of most land animals; narrow connections, with their limited range of environments, act as filters and are commonly known as **filter barriers.** The modern vertebrate fauna of western Europe, for instance, is very similar to that of China, several thousand miles away, but quite unlike that of Africa, a mere thousand miles away. If we look at the map, we see that migration was fairly easy in the first instance but difficult, if not impossible, in the second. Extensive bodies of water prevent the dispersion of most land animals. The exception—the accidental rafting of small animals through a sea with scattered islands—is commonly called, because of its uncertainty, the **sweepstakes route.**

We need a background of historical and paleontological data to understand the significance of the zoogeographic provinces of both land animals and marine animals. In Chapter 14 (on Cenozoic life), once we get a better perspective of the physical and biological history leading up to the modern world, we shall have a further discussion of Figure 2-19.

From a study of ancient faunas and floras and the nature of the enclosing strata we can make maps of particular times in the

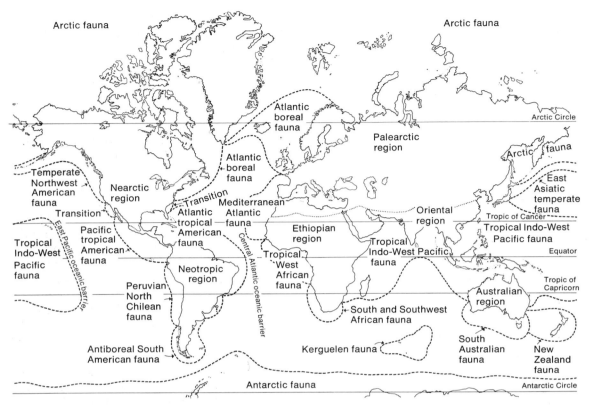

FIGURE 2-19

Zoogeographic provinces of shallow-water marine faunas and of land animals, primarily mammals. (Data from Ekman, 1953, and Simpson, 1953.)

earth's history. This study, **paleogeography,** is an extremely complex phase of historical geology; it requires a synthesis of all available data from both fossils and their entombing strata. Our study of the geography of modern plants and animals establishes a framework for the study of ancient geography.

Fossils as Documents of Evolution

William Smith's discovery of the usefulness of fossils in the identification and correlation of rock strata was a great stimulant to biologists. Before that discovery the age of the earth and the chronological succession of strata were imperfectly understood, and the concept of geologic time, as we know it, was almost completely lacking. Out of that discovery came new methods of studying the processes and history of life, and the geologist and the paleontologist became active contributors to the study of evolution.

There is no longer any controversy as to whether or not evolution has taken place. The huge and growing mass of data from paleontology, comparative anatomy, embryology, taxonomy, and biochemistry all demonstrate decisively the fact of evolution. The mechanism of evolution, however, is not yet completely understood. The bare outline of evolutionary theories given here cannot do justice to an enormously complex subject, but a cursory acquaintance with the major theories and with some of the problems of evolution will help us to understand the geologic and biologic history of the earth that is related in the following chapters.

Two men—the Chevalier de Lamarck (1744–1829) and Charles Darwin (1809–1882)—removed the study of the evolution of life from a theological and metaphysical frame to its proper place in biological philosophy. Lamarck was the first really scientific student of evolution. His book *Zoological Philosophy*, published in 1809, aroused a controversy that is still not completely resolved. Lamarck recognized, in the successive organisms that he studied, a progressive complication of organization, and to him that meant evolution; he saw, however, that living organisms could not be classified into a simple, continuous series of ever increasing complexity. Environment, he believed, was a factor in, but not a cause of, evolution. An important concept in his thesis was the use or disuse of organs: any organ that is constantly used will, he believed, increase in size or efficiency in succeeding generations, and any that is not used will decrease. This belief led to the idea that a characteristic acquired by an individual organism during its life could be inherited by its descendants. The idea is appealing, and it is not surprising that Lamarck had a great following, but most biologists today agree that the idea is wrong. The mechanism of biological heredity, as revealed by the science of genetics (to be mentioned later in this section), seems to make the inheritance of acquired characteristics impossible. For many years, however, the idea was a guide to zoologists and paleontologists, and only in the last two generations has it been pretty well weeded out of scientific literature.

Darwin's *The Origin of Species*, which appeared in 1859 and immediately stirred up a violent controversy, is generally recognized as one of the most important works ever produced by the mind of man. The idea of evolution had been known long before Darwin, but he was the first to demonstrate convincingly the general principle of evolution, to show (incompletely) how it works, and—by his theory of natural selection—to explain adaptation. Briefly, Darwin's thesis is as follows: Organisms are enormously variable; that is, no two individuals are exactly alike. Some of the variability in animals is hereditary. Most animals are enormously fertile and produce large numbers of eggs or offspring. In spite of this high fertility, the number of individuals of a species remains approximately constant. The individuals that are best adapted to their environmental niches are the ones that are most likely to mature and reproduce—that is, to survive.

The process by which all this takes place is known as **natural selection.** In the selective process the individuals that have characters or traits that give them even a slight advantage over other individuals, of the same species or of other species competing for the same environmental niche, tend to live longer and hence to produce more offspring. They pass on to many of their descendants the characteristics that have enabled them to live longer than other individuals in the same environment. Their descendants pass on such characteristics to some of their own offspring; and in the repetition of the process more and more of the better-adapted individuals will occupy the environment, while fewer and fewer of the less well-adapted will survive. Though Darwin himself did not believe that natural selection was the complete explanation of animal evolution, some of his supporters did. Their school of thought is generally called **neo-Darwinian.**

The study of evolution was revitalized at the beginning of the twentieth century with the introduction of the science of **genetics,** which deals with the mechanism of heredity. The controversy over Darwin's theory of natural selection centered on two questions

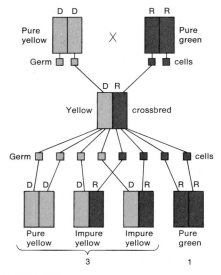

FIGURE 2-20

Diagram showing the segregation of yellow and green pea seeds, yellow being dominant (*D*) and green recessive (*R*).

that he had not answered: "How do variations arise?" and "How does heredity operate?" While pro-Darwin and anti-Darwin factions carried on their arguments, a clue to the answers was published by Gregor Johann Mendel, an Augustinian monk of Moravia (now a part of Czechoslovakia) in 1866, only seven years after Darwin's *Origin of Species.* Mendel's contribution was published by a local natural history society and went unnoticed until 1900. He was interested in heredity and especially in the inheritance of variations. He experimented with the ordinary culinary pea, which has two types of seeds—a green and a yellow. Upon crossing pure-green and pure-yellow seed types, he obtained nothing but yellow-seeded peas. Numerous repetitions of this experiment, always beginning with pure-breeding green and yellow seeds, produced the same result. When he bred the first generation of the crossbreds, however, he obtained pure-yellow, impure-yellow, and pure-green peas. This established the fact that, whereas the first-generation cross-

breds did not produce any green peas, they must have carried the green color without showing it. Another surprising result was the ratio of colors: three yellow to one green. Upon breeding the second-generation green seeds, Mendel got only green seeds, and repeated generations continued to produce only green seeds. The story was quite different with the yellow seeds. Though some of the yellow seeds bred true, in the same way as the green seeds, the rest of the yellows produced the same mixture of pure yellows, impure yellows, and pure greens (Fig. 2-20). The nature of the yellow seeds could be determined only by breeding.

Mendel's experiments led him to four conclusions: (1) the seed color is controlled by a pair of hereditary factors; (2) the factors of each pair are derived from the plant's parents, one member of the pair from each parent; (3) each germ cell (reproductive cell) bears only one factor, and on fertilization two factors are brought together; (4) the factors for yellow seed and green seed are alternative forms of the factor for seed color, the yellow being **dominant** over the **recessive** green.

Mendel's contribution was discovered in 1900 by three European scientists and soon caused a revolution in biologic thought. The simple relations discovered by him were verified and have since been greatly elaborated into a new science, **genetics.** It is now known that within the nuclei of cells are small bodies called chromosomes (so called because they readily take up dyes). The chromosomes contain the actual units of inheritance, called **genes**—the paired hereditary factors of Mendel. The alternative forms of each factor are called **alleles;** the gene for yellow seeds and the gene for green seeds are alleles of each other. Spontaneous changes in the chromosomes or genes, called **mutations,** are the source of new hereditary forms. The causes of mutations are not understood; their appearance can be accelerated by X-rays and certain chemicals, but their nature seems to be completely random and is not predictable. Genetics, an extremely complex science, is still developing. It is to the geneticist that we now turn for data on the actual mechanism of evolution.

Students of evolution, early in this century, were faced with two strong schools of thought, the Darwinian and the Mendelian. It turned out, as it usually does, that neither of the schools was completely correct. The modern theory of organic evolution, combining the two schools, is called the **synthetic theory.** This theory regards random mutations as the materials of evolution and natural selection as the guiding process.

Darwin devoted two full chapters of *The Origin of Species* to the geologic record and paleontology, and even before his time the contributions of paleontology to the study of evolution had been significant. Paleontology cannot tell how changes take place, but it adds the third dimension, time, which the zoologist and geneticist alone cannot supply. The fossil record—a chronological record of past life—furnishes clues to the patterns of evolution. Relations between fossil animals and plants are determined by morphological similarity (that is, likeness in form and structure) in conjunction with stratigraphic and geographic position. If we take any small group of closely related or morphologically similar animals and arrange them according to their stratigraphic position, we can generally trace their changes in structure. Figure 2-21 illustrates the appearance of four successive species of oysters of the genus *Cubitostrea,* which lived in Middle Eocene time in the ancient Gulf of Mexico. These species are separated by stratigraphic gaps in the geologic record, and intergrading populations are therefore not known. The structures and life histories of these species, when seen in their stratigraphic relations, demonstrate clearly that the progressive changes they have undergone include (1) an increase in size, (2) an increase in the

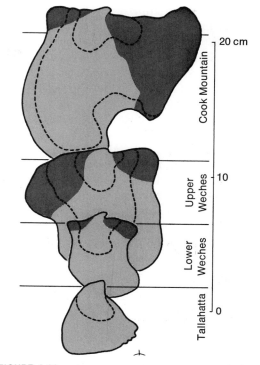

FIGURE 2-21

Left valves of species composing the *Cubitostrea sellaeformis* (Conrad) stock, ½ natural size. These are outlines of the left valves of large specimens of each species. They show the development of auricles (projections of shell along hinge line) and the increase in shell size from species to species. The species are arranged with the earliest at the bottom and the latest at the top. Horizontal bars are extensions of the hinge axis. Auricles are shaded differently from the main shell body. Some growth lines are indicated by broken lines. (After Stenzel, 1949.)

size of the auricles, (3) a decrease in the strength of the ribs, (4) the disappearance of the triangular outline, and (5) the appearance of a twist in the shell.

Progressive increase in size is commonly observed in a group of closely related animals, whether invertebrate or vertebrate. The large dinosaurs of the Jurassic and Cretaceous were the culmination of dinosaur lines that had been gradually increasing in size for several million years.

Many orders of Cenozoic mammals underwent striking increases in size during their evolution. In the invertebrate phyla such increase is common among the brachiopods, echinoids, corals, and mollusks—to name just a few. The crinoid genus *Calceolispongia* is found throughout 7,000 feet of Middle Permian strata in Western Australia. The preservation of the specimens is not good enough to give us data on the size of the complete animal, but one structure, the basal plate, increased in volume 40,000 percent during a time estimated as 6,000,000 years. The largest known representatives of the vertebrates, crustaceans, echinoderms, pelecypods, gastropods, cephalopods, coelenterates, and annelids—all appreciably larger than their largest known fossil relatives—are living forms. Increase in size is not, however, universal among animals; many groups show no such increase in their evolution.

Evolutionary changes in animals can be convergent, divergent, or parallel. If unrelated groups come to have a similar appearance, we say that their evolution is convergent. Certain genera of fossil brachiopods and pelecypods, for example, had a cone-shaped shell that was cemented to the sea bottom like that of the solitary corals—the brachiopods and pelecypods had become adapted to an environmental niche similar to that of the corals. If closely related groups become less similar as they evolve, their evolution is said to be divergent. Evolution that is neither convergent nor divergent is said to be parallel. Parallelism is common in the fossil record.

One interesting aspect of animal evolution is the apparent orientation of trends—a phenomenon called **orthogenesis**. Oriented trends have been described for many groups of animals and are believed by some to demonstrate the existence of an internal force, an inherent tendency. The example most commonly cited is the evolution of the horse. The earliest known horse, eohippus, lived in Eocene time, was about a foot and a half high, and had four toes on the fore-

feet and three complete toes and a remnant of a fourth on the hind feet. The teeth were not fit for grazing but only for browsing on soft vegetation. Popular discussions interpret the evolutionary history of the horse as a steady, gradual change, in a single line, from Eohippus to *Equus*, the modern horse. Detailed analysis of the abundant fossil data shows, however, that the evolutionary changes in limb proportions, teeth, number of toes, etc. were not constant, and that there were dozens of lines rather than one. Figure 2-22 is a highly simplified diagram of the evolutionary history of the horse. The early three-toed horses, browsing animals with low-crowned teeth, underwent several episodes of evolutionary **radiation**—that is, produced several distinct lines from a common ancestor. One radiation gave rise to three-toed horses with high-crowned teeth adapted to eating grass. Such horses underwent evolutionary radiation into several different lines, one of which produced the one-toed horses with high-crowned teeth that are ancestral to our modern horse. Instead of a single orientation there were, we see, several periods of radiation, from which certain adaptive types survived and others did not. Many examples of so-called oriented evolution break down thus upon detailed analysis. Evolution is sometimes oriented; when it is, the orienting factor is not something inherent in life, but adaptation—the interplay between organisms and the environment.

Fossils as Paleontological Clocks

Have the number of days and months per year always been what they are now? The question is extremely important, as the answer is needed for our understanding of the earth's rotation and the factors affecting it. The earth's rotation has been slowed by the friction of the tides, and calculations lead to the conclusion that the length of the day has increased slowly throughout geologic time. These calculations indicate

FIGURE 2-22

Evolution of the horse family. The animals shown
are only a few of the many in the family, and lines
of evolution were more numerous and complex than
indicated here. The various forms are drawn to
the same scale. (After Simpson, 1949.)

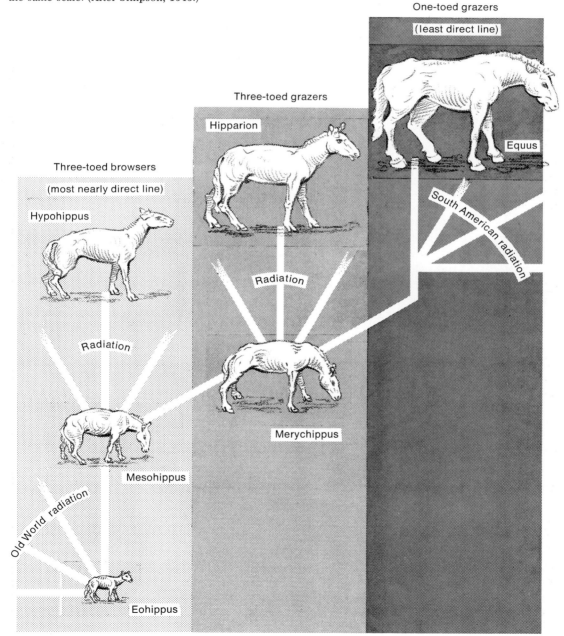

One-toed grazers

(least direct line)

Equus

Three-toed grazers

Hipparion

Three-toed browsers

(most nearly direct line)

Hypohippus

South American radiation

Radiation

Radiation

Merychippus

Mesohippus

Old World radiation

Eohippus

that the year had about 428 days at the beginning of the Cambrian and about 400 days in the middle of the Devonian. What has been needed is a check on these calculations. The study of growth lines of marine invertebrates has shed light on this problem.

Professor John W. Wells of Cornell University was the first to suggest that daily growth lines were preserved on fossil corals. These organisms have long been known to have distinct bands that represent annual growth. Wells counted the fine bands on several Middle Devonian corals and found that they range in number from 385 to 410, with an average of 400. This was an encouraging beginning, as it agreed with calculations based on tidal friction. Work is continuing to establish the data more firmly. Surprisingly, data on the growth of living corals are still few. Visual counting of these minute lines needs to be mechanized to eliminate error.

This same general line of research has been applied to Mollusca, mainly pelecypods, with some very interesting, though preliminary results. The results of a systematic survey of such mollusks of Upper Cambrian to Recent age are shown in Figure 2-23. According to this study the lunar month in the Upper Cambrian was 31.56 days, in the Upper Cretaceous, 29.92 days, etc. The most interesting feature of this graph is the break in slope between the Pennsylvanian and the Upper Cretaceous. If the data are correct the graph clearly indicates that the deceleration rate of the earth's rotation has not been constant. It must be emphasized that the data on the corals and mollusca are as yet preliminary and that much additional work is needed. The data assembled to date, however, are very promising.

Fossils as Documents of Chronology

The one unifying factor in the study of historical geology is the change of organisms with time. Even before Darwin popularized

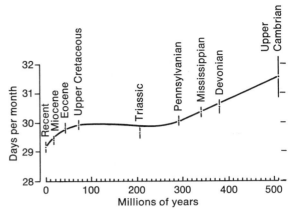

FIGURE 2-23

Variation in the length of the synodic month through geologic time as determined from count of growth lines on fossil Mollusca (mainly pelecypods). The bars for each point indicate the standard error. (After G. Pennella, C. MacClintock, and M. N. Thompson, *Science*, **162**, Nov. 15, 1968. Copyright 1968 by the American Association for the Advancement of Science.)

the theory of evolution, William Smith, as we have seen, had recognized the value of fossils for the identification and correlation of rock strata, and a standard geologic column had been established. Field observations in the type areas of the major divisions of the geologic column yielded an empirical sequence of fossil forms. Each geologic system was recognized as having its own fossil organisms, distinct from those of other systems, and within each system subdivisions based on fossil content were recognized.

Since animal evolution is the framework of geological chronology, and since the physical history and the biological history of the earth are intimately related, it is appropriate here, before we take up specific aspects of the earth's physical history, to consider the broad aspects of animal evolution. For the moment let us consider only the larger groups of animals, the phyla and the classes. When do the various phyla first appear in the earth's record? What was their ancestry? What particular patterns or modes of evolution do they display? Did they become extinct? Is there any relation between the evolutionary patterns of animals and plants and the time scale adopted early in the nineteenth century? Only partial answers can be given to most of these questions.

2-4 INTERRELATIONSHIPS OF THE ANIMAL PHYLA

Since most phyla common in the fossil record are present in Cambrian or Ordovician formations as distinct entities, their evolutionary diversification must have taken place during the Precambrian. The fossil animals of the Precambrian, unfortunately, are all but unknown, and we must therefore rely on embryology and comparative zoology for clues to the interrelationships of the phyla.

The degree of structural complexity does throw some light on this matter. The simplest (most primitive) organisms—capable, however, of all the functions of living and reproducing—are the one-celled members of the kingdom Protista. We are fairly sure that life began in some such one-celled structure. A grade above the one-celled organism is one in which a number of cells, joined together and differentiated, have a more definite outline. The sponges represent this grade. All multicellular animals belong to the kingdom Animalia. This kingdom is divided into animals whose cells are not organized into tissues and organs (the Porifera, or sponges) and animals whose cells are organized into tissues and organs (all the remaining phyla).

The sponges are not closely related to any other phyla of animals; they constitute a branch that split off from the main course of development early in the history of animals and has persisted until the present with little fundamental modification of structure.

The Coelenterata, including the corals and the jellyfish, are morphologically advanced over the sponges by having radial symmetry, a definite mouth, a digestive tract (the sole cavity of the body), and a body wall consisting of two layers, the cells of which are organized into tissues and incipient organs.

All the remaining phyla have bilateral symmetry, three-layered body walls (a mesoderm being inserted between the outer ectoderm and the inner endoderm), and body spaces other than the digestive tract. Tissues and organs are more highly developed. The two main divisions of these phyla, the chordate line and the arthropod line, are divided by origin of the mesoderm. In the chordates and echinoderms the origin follows nearly identical patterns, but in all the other phyla it follows a different pattern (Fig. 2-24). This similarity, plus data from embryology, indicates that the vertebrates and the echinoderms are more closely related to each other than to any other phyla. The Annelida, Mollusca, Brachiopoda, and Bryozoa all have a similar free-swimming larval stage called trochophore—a similarity taken to indicate that they all originated in some unknown common ancestor. The Arthropoda, structurally,

FIGURE 2-24

Origin of the mesoderm in arthropod and chordate lines. In the chordate line the mesoderm originates as pouches from the sides of the primitive endoderm. In the arthropod line primitive mesoderm cells bud off from the primitive endoderm. (After *Biology: Its Principles and Implications*, by Garrett Hardin. W. H. Freeman and Company. Copyright © 1966.)

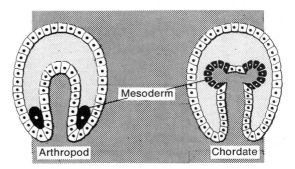

are closely related to the Annelida. The interrelationships of the main phyla are shown in Figure 2-25.

2-5 PRECAMBRIAN LIFE

The most dramatic breakthrough in the study of ancient life in the past decade has been the discovery, through refined instrumentation not previously available, of a great variety of microorganisms of Precambrian age. In addition, significant advances in organic geochemistry have made it possible to isolate and characterize many biologically important organic compounds preserved in early sediments. Further, electron microscopy has made possible the detection and study of fossil bacteria less than a micron in size.

Paleontologists have long been interested in Precambrian life. One of the most common of Precambrian fossils are stromatolites, which are biogenic sedimentary structures produced by the calcium-carbonate-precipitating and sediment-binding activities of successive mats of algae, predominantly blue-green algae. They are found in all the principal Precambrian formations. Among the animal fossils reported to have been found in these ancient rocks are jellyfish, sponge spicules, burrows and trails, radiolarians, bacteria, foraminifers, brachiopods, and arthropods. Only the burrows and trails of worm-like animals have been verified; the other reports, on re-examination of the evidence, have been refuted. The stromatolites and the burrows and trails are not an impressive record for a span of time covering more than three quarters of earth's history, but now let's look at the new data. These we shall discuss chronologically from oldest to youngest.

The oldest direct evidence of life has come from the Fig Tree Series, a sedimentary formation near Barberton, South Africa, that contains black, carbon-rich cherts. The radiometric dates for this series of rocks indicate an age of more than 3.1 billion years. The fossils of the Fig Tree consist of spheroidal alga-like organisms (Fig. 2-26, *41–45*) and bacterium-like cells (Fig. 2-26, *46–48*). That the organisms in the Fig Tree Series were photosynthetic is suggested by the presence of certain organic compounds, but the data as yet are too incomplete to establish this for certain. Nevertheless, the presence of photosynthetic organisms would be consistent with the occurrence of stromatolites, probably produced by blue-green algae, in a limestone near Bulawayo, Rhodesia, that is about 2.7 billion years old.

The most significant discovery of Precambrian fossils was made in black cherts of the Gunflint Iron Formation near Thunder Bay, Ontario, on the northern shore of Lake Superior. Radiometric age determinations indicate that this formation is approximately 1.9 billion years of age. The microorganisms of the Gunflint are abundant and quite diverse. The most abundant are unbranched filaments, some septate, which are probably related to modern blue-green algae and have been given the name *Gunflintia* (Fig. 2-26, *13–16*). Certain finely septate types named *Animikiea* (Fig. 2-26, *24–25*) exhibit a basic morphology comparable to that of living filamentous blue-green algae, such as *Oscillatoria* and *Lyngbya*. Certain of the filaments contain spheroidal spore-like bodies (Fig. 2-26, *26*) that are similar to spore-containing filaments in certain modern blue-green algae and iron bacteria. Other forms, such as *Eoastrion* (Fig. 2-26, *17–18*), resemble modern iron- and manganese-oxidizing bacteria found in lakes of northwestern U.S.S.R. Finally, some forms are so unlike any known organisms that their biologic relations are uncertain (*Eosphaera*, Fig. 2-26, *19*; and *Kakabekia*, Fig. 2-26, *20–23*).

This assemblage of microorganisms in the Gunflint Formation appears to consist predominantly of photosynthetic algae that

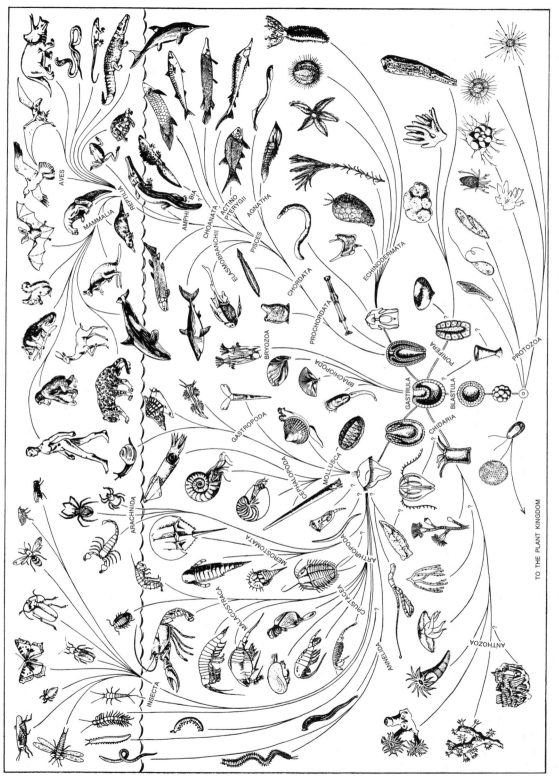

FIGURE 2-25

Relationships of the animal kingdom. (Reproduced with the permission of A. Heintz and L. Størmer, Paleontological Museum, Oslo, Norway.)

FIGURE 2-26 (*see following two pages*)

Miscellaneous Precambrian microorganisms and "geologic clock" in billions of years, showing major fossiliferous Precambrian sediments and time ranges of primitive organisms.

Late Precambrian microorganisms
(Bitter Springs Chert, Central Australia, approximately 1,000 million years old)

1 through 3: *Palaeolyngbya barghoorniana* Schopf; scales = 10 μ; a blue-green alga very similar to members of the modern genus *Lyngbya*. These microorganisms have evolved little or not at all since Bitter Springs time.

4: *Contortothrix vermiformis* Schopf; scale = 10 μ; a coiled blue-green alga.

5: *Archaeonema longicellularis* Schopf; scale = 10 μ; a well-preserved blue-green alga, also known from the late Precambrian of South Australia.

6: *Oscillatoriopsis obtusa* Schopf; scale = 10 μ; a blue-green alga quite comparable with members of the extant genus *Oscillatoria*.

7: *Caudiculophycus rivularioides* Schopf; scale = 10 μ; this blue-green alga is similar to some members of the Rivulariaceae.

8: *Halythrix nodosa* Schopf; scale = 10 μ; a distinctive filamentous alga composed of spool-shaped cells; although probably related to modern blue-green alga, also known from the late Precambrian of South Australia.

9 and 10: *Cephalophytarion grande* Schopf; scales = 10 μ; the constricted "neck" shown in these filaments is typical of some species of the modern Cyanophycean genus *Microcoleus*. Like *Paleolyngbya* (see 1 through 3) and *Oscillatoriopsis* (see 6), this blue-green alga has apparently exhibited little morphological change since Bitter Springs time.

11: *Heliconema australiensis* Schopf; scale = 10 μ; this Bitter Springs alga is comparable with twisted filamentous Cyanophytes of the extant genus *Spirulina*.

12: *Siphonophycus kestron* Schopf; scale = 10 μ; this microfossil probably represents the remnants of a gelatinous sheath that originally enclosed a blue-green algal filament.

Middle Precambrian microorganisms
(Gunflint Chert, Ontario, Canada, 1,900 million years old)

13: *Gunflintia minuta* Barghoorn; scale = 10 μ; a well-preserved, slender, filamentous blue-green alga.

14 and 15: *Gunflintia grandis* Barghoorn; scales = 10 μ; algal filaments, probably related to modern blue-green algae, with well-defined septa and cells of rather irregular dimensions.

16: Typical assemblage of tangled algal filaments (*Gunflintia*) and spheroidal algae (*Huroniospora*); scale = 10 μ; in densely packed areas such as this one, a single cubic inch of chert may contain several thousand microorganisms.

17 and 18: *Eoastrion simplex* Barghoorn (17) and *E. bifurcatum* Barghoorn (18); scales = 10 μ for 17, 5 μ for 18. These microorganisms seem quite comparable in morphology with a modern iron- and manganese-oxidizing bacterium known from several Karelian lakes.

19: *Eosphaera tyleri* Barghoorn; scale = 10 μ; although probably algal in affinity, the complete lack of modern morphological counterparts of this distinctive microfossil makes its relationship to living organisms quite problematical.

20 through 23: *Kakabekia umbellata* Barghoorn; scales = 10 μ; the basic morphology of this organism consists of three parts: bulb, stipe, and umbrella-like crown. The biological relationships of this intricate microorganism are uncertain.

24 and 25: *Animikiea septata* Barghoorn; scales = 10 μ; in general appearance, size and form, these filamentous microfossils resemble species of the living blue-green algal genera *Lyngbya* and *Oscillatoria*.

26: *Entosphaeroides amplus* Barghoorn; scale = 10 μ; with internally contained spheroidal sporelike bodies; similar spore-containing filaments are known in certain modern blue-green algae and iron bacteria.

27 through 29: *Huroniospora psilata* Barghoorn (27), *H. macroreticulata* Barghoorn (28) and *H. microreticulata* Barghoorn (29); scales = 5 μ for 27 and 28, 10 μ for 29. These organic spheroids, of probable algal affinity, are ubiquitous in the Gunflint chert (e.g., 16 and 26); similar forms are widespread throughout the Precambrian (34 to 36, 41 to 45).

Late Precambrian Microorganisms
(Bitter Springs Chert, Central Australia, approximately 1,000 million years old)

30 through 32: *Caryosphaeroides pristina* Schopf; scales = 10 μ; these spheroidal green algae contain remnants of their original nuclei and are the oldest demonstrably nucleated organisms now known.

33: *Sphaerophycus parvum* Schopf; scale = 10 μ; these spheroidal blue-green algae were apparently preserved while the organism was undergoing the process of cellular fission and before the resulting daughter cells had separated.

34 and 35: *Myxococcoides minor* Schopf; scale = 10 μ; well-preserved colonies of spheroidal blue-green algal cells. Note that some of the cells are somewhat angular because of compression against adjacent cells.

36: *Palaeoanacystis vulgaris* Schopf; scale = 10 μ; unlike most of the other Precambrian microorganisms shown here (which were photographed in thin sections of the rock), this blue-green algal colony was freed from the mineral matrix by dissolving the chert in hydrofluoric acid.

37: *Globophycus rugosum* Schopf; scale = 10 μ; the biological affinity of this coarsely reticulate alga-like microfossil is uncertain.

38: *Gloeodiniopsis lamellosa* Schopf; scale = 10 μ; although probably algal in affinity, the relationship of this ellipsoidal, sheath-enclosed cell to modern organisms is uncertain.

39: *Glenobotrydion aenigmatis* Schopf; scale = 10 μ; a well-preserved, sheath-enclosed colony of green algae. The circular bodies (denoted by the arrow) in each of the spheroidal cells may represent the remnants of starch storage structures known as pyrenoids.

Middle Precambrian Bacteria
(Gunflint Chert, Ontario, Canada 1,900 million years old)

40: Electron micrograph showing rod-shaped bacteria in a platinum-shadowed carbon replica of the chert surface; scale = 1 μ; these well-preserved bacilli are morphologically similar to certain extant iron bacteria.

Early Precambrian Microorganisms
(Fig Tree Chert, South Africa, approximately 3,200 million years old)

41 through 45: *Archaeosphaeroides barbertonensis* Schopf and Barghoorn; scales = 10 μ; these spheroidal, organic, alga-like microfossils, together with the bacterium-like rods shown in 46 to 48, comprise the oldest fossil evidence of life now known from the geologic record. Note the similarity between these early Precambrian microorganisms and *Huroniospora* (27 to 29) from the Gunflint chert (middle Precambrian).

46 through 48: *Eobacterium isolatum* Barghoorn and Schopf; scales = 1 μ; well-preserved, rod-shaped bacterium-like cells shown in longitudinal (46 and 48) and transverse sections (47) in electron micrographs of platinum-shadowed carbon replicas of the chert surface. The organically preserved cell wall (arrow, 47) is composed of two layers, as are the walls of many modern bacteria. The occurrence of cellular imprints in the rock surface, such as the one shown in 48 transgressing a polishing scratch in the mineral matrix, establishes that these rod-shaped cells are indigenous to this early Precambrian chert.

(Photographs from E. S. Barghoorn and J. William Schopf; "Geologic Clock" modified and adapted from J. W. Schopf, McGraw-Hill Yearbook of Science and Technology, 1967.)

Late Precambrian Microorganisms

Bitter Springs Chert,
Central Australia,
approx. 1,000 million years old

Middle Precambrian Microorganisms

Gunflint Chert, Ontario, Canada,
1,900 million years old

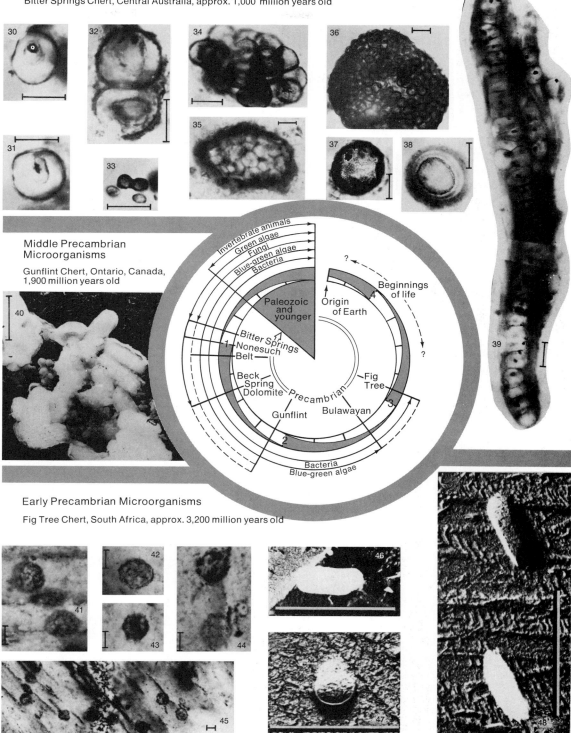

Late Precambrian Microorganisms

Bitter Springs Chert, Central Australia, approx. 1,000 million years old

Middle Precambrian Microorganisms

Gunflint Chert, Ontario, Canada, 1,900 million years old

Early Precambrian Microorganisms

Fig Tree Chert, South Africa, approx. 3,200 million years old

Invertebrate animals
Green algae
Fungi
Blue-green algae
Bacteria

Paleozoic and younger

Origin of Earth

Beginnings of life

Bitter Springs
Nonesuch
Belt

Beck Spring Dolomite

Precambrian

Fig Tree

Gunflint

Bulawayan

Bacteria
Blue-green algae

grew in laminar sheets or mats in shallow, agitated waters. The organisms were entrapped and preserved in a relatively unaltered state by their encasement in colloidal silica, which became lithified to chert.

The best-preserved and most diverse Precambrian flora known to date is from carbonaceous chert of the Bitter Springs Formation of central Australia, which is approximately one billion years old. The microfossils of the Bitter Springs Formation grew in the form of laminar sheets or mats near the sediment-water interface in a shallow, apparently marine environment. Silica-laden solutions, possibly derived from contemporary volcanic sources, permeated the sediments and entrapped the microorganisms. Thirty distinct microfossils have been described from this assemblage: nearly half are species of blue-green algae, many related to modern forms (Fig. 2-26, 1–12). Some appear to have evolved little or not at all since Bitter Springs time. In addition to the blue-green algae the Bitter Springs assemblage contains colonial bacteria, fungus-like filaments, and spheroidal green algae (Fig. 2-26, 30–39. Perhaps the most significant feature of the Bitter Springs flora is the preservation of nuclei in some of the spheroidal green algae (Fig. 2-26, 30–32). These are the oldest demonstrably nucleated organisms known.

Our discussion so far has mentioned only plant microfossils. What about the record for animals? Very late Precambrian formations in Australia, Siberia, England, and

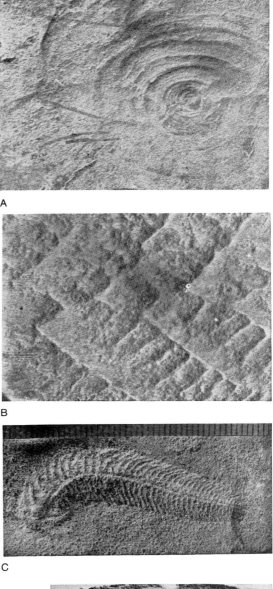

A

B

C

FIGURE 2-27

Fossil impressions from the late Precambrian of the Ediacara Hills, South Australia. (From Glaessner, 1961.)

A. *Cyclomedusa*, probably a jellyfish.
B. *Charnia*, probably a sea-pen, a kind of coral.
C. *Spriggina*, an annelid worm (scale in mm).
D. *Dickinsonia*, a flatworm (?) similar to the modern *Spinther* (scale in mm).

D

Southwest Africa have yielded an intriguing assortment of impressions of soft-bodied metazoans. The best known of these faunas is from the Pound Quartzite (Fig. 6-34), which crops out in the Ediacara Hills, 280 miles north of Adelaide, South Australia. The Ediacara fauna consists of jellyfish, sea pens, polychaet worms, and a number of forms whose biological affinities are still uncertain. A few of these strange fossils are illustrated in Figure 2-27. All these organisms were soft bodied and very different from anything we know from Cambrian formations.

The absence of metazoan animals in the Precambrian, except for the Ediacara and related faunas, has long been a puzzle. An examination of earliest Cambrian life, however, will help us to attain a better perspective of Precambrian life. The earliest Cambrian record includes archaeocyathids,

worm trails and burrows, brachiopods, gastropods, sponges, trilobites, and echinoderms. From this list it is apparent that the earliest Cambrian faunas were quite diversified but were composed mainly of rather primitive representatives of the various phyla. Each representative, however, was a highly complex organism. Moreover, all but one clearly belong in one or another of our modern phyla; the archaeocyathids are the exception, as their biologic position is not satisfactorily known.

What is of particular interest is that the first appearance of all these forms took place within a relatively short period of time. Figure 2-28 shows the time of first appearance of animal fossils in the earliest Cambrian strata of the western United States and of the U.S.S.R. The range of time over which these fossil forms first appear in the record from the western

FIGURE 2-28

Approximate time of first appearance of metazoan fossils in the Lower Cambrian of the United States and the U.S.S.R. (Data supplied by A. R. Palmer.)

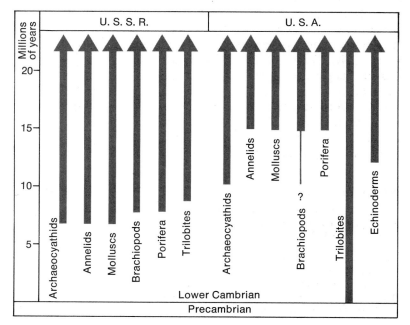

United States is approximately 15 million years, but in the U.S.S.R. they all make their first appearance within a few million years.

The question "Why are Precambrian metazoan fossils scarce?" has been asked many times, but we have not yet found a satisfactory answer. Most of the early answers are so improbable that a review of them is not worthwhile. Much progress has been made in recent years regarding the evolution of the earth's atmosphere, and these new data provide at least a reasonable working hypothesis. Approximately 3.5 billion years ago, the age of the oldest dated rocks, the atmosphere consisted of juvenile gases but no free oxygen. Photosynthesizing plants first appeared about 3 billion years ago. From that time until about 2 billion years ago the oxygen produced by photosynthesis was taken up by ferrous iron, but by about 2 billion years ago plants had evolved sufficiently to acquire oxygen-mediating enzymes, and free oxygen slowly began to be released to the atmosphere. This slow build-up of oxygen in the atmosphere continued until late Precambrian time, when the atmosphere contained perhaps as much as 3 percent of the present atmospheric content of oxygen, an amount sufficient for Metazoa of low oxygen requirements. At this stage the evolutionary radiation of the Metazoa could take place. This whole story is as yet speculation, but at least it seems to be a reasonable working hypothesis.

2-6   THE PHANEROZOIC FOSSIL RECORD

Once we enter the Cambrian Period, our evidence on the evolution of life becomes more satisfactory, and the nearer we approach the Recent the more detailed and complete this evidence becomes. The history of any phylum of animals is based on data painstakingly assembled in countless paleontological publications for more than 150 years. The gathering of these data has followed a logical pattern. The first contributions, those of the early nineteenth century, were monographs describing the fossils of large geographic areas. Such works were especially common in England, France, Italy, and Germany. They were followed by descriptions of fossil faunas in stratigraphic units—that is, the faunas of single formations. Monographs of this type were produced by the hundreds during the latter half of the nineteenth century and later. From the treatment of a formation as a unit of faunal description there developed the **paleobiological** treatment—that of a biological group such as the brachiopods of a particular time unit or of the time range of particular brachiopod genera and families. The next stage was the **synthetic** stage, in which the fossil records of whole families, orders, or even classes of animals were the units of study.

The description of fossils, either by formational units or by biologic units, is still going on, and it must, of course, continue if geologic knowledge is to advance. We already have enough data to try to answer some of the broader questions of evolution and historical geology. Paleontology has become so complex, however, that one person can no longer master all the phyla. The best that one worker can hope for is to attain competence in one or two large groups as they existed during a few of the geologic periods. To each such specialist the group he works on is merely the "tool" or medium by which he hopes to solve the broader problems.

Let us briefly review the geologic record of each of the major phyla (Fig. 2-29). This broad fossil record should be kept in mind as we systematically discuss the physical history of the earth in later chapters. If you are not familiar with the basic morphological features of the main animal and plant phyla, be sure to read Appendix 1 before you continue with this chapter.

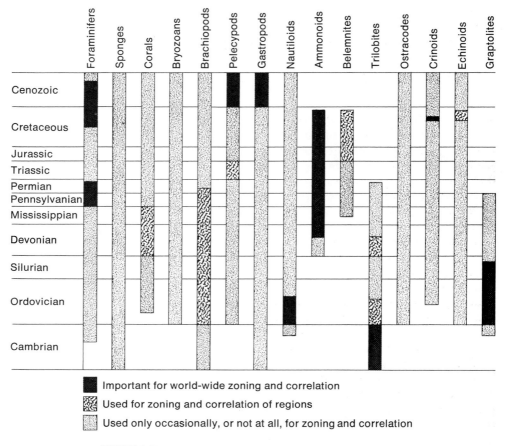

Important for world-wide zoning and correlation

Used for zoning and correlation of regions

Used only occasionally, or not at all, for zoning and correlation

FIGURE 2-29

Geologic range and relative biochronologic significance of the major groups of marine invertebrates during geologic time. (After Teichert, 1958.)

Protista

This very complex kingdom of one-celled organisms includes some that were once variously classed as animals and as plants. Some groups of Protista combine significant features of both plants and animals. For our purposes, however, only a few groups are geologically important. By far the most important group is the Foraminifera. Because of its importance in petroleum exploration, more paleontologists concentrate their studies on this group than on any other group of fossil organisms. The foraminifers are excellent as both stratigraphic and ecological indicators and are easily recovered from drill cuttings. The first unquestioned Foraminifera appear in the Cambrian. It is not until the late Paleozoic that they are important to the historical geologist. A special group of Foraminifera, the fusulinids, are extremely important guide fossils for the Pennsylvanian and Permian Systems. The remains of a foraminifer called *Globigerina* make up the bottom sediment of nearly 50 percent of the modern ocean floor, and some limestone formations of great geographic extent are made up almost entirely of remains of similar organisms.

Porifera

This ancient branch of the animal kingdom does not have an abundant fossil record, but it surely has a long one. Several occurrences of Precambrian sponge spicules have been recorded, but none seem to have been verified. Sponges are sporadically distributed from the Cambrian to the Recent and at times become important parts of the rock record. The earliest reefs formed by animals were Cambrian sponge reefs, but sponges have not contributed significantly to geochronology.

Coelenterata

Massive coral reefs are as common in Paleozoic and Mesozoic strata as they are in present-day equatorial oceans. The corals have been outstanding rock-builders from the Ordovician to the Recent. Interest in fossil coral reefs has increased greatly in recent years with the discovery that some are important oil reservoirs. Of the three principal orders of corals—Rugosa, Tabulata, and Scleractinia—the first two are confined to the Paleozoic and the third appears first in the Middle Triassic. The low point in the evolutionary history of the corals came in the Triassic; no Lower Triassic corals are yet known, in fact, though they were, of course, in existence. The geochronologic value of corals has not been great, but they have been of tremendous importance as rock-builders and, because of their temperature and depth restrictions, to paleoecology.

Bryozoa

These small, structurally complex colonial animals are not yet well understood. Because special techniques are required for their study, progress is slow. The group is biologically and geologically important, however, and deserves more attention. The Bryozoa first appear in the Cambrian, and their geologic history indicates a rather even, low evolutionary rate over a long time span.

Brachiopoda

The brachiopod faunas of our modern seas are but a poor remainder of a once flourishing and diverse phylum with a history reading back to the Cambrian. During the first two-thirds of the Paleozoic Era—that is, up to the Mississippian—the brachiopods were extremely important elements of marine faunas. The two classes of brachiopods, the Inarticulata and the Articulata, have contrasting evolutionary histories. The Inarticulata had their period of maximum expansion in the Ordovician and declined rapidly thereafter. The inarticulate brachiopod *Lingula* is one of the modern "living fossils"; this genus has existed since the Ordovician. The articulate brachiopods are a much more important and larger group than the inarticulates, and evolved faster. The two groups appear to replace each other in time; that is, as the inarticulates declined in evolutionary activity, the articulates increased.

The brachiopods are, in general, good fossils for correlation and are especially useful for early Paleozoic systems. In the Mesozoic, and especially in the Jurassic, brachiopods are abundant in some provinces. Their study has already provided some useful stratigraphic and ecologic results, even though they are overshadowed by the ammonoids.

Mollusca

The Pelecypoda, Gastropoda, Cephalopoda, Scaphopoda, and Amphineura constitute this paleontologically important phylum. Their abundance, diversity, and great beauty, as well as their excellence as stratigraphic tools, have commanded the attention of many students. The marine strata of the Mesozoic Era throughout the world are zoned principally by means of fossil cephalopods. The phylum ranges from the Cambrian to the Recent. The Amphineura and Scaphopoda are very small groups. Only a hundred species of fossil Amphineura, distributed from the Cambrian to the Recent, are known, and approximately three hundred species of Scaphopoda are known, the first appearing in Ordovician strata. These two classes of the Mollusca, because of their general rarity, simple morphology, and unspectacular evolution, are not important for stratigraphic purposes.

The oldest pelecypods known are from Ordovician formations. Their number and diversity increase steadily after the Ordovician, and the class now appears to be at the climax of its history. The evolutionary relations of the main stocks of pelecypods are not well known. Most span a long time, but many genera and species are excellent guide fossils. Because most pelecypods are adapted to benthonic life and are influenced in their distribution by the bottom sediments, they are excellent paleoecologic tools. The history of the class can be summarized as a general and continuous development of new forms through a slow evolution.

Few animal groups are more successfully adapted or more widely distributed than modern gastropods. Like the pelecypods, the gastropods now appear to be at the height of their development. After the first snails appear in Lower Cambrian formations, the class shows an ever increasing diversity and abundance. It is important for stratigraphic purposes in the Cenozoic; in the Mesozoic and Paleozoic, however, other groups are more useful—perhaps because the gastropods in those formations have not been sufficiently studied.

The cephalopods evolved more rapidly than most other invertebrate groups and are therefore more successfully used for stratigraphic correlation. About four thousand genera are known, ranging from the Cambrian to the Recent. The modern *Nautilus* is the last surviving genus of a group that first appeared in the Late Cambrian. In Ordovician time the nautiloids underwent a tremendous radiation and developed numerous distinct morphological types. Nearly half of the 700 known nautiloid genera occur in the Ordovician. From that period until the Recent the nautiloids declined rapidly in numbers of genera. The group was reduced greatly in the Late Triassic; and, although new radiation took place therafter, the group never regained either its numbers or its diversity, and by the Cenozoic its decline was rapid.

The ammonoid group first appeared in the Early Devonian and expanded slowly during the remaining Paleozoic periods. In the Permian, however, most stocks became extinct; only one survived to give rise, in the Triassic, to a very large radiation of ammonoids. Again, near the end of the Triassic, the ammonoids suffered severely; all except one of the many stocks died out. The one stock that did survive gave rise to a third and last great radiation, which flourished throughout the Jurassic and Cretaceous. In these two geologic periods approximately 1,500 genera of ammonoids are recognized. The ammonoids became extinct by the end of the Cretaceous.

Throughout the Mesozoic the ammonoids are our best "tools" of stratigraphy. The time-rock divisions of the Mesozoic systems are largely defined on the basis of ammonoid zones.

The dibranchiate, or two-gilled, cephalopods, including the belemnoids, squids, and octopods, are the most recent members of the Cephalopoda. Of these only the belemnoids have geological importance. This group first appeared in the Mississippian and died out in the Eocene. They were derived from straight-shelled nautiloids. The climax of their evolution came in the Jurassic.

Arthropoda

Species of Arthropoda comprise three-quarters of the modern animal kingdom. The phylum is varied and complex, but, because the fossil record is fragmentary, the exact relations of many groups are in doubt. The oldest fossil arthropods are trilobites, which range from the Lower Cambrian to the Permian and there become extinct. Trilobites are the fossils most important for geochronology, not only in the Cambrian but also in the Ordovician. By Silurian time they were on the decline, and they become gradually less and less conspicuous in the record until their final extinction in the Permian.

The ostracodes also had an active and varied evolutionary history and played an important part in geochronology. Insects, however, which appear first, in Lower Carboniferous strata, in highly developed forms that indicate a long previous evolutionary history, are not abundant enough in the fossil record to play any part in our chronology. This statement applies to practically all the other major groups of arthropods.

Echinodermata

Few invertebrate phyla are more complex or diverse than the Echinodermata. It has long been customary to group the stalked or attached forms—for example, the crinoids and blastoids—in one subphylum and the free-living forms, like the echinoids and starfish, in another subphylum. In the past decade this phylum has been intensely studied, and many new and highly interesting fossil forms have been discovered, leading to a completely new concept of relationships in this phylum.

The echinoderms first appear in the fossil record in the Lower Cambrian, where eight distinct classes of primitive forms are present. By the Ordovician there were sixteen classes, and evolutionary radiation was well underway. Many early Paleozoic classes are bizarre forms; by the close of the Devonian many had become extinct.

The living representations of this phylum, the crinoids, echinoids, asteroids, ophiuroids, and holothurians, all originated in the Ordovician. The crinoids are a large class that underwent a very diverse evolution. Of the four subclasses, three were confined to the Paleozoic. The fourth appeared in the Triassic and survives today. The crinoids were common during the Paleozoic and are generally good stratigraphic tools. Since, however, like all the Echinodermata, they were gregarious, they are generally found either in large numbers or not at all.

The starfish, the ophiuroids, and the rare extinct forms related to them are not abundant in the geologic record. The class Stelleroidea, which includes these forms, ranges from the Ordovician to the Recent. The relative scarcity of fossil stelleroids is a serious obstacle to understanding their evolution and to using them as stratigraphic tools.

The only parts of the sea cucumber (holothurians) that have been preserved as fossils are the calcareous plates that lie within its leathery skin. These plates, which are generally rare, have been identified in strata from Ordovician to Recent.

The sea urchins (echinoids) are geologically and paleontologically the most important class of Eleutherozoa. They first appeared in the Ordovician; they had a modest record in the Paleozoic; in the Jurassic they became abundant, and they have continued in great diversity until the present day. In the Jurassic, Cretaceous, and Cenozoic they are excellent stratigraphic tools.

Graptolithina

For Ordovician and Silurian strata no fossil groups have been more useful as tools of stratigraphy than the graptolites. These small colonial animals, which at first look like mere pencil marks on a rock, form an extremely complex group. Their classification has been a source of great confusion. Excellently preserved specimens now suggest that they may really be very primitive chordates, and some leading specialists place them in a subphylum of the Chordata.

The group first appeared in the Middle Cambrian and became extinct in the Early Carboniferous. Many were planktonic or pseudoplanktonic and thus had a wide geographic distribution. This fact, coupled with their exceptionally high rate of evolution during the Ordovician and Silurian, makes the graptolites excellent guides to synchroneity for the historical geologist.

Chordata

Fossil vertebrates are not nearly as abundant or as widely distributed as marine invertebrates; nevertheless, they are often excellent tools for geochronology. The fossil vertebrates have also contributed more than most invertebrate phyla to our understanding of evolution. The study of fossil vertebrates has so progressed that the broad framework of relations among the major groups is now fairly well known and understood (Fig. 2-30). In general, we know the evolutionary progression from the earliest vertebrates, the jawless fish, in the Ordovician, through the jawed fish to the bony fishes, to the amphibians, then to the

FIGURE 2-30

Geologic range of the chordates. The width of the shaded band indicates the relative importance of the group. The broken lines indicate discontinuities in the geological record—that is, gaps in the discovery of fossils in the rocks of the period. (After *Biology: Its Human Implications,* by Garrett Hardin. W. H. Freeman and Company. Copyright © 1949.)

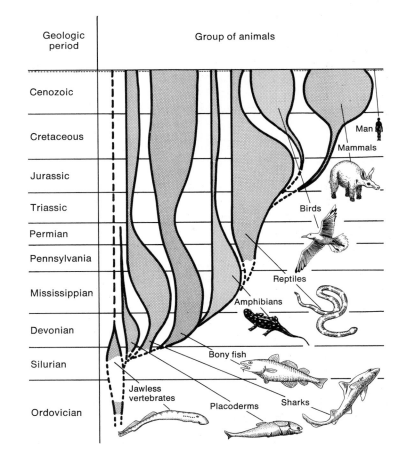

reptiles, and finally to the mammals and birds.

The fossil record of the vertebrates for the Ordovician and Silurian is poor. Not until the Devonian do fossil vertebrates (fish) attain a respectable representation in the record. The first land vertebrates, the amphibians, appear in Late Devonian or earliest Mississippian time. From the amphibians evolved the reptiles in Early Pennsylvanian time. Late Paleozoic time saw a great diversity of amphibians and reptiles. It was in the Mesozoic, however, that the great evolutionary radiation of the reptiles took place; we commonly refer to the Mesozoic, in fact, as the "Age of Rep-

FIGURE 2-31

Geologic distribution of the components of the three major terrestrial floras. The width of each labeled band is proportional to the known variety of the group represented, primarily in terms of families, secondarily in terms of genera. (After Dorf, 1955.)

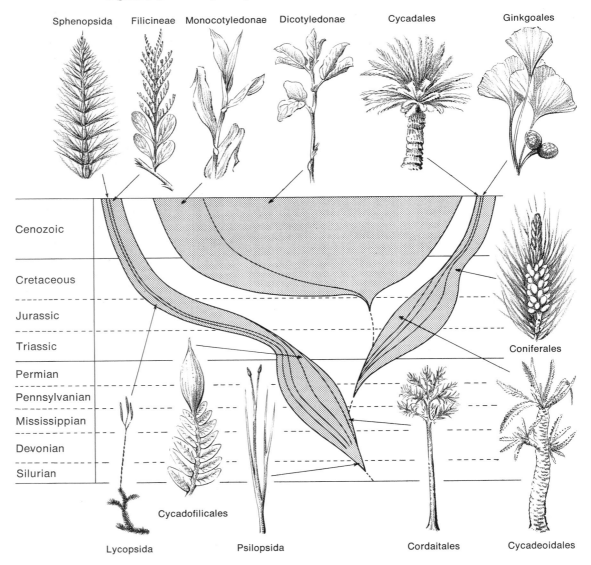

tiles." In addition to adapting themselves to many different forms of land life, reptiles also invaded the sea and the air. The reptiles are a substantial aid in the geochronology of continental Mesozoic strata. Most became extinct at the end of the Mesozoic, and only small lines continued through to the Recent.

As the reptiles were first beginning to expand, in numbers of both individuals and kinds, in the early Mesozoic, two new major groups arose from reptilian ancestry, the mammals and the birds. Both groups first appeared in the Jurassic. In the Cenozoic the mammals became so dominant in the evolutionary scene that the era is commonly called the "Age of Mammals." In rapidity and extent of evolution the mammals are second to no other animal group. They are the principal fossil group in geochronologic studies of continental Cenozoic formations. The culmination of mammalian evolution, at least to us, was the appearance of man approximately 2 million years ago. The rise and evolution of the mammals are best understood against the background of the physical evolution of the earth in Cenozoic time, a topic that will be discussed in Chapters 12 and 13.

Plants

The record of fossil plants, as now interpreted, has brought to light a complete and orderly history from the simple aquatic plants of the late Precambrian and early Paleozoic to the progressively more com-plex terrestrial plants of the later Paleozoic, the Mesozoic, and the Cenozoic. Each of the three major land floras shown in Figure 2-31 was characterized by the dominance of one or several major groups, which largely gave way in time to other groups, each more advanced in its characteristics than its predecessors. Each of the major floras appears to have been ushered in by a short episode of rapid evolution, which was followed by a longer episode of evolutionary stability.

The first land plants, the Psilopsida, of Late Silurian age, were very simple vascular plants without true roots. By the Middle Devonian real forests became established; in the Pennsylvanian the evolutionary development of the Paleozoic land flora reached its culmination. The predominant plants were the club mosses (Lycopsida), the horsetails (Sphenopsida), the ferns (Filicineae), the gymnosperms (plants with naked seeds), the seed ferns (Pteridospermae), and the cordaites (Cordaitales). In the Mesozoic flora the gymnosperms were conspicuous, especially the cycads, conifers, and ginkgos. A major change came in the Early Cretaceous with the appearance of the angiosperms, the plants with flowers and encased seeds. These are by far the most successful of land plants today, and they represent the highest level of plant evolution achieved up to now. The greater part of our food is of angiosperm origin, directly or indirectly. The expansion of the angiosperms coincided with the decline or extinction of most groups of the gymnosperms.

SUGGESTED READINGS

Allee, W. C., and K. P. Schmidt, *Ecological Animal Geography* (Wiley, New York, 1951).

Andrews, Jr., H. N., *Studies in Paleobotany* (Wiley, New York, 1961).

————, "Early Seed Plants," *Science,* **142** (1963).

Arnold, C. A., *An Introduction to Paleobotany* (McGraw-Hill, New York, 1947).

Banks, H. P., "The Early History of Land Plants," *Evolution and Environment,* Symposium for the 100th Anniversary of the Peabody Museum of Natural History, E. T. Drake, ed. (Yale Univ. Press, New Haven, 1968).

Beerbower, J. R., *Search for the Past,* 2nd ed. (Prentice-Hall, New Jersey, 1968).

Brouwer, A., *General Paleontology* (Oliver & Boyd, Edinburgh, 1967).

Colbert, E. H., *Evolution of the Vertebrates* (Wiley, New York, 1955).

Darwin, Charles, *The Origin of Species by Means of Natural Selection, or The Preservation of Favored Races in the Struggle for Life* (Murray, London, 1859). A facsimile of the 1st ed. (Harvard Univ. Press, Cambridge, 1964).

Delevoryas, T., *Morphology and Evolution of Fossil Plants* (Holt, Rinehart & Winston, New York, 1962).

Dobzhansky, Theodosius, *Evolution, Genetics, and Man* (Wiley, New York, 1955).

Ekman, Sven, *Zoogeography of the Sea* (Sidgwick & Jackson, London, 1953).

Halstead, L. B., *The Pattern of Vertebrate Evolution* (W. H. Freeman and Company, San Francisco, 1969).

Harland, W. B., ed., *The Fossil Record* (Geol. Soc. London, 1967).

Jensen, W. A., and L. G. Kavaljian, eds., *An Evolutionary Survey of the Plant Kingdom* (Wadsworth, California, 1965).

Moore, R. C., C. G. Lalicker, and A. G. Fischer, *Invertebrate Fossils* (McGraw-Hill, New York, 1952).

Olson, E. C., *The Evolution of Life* (Mentor Books, New York, 1965).

Rhodes, F. H. T., *The Evolution of Life* (Penguin Books, Baltimore, 1962).

Romer, A. S., *The Vertebrate Story* (Univ. of Chicago Press, Chicago, 1959).

———, *Vertebrate Paleontology,* 3rd ed. (Univ. of Chicago Press, Chicago, 1966).

Shrock, R. R., and W. H. Twenhofel, *Principles of Invertebrate Paleontology* (McGraw-Hill, New York, 1953).

Simpson, G. G., *The Meaning of Evolution: A Study of the History of Life and of Its Significance for Man* (Yale Univ. Press, New Haven, 1949; Mentor Books, New York, 1951).

———, *Life of the Past: An Introduction to Paleontology* (Yale Univ. Press, New Haven, 1953).

———, *The Geography of Evolution* (Capricorn Books, New York, 1965).

Stebbins, G. L., *Processes of Organic Evolution* (Prentice-Hall, New Jersey, 1966).

Sverdrup, H. U., M. W. Johnson, and R. H. Fleming, *The Oceans: Their Physics, Chemistry, and General Biology* (Prentice-Hall, New York, 1946).

Precambrian Life

Barghoorn, E. S., and J. William Schopf, "Microorganisms Three Billion Years Old from the Precambrian of South Africa," *Science,* **152** (1966).

———, "Alga-like Fossils from the Early Precambrian of South Africa," *Science,* **156** (1967).

Barghoorn, E. S., and S. A. Tyler, "Microorganisms from the Gunflint Chert," *Science,* **147** (1965).

Cloud, Jr., P. E., "Significance of the Gunflint (Precambrian) Microflora," *Science,* **148** (1965).

———, "Pre-Metazoan Evolution and the Origins of the Metazoa," In E. T. Drake, ed., *Evolution and Environment* (Yale Univ. Press, New Haven, 1968).

Glaessner, M. F., and M. Wade, "The Late Precambrian Fossils from Ediacara, South Australia," *Paleontol.,* **9** (1966).

Schopf, J. William, *Antiquity and Evolution of Precambrian Life* (McGraw-Hill Yearbook of Science and Technology, 1968).

———, "Microflora of the Bitter Springs Formation, Late Precambrian, Central Australia," *J. Paleontol.,* **42** (1968).

Scientific American Offprints

The *Scientific American* Offprints listed below and after the other Suggested Readings in this book are available from your bookstore or from W. H. Freeman and Company, 660 Market Street, San Francisco, California, 94104, and Warner House, Folkestone, Kent. Please order by the number preceding the author's name. The month and year in parentheses following the title of an article is the issue of the magazine in which the article was originally published.

6 Theodosius Dobzhansky, *The Genetic Basis of Evolution* (January, 1950).

47 George Wald, *The Origin of Life* (August, 1954).

837 Martin F. Glaessner, *Pre-Cambrian Animals* (March, 1961).

871 S. K. Runcorn, *Corals as Paleontological Clocks* (October, 1966).

872 Adolf Seilacher, *Fossil Behavior* (August, 1967).

3

THE RECORD OF THE SEDIMENTARY ROCKS

The fundamental working units of the historical geologist are rocks: those that are exposed at the surface of the earth and those that can be reached by drilling. The physical and chemical properties of a rock, its structural and geographic relations to other rocks, all give us clues to the history of the earth.

From your study of physical geology you will remember that there are three major types of rock: igneous, metamorphic, and sedimentary. Our attention will be directed chiefly to the sedimentary rocks, since at the very surface of the earth sedimentary rocks make up fully 75 percent of the outcrop area and igneous rocks only 25 percent. But it is not only the accessibility of the outcropping that makes sedimentary rocks so important to us; they are the rocks that contain the fossils from which the chronology of the earth's history has been estab-

lished. Even igneous rocks are dated by their structural relations to sedimentary rocks; until the development of radiometric dating methods (discussed in §1-5) this was, in fact, the only method of dating igneous rocks.

In this chapter we shall investigate briefly the chemical, physical, and structural characteristics of sedimentary rocks and their historical meaning. After reviewing the classes of sedimentary rocks, we shall start with the hand specimen of a rock and study the number of data it can yield us. Next we shall look at the rock's natural setting in the field—that is, the internal and external features of the sedimentary body. Finally we shall stand far off and look at the physical and historical relations of large sedimentary bodies within the frameworks of continents.

3-1 CLASSES OF SEDIMENTARY ROCKS

Sedimentary rocks, which are deposited layer upon layer, at the surface (that is, on top of underlying rocks) at fairly low temperatures and pressures, are classified as (1) clastic rocks, which are mechanical accumulations of detrital material and rock fragments, and (2) nonclastic rocks, which are of chemical or biological origin. The detrital particles of clastic rocks have been broken from their parent rock by weathering, have been transported by water or wind, or possibly by both, and have finally come to a resting place, where natural processes cement them together into a sedimentary rock (Fig. 3-1). Nonclastic sedimentary rocks, on the other hand, are either formed by chemical action and (usually) precipitated out of solution (evaporites: gypsum, anhydrite, some limestones) or formed by living organisms that secrete shells of calcium carbonate (many limestones were formed from accumulations of such shells). The differences that clastic and nonclastic rocks exhibit in characteristics and composition, imply that they have quite different histories.

3-2 CLASTIC SEDIMENTARY ROCKS

The detrital particles composing a clastic rock suggest three questions: Where did the particles come from? How far did they travel? How were they carried to the basin of deposition? What were the conditions in that basin? Answers to these questions may be deduced from the chemical and physical characteristics of the rock and from the three-dimensional distribution of the sedimentary body from which the sample was taken.

To study the physical characteristics of

FIGURE 3-1

Any sediment is a mechanical mixture of a detrital fraction, brought as a solid into the basin of deposition from the source area, and a chemical fraction, precipitated from solution within that basin. (After P. D. Krynine, *J. Geol.*, **56**, 1948.)

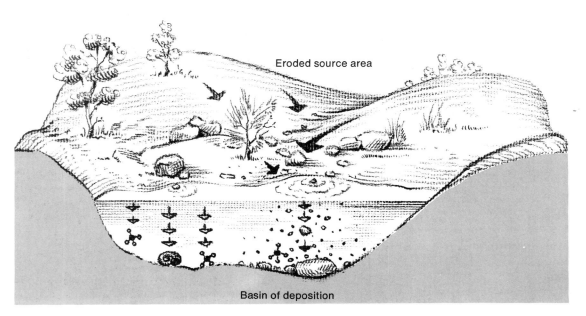

Eroded source area

Basin of deposition

a rock, we normally begin with the examination of a hand sample. Merely by looking at it we can tell something of its texture, its color, and its composition. Let us see what we can discover about our specimen from each of these features.

Texture

The **texture** of a clastic rock is determined by the size, shape, and spatial relations of its particles. Such a rock, being an aggregate of mechanically deposited fragments, generally consists of two main components, the particles themselves and the cement that holds the particles together. In some sandstones and in practically all conglomerates there is also a **matrix,** composed of fine material, that may also act as a cement.

The sizes of the particles, which can be estimated in the field or accurately measured in the laboratory, are named according to a standard scale (Table 3-1). A rock composed of clay-size particles is called a **shale;** a rock composed of sand-size particles is called a **sandstone;** and a rock composed of pebbles, cobbles, and boulders—generally in a clay-and-sand matrix—is called a **conglomerate.** Particle size reflects the mode and extent of transportation. The larger the particle the greater the force needed for

TABLE 3-1

Particle sizes most commonly used by geologists

Diameter (mm)	Name
Above 256	Boulder
64−256	Cobble
4−64	Pebble
2−4	Granule
1/2−2	Coarse sand
1/2−1/4	Medium sand
1/4−1/8	Fine sand
1/8−1/16	Very fine sand
1/16−1/256	Silt
1/256−1/2,048	Clay

its movement. Study of the movement of particles in natural streams and in laboratory flumes has yielded a great body of data that aids in the interpretation of ancient sedimentary deposits. Wind-transported sediments have been studied by similar methods.

One of the most useful ways of characterizing the variation of particle size is to specify the sediment's **sorting**—that is, the range of sizes deviating from the average size. Rocks composed of particles that deviate little from the average size are said to be well sorted; those that include particles of a wide range of sizes are poorly sorted. Poor sorting indicates that the particles were rapidly deposited without being selectively redistributed, by currents, according to size. A large quantity of sediment entering a rapidly subsiding basin is not likely to remain unburied long enough to be sorted by currents, and such a sediment has, at times, a "poured in" appearance. Off a stable coast, however, currents are continually shifting the bottom deposits, which therefore, if the currents are strong enough, are likely to be well sorted. Sorting, then, reflects the conditions of transportation and the environment of deposition. However, no matter what the mechanism of transportation or the environment of deposition may be, if the source of the particles is an ancient, well-sorted sandstone, the new deposit also will be well sorted.

The shape and the roundness of particles also have historical meaning. **Shape** is measured with reference to a sphere. **Roundness** is acquired as a particle loses the sharpness of its edges and corners, and we express degrees of roundness by saying that a particle is angular, subangular, subrounded, rounded, or well rounded. Studies of streams and laboratory experiments show that particles become more rounded and more nearly spherical in transport. Thus the roundness of a particle sums up the history of its abrasion and reflects the conditions and extent of its transportation. A gravel, for instance, is soon rounded, but a sand-size particle requires prolonged abrasion. In general, other conditions being the same, the smaller the particle the longer it takes to round it off. Recent field and laboratory studies of the shape and roundness of particles have accumulated an immense amount of data and have greatly aided the historical geologist in deciphering the meaning of ancient rocks.

Color

The most obvious characteristic of a sedimentary rock is probably its color. The apparent **color** of a rock is generally dependent on the colors of the component mineral grains, the color of the finer-grained matrix or of the cement, the color of any coating on the grains, such as iron oxide on quartz, and the size of the grains. Two colors, red and black, are of special significance.

Fine-grained black rocks are common in the geologic record. The blackness is due mainly to organic matter or to fine grain size or to both. The fossil fauna of marine black shales generally consists of nektonic or planktonic organisms; benthonic forms, if present, are dwarfed or aberrant. Black shales are formed only if the organic material is preserved, as it is if the local circulation of oxygenated water is poor or if it is quickly buried below the zone of decomposition at the surface of the sea bottom.

Redness presents a more complex problem than blackness. Sedimentary rocks with red coloration are called redbeds. The redness is due usually to the presence of anhydrous ferric oxide (hematite), occasionally to an abundance of pigmented orthoclase (a type of feldspar rich in potassium). Redbeds are frequently associated with such evaporite deposits as gypsum, anhydrite, and salt. This and other evidence has led some to conclude that redbeds were formed under semi-arid, oxidizing, and evaporative conditions and where there were alternate wet and dry seasons; others have concluded, however, that redbeds were formed in warm, humid climates and in places where red soils were formed. Whichever theory comes closer to the truth, redbeds could be formed wherever the sediments were buried in an oxidizing environment.

Composition

The mineral **composition** of a clastic rock is of special importance to us because it gives us clues to the history of the rock: the source of the sediments, the mode of transportation, the environment of deposition, and even the degree of deformation of the earth's crust that was taking place during the deposition of the rock series.

Among the clastic rocks—sandstones, shales, and conglomerates—the sandstones have been studied the most thoroughly and are therefore the best understood. The most significant characteristics of a sandstone are those that reflect the nature of the source area, the "maturity" of the sandstone, and the mode of transport.

The weathering of a plutonic terrane of quartz-bearing rocks such as granite yields sand-size particles, notably of quartz and feldspar. If the terrane is an ancient one, containing sedimentary, metamorphic, and volcanic rocks, the weathered particles of sand size will be mostly rock particles rather than monomineralic particles. Thus the ratio of rock particles to feldspar particles reflects the nature of the source area.

The stablest of the rock-forming minerals is quartz. Sandstones containing a very high percentage of quartz particles have necessarily gone through more than one cycle of weathering, transport, and deposition, during which the less stable minerals have gradually disappeared. The ultimate sandstone is one composed only of quartz particles; the percentage of quartz particles is therefore an excellent indication of the maturity of a sandstone.

Many sandstones are compact assemblages in which the sand grains are in contact with one another; voids between the sand grains either are empty spaces or are filled with a mineral cement. In other sandstones, however, the sand grains are generally not in contact with one another; the voids are filled with a fine matrix, the particles of which are of clay and silt size. These two types of sandstone reflect quite different media of transport. With reference to the first type, there must have been a great difference in density between the particle and the transporting medium, so that separation was rapid and complete, the clay-size materials remaining suspended and being carried to some other site of deposition. For the second type, the transporting medium must have been highly viscous, as in a mud flow or a subaqueous turbidity flow. There is always the possibility, however, that the clay-and-silt matrix was, not discrete particles, but soft rock fragments, which, on compaction and lithification, lost their identity as rock fragments. If this was so, the sand particles and the rock fragments of clay or silt would have approximately the same specific gravity. It is often difficult to tell whether the matrix was originally discrete particles or rock fragments; if one can tell, one knows something about the viscosity of the transporting medium.

There are four basic types of sandstone — graywacke, subgraywacke, arkose, and orthoquartzite.

The **graywackes** are distinctive for their high content of matrix, which consists largely of clay minerals, chlorite, and sericite. The sand-size particles consist of quartz, rock fragments, feldspar, and various other minerals in minor amounts (Fig. 3-2). The sorting of graywackes is poor, and the particles are often angular. The heterogeneous composition and poor sorting indicate instability both in the source area and in the depositional area (Fig. 3-3). Deformation and rapid erosion in the source area, and rapid burial in the depositional area, often give graywackes their "poured in" appearance.

Beds of graywackes usually vary from a few inches to a few feet in thickness and seldom show any lamination. On close examination, however, we can often observe that within a particular bed the particles are graded, the larger particles being in the lower part of the bed and the smaller particles in the upper part—a feature called **graded bedding** (Fig. 3-4). The matrix is not graded. Bedding can be graded only by the settling out of particles from a liquid. Laboratory experiments have demonstrated that deposits from turbidity currents are usually graded. In nature any mass of clay, silt, and sand with a very high water content may be set in motion down a slight slope by a seaquake or a surface slumping; as soon as the flow slows down sufficiently, the sediments will begin to settle out, and the coarse particles, obviously, will settle first. The fairly common association of graywackes with radiolarian cherts, submarine volcanic deposits, and shales containing deep-water faunas suggests that they were deposited in deep water in unstable environments. In support of this conclusion, such structures as cross-bedding and ripple marks, which are characteristic of shallow shelf environments, are seldom found in graywackes. Another significant feature of graywackes is that the formations they make up are generally thousands of feet thick.

Subgraywackes are the commonest sandstones; they comprise, in fact, more than a third of all sandstones. Their content of quartz and chert is high, and they contain more rock fragments than feldspar, which makes up only 0–10 percent; the fine-grained detrital matrix is 15 percent or less; the voids are filled chiefly with mineral cement but also with some clay materials

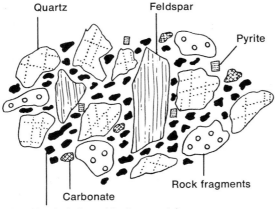

FIGURE 3-2

Schematic representation of a typical graywacke (× 125). The diagram is an "exploded" one; that is, the constituents have been slightly pulled apart for the sake of clarity. (After P. D. Krynine, *J. Geol.,* **56,** 1948.)

(Fig. 3-5). Such sandstones were derived mostly from a sedimentary or low-rank metamorphic terrane. Because of their content of rock particles they are often gray. They are usually much better sorted than graywackes, and the particles are gener-

ally better rounded. Subgraywackes are usually well stratified, and they often show cross-bedding and ripple marks. They are commonly found with coal beds, in both the subaerial and the subaqueous parts of deltas, and in closely associated marine environments.

Arkoses are sandstones that contain 15 percent or less detrital matrix and more feldspar than rock fragments (Fig. 3-6). They are formed largely of particles derived from a granitic terrane and have, in fact, a composition very similar to that of a granite. Quartz and feldspar are the main minerals, but clay minerals (mostly kaolin and mica) are also important. Arkoses are generally moderately well sorted, and the particles generally vary from coarse and angular to subangular. The rocks commonly occur as thin, sheet-like deposits or in very thick wedges. If thin sheets, they are generally the basal sedimentary unit overlying a granitic terrane; if thick wedges, they were deposited in structural troughs or basins near highlands composed largely of granite (Fig. 3-7).

Orthoquartzites, because of their very

FIGURE 3-3

Sketch of idealized geologic conditions for the deposition of graywackes. (After S. J. Pirson, *Elements of Oil Reservoir Engineering,* McGraw-Hill, New York, 1950.)

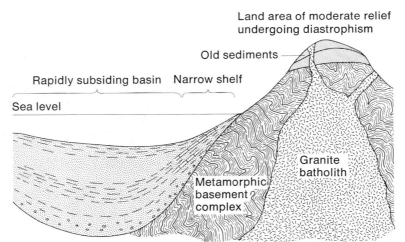

FIGURE 3-4

Graded bedding in the Coalido Formation (upper Eocene), Coos Bay, Oregon. The coarser particles form the unstratified part, the finer particles the darker laminations at the top of each bed. The irregular portion (load casts) at the base of each bed is formed by the intrusion of the underlying fine sediment into the overlying coarse sediment and is caused by slight inequalities of loading or of sinking of the denser overlying sediment into the less dense and somewhat plastic underbed. (Photograph by R. H. Dott, Jr.)

high content of quartz, are sometimes called "pure-quartz sandstones"; they have little or no clay matrix, and their voids may or may not be filled with a mineral cement (Fig. 3-8). They also contain, however, very small amounts of stable heavy minerals such as zircon and tourmaline. Orthoquartzites are usually very well sorted and the grains well rounded. Cross-bedding and ripple marks are particularly common in these rocks, but calcareous fossils are not usually found in them. Orthoquartzites occur oftenest as blanket deposits only a few hundred feet thick but covering wide areas. They are commonly, but by no means always, associated with, or grade into, limestones and dolomites. Their mineral composition and texture clearly indicate that they are mature sandstones. They may have been deposited in fairly stable environments (Fig. 3-9), and there may have been time for weathering and abrasion to take out all the clay and unstable minerals, leaving only sandsize grains of quartz; or

they may have gone through the cycle of weathering, transport, and deposition more than once.

The principal characteristics of the main types of sandstone are summarized in Table 3-2.

3-3 NONCLASTIC SEDIMENTARY ROCKS

Nonclastic sedimentary rocks—limestones, dolomites, gypsum, salt, phosphates, and others—are of biological or chemical origin. The commonest are the carbonates, limestone and dolomite.

The genesis of limestones is an extremely complicated subject and one upon which there is a wide diversity of opinion. Nevertheless, many limestone bodies can definitely be shown to consist primarily of the remains of plants and animals. Organic reefs are common from the Cambrian to the Recent and persist in the modern equatorial belt. Many limestone bodies, however, show no reef structure whatever, but are well-stratified, generally fine-grained rocks.

Examination of thin or polished sections shows that many limestones are made up of

FIGURE 3-5

Exploded schematic representation of a typical subgraywacke (× 125). (After P. D. Krynine, *J. Geol.*, **56**, 1948.)

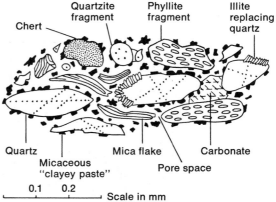

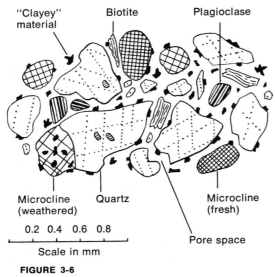

"Clayey" material

Biotite

Plagioclase

Microcline (weathered)

Quartz

Microcline (fresh)

0.2 0.4 0.6 0.8

Scale in mm

Pore space

FIGURE 3-6

Exploded schematic representation ($\times$ 40) of an arkose. (After P. D. Krynine, *J. Geol.*, **56**, 1948.)

shells and fragments of shells cemented by a finely crystalline calcite. Figure 3-10, A, B, C shows thin sections of limestones composed mainly of recognizable skeletal remains in a fine-grained calcite matrix. We usually designate these limestones by their predominant constituents: A is a foraminiferal limestone, B is a gastropod limestone, and C is a trilobite limestone. Limestones made up of small fragments of skeletal materials are illustrated in Figure 3-10, D, E, F; D consists largely of bits of brachiopod and mollusk shells, E is made up mostly of crinoid remains, and F consists of cephalopod shells in a fine-grained matrix. Often, however, it is impossible to recognize the organisms represented.

Other limestones, which do not show evidence of organic origin, are thought to be of purely chemical origin. Some limestones (Fig. 3-10, G, H, I), for instance, are made up largely of small, spherical concretions called **oolites**. Oolites are formed in agitated saline waters under agitated conditions. Calcareous oolites are formed by more or less even deposition of calcite on a nucleus being tossed about by the water. Many oolites result from accretion around a shell fragment or a grain of calcite or quartz, and others have no nucleus whatever. Most dolomites result from post-depositional alteration of limestone, but some dolomites associated with evaporites

FIGURE 3-7

Sketch of idealized geologic conditions for the deposition of arkose. (After S. J. Pirson, *Elements of Oil Reservoir Engineering*, McGraw-Hill, New York, 1950.)

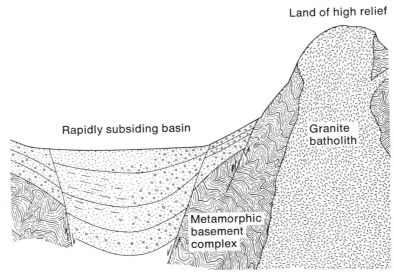

Land of high relief

Rapidly subsiding basin

Granite batholith

Metamorphic basement complex

appear to be primary chemical deposits. Gypsum and salt are typical evaporites.

The presence of nonclastic rocks indicates an environment in which large quantities of detritus were not available. This lack could mean either that the depositional area was far from land or that the adjoining land was of such low relief that the amount of material weathered and transported into the depositional area was negligible.

3-4 STRATIGRAPHIC RELATIONS OF SEDIMENTARY FORMATIONS

The boundaries of sedimentary formations are extremely variable. Laterally, a formation may wedge out or may grade or intertongue with an adjoining formation; vertically, formations may be conformable or nonconformable.

A discontinuity between two successive formations indicates a break in sedimentation. The surface representing this break is an **unconformity.** There are several kinds of unconformity. If the formations on the two sides of the break are parallel, the unconformity is called a **disconformity.** If there is a difference in dip and strike between the two formations, we have an **angular unconformity.** If bedded rocks rest on an eroded surface of plutonic or metamorphic rocks, we have a **nonconformity.** Very minor breaks in the sedimentary record are called **diastems.**

The various types of unconformity are

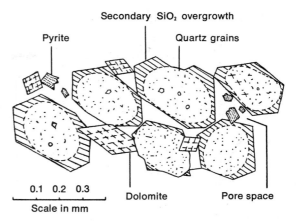

FIGURE 3-8

Schematic representation ($\times$ 80) of an orthoquartzite, showing secondary quartz overgrowths on the originally rounded quartz grains. (After P. D. Krynine, *J. Geol.,* **56,** 1948.)

tectonic relations; that is, they result from movements of the earth's crust, generally followed by erosion. Movements that result in major deformation of the crust are called **diastrophic.** Slow or slight diastrophic movements producing broad uplifts or downwarps are **epeirogenic;** greater diastrophic movements producing faults and folds are **orogenic.**

3-5 SEDIMENTARY FACIES

A **sedimentary facies** is defined as the sum of all the primary characteristics of a sedimentary rock from which the environment

FIGURE 3-9

Sketch of idealized geologic conditions for the deposition of orthoquartzite. (After S. J. Pirson, *Elements of Oil Reservoir Engineering,* McGraw-Hill, New York, 1950.)

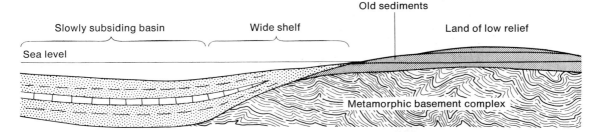

TABLE 3-2

Summary of principal characteristics of the main types of sandstone

Property	Orthoquartzite	Subgraywacke	Graywacke	Arkose
Mineral composition	Quartz ($> 95\%$)	Quartz, little feldspar, mica, clay	Quartz, feldspar, mica, clay	Quartz, feldspar
Texture	Clean, well sorted	5–15% matrix, sand fraction well sorted	Up to 50% matrix, sand fraction not well sorted	Little matrix, sorting poor or good
Cement	Silica, carbonate	Silica, carbonate	None	Carbonate
Roundness	High	Fairly high	Low	Low or intermediate
Thickness of beds	Thin as a rule but occasional great thickness of single bed	Thin or medium	Thin	May be very thick
Thickness of formations	Thin	Thin or intermediate	Thick	Thick
Sedimentary structures	Cross-bedding, ripple marks, etc.	Cross-bedding, ripple marks, slumping	Graded bedding	Cross-bedding, ripple marks, mudcracks, footprints
Associated rocks	Limestone, calcareous shale	Shale, thin limestones, coal	Siltstone, shale, bedded cherts, volcanic deposits	Red shales, siltstones, volcanics

Table compiled by Raymond Siever.

of its deposition may be induced. To illustrate the concept of facies, let us look for a moment at a modern diverse sedimentary environment.

In Batavia Bay, on the northern coast of Java, a small island rising a mere few feet above sea level is composed mainly of finely broken fragments of corals—a calcareous sand; and this material is nearly identical with the sediment of the adjacent sea bottom, especially on the leeward side of the island (Fig. 3-11, A). In the windward direction from the island, in water of moderate depth, is a thriving coral reef, beyond which the bottom sediment is mud. Between the coral reef and the sand island is a linear zone of rubble, made up of fragments of coral and algae. Leeward of the sand island lies an impoverished coral reef. Each of these divisions is a distinct facies. If, by a large number of bore holes, we were able to reconstruct a portion of the depositional history of this small area, we would

probably find that the boundaries between facies that are adjacent on the surface shift laterally with depth (Fig. 3-11, B).

The maps of the bottom sediments and animal communities of the Kattegat, between Denmark and Sweden (Fig. 2-17, pp. 34 and 35), are excellent illustrations of the present facies of that area. If we uplifted this area and permitted erosion to expose a continuous profile from Denmark to Sweden, it might look somewhat like the cross section in Figure 3-12. For the sake of simplicity we assume that all the rocks exposed are of the same age and therefore represent a very short span of time. Note the great variation in type of sediment across this profile.

In our study of the maps in Figure 2-17 we saw that each type of sediment had a different bottom fauna. This difference would be reflected by distinctive fossil faunas in the facies of our profile. In the coarse sand-and-conglomerate facies of the margins we

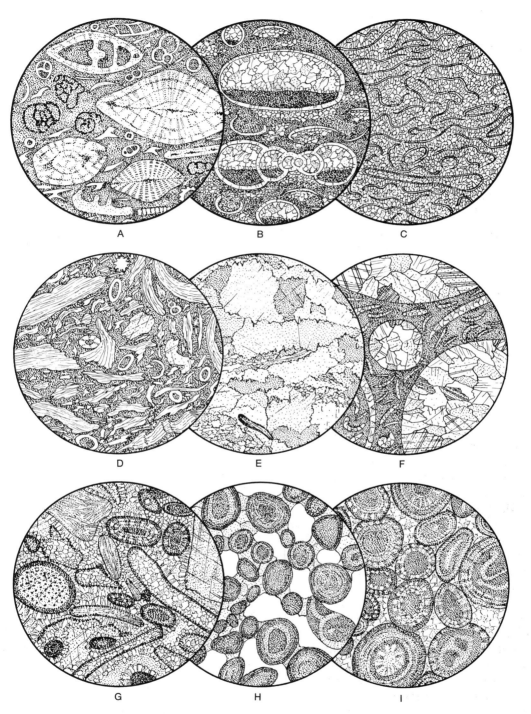

FIGURE 3-10

Various limestone types in thin section. *A, B,* and *C* are limestones made up largely of skeletal matter; *D, E,* and *F* are limestones composed of small shell fragments; *G, H,* and *I* are oolitic limestones. The diameters represented vary from 2.5 to 4 mm. (From *Petrography*, by Williams, Turner, and Gilbert. W. H. Freeman and Company. Copyright © 1954.)

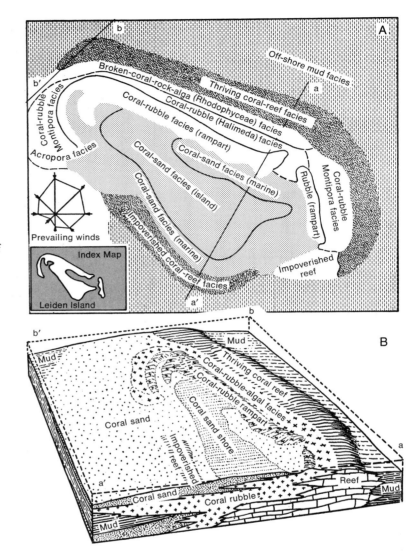

FIGURE 3-11

A. Present variation in types of sediment in Batavia Bay, Java, illustrating the relations of facies.

B. Block diagram of part of Batavia Bay, showing the inferred relations of facies in cross section. (After R. C. Moore, *Geol., Soc. Am.,* Mem. 39, 1949.)

should find fossils of the *Venus* community illustrated in Figure 2-17. The shale facies (clay bottoms) would have the *Brissopsis* community and the *Haploops* community. The facies intermediate between the coarse and the fine facies — that is, the one underlying a mixed clay-and-sand bottom — would have the *Echinocardium* community. Each of these fossil faunas would be quite distinctive.

If, now, structural complications and periods of erosion leave us with only isolated outcrops of various segments of our profile, it becomes difficult to establish the synchroneity of these rocks. We can solve such a problem only if we thoroughly understand the physical aspects and interrelations of the sediments and the biological factors of each facies.

Not all changes in facies are as distinct or abrupt as those we have mentioned. Sometimes we recognize no difference in the rock formations, but the fossil assemblages tell us that there has been a change in environment. The Wakarusa Limestone of Kansas and Oklahoma, of Pennsylvanian age, is from two to four feet thick and exhibits little lithologic change for more than two hundred miles along the outcrop, but the organic changes define facies subdivisions that have paleogeographic significance. In central Kansas the fossil community contains brachiopods, crinoids, fusulinids (foraminifers), and bryozoans. In southern Kansas the fusulinids become rare, and a calcareous alga (*Ottonosia*) appears as a new element in the fauna. In northern Oklahoma the fusulinids are completely lacking, and the algae become abundant.

A striking and paleogeographically important type of facies relation is that formed when the sea transgresses on the land. The bottom sediments are, successively, a near-shore coarse clastic facies, an off-shore fine clastic facies, and a carbonate facies. As the shore line moves inland, the boundaries of these facies shift in the same direction. This sequence of events, illustrated in Figure 3-13, A, is called **onlap**. When the seas retreat, the facies boundaries again move in the same direction as the shore line (Fig. 3-13, B). This is known as **offlap**. In onlap successively younger beds have wider geographic extent; in offlap the younger beds have narrower extent.

3-6 CORRELATION OF ROCK UNITS

We have already seen how the geologic time scale, with its various subdivisions is based on fossils. Systems, series, and stages are recognized round the world by fossil criteria and not by the physical characteristics of the strata. In the early nineteenth century the primary basis of paleontological correlation was the superposition of strata (that is, the succession of faunas was an empirical fact, and, although present knowledge enables us to determine the relative ages of many animal and plant groups by

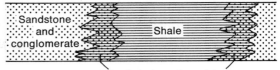

Mixed shale and sandstone

FIGURE 3-12

Hypothetical cross section, along the lines indicated in Fig. 2-17, of the sediments of the Kattegat.

their morphology and evolutionary status, without the corollary evidence of stratigraphic position, the superposition of strata is still the primary basis of paleontological correlation.

The only practical means of establishing the synchroneity of rock units from one continent to another—or even, in many cases, within continents—is fossils. It is impossible to correlate the formations of two continents by purely physical characteristics.

In the detailed study of a local area, however (which generally involves mapping), correlation of local rock units by fossils is often not possible or practicable. We are interested in establishing the distribution of rock units, without regard to relative times, and the fundamental unit for this purpose is the **formation**. When outcrops

FIGURE 3-13

Diagram illustrating the relations of facies in (A) overlap in a transgressing sea and (B) offlap in a regressing sea. Time lines are parallel to the bottom of the diagram.

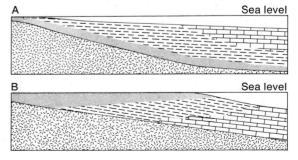

are abundant, it is easy enough to trace, or "walk out," a formation's boundaries. There are many methods by which a formation's continuity or its equivalence to another formation can be established. Lithologic identity and constant position in a geologic sequence are the commonest means of correlation. Most formations have identifying lithologic characteristics that permit correlation between widely spaced exposures. Such characteristics could be color, bedding, primary structures, mineralogic constitution, even peculiarities of weathering. Intense exploration and drilling for petroleum, in the last few decades, have developed many techniques for the correlation of subsurface formations penetrated by the drill.

Figure 3-14 illustrates the relations among Cambrian formations in the Grand Canyon. The lower faunal horizon of the Bright Angel Shale at the left of the diagram contains Lower Cambrian guide fossils. In the right-hand portion fossils at about the same distance from the underlying Tapeats Sandstone indicate Middle Cambrian age. Here, clearly, the Bright Angel Shale comprises a succession of facies of different ages while maintaining the same relations to the limestones facies above and the sandstone facies below. In this area it has been possible to establish correlation by the positions of the faunas in the rock units as well as by the sequence and positions of particular conglomerate beds and certain other lithologic markers.

FIGURE 3-14

Relations of Cambrian formations in the Grand Canyon, showing the principal key beds and horizon markers considered to be approximate time planes. (After E. D. McKee, *Geol. Soc. Am.*, Mem. 39, 1949.)

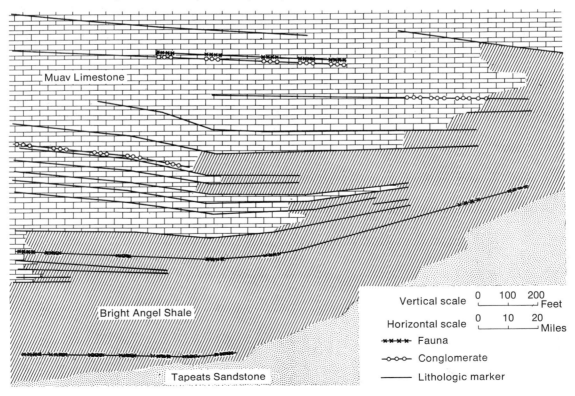

Correlation by lithology can be applied only to single sedimentary basins or even more restricted areas, but for the daily mapping of any new area it is the principal method of correlation. Correlation by fossils is used primarily to place a unit in the proper time-rock division.

3-7 THE SEDIMENTARY FRAMEWORK OF CONTINENTS

The year Darwin's *Origin of Species* appeared, 1859, saw the introduction of another concept that is fundamental to geology. James Hall, State Geologist of New York, in that year published a large monograph in which he summarized the results of his work on the stratigraphy and structure of the Appalachian Mountains of eastern North America. The northern Appalachians, particularly in the State of New York, where Hall worked, were then without doubt the geologically best-known mountain system in the world, and the generalizations that Hall drew from his study of field relations and stratigraphy soon caused widespread discussion.

The Concept of Geosynclines

Hall pointed out that the Paleozoic strata in the Appalachian Mountains were much thicker than beds of the same age in the Mississippi Valley—in other words, that the crust had subsided much more in the former area than in the latter. The greater bulk of the sediments in the Appalachian region were predominantly clastic, and all had been laid down in shallow water. Such a pattern of deposition could have persisted only if the subsidence of the area had kept pace with the sedimentation. This, obviously, was the way the great synclinal structure was formed. Along the axis of the syncline the beds were folded into smaller synclines and anticlines. As they

were fractured, magma could intrude into them. Metamorphism, due to folding and the rise in temperature at depth, was also strongest along the axis.

Hall concluded that all great mountain chains were belts of greatest accumulation of sediments. Such belts have come to be known as **geosynclines.** Even though the concept of geosynclines is one of the fundamental concepts of geology, it is not possible, strange as it may seem, to give a precise, generally acceptable definition. We do need a working understanding of the term, and we shall define it, for the moment, as a surface of regional extent subsiding deeply during the accumulation of succeeding layers of rocks. The full meaning of the concept will become clearer as we proceed to a systematic survey of the earth's history.

As more and more geologic data accumulated, the source of the sediments and the nature of the borders of geosynclines became centers of interest. One of the earliest theories intended to explain the source of sediments held that a series of Precambrian masses or islands existed along the axial area of the geosyncline and supplied the sediments deposited in the surrounding subsiding basin. These masses were thought to have persisted throughout the history of the geosyncline. This explanation soon had to be discarded, however, for many of the so-called Precambrian areas turned out to be metamorphic complexes of Paleozoic age.

The fact that several early Paleozoic formations in the Appalachian Mountains clearly indicate a source to the east led to the idea of borderlands. This theory, developed in the past half century and widely accepted by North American geologists, holds that geosynclines along the margins of continents were bounded on their outer margin by an extensive land mass, or borderland. The borderland adjacent to the Appalachian geosyncline was named Appalachia. A comparable geosyncline, the

Cordilleran, is found in western North America, and its borderland was named Cascadia. The borderlands of Pearya and Llanoria fill in the picture for North America (Fig. 3-15). The sediments in the geosynclines came mainly from these borderlands. The vertical variation in the lithologies of the sedimentary rocks of the geosynclines reflected changing rates of uplift and erosion in the borderlands. When the development of a particular geosyncline came to an end, and the sediments folded, beginning the mountain-making process, the adjacent borderland subsided beneath the ocean.

In recent years another concept of geosynclinal margins, derived from analysis of the gross distribution of rock types, has begun to attract a wide following. In the northern part of the western North American geosyncline there was an outer belt, with extrusive volcanic materials and intrusive magmatic bodies, and an inner, amagmatic belt. In the early Paleozoic such conditions were also found in the Appalachian geosyncline. This had led to the conclusion that active volcanic islands and tectonic welts of older rock supplied the sediments and volcanics found in the volcanic belts. A reconstruction of the North American continent during Early Ordovician time on the basis of this theory is shown in Fig. 3-16. The long geosynclines on the margins of the continent each consist of an outer, magmatic segment called a **eugeosyncline** and an inner, amagmatic segment called a **miogeosyncline**. The broad, stabler area between the geosynclines, where sediment accumulation is much smaller, is called the **continental plate**. Within the plate area are local basins and complementary stable areas.

Sediments in Geosynclines

The nature of geosynclines is largely determined by study of their rocks.

Because eugeosynclines are in volcanic belts, they contain lavas and volcanic detritus in some abundance. The volcanic rocks generally present include pillow lavas, lava flows, tuffs, accumulations of coarser fragmental materials, and derivatives of the weathering of volcanic islands.

The clastic rocks of a eugeosyncline are predominantly of the graywacke type and are derived from the erosion of tectonic land within the eugeosyncline. Such graywackes are generally poorly sorted and frequently show graded bedding, but cross-bedding and ripple marks are rare; in particle size they range from coarse conglomerates to fine sandstones. Eugeosynclines are said to be **cannibalistic** because the sediments of each come from within the eugeosyncline itself, from the tectonic lands (islands) consisting of ancient rocks and volcanoes.

Though graywacke and volcanic sediments prevail in eugeosynclines, chemical deposits—limestones, dolomites, and cherts—are not uncommon in areas where no terrigenous material entered to mask them. Orthoquartzites and arkoses also may be abundant in some places but are not common. The lands within the eugeosynclines did not have extensive areas of rocks that could yield sorted quartz sand and fresh feldspar in abundance; and stable shallow-water shelf areas, where sediment accumulation would be slow and conditions would favor good sorting, were apparently uncommon.

The main impression we obtain from study of the sediments of eugeosynclines is one of great mobility in these regions. It is here that much of the earth's diastrophic activity is concentrated and that granitic batholiths frequently intrude.

In miogeosynclines the lithologic types are quite different. There are, by definition, no volcanic rocks. The dominant types are orthoquartzites and limestones. The source of most of the clastic rocks is the interior of the continent. When tectonic lands arose

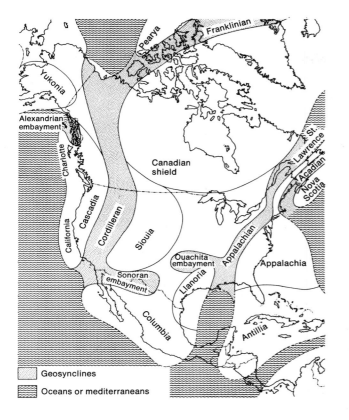

FIGURE 3-15

Paleozoic tectonic framework of
North America, showing the relations
of geosynclines to borderlands.
(After Schuchert, 1923.)

along the borders of a eugeosyncline, however, they did contribute subgraywackes
and shales to the adjoining miogeosyncline.
Detailed studies of rocks within miogeosynclines clearly show that cannibalism was
common there also: island areas, periodically raised within the belt, contributed
pre-existing sediments.

3-8 THE TECTONIC FRAMEWORK OF THE EARTH

Seismic and gravity studies of the earth's
crust, both beneath the oceans and on the
continents, have added interesting data to
our consideration of continental development. The first startling conclusion is that
the crust beneath the ocean basins is different from that beneath the continents.

In their study of earthquake waves the
seismologists detect a layering of the earth's
crust. The velocity of the waves in the outer
layer closely approaches that computed
from the elastic properties of granite. Although this layer is often called the **granitic
layer**, it is known to be heterogeneous and
largely not true granite: but it resembles
granite in its minerals. Beneath this upper
layer is another, often called the **basaltic
layer**, in which wave velocities are somewhat like those in gabbro or basalt. A significant change in wave velocities takes place
at the boundary between the basaltic layer
and an underlying layer that is thought to
consist of ultramafic rocks (dark rocks composed of ferromagnesian minerals). This
boundary, the **Mohorovičić discontinuity**,
marks the base of the crust.

The crust is very thick beneath the con-

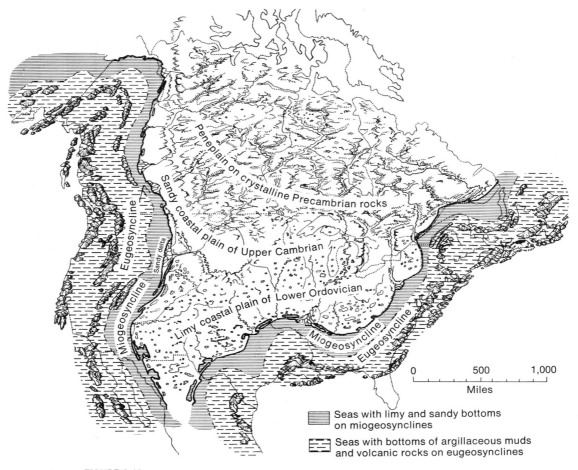

Seas with limy and sandy bottoms on miogeosynclines

Seas with bottoms of argillaceous muds and volcanic rocks on eugeosynclines

FIGURE 3-16

Paleogeography of North America during Early Ordovician time according to the island-arc hypothesis. Each geosyncline consists of an inner miogeosyncline and an outer eugeosyncline, the latter with islands and volcanoes. (After Kay, 1951.)

tinents, but thins toward the continental borders, and the Mohorovičić discontinuity is consequently deeper under the continents than under the oceans. Under the oceans the crust is thin, below the deep-sea sediments. The wave velocities of oceanic crust are similar to those in the basaltic layer of continental crust. The crust is about thirty kilometers thick under the continents and from six to eight kilometers thick under the oceanic basins.

Some areas of the earth's crust present interesting exceptions. In island archipel-

agos, such as the East Indies, the Philippines, Japan, and the West Indies, we find both continental and oceanic crust. The larger islands clearly have a continental crust. Within each of these areas, however, are portions underlain by oceanic crust only. Such areas are thought to be, in a sense, embryonic continents, which in the future may merge with other continents or become large continents themselves.

Extensive seismic and gravity surveys have been made in Nevada, California, and the adjacent areas of the Pacific Ocean. The relations of the crust to the mantle along a section of this region is shown in Figure 3-17.

The Continents

A tectonic framework like that of North America, with the stable interior (the continental plate) bordered by mobile geosynclines, can be demonstrated to have existed for most of the continental masses of the earth. The geosynclines of the world have varied histories and ages, and the later chapters of this book will deal at length with their relations in time and space. For the moment we shall briefly generalize their history with the aid of Figure 3-18. The period from the birth of a geosyncline to its ultimate stabilization and the final retreat of the sea may be called a major diastrophic sedimentation cycle. The geosyncline subsides faster than the adjoining continental plate and receives a larger supply of sediment. The sediments of the miogeosyncline come mostly from the foreland, the adjacent part of the continental plate. The sediments of the eugeosyncline come from its own volcanic islands and tectonic lands. The eugeosyncline—the site of much tectonic instability, in contrast to the more stable conditions of the miogeosyncline and the continental plate—undergoes one or more diastrophic phases. Later tectonic

FIGURE 3-17

Cross section of the earth's crust and upper mantle across Nevada, California, and part of the Pacific Ocean. Dot pattern = sedimentary rocks; check marks = granitic rocks; black pattern = greenstone; ρ = density of rocks in g/cm^3. (After G. A. Thompson and M. Talwani, *Science*, **146**, Dec. 18, 1964. Copyright by the American Association for the Advancement of Science.)

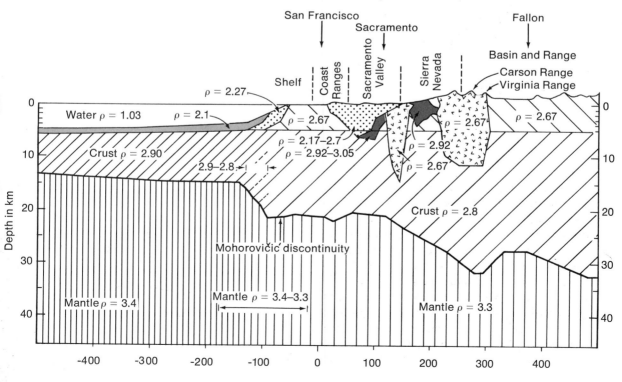

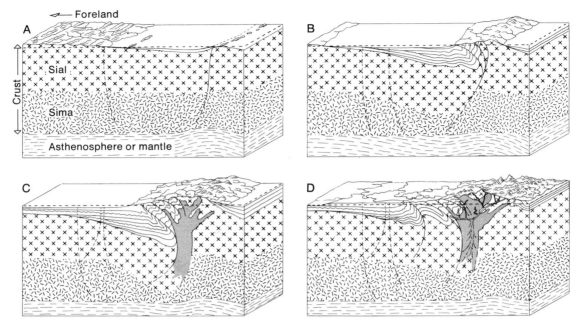

FIGURE 3-18

Schematic representation of the evolution of a geosyncline: (A) Initial subsidence of the continental margin, the miogeosyncline receiving most of its sediments from the foreland, the eugeosyncline from volcanic islands and tectonic welts. (B) Orogeny in the eugeosyncline creates larger tectonic lands, which supply much of the sediment carried into the miogeosyncline. (C) Orogeny in the eugeosyncline is intensified and is accompanied by granitic intrusions; large quantities of sediments are contributed by these tectonic lands to adjoining basins. (D) The final orogenic phase, with further plutonic intrusions, converts the former mobile belt into a rigid segment of the continental framework. (After L. G. Weeks, *Bull. Am. Assoc. Petrol. Geol.*, 1952.)

phases involve both the miogeosyncline and the eugeosyncline in the mountain-making deformation that folds and faults the contained sediments. These areas then become consolidated and attached as a rigid unit to the continental plate.

The stable interior of a continent is composed mainly of Precambrian rocks, which may or may not have a thin cover of younger strata. Such stable areas have been named **shields,** and will be discussed in more detail in the next chapter. At the moment let us examine shields in their relations to mobile belts and for the part they have played in the last half-billion years of the earth's history. The very generalized map of Figure 3-19 shows the major shields of the world and the belts of geosynclines and orogenies.

We can see immediately that the belts lie around or between the shields. Each belt has gone through a very mobile period of subsidence, deposition, and diastrophism. The exact sequence and time of the major events in the growth of these belts have, of course, varied from one region to another. Most of the belts underwent final diastrophic phases and subsequently acted as rigid, welded masses, which, with the adjacent shields, form our present continents.

THE OCEAN BASINS

The ocean basins have had a very different history from that of the continents. As was mentioned above, it has long been known

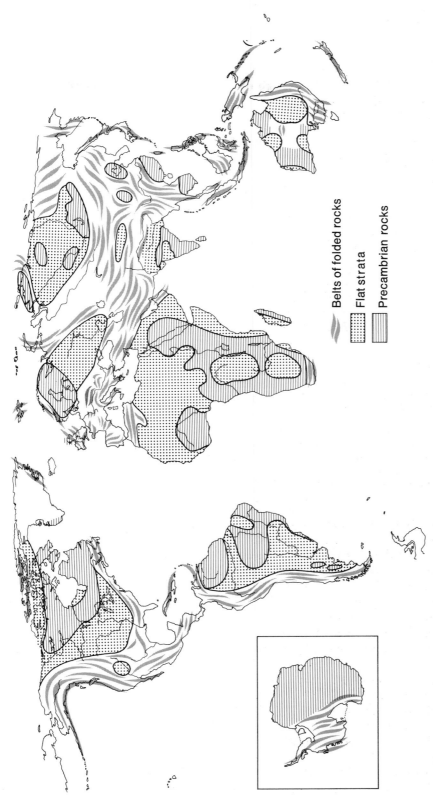

Belts of folded rocks

Flat strata

Precambrian rocks

FIGURE 3-19

Structural framework of the continents since Precambrian time, showing the areas of exposed Precambrian rocks (shields) and the former mobile belts, now with folded strata. (After Umbgrove, 1946.)

that one of the primary differences between the continental and oceanic crusts is in their thickness. The continents have a thick crust; the oceans, a thin crust. The oceanic crust appears, furthermore, to be relatively young. According to preliminary reports on the drilling in the Atlantic and Pacific by the JOIDES* project (a joint enterprise of five American Oceanographic Institutes supported by the National Science Foundation), no sediments older than 150 million years have been encountered, and most are younger than 80 million years. The ocean basins are not featureless plains but include ranges of great relief. These mountain systems are quite unlike the mountain systems on the continents, which are built up of thick folded sequences of sedimentary and igneous rocks. The oceanic mountains are formed almost entirely of basalts that have emerged from the interior of the earth. These mountains likewise occupy approximately mid-oceanic positions and have cracklike valleys along their crests. These data lead to the conclusions that the oceanic crust is being added to along the cracklike valleys of the mountain systems and that other areas of the oceanic crust disappear into the mantle elsewhere. This is a revolutionary new theory on the history of the ocean basins. It has profound bearing on the most fundamental problem of the earth's history: Has the relative position of the continents and the oceans always been the same? The answer to this question may lie in the theory of continental drift. These new ideas on the history of the oceans have developed only in the past decade. Intensive research on the oceans continues, and new data are accumulating at an astonishing rate.

This new theory on the oceans will, when further developed and substantiated, surely change our understanding of and approach to the earth's history profoundly, but I feel that it is premature to synthesize the history of the oceanic areas with that of the continents. For this reason the main focus of this book is on the geological evolution of the continents. The bearing of this new theory of oceanic history on the earth's history is discussed in more detail in Chapter 11.

SUGGESTED READINGS

Aubouin, Jean *Geosynclines* (Elsevier, Amsterdam, 1965).

Beloussov, V. V., *Basic Problems in Geotectonics* (McGraw-Hill, New York, 1962).

Bucher, W. H., *The Deformation of the Earth's Crust* (Princeton Univ. Press, Princeton, 1933).

Dunbar, C. O., and John Rodgers, *Principles of Stratigraphy* (Wiley, New York, 1957).

Goguel, J., *Tectonics* (W. H. Freeman and Company, San Francisco, 1962).

Kay, Marshall, "North American Geosynclines," *Geol. Soc. Am.*, Mem. 48 (1951).

Krumbein, W. C., and L. L. Sloss, *Stratigraphy and Sedimentation*, 2nd ed. (W. H. Freeman and Company, San Francisco, 1963).

Pettijohn, F. J., *Sedimentary Rocks*, 2nd ed. (Harper & Brothers, New York, 1957).

Schuchert, C., "Sites and Nature of the North American Geosynclines," *Geol. Soc. Am. Bull.*, **34** (1923).

Umbgrove, J. H. F., *The Pulse of the Earth* Nijhoff, The Hague, 1947).

Weller, J. M., *Stratigraphic Principles and Practice* (Harper & Brothers, New York, 1960).

Williams, H., F. J. Turner, and C. M. Gilbert, *Petrography: An Introduction to the Study of Rocks in Thin Sections* (W. H. Freeman and Company, San Francisco, 1954).

Scientific American Offprints

803 Ph. H. Kuenen, *Sand* (April, 1960).

808 Henry C. Stetson, *The Continental Shelf* (March, 1955).

816 Marshall Kay, *The Origin of Continents* (September, 1955).

882 K. O. Emery, *The Continental Shelves* (September, 1969).

* Joint Oceanographic Institutions for Deep Earth Sampling.

THE PRECAMBRIAN ERAS

The record of rocks and fossils from which we deduce the earth's history becomes sparser and sparser as we go back in time. Since each rock system covered vast areas of the preceding systems, the younger the system the greater the area exposed, and the older the system the greater the chance of its being covered by younger rocks or of being eroded to produce them. This point is illustrated by Figure 4-1, which shows, for the United States, the area of exposure of each geologic system plotted against time. For other continents the chart would be somewhat different in its percentages, but it presents a reasonable estimate for the world as a whole. This aspect of the rock record is also reflected in the development of the geologic time scale more than a century ago.

With William Smith's inspired recogni-

tion of the usefulness of fossils for purposes of correlation, the first segments of the geologic column to be defined were the systems included in the Cenozoic, Mesozoic, and late Paleozoic. In Europe these systems crop out abundantly and are commonly fossiliferous. The great challenge in the first part of the nineteenth century, therefore, was Werner's Transition class and others that we discussed in Chapter 1. These classes included the ancient sedimentary formations, which were usually strongly folded and faulted and not rich in fossils. The conquest of these formations by Sedgwick and Murchison was one of the great milestones of geology. Subsequent work has greatly enriched our knowledge of those formations, and they are no longer considered a part of the geological no-man's-land Today the Precambrian rocks

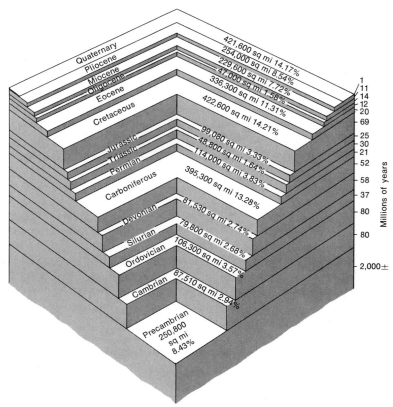

FIGURE 4-1

Block diagram showing relative areas of outcrop of systems and series.
The area of the "tread" on top of each unit is proportional to the
outcrop area in the United States; the height of the "riser" is proportional
to the duration of the period or epoch. The outcrop area per million
years' duration is greater the younger the rocks involved. (After
J. Gilluly, *Bull. Geol. Soc. Am.*, 1949.)

are the most difficult portion of the earth's
record to interpret.

Precambrian rocks represent more than
three-quarters of the time span of the record
observable in the earth's crust. They include
our most ancient rock formations, many so
intensely metamorphosed that we can only
guess at their original character. In Precambrian rocks that are not intensely metamorphosed, however, the lithologies, sedimentary structures, and facies resemble
those found in younger rocks. We therefore have no reason to believe that the
agencies operating in Precambrian time
were greatly different from those in all the
following eras. In other words, the uniformi-

tarian doctrine of Hutton applies in a rough,
general way to all geologic ages.

As was pointed out in Chapter 2, the great
difficulty in studying Precambrian formations is their almost complete lack of fossils.
By observing physical properties (such as
lithologic similarity and sequence of strata)
and structural relations (such as unconformities and the relations of intrusive bodies)
we can sometimes interpret the rocks of
small areas, but it is practically impossible,
by such methods, to extend correlations to
other geologic provinces or to other continents. The radiometric method of dating,
however, lends some encouragement to
those who are trying to correlate Precam-

brian events. Once this method becomes workable, we can expect great advances in our knowledge of Precambrian rocks. Whatever the means employed, all we need is a method of establishing synchroneity of events.

About 80 percent of the present land surface is covered by Phanerozoic rocks (Fig. 4-1) and 20 percent by the Precambrian formations. The latter are exposed in many major mountain ranges and in large continental areas, where they are called shields. The shields are generally areas of low relief and have been very stable since Precambrian time. Only thin, sporadic Phanerozoic formations are found on the shields, the locations and extent of which are shown in Figure 3-19. The shields, we have seen, form the cores of the continents, and Phanerozoic geosynclines generally lie parallel to, lie between, or are wrapped round these shields. Each continent has one or more Precambrian shields. Much of the geological history of the continents is revealed in the interrelations between these stable masses and the adjoining mobile belts.

Because Precambrian history is so complex and our knowledge incomplete, we shall direct our attention to the gross aspects of each shield in its relation to the Phanerozoic rocks and orogenic belts. At the same time we shall touch briefly on the rock record we have for each shield.

4-1 THE CANADIAN SHIELD

The nearly 3,000,000 square miles of Precambrian outcrop in the Canadian shield form the largest area of its kind in the world. It includes eastern, central, and northwestern Canada and much of Greenland. The shield is composed mainly of granite and gneiss, including some of the most ancient rocks known.

The extremely rich mineral deposits of the Canadian shield have attracted geologists for nearly a hundred years, but its vast area, the extreme complexity of its geology, and the difficulties of travel have all hindered progress in deciphering its history. We know that there were several episodes of diastrophism and granite intrusions and several of geosynclinal accumulation of sedimentary rocks, usually accompanied by the extrusion of volcanic rocks. Within particular areas the sequence of geosynclinal deposition, volcanism, diastrophism, and granite intrusions has been approximately worked out, but their correlation with other areas is difficult and often confusing.

The geographic and historical relations of the Canadian shield with the rest of the North American continent were discussed briefly in Chapter 3. As we go into the history of the shield itself, study Figure 4-2, and keep in mind particularly the position of the folded rocks in relation to the shield. The great stability of the shield since the Precambrian is clearly demonstrated by this map.

The history of the Canadian shield can be divided into two eras—an older, the Archeozoic, and a younger, the Proterozoic. The rocks of the two eras are separated by a very pronounced unconformity over most of the shield. The stratigraphic and historical aspects of the Proterozoic are much better understood than those of the Archeozoic. This twofold classification of Precambrian rocks is generally applicable to the Canadian shield, where the terms were first applied; but correlation of these divisions with the Precambrian formations of the other shields is not yet possible.

By far most of the Archeozoic rocks of the Canadian shield are little-known granites and gneisses. Among them, however, are long, isolated stretches of sedimentary and volcanic rocks. Detailed examination shows that these sedimentary rocks are not all as highly metamorphosed as has been commonly believed, and that the composition, sedimentary structures, and stratigraphic

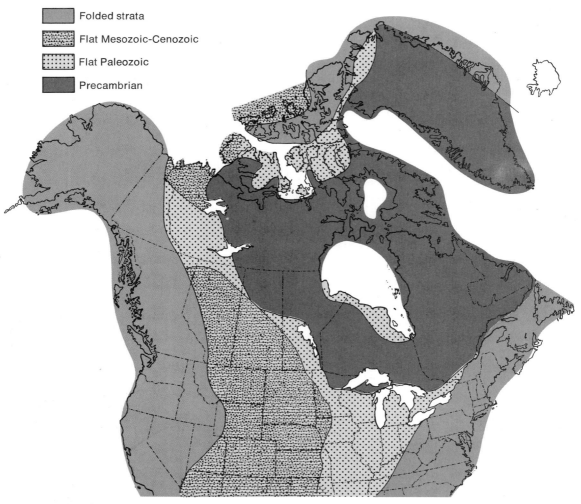

Folded strata

Flat Mesozoic-Cenozoic

Flat Paleozoic

Precambrian

FIGURE 4-2

Generalized geologic sketch map of northern North America, showing the Precambrian shield and the disposition of Phanerozoic formations.

relations are like those in younger rocks. A striking lithologic unity in these early Precambrian rocks, when taken to mean identity in age, has led to many erroneous correlations. The unity of lithology means unity only in mode of origin and has nothing to do with time.

Studies of the Archeozoic sedimentary rocks just north of Lake Superior demonstrate how much can be read from the rock types, compositions, and structures (Fig. 4-3). The Archeozoic rocks of this area con-

sist mainly of conglomerates, graywackes, and greenstones (lavas, pyroclastics, etc.).

Conglomerates are most conspicuous; generally, but not always, they are at the base of a sedimentary series. Many are more than 1,000 feet thick, and one is nearly 9,000 feet thick. They are composed mainly of wellrounded pebbles, boulders, and cobbles, generally of granite and greenstones, in places crudely sorted. Most conglomerate formations are wedge-shaped. Their length along strike may be more than

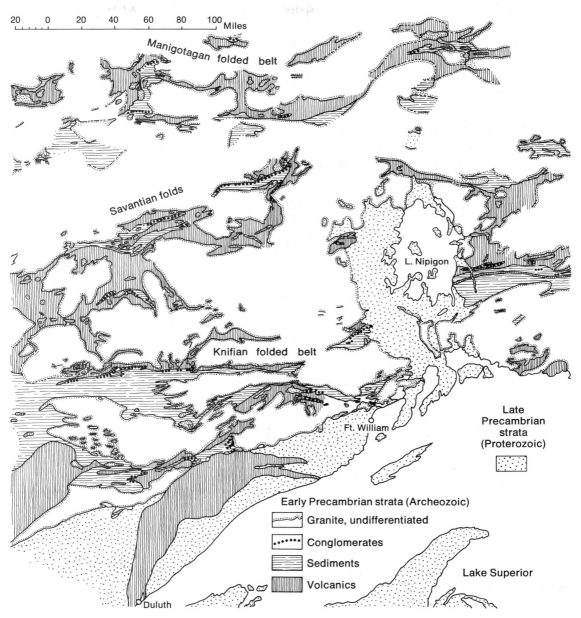

FIGURE 4-3
Geologic map of a portion of the southern part of the Canadian shield, showing the
areas of Archeozoic sediments. (After F. J. Pettijohn, *Bull. Geol. Soc. Am.*, 1943.)

sixty miles, but across strike they wedge out fairly rapidly. One thins from nearly 2,500 feet to nothing in a distance of about three miles across the strike.

The thickness and wedge shape of the conglomerate formations indicate that they are basin-margin deposits, recording fairly well the boundary between areas of erosion and of sedimentation during their deposition. The linear pattern of the Archeozoic patches in the area north of Lake Superior, and the linear wedge-shaped conglomerate bodies, which required local sources, preclude the notion that all the material of these conglomerates came from a common source; there were evidently many independent basins or geosynclines.

Most of the Archeozoic sedimentary rocks are graywackes, generally dark sandstones of quartz, feldspar, and rock fragments set in a fine-grained clay matrix. Graded bedding is common, but cross-bedding and ripple marks are rare. The graphite in slate beds, associated with pyrite, is probably of organic origin. The carbon content of the average modern marine sediment is about 1 percent; some Archeozoic sediments have yielded 1.92 percent. Since the preservation of the carbon and associated pyrite demands reducing conditions, a marine environment is favored for the deposition of these rocks. To establish, in the absence of fossils, whether a sedimentary formation was of marine or nonmarine origin is often very difficult. The carbon content of these Precambrian formations is therefore an important clue to their depositional environment.

The composition and structure of graywackes require certain sources and modes of deposition. The composition implies source areas undergoing rapid erosion. The structure, characterized by graded bedding and the general absence of cross-bedding and ripple marks, implies rapid deposition, one result of which is enormous thickness. Some Archeozoic graywackes are more than 20,000 feet thick.

Associated with the graywacke formations, underlying, overlying, and interbedded, are greenstones. This association of both intrusive and extrusive volcanic rocks with graywackes is found in many geosynclines of all ages. The greenstones consist of lavas, pyroclastic materials, and various intrusive bodies. Pillow lavas, which show a subaqueous origin, are rather common. Some of the Archeozoic greenstones are more than 30,000 feet thick.

How shall we interpret our data on the history and origin of these Archeozoic rocks? First we note the assemblage of rock types — graywackes, conglomerates, and volcanics. "Why no limestone or orthoquartzite?" we may ask. The general character of the graywackes gives us some clues: their composition, poor sorting, and graded bedding testify to rapid denudation and rapid deposition; their thickness bears witness to great subsidence; these factors, along with the association of volcanic rocks, strongly suggest deposition in a mobile eugeosyncline. Limestones and orthoquartzites are usually accumulated on stable shelf areas in thin, plate-like deposits, which reflect much slighter and slower subsidence than we find in eugeosynclinal belts. These Archeozoic rocks underwent a severe orogenic phase and then a long period of erosion. The record that remains for us to study represents merely the roots of the former depositional basins. The shelf sediments have long since been removed by erosion, and only the sediments deposited in the deeper, mobile parts of the geosyncline have been preserved.

The sedimentary and igneous rocks discussed above form only a small part of the total area of Archeozoic rocks of the Canadian shield. The remaining area is made up of granites and gneisses. During the Archeozoic there were at least two, and probably many more, great episodes of granitic intrusion.

Proterozoic rocks unconformably overlie the Archeozoic. They consist of quartzite, dolomite, limestone, slate, graywacke, arkose, conglomerate, and iron-bearing formations. Volcanic rocks, though present, are generally less abundant than in the Archeozoic. The aggregate thickness of Proterozoic rocks in Canada is about 65,000 feet.

Within the Proterozoic rocks several unconformities are recognized, each thought, however, to represent a much shorter time than represented by the Archeozoic-Proterozoic unconformity.

Analysis of the orogenic belts and the available radiometric ages have enabled a subdivision of the shield into structural provinces (Fig. 4-4). Correlation from one province to the other is generally not possible, but radiometric ages establish a general framework for correlation. The K-Ar ages determined so far show, for the most part, a rather impressive grouping around 2,500, 1,700, 1,350, and 1,000 million years, presumably indicating the main orogenic periods. These are called respectively the Kenoran, the Hudsonian, the Elsonian, and the Grenville orogenies. Because of analytical inexactness and geological uncertainties, such as possible loss of argon, a considerable latitude of say $\pm$ 300 m.y. must be allowed in estimating the age of the peak or of the time interval from the beginning to the end of an orogeny.

A schematic interpretation of the Precambrian evolution of North America is shown in Figure 4-5. The oldest areas of the continent, which underwent final orogeny about 2,500 m.y. ago or earlier, include the Superior and Slave provinces of the Canadian shield and Wyoming (Fig. 4-5, A).

FIGURE 4-4

Structural provinces of the Canadian Shield. (After C. H. Stockwell, *Geol. Surv. Canada*, Paper 64–17, 1964.)

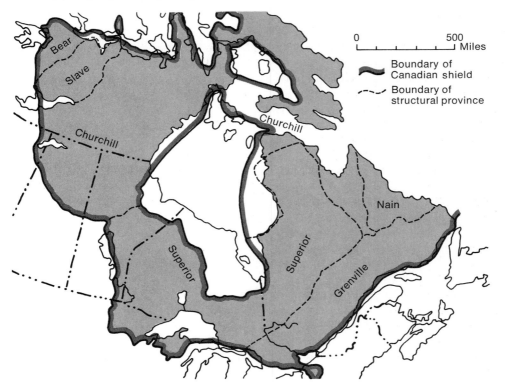

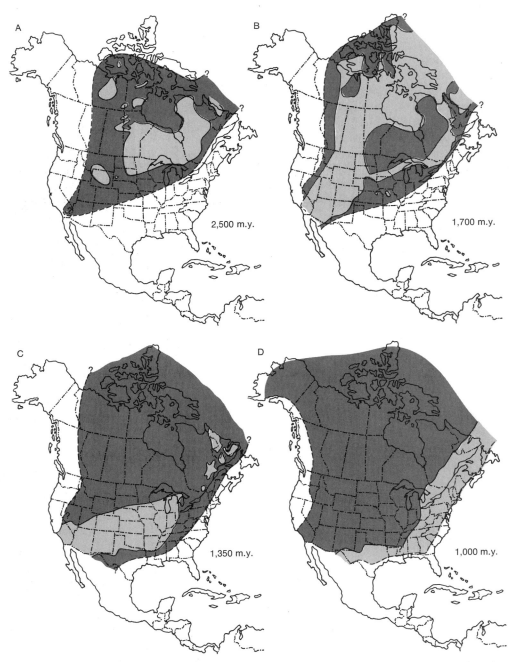

FIGURE 4-5

Schematic diagram showing evolution of North American continent during the Precambrian. (A) 2,500 m.y. ago. (B) 1,700 m.y. ago. (C) 1,350 m.y. ago. (D) 1,000 m.y. ago. Actual size of continent is unknown for any of these intervals because of lack of data on older rocks in marginal regions. Ruled pattern = area known to be underlain by rocks giving radiometric ages equal to or greater than that indicated in figure (± 100 m.y.). Shaded area = probable minimum size of continent (or continental crust) based on the assumption that areas dated are in their same relative position. (After W. R. Muehlberger, R. E. Denison, and E. C. Lidiak, *Bull. Am. Assoc. Petrol. Geol.*, 1967.)

That these rocks 2,500 m.y. old or older in intermediate locations were involved in later orogeny suggests that these three areas were not separate but were joined by continental crust. Thus in Figure 4-5, A the shaded area surrounding and between the three ancient cratons indicates the probable minimum size of North America 2,500 m.y. ago. The successive parts of Figure 4-5 show the gradual enlargement of the North American continent to about its present size. Much of this enlargement, however, is actually stabilization of mobile regions.

Many geologists believe continental accretion was important in the geological evolution of North America—that is successive mobile belts around the ancient cratons oceanic crust was transformed into continental crust through complex geosynclinal processes. One difficulty in applying this concept is that little is known about the composition or age of the crystalline basement for the entire Cordilleran belt of western North America, the Canadian Arctic Islands, and the Gulf and Atlantic coastal plains of the United States. Thus the maps of Figure 4-5 can only indicate a suggested minimum size for each time period.

4-2 THE SOUTH AMERICAN SHIELDS

The eastern two-thirds of South America is a vast, stable complex of Precambrian rocks and mildly disturbed sedimentary and volcanic Phanerozoic rocks. The western margin of South America is a highly folded zone of Phanerozoic formations with a long and interesting geosynclinal history. The shield area of South America (Fig. 4-6) is broken into three masses by troughs filled mainly with undisturbed Paleozoic sediments. The northern trough is along the course of the Amazon River; the southern cuts across central Argentina. It is customary to recognize three distinct shields: the Guianan

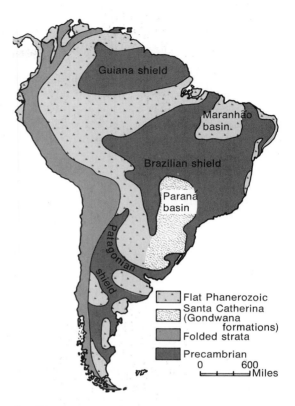

FIGURE 4-6

Generalized geologic map of South America, showing Precambrian shields and disposition of Phanerozoic formations. (After V. Benavides, *Geol. Soc. Am.*, Special Paper 88, 1968.)

shield north of the Amazon, the Brazilian shield between the two troughs, and the Patagonian shield in southern Argentina.

The Precambrian strata of South America are still little known. Relatively few radiometric ages are available, though extensive programs are underway. Data are insufficient to allow even the kind of synthesis that is now possible for North America and Africa. The Brazilian shield, the largest and best-known, is made up of poorly understood Early Precambrian gneiss and granites. Middle Precambrian rocks are found in a northeast-southwest trend on the western and eastern sides of the Paraná and Maranhão basins (Fig. 4-6). The present outcrop belt is approximately 2,000 miles long and 600 miles wide. These Middle Precambrian strata accumulated in a eugeosyncline that was intensely folded and intruded by a variety of plutonic rocks. After extensive erosion, approximately the same area again became geosynclinal in Late Precambrian time, though with little or no volcanic activity in contrast to that of the Middle Precambrian geosyncline. Toward the close of Precambrian time the area again underwent folding but with little plutonism. This phase of folding was much less intense than that of Middle Precambrian time. Much of the area of this Precambrian geosynclinal region is now covered by flat-lying Phanerozoic sedimentary deposits. It is interesting to speculate that perhaps the "tectonic energy" of this area was slowly dissipated during the long evolution of the region. Eugeosynclinal in character during the Middle Precambrian, miogeosynclinal in the Late Precambrian, it ended by being basinal in the Early Paleozoic, the subnegative tendency of the region becoming less and less marked in the periods that followed.

The Guianan shield occupies most of French and British Guiana and Surinam. Here, as in the Brazilian shield, three main divisions are generally recognized, each representing a long, complex depositional cycle followed by orogeny and granitic intrusions. The oldest Precambrian rocks are gneisses and schists of sedimentary origin. The Middle Precambrian is mainly volcanic but includes some sedimentary deposits, all generally highly metamorphosed, folded, and cut by granitic bodies. The Late Precambrian consists mainly of clastic and slaty rocks with volcanic flows and tuff. There is a suggestion in the rock record that there was a progressive migration of the geosyncline from south to north during Precambrian time.

On the Patagonian shield only small, isolated areas of Precambrian rocks are exposed. A blanket of Late Cretaceous and Cenozoic formations masks this old terrane. For much of Phanerozoic time this area has been a stable block, which, late in its history, subsided a little. The basement strata of the Patagonian shield are thought to be late Precambrian in age.

4-3   THE BALTIC SHIELD

Most of Sweden and Finland consist of Precambrian rocks of the Baltic shield, whose western edge is bounded by an early Paleozoic orogenic belt; its eastern and southern margins are overlain by thin, flat, Phanerozoic sediments that gradually thicken away from the shield. The relations of these main geologic units are shown by the generalized geologic map of Figure 4-7. Solely from a study of this map we can see that the Baltic shield has been remarkably stable since Precambrian time.

The extremely complex Precambrian history of the Baltic region has long been studied intensively by Swedish and Finnish geologists, who recognize a succession of sedimentary cycles and epochs of diastrophism and granitic intrusion. We now recognize four distinct sedimentary and volcanic cycles separated by diastrophic movements and granitic intrusions. The two oldest cycles, the Katarchean and the Archean, are characterized by schistose and gneissic rocks, originally clastic sedimentary or volcanic rocks such as lavas and tuffs. At the end of each cycle widespread granites were intruded. The next cycle, the Karelidic, is represented by slightly metamorphosed quartzitic sandstones (many with ripple marks), conglomerates, dolomites, iron ores, slates, and greenstones. The youngest Precambrian cycle (not formally named) produced little but unmetamorphosed volcanic rocks and continental sandstones, into which granites have intruded.

Correlation of complex granitic and metamorphic bodies over wide areas is extremely difficult, but many radiometric studies have recently added to our understanding of the shield's geologic history. By geologic and radiometric studies the Baltic shield can be subdivided into three zones (Fig. 4-8). In the Saamo-Karelian zone radiometric dates range from 3,000 m.y. to 1,900 m.y. The four subdivisions of the Saamo-Karelian zone are based on some differences in radiometric ages, but the details need not concern us. Southwestern Finland and most of Sweden make up the second zone, the Sveco-Fennian. In this zone the radiometric ages range from 2,300 m.y. to 1,500 m.y. A third zone, the Sveco-Norwegian, occupies southwestern Sweden and southern Norway; ages in this zone range from 1,200 m.y. to 900 m.y. The rocks of these three zones are now largely granites and highly metamorphosed rocks of geosynclinal origin. There are also post-orogenic Precambrian complexes of essentially flat-lying sedimentary formations, volcanic rocks, and igneous intrusions with ages of 1,800 m.y. to 1,000 m.y. (Fig. 4-8).

Most K-Ar age values date the latest local plutonic processes; these values, however, are often influenced by rejuvenation. A

FIGURE 4-7

Generalized geologic sketch map of northern Europe, showing Precambrian shield and disposition of Phanerozoic formations. (Data from *Europa, Geologische Übersicht*, Geotektonischen Institut der Deutschen Akad. Wiss., Berlin, 1963.)

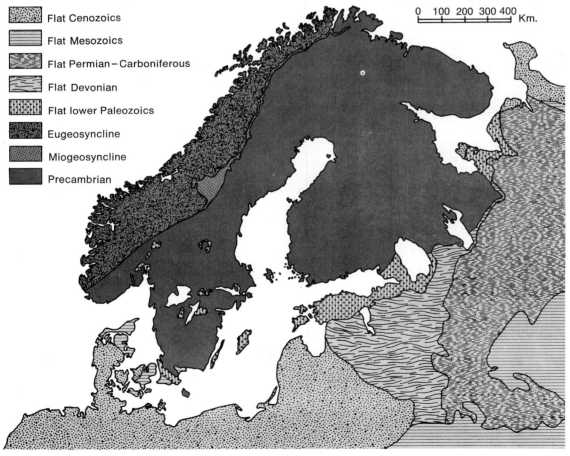

Flat Cenozoics

Flat Mesozoics

Flat Permian – Carboniferous

Flat Devonian

Flat lower Paleozoics

Eugeosyncline

Miogeosyncline

Precambrian

0 100 200 300 400 Km.

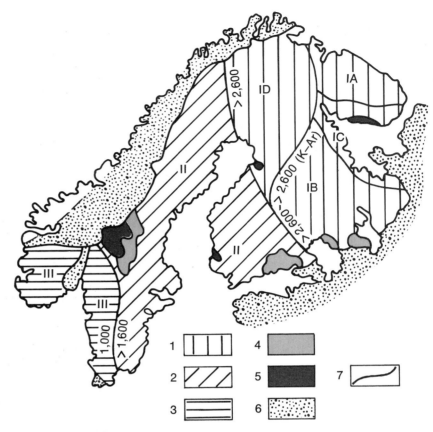

FIGURE 4-8

Schematic geochronological map of the Precambrian of the
Baltic shield. I = Saamo-Karelian zone; II = Sveco-Fennian
zone; III = Sveco-Norwegian zone. Geosynclinal associations:
1 = age 3,000–1,900 m.y.; 2 = age 2,300–1,500 m.y.; 3 = age
1,200–900 m.y. and partly older. Platform associations: 4 = age
1,800–1,600 m.y.; 5 = age 1,450–1,000 m.y.; 6 = latest Precambrian
and Paleozoic; 7 = boundary of geologic-geochronological zones
and subzones. (After E. Gerling, K. Kratz, and S. Lobach-Zhuchanko,
Report 23rd. Intern. Geol. Cong., 1968.)

summary of the K-Ar dates for the Baltic
shield is given in Fig. 4-9.

In a very generalized way the tectonic
evolution of the Baltic shield is one of progressive stabilization of the region. The
oldest and first stable portion is that to
the northeast; the intermediate region was
the next to become immobile and the most
southern portion the last.

4-4 THE ETHIOPIAN SHIELD

Probably the stablest continental block in
the Phanerozoic history of the world is the
great continent of Africa. An inspection of
Figure 4-10 will help to explain this stability. Outcrops of Precambrian rocks cover
more than half the surface. There are only
two orogenic belts with folded strata, a

narrow one at the northwestern corner and a small one at the southern tip. The rest of the continent consists either of exposed Precambrian formations or of only mildly disturbed layers of Phanerozoic sediments and volcanics. These younger formations are generally marine at the margins of the continent and nonmarine in the interior.

The Precambrian formations are an extremely complex array of metamorphosed sedimentary and igneous rocks including large masses of granite. The character of the rocks and their structural relations testify to a long and very complex geological history. The use of local stratigraphic names in each country and the difficulty of correlation make synthesis difficult, but the recognition of orogenic belts in the Precambrian of central and southern Africa and radiometric dating have recently established at least a rough outline of the orogenic history.

FIGURE 4-9

Schematic map of K-Ar age dates of orogenic formations of the Baltic shield. Areal distribution of rocks, 1 = age exceeding 2,600 m.y.; 2 = age 2,300–1,650 m.y.; 3 = intense rejuvenation of an age 1,950–1,650 m.y.; 4 = partial rejuvenation of an age 1,950–1,650 m.y.; 5 = age 800–1,200 m.y.; 6 = intense rejuvenation age 1,000 m.y.; 7 = boundaries of zones; 8 = late Precambrian-Paleozoic. (After E. Gerling, K. Kratz, and S. Loback-Zuchenko, *Report 23rd. Intern. Geol. Cong.*, 1968.)

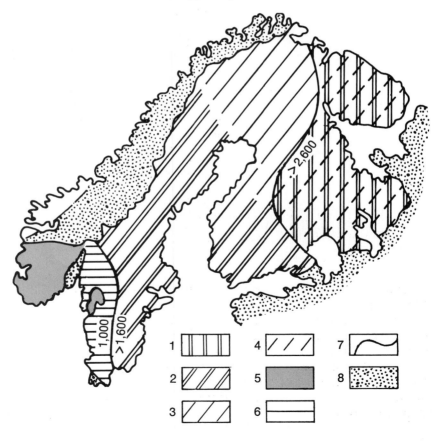

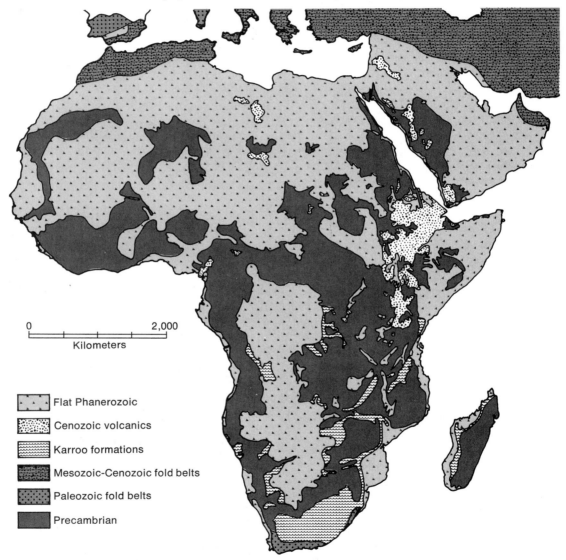

FIGURE 4-10

Generalized geologic map of Africa and Arabia and disposition of Phanerozoic formations. (Data from Esquisse Structurale Provisoire de L'Afrique, Assoc. des Ser. Géol. Africans, Paris, 1958.)

Flat Phanerozoic

Cenozoic volcanics

Karroo formations

Mesozoic-Cenozoic fold belts

Paleozoic fold belts

Precambrian

During the past two decades many hundreds of radiometric age determinations have become available. These ages tend to cluster, making possible the following subdivision of the Precambrian:

Dated Events (m.y.)	Stratigraphic Subdivisions
450–680	Early Paleozoic and Late Precambrian
1,100 ± 200	Middle Precambrian
1,850 ± 250	
ca. 2,600	Early Precambrian
ca. 3,000	

These clusters broadly identify orogenic episodes. The analysis of these events, however, is highly complicated. Consider, for example, Figure 4-11 which shows a series of highly idealized sections of a fold belt. The deformation of the geosynclinal trough, accompanied by igneous intrusions, is the thermal event that is radiometrically dated. This thermal event not only affects the geosynclinal deposits but also to varying degrees the older basement rocks. Suppose that erosion has removed most or all of the deformed geosynclinal sequence, leaving only the older basement complex, upon which a younger thermal episode has been imprinted. Africa underwent many lengthy periods of erosion during the Precambrian and Phanerozoic and has remained largely continental since Lower Paleozoic times, so that many levels in Precambrian to early Paleozoic orogenic zones are exposed. Another complicating factor is that thermal imprints can be completely unrelated to orogenic episodes.

A plot of ages greater than 2,100 m.y. is shown in Figure 4-12, A. They tend to be concentrated mainly in the southern half of Africa, except for a few in West Africa. The oldest dates are in South Africa, where several formations yield ages of more than 3,000 m.y. The areas of these very ancient dates have not been orogenically deformed since at least 1,500 m.y. ago. Ages in the range 1,850 ± 250 m.y. are widespread in southern and West Africa (Fig. 4-12, B) and clearly reflect important orogenic events. The primary result of this orogenic episode was the stabilization of four cratonic areas as indicated in Figure 4-12, B. The distribution of ages ranging from 800 m.y. to 1,300 m.y. is largely concentrated in the southern half of Africa, and reflects an important orogenic episode known as the Kibaran

FIGURE 4-11

Idealized sections of a fold belt. (A) The depositional phase of the mobile zone. (B) The deformed geosynclinal volcanic-sedimentary pile. (C) The imprint of thermal effects on the older basement. (After T. N. Clifford, *Radiometric Dating for Geologists*, Wiley, New York, 1968.)

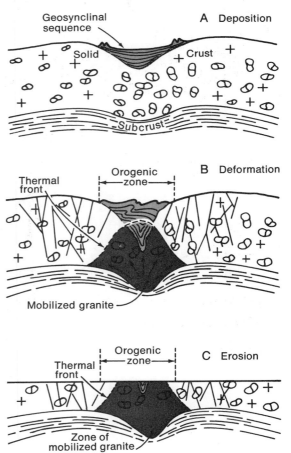

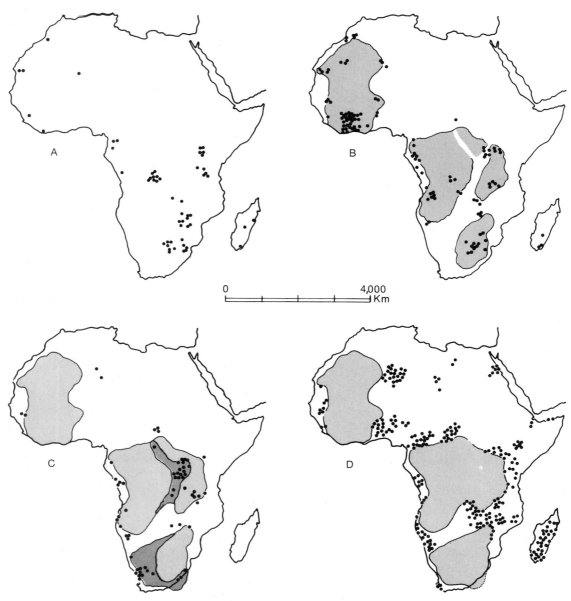

FIGURE 4-12

Distribution of radiometric ages for Africa. (A) Ages greater than 2,100 m.y. (B) Ages 1,300–2,100 m.y.; ruled pattern indicates regions that have remained stable since the end of the 1,850 ± 250 m.y. orogenesis. (C) Ages 800–1,300 m.y.; ruled pattern same as in B; stippled pattern indicates areas of Kibaran orogenesis (1,100 ± 200 m.y.). (D) Ages 450–680 m.y.; ruled pattern indicates regions that have been stable since end of Kibaran orogenesis (1,100 ± 200 m.y.). (After T. N. Clifford, *Radiometric Dating for Geologists*, Wiley, New York, 1968.)

orogenesis (Fig. 4-12, C). By far the greatest concentration of ages in Africa lies in the range 450–680 m.y. (Fig. 4-12, D). This range of ages probably reflects more than one orogenic episode. Some of the ages are of geosynclinal sequences, but others are clearly thermal imprints on older terranes. As can be seen from Figure 4-12, D, the mobile belts reflected by these ages are around and between three major ancient cratons.

The geological evolution of Africa now appears to be characterized by enlargement of cratons as a result of the stabilization of zones of mobility. This concept is schematically shown in Figure 4-13. At a period of approximately 1,500 m.y. ago four major, stable cratonic areas were present in south and western Africa. With the conclusion of the Kibaran orogeny (approximately 800–900 m.y. ago) these cratons enlarged slightly, and the two central ones coalesced to form the Congo craton. In the period 400–500 m.y. ago another orogenic episode (Pan-African) stabilized most of the remaining portion of Africa (Fig. 4–13, D). The remaining mobile areas are limited to the Atlas Mountain region in the northwest and to the Cape region in the south. These areas underwent folding in the period between the mid-Paleozoic and the early Mesozoic.

4-5 THE INDIAN SHIELD

Peninsular India is another continental block that has been remarkably stable since the beginning of Cambrian time. Fully two-thirds of the peninsula has outcropping Precambrian rocks (Fig. 4-14), whose depositional and structural records, like those in most Precambrian shields, is extremely complex.

The main geological features of India are shown in Figure 4-14. Peninsular India is bounded on the north by folded geosynclinal complexes of Phanerozoic age. The shield itself has wide areas of flat lavas (the Deccan traps of latest Cretaceous and early Cenozoic age) and sedimentary formations (the Gondwana strata, from Early Permian to Early Cretaceous in age, which will be discussed in § 11-1). Of interest is the fact that these later deposits are terrestrial or extrusive (with very few exceptions) and have never been involved in any orogenies. Their character and distribution testify to the stability of the shield since Precambrian time.

Even though these Precambrian rocks have been studied for more than a hundred years, many fundamental problems regarding the sequence of the main rock bodies remain. Radiometric dates are still too few to allow a realistic synthesis of Precambrian history. It has been possible, however, to recognize a few major regional trends in the Precambrian of peninsular India from which some of the principal features of this history can be outlined. These major trends are the Dharwar schist belt (N.N.W.–S.S.E.), the Eastern Ghats belt (N.E.–S.W.), the Satpura belt (W.S.W.–E.N.E.), and the Aravalli and Delhi belts (N.E.–S.W.), and shown in Figure 4–14.

The Dharwar belt of southwest peninsular India includes some of the oldest rocks (2,400 m.y.) known on the subcontinent, an extremely complex array of gneisses and granite. The depositional history covers a very long time: deposition of volcanic and sedimentary rocks was followed by three or more phases of orogeny accompanied by granitic intrusions. The Dharwar belt is one of the older stable cratonic nuclei of peninsular India. The Bundelkhand granite of north-central peninsular India (Fig. 4-14) has yielded an age of 2,500 m.y. and is thus the approximate time equivalent of the Dharwar belt.

The Eastern Ghats belt parallels much of the eastern coast of peninsular India and strikes almost normal to the Dharwar trends. The rocks of the Eastern Ghats are highly metamorphic and igneous complexes like those of many Precambrian complexes.

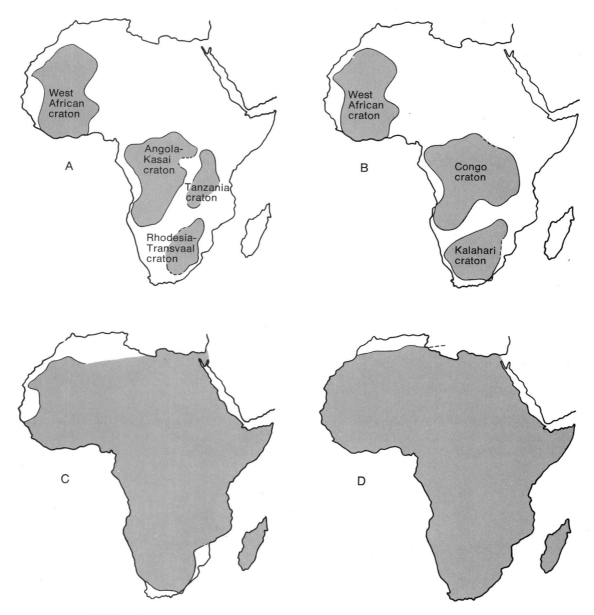

FIGURE 4-13

Stages in the structural consolidation of Africa. (A) Regions stable since at least 1,500 m.y. ago. (B) Regions stable since the end of the Kibaran orogeny (800–900 m.y. ago). (C) Regions stable since (Pan-African) orogenesis (400–500 m.y. ago). (D) Regions stable since end of middle Paleozoic-early Mesozoic orogenesis. (After T. N. Clifford, *Radiometric Dating for Geologists*, Wiley, New York, 1968.)

FIGURE 4-14
Generalized geologic sketch map of India showing Precambrian shield, tectonic trends, and disposition of Phanerozoic formations. (Data from *Geological Map of India*, Geol. Surv. India, 1962, and A. Holmes, *Proc. Geol. Assoc. Canada,* 1955.)

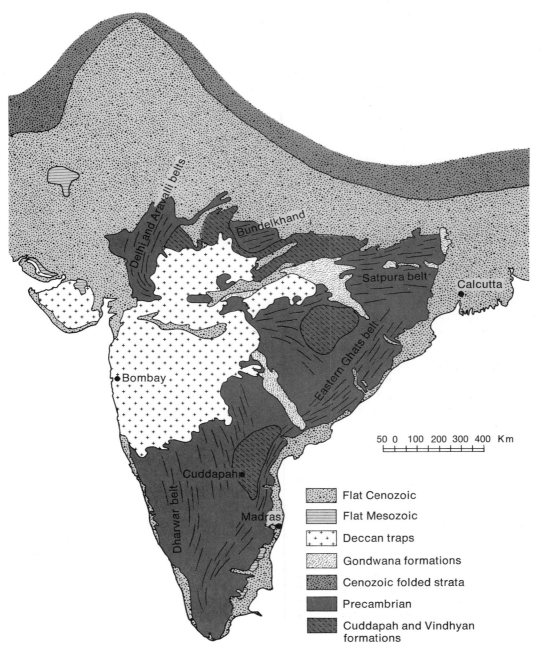

The few radiometric ages from this belt cluster around 1,600 m.y.

The Satpura belt is south of the Ganges alluvial plain and has a E.N.E.–W.S.W. trend. The western extension of this belt is buried beneath the Deccan traps. The original deposits of the Satpura belt were sedimentary, some rich in manganese, others in hematite. These sedimentary formations were subjected to severe orogeny and extensive plutonic intrusions approximately 1,000 m.y. ago, giving rise to the complex metamorphic and granitic rocks of this belt.

The Aravalli-Delhi belt of northwest peninsular India includes three major orogenic cycles, each with its own sedimentary depositional history, orogeny, metamorphism, and plutonic intrusions (Fig. 4-14). The Delhi belt is the youngest and is directly superimposed on the Aravalli belt. The latter is underlain by a little-known gneiss complex. The relationship of these Precambrian rocks to others on peninsular India is uncertain. The trend of the Aravalli belt as it approaches the Deccan traps suggests a link-up under the traps with the Dharwar belt. At the same time the more south-easterly trend close to the Deccan traps may continue in a more easterly trend and thus link with the Satpura belt. Data as yet are insufficient to solve this problem, but current consensus is that the latter interpretation is more likely. A few radiometric ages in the Aravalli-Delhi trend give dates of approximately 750 m.y., probably dating the youngest plutonic events in this belt.

These Precambrian belts are all characterized by highly deformed metamorphic and granitic terranes. There are also mildly deformed or metamorphosed Precambrian formations. One of these, the Cuddapah group, outcrops mainly in a large crescent-shaped area northwest of Madras (Fig. 4-14), where it rests on the Dharwar rocks with striking unconformity. The rocks of this group consist of slates, quartzites, and limestones with numerous igneous intrusions.

Over most of this area the Cuddapah group is little disturbed but along the eastern margin it is folded roughly parallel to the Eastern Ghats belt. On this basis it has been suggested that the group may belong to the end of the Eastern Ghats cycle and the period immediately following. The few radiometric ages available tend to support this suggestion. It is not surprising to note that this slightly deformed Precambrian complex is much older than the highly deformed Satpura and Aravalli-Delhi belts.

In the north-central margin of the Indian shield, just south of the Ganges River, is a vast area (40,000 square miles) of stratified sandstones, shales, and limestones that rest unconformably on older Precambrian rocks (e.g., Delhi belt, Bundelkhand granites, etc.). These strata—the Vindhyans—reach a thickness of 14,000 feet and, over wide areas, are horizontal. The lower beds consist of marine calcareous and argillaceous facies and the upper ones of fluvial and esturine deposits. We are still uncertain of the age of the Vindhyans. Some paleontologists believe that minute, disk-like impressions discovered in the upper beds were made by primitive brachiopods; others doubt even the organic nature of these fossils. The Geological Survey of India places the Vindhyans in the late Precambrian; it is possible that part of the Upper Vindhyans may be early Paleozoic in age.

Tectonic forces slowly lifted the Vindhyan deposits above sea level and formed of them a continental land area. This was the last significant earth movement recorded in the history of the peninsula; no other disturbance of a similar nature has affected the stability of this land mass since late Precambrian time.

4-6 THE AUSTRALIAN SHIELD

Australia is a continent without young mountain ranges. The eroded roots of Paleozoic mountains are found in eastern Australia,

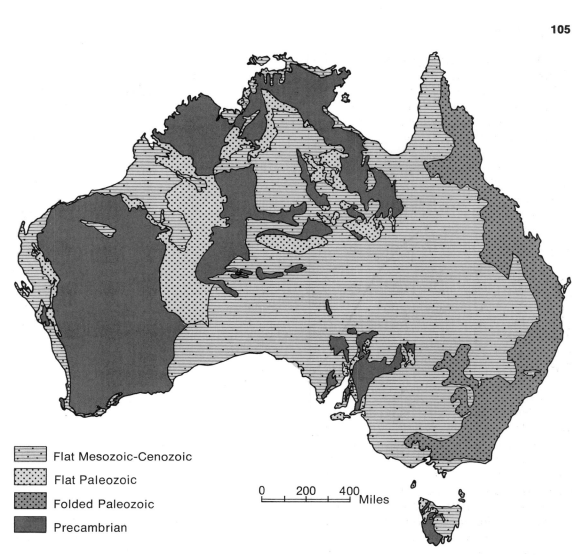

Flat Mesozoic-Cenozoic

Flat Paleozoic

Folded Paleozoic

Precambrian

0   200   400 Miles

FIGURE 4-15
Generalized geologic map of Australia showing Precambrian shield and disposition
of Phanerozoic formations. (Data from Geologic Map, *Atlas of Australian Resources*,
Dept. of National Development, Canberra, 1958.)

but in the central and western part of the island continent Precambrian rocks, often covered by essentially flat-lying Phanerozoic strata, predominate and make up the Australian shield (Fig. 4-15). Folding virtually ceased in Australia with the end of Paleozoic time.

The Precambrian of Australia includes much highly deformed schist and gneiss of sedimentary and igneous origin, but the most striking feature is the widespread occurrence of thick, unmetamorphosed sedimentary sequences. Over the past several years many radiometric age determinations have been made, enabling geologists to make a reasonable correlation of strata and events. The correlation now used is as follows:

	Cambrian System	
	Adelaidean System	
		1,400 m.y.
PROTEROZOIC	Carpentarian System	
		1,800 m.y.
	Nullaginian System	
		2,300 m.y.
ARCHEAN		

The terms Archean and Proterozoic are the same as used for the Canadian shield and essentially reflect a two-fold division into older and younger phases of Precambrian history. The Archean rocks are older than 2,300 m.y. and crop out over much of southwestern and central Australia. These Archean rocks consist mainly of gneiss and granite that reflect two or more phases of folding and plutonic intrusions, spanning at least from 2,300 to 3,000 m.y.

The most unusual feature of the Proterozoic formations is the predominance of quartz sandstone over graywacke. Some graywacke is present in the Lower Proterozoic (Nullaginian) formations, much less in the Middle Proterozoic (Carpentarian), and none in the Upper Proterozoic (Adelaidean). A schematic paleogeographic map for the Nullaginian is shown on Figure 4-16, A. One must keep in mind that such a map is highly generalized, as it covers a span of time equal to almost the whole Phanerozoic. Even so, it is useful. The Archean outcrop areas of southwest Australia are now a stable cratonic nucleus with the Nullaginian geosynclines wrapped around the north, east, and, to a lesser extent, the west sides. The sedimentary facies include quartz sandstone, limestone, dolomite, and some graywacke. Nullaginian deposition came to an end about 1,800 m.y. ago as a result of orogenesis accompanied by granitic intrusions.

The paleogeography for Carpentarian time is roughly similar to that for the preceeding Nullaginian (Fig. 4-16, B). Quartz sandstone, limestone and dolomite predominate. This, like the preceeding phase, ended in orogenesis accompanied by granitic intrusions in some areas.

The final phase of the Precambrian history of Australia (the Adelaidean) is probably the most interesting. Strata of this period crop out mainly in a N–S belt through central Australia. A schematic paleogeographic map is shown on Figure 4-16, C. The dominant facies are quartz sandstone and carbonates. Of special interest is the presence of at least two glacial episodes in the upper half of the sequence. Evidence for this Late Precambrian glaciation in Australia is widespread. Another interesting aspect is the occurrence of fossil metazoans in the upper part of the sequence (Pound Quartzite) as discussed in Chapter 2 and illustrated in Figures 2-27, 6–34. In the Adelaide geosyncline of South Australia this series is 45,000 feet thick. The Adelaidean, unlike the previous Proterozoic episodes, did not end with orogenesis and igneous activity except in parts of central Australia. The thick sequences of the Adeleide geosyncline of south Australia were folded later in the Cambrian accompanied by granite emplacement.

4-7 THE ANGARAN SHIELD

This stable block, commonly thought of as the nucleus of Asia, is, of all the shields, the least known to geologists of the western hemisphere. It includes the area between the Yenisei River on the west, the Lena on the east, the Arctic Ocean to the north, and the general latitude of Lake Baikal to the south. The Angaran shield is separated from the Baltic shield by a mobile belt of Paleozoic age, now comprising the Ural Mountains. It is separated from the Indian shield by a vast mobile belt, the Himalayan

Mountains and folded zones in Tibet and China. Figure 4-17 presents the relations of this stable segment to the surrounding belts of folded strata. One of the unusual features of the Angaran shield is the relatively small outcrop of Precambrian rocks. Most of the shield is covered by flat Paleozoic, some by Mesozoic, rocks. These younger formations are thin, shallow-water deposits, which reflect subsidence and emergence of the stable plate in Phanerozoic time.

The rocks of Precambrian age exposed in the shield are like those of the other shields — an older complex of gneiss, schist, and granites, and a younger complex of slightly metamorphosed sedimentary series, including volcanics and granite bodies. The older complexes are poorly understood as yet, but the late Precambrian is better known. The late Precambrian includes strata ranging in age from 1,600 m.y. to earliest Cambrian, and is referred to as the Riphean. The maps in Figure 4-18 summarize the available data for four subdivisions of the Riphean. The major feature of these maps clearly shows that the Angaran shield was already stable more than 1,600 m.y. ago, with only minor transgressions on it during most of the late Precambrian. A widespread transgression came in latest Precambrian time, setting a pattern that was to continue through the early Paleozoic.

In early Riphean time the shield area was emergent and surrounded by seas wherein accumulated clastic sediments with minor amounts of limestone (Fig. 4-18, A). Transgression characterizes medial Riphean

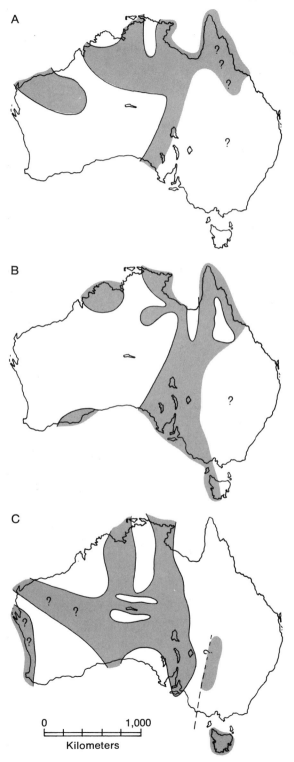

A

B

C

0   1,000

Kilometers

FIGURE 4-16

Schematic paleogeographic maps of Australia for three episodes of the Proterozoic. Shaded areas represent probable distribution of largely marine deposits. (A) Lower Proterozoic (Nullaginian) 1,800–2,300 m.y. (B) Middle Proterozoic (Carpentarian) 1,400–1,800 m.y. (C) Upper Proterozoic (Adelaidean) 600–1,400 m.y. (After D. A. Brown, K. S. W. Campbell, and K. A. W. Crook, *The Geological Evolution of Australia and New Zealand*, Pergamon Press, 1968.)

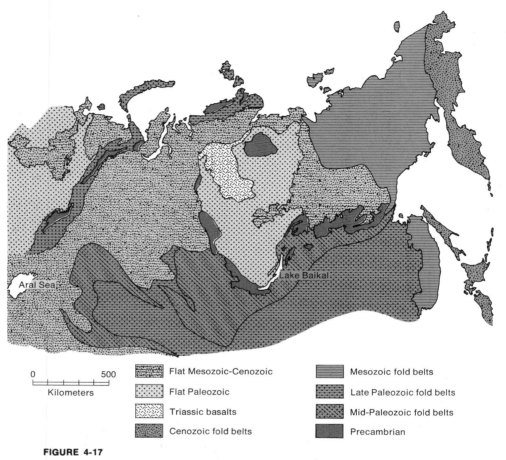

FIGURE 4-17

Generalized geologic map of eastern U.S.S.R., showing Precambrian shield and disposition of Phanerozoic formations. (Data from *Geologic Map of U.S.S.R.*, 1960.)

time, and some of the shield was covered by seas. Clastic sediments are prominent, but there was also widespread limestone deposition (Fig. 4-18, B). The late Riphean was a time of some regression off the shield area and the uplift of linear belts to the west and south of the shield. These uplifted belts were the source of much clastic sediment and the center of much volcanic activity (Fig. 4-18, C). Latest Riphean time was marked by broad transgression of very shallow seas across the shield, on which dolomites and evaporites accumulated (Fig. 4-18 D). Linear uplifted belts characterized the southwestern, southern, and north-

eastern marginal areas of the shield, and in these areas clastic sedimentation predominated. The transgressive pattern of the latest Riphean continued through part of early Paleozoic time.

4-8 THE ANTARCTIC SHIELD

The least known of all the continents is the Antarctic. Geological observations made there during the past 60 years reveal a structural division into East and West Antarctica. These were thought to be, respectively, an ancient stable shield and a much younger

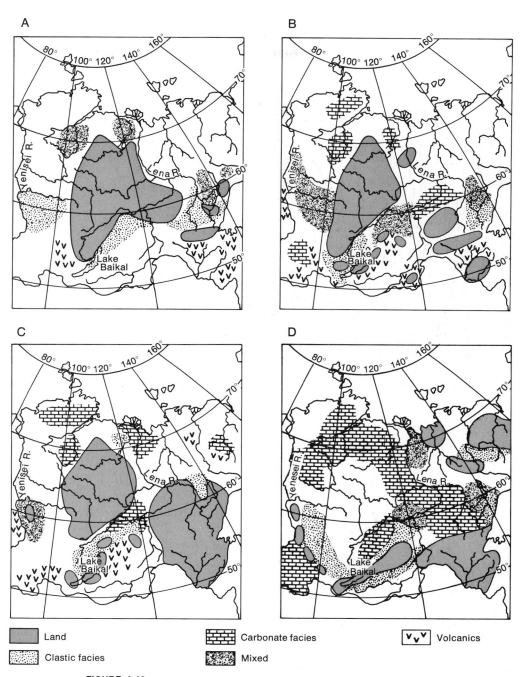

FIGURE 4-18

Schematic paleogeographic maps for the late Precambrian of Siberia.
(A) Early Riphean (1,600 ± 50–1,350 ± 50 m.y.). (B) Middle Riphean
(1,350 ± 50–950 ± 50 m.y.). (C) Late Riphean (950 ± 50–675 ± 25 m.y.).
(D) Latest Riphean (675 ± 25–570 ± 10 m.y.). (After B. M. Keller *et al.*,
Report 23rd Intern. Geol. Cong., 1968.)

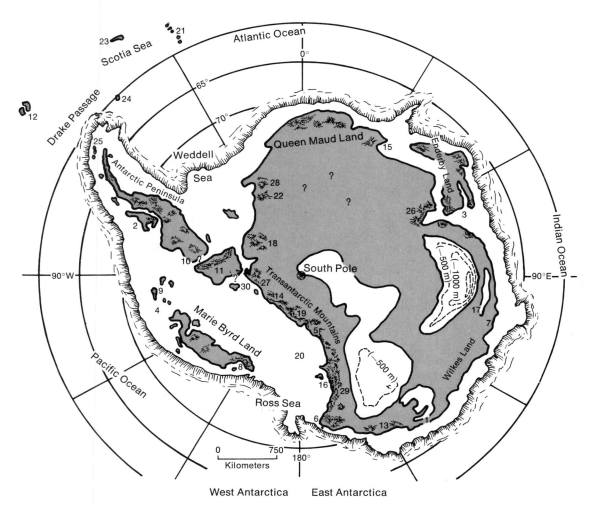

FIGURE 4-19

Physiographic diagram of Antarctica with ice cover removed, unadjusted for isostatic uplift. Hachures show continental slope and major mountain areas; stippled regions are above sea level. In alphabetical order, numbered features are as follows: (1) Adelie coast, (2) Alexander Island, (3) Amery ice shelf, (4) Amundsen Sea, (5) Beardmore Glacier, (6) Cape Adare, (7) Denman Glacier, (8) Edsel Ford Ranges, (9) Eights coast, (10) Ellsworth Land, (11) Ellsworth Mountains, (12) Falkland Islands, (13) George V coast, (14) Horlick Mountains, (15) Lutzow-Holm Bay, (16) McMurdo Sound, (17) Mount Sandow, (18) Pensacola Mountains, (19) Queen Maud Range, (20) Ross ice shelf, (21) Scotia Arc, (22) Shackleton Range, (23) South Georgia, (24) South Orkney Islands, (25) South Shetland Islands, (26) Southern Prince Charles Mountains, (27) Thiel Mountains, (28) Theron Mountains, (29) Victoria Land, and (30) Whitmore Mountains. (After Ford, 1964.)

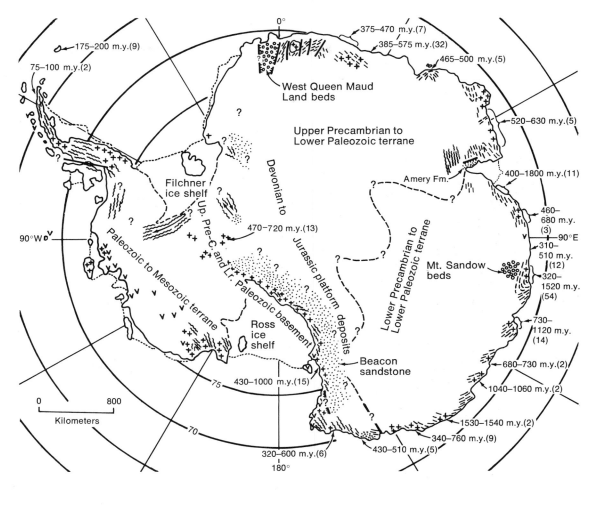

FIGURE 4-20

Generalized geologic map of Antarctica. Numbers show approximate ranges
of radiometric ages, with the numbers of dates within each range given in
parenthesis. (After Ford, 1964.)

alpine-type, fold-mountain system of late Mesozoic to Cenozoic age. More recent data suggest that this twofold division is an over-simplification.

The average thickness of the Antarctic ice sheet is approximately 3,000 feet, but in places the thickness appears to be as much as 9,000 feet. Extensive geophysical studies on the continent have yielded data on the bedrock surface of the Antarctic land mass (Fig. 4-19). Most of the bedrock surface of East Antarctica is above sea level, except in three relatively extensive depressions; in at least one of these the bedrock surface is 4,500 feet below sea level. It has not been calculated what parts of these areas would remain below sea level if the overlying ice were removed, resulting in isostatic rise of the continent and, at the same time, allowing for a nearly 180-foot rise in sea level resulting from the great volume of melted ice. There are several areas in East Antarctica with mountains reaching altitudes of approximately 10,000 feet. This great relief is certainly not typical of Precambrian shield regions elsewhere in the world.

Most geological observations in East Antarctica have been confined to narrow coastal strips. A generalized geological map of Antarctica is shown in Figure 4-20. The few scattered coastal outcrops suggest that in the region from the Amery Ice Shelf to the Adelie Coast ancient shield rocks are exposed, whose extent inland is uncertain. Igneous and metamorphic rocks along this coastal region have yielded radiometric age dates as old as 1,800 m.y. Besides the older Precambrian dates, some as young as middle Paleozoic are abundant throughout the Amery Ice Shelf-Adelie Coast area, suggesting a long and doubtless complex history of deformation, regional metamorphism, and plutonic intrusions. Most details of the geologic history have yet to be worked out, but it is already obvious that East Antarctica is not merely a simple, longstable Precambrian shield, as was heretofore commonly believed.

In addition to the intensely deformed strata that yield ancient Precambrian dates there are sedimentary and low-grade metamorphosed sedimentary rocks that appear to be Late Precambrian and possibly early Paleozoic in age. In eastern Queen Maud Land widespread plutonic rocks have yielded radiometric ages of 400 to 600 m.y.— early Paleozoic. Older Precambrian rocks have not as yet been identified in the coastal belt from Queen Maud Land to the Amery Ice Shelf.

West Antarctica is a mobile belt of Paleozoic and Mesozoic formations. The folded belt trends northward through the Antarctic Peninsula, swings sharply eastward to include the South Orkney and South Sandwich islands, then curves north and west to the South Georgia Islands and joins the Andean fold belt. Beyond the Antarctic continent is a sparse ring of subantarctic islands, mostly on the threshold of the Atlantic and Indian Oceans. Aside from the Palmer Archipelago and other islands closely associated with western Antarctica, only Macquarie Island stands on the threshold of the Pacific. All these subantarctic islands consist exclusively of volcanic, shallow intrusive, and sedimentary rocks going back to the Cenozoic in age.

SUGGESTED READINGS

The literature on geology is extremely plentiful and is increasing at a phenomenal rate. Throughout the remainder of this book we shall discuss each continental area in more and more detail in terms of the broad pattern of the earth's history. The references listed here—volumes on the general geology of particular countries, regions, or geologic periods—are recommended for those wishing to pursue their studies one step further.

Regional Studies

North America

Cady, W. M., "Regional Tectonic Synthesis of Northwestern New England and Adjacent Quebec," Geol. Soc. Am., Mem. 120 (1969).

Childs, O. E., and B. W. Beebe, eds., "Backbone of the Americas, A Symposium," Am. Assoc. Petrol. Geol., Mem. 2 (1963).

Clark, T. H., ed., Appalachian Tectonics (Royal Soc. Canada, Spec. Publ. 10, 1967).

Eardley, A. J., Structural Geology of North America, 2nd ed. (Harper & Brothers, New York, 1962).

Fisher, G. W., et al., eds., Studies of Appalachian Geology, Central and Southern (Wiley, New York, 1970).

Fortier, Y. O., et al., "Geology of the North-Central Part of the Arctic Archipelago, Northwest Territories (Operation Franklin)," Geol. Surv. Canada, Mem. 320 (1963).

King, P. B., The Tectonics of Middle North America (Princeton Univ. Press, Princeton, 1951).

———, The Evolution of North America (Princeton Univ. Press, Princeton, 1959).

McCrossen, R. G., and R. P. Glaister, eds., Geological History of Western Canada (Alberta Soc. Petrol. Geol., Alberta, 1964).

Murray, G. E. Geology of the Atlantic and Gulf Coastal Province of North America (Harper & Brothers, New York, 1961).

Neale, E. R. W., and H. Williams, eds., Geology of the Atlantic Region (Geol. Assoc. Canada, Spec. Paper 4, 1967).

Proctor, P. D., ed., A Coast to Coast Tectonic Study of the United States, UMR J., No. 1 (1968).

Ruedemann, R., and R. Balk, eds., Geology of North America (Gebrüder Borntraeger, Berlin, 1939).

Smith, P. S., Areal Geology of Alaska (U.S. Geol. Surv., Prof. Paper 192, 1939).

Stockwell, C. H., ed., Geology and Economic Minerals of Canada, 4th ed. (Geol. Surv. Canada, Ec. Geol. Ser., 1957).

Zen, E-an, et al., eds., Studies of Appalachian Geology, Northern and Maritime (Wiley, New York, 1968).

Antilles

Butterlin, J., La Constitution Géologique et la Structure des Antilles (Centre National de la Recherche Scientifique, Paris, 1956).

———, La Géologie de la Républic D'Haiti et Ses Rapports Avec Celle des Régiones Voisines (Imprimerie Jouve, Paris, 1964).

Schuchert, C., Historical Geology of the Antillean Caribbean Region (Wiley, New York, 1935).

Weyl, R., Geologie der Antillen (Gebrüder Borntraeger, Berlin, 1966).

Woodring, W. P., "Caribbean Land and Sea Through the Ages," Geol. Soc. Am. Bull., 65, No. 8 (1954).

Central and South America

Ahlfeld, F., "Geología de Bolivia," Revista ael Museo de la Plata, (n.s.) 3 (1946).

Ahlfeld, F., and L. Branisa, Geología de Bolivia (Inst. Boliviano Petroleo, La Paz, 1960).

Beurlen, K., Geologie von Brasílien (Gebrüder Borntraeger, Berlin, 1970).

Bossi, J., Geología del Uruguay (Univ. of Montevideo, Uruguay, 1966).

Bruggen, Juan, Fundamentos de la Geología de Chile (Inst. Geogr. Militar, Santiago de Chile, 1950).

Childs, O. E., and B. W. Beebe, eds., "Backbone of the Americas, A Symposium," Am. Assoc. Petrol. Geol., Mem. 2 (1963).

Furrazola-Bermudez, G., et al., Geología de Cuba (Inst. Cubano de Recursos Min. Havana, 1964).

Garfias, V. R., and T. C. Chapin, *Geología de México* (Jus, Mexico, 1949).

Gerth, H., *Geologie Südamerikas*, 3 vol. (Gebrüder Borntraeger, Berlin, 1932, 1935, 1941).

———, *Der geologische Bau der südamerikanischen Kordillere* (Gebrüder Borntraeger, Berlin, 1955).

Guimarães, D., "Geologia do Brasil," *Divisão de Fomento da Producão Mineral*, Mem. 1, Rio de Janeiro (1964).

Harrington, H. J., "Paleogeographic Development of South America," *Bull. Am. Assoc. Petrol. Geol.*, **46** (1962).

Ijzerman, R., *Outline of the Geology and Petrology of Surinam (Dutch Guiana)* (Kemink & Zoon, Utrecht, 1931).

Jenks, W. F., ed., "Handbook of South American Geology: An Explanation of the Geologic Map of South America," *Geol. Soc. Am.*, Mem. 65 (1956).

Liddle, R. A., *The Geology of Venezuela and Trinidad*, 2nd ed. (Paleontological Research Institution, Ithaca, New York, 1946).

Morrison, R. P., *A Resume of the Geology of South America* (Univ. Inst. Earth Sci., Toronto, 1962).

Oliveira, A. I. de, and O. H. Leonardos, *Geologia do Brasil*, 2nd ed. (Serviço de Informaçao Agricola, Ministério de Agrícultura, Rio de Janeiro, 1943).

Putzer, H., *Geologie von Paraguay* (Gebrüder Borntraeger, Berlin, 1962).

Sheppard, G., *The Geology of Southwestern Ecuador* (Murby, London, 1937).

Steinmann, G., *Geologie von Peru* (Winter, Heidelberg, 1929).

Weyl, R., *Die Geologie Mittelamerikas* (Gebrüder Borntraeger, Berlin, 1961).

Windhausen, A., *Geología Argentina* (Peuser, Buenos Aires, 1931).

Zeil, W., *Geologie von Chile* (Gebrüder Borntraeger, Berlin, 1964).

Europe

Abrard, A., *Géologie de la France* (Payot, Paris, 1948).

Acciaiuoli, L. de M., *Geologia de Portugal* (Servicos Geologicos, Lisbon, 1957).

Anderson, J. G. C., and T. R. Owen, *The Structure of the British Isles* (Pergamon Press, New York, 1968).

Baluk, W., M. Kamieniecka, and Fr. Szczepanski, *Geologia Polski* (Panstwowe Wyd. Szkol. Zawod., Warsaw, 1960).

Bennison, G. M., and A. E. Wright, *The Geological History of the British Isles* (St. Martin's Press, New York, 1969).

Brinkmann, R., *Abriss der Geologie, Historische Geologie* (Enke, Stuttgart, 1954). *Geologic Evolution of Europe,* A condensed version translated into English by J. E. Sanders (Enke, Stuttgart, Hafner, New York, 1960).

Bubnoff, S. von, *Geologie von Europa*, 2 vol. (Gebrüder Borntraeger, Berlin, 1926–1936).

Charlesworth, J. K., *The Geology of Ireland* (Oliver & Boyd, Edinburgh, 1953).

———, *Historical Geology of Ireland* (Oliver & Boyd, Edinburgh, 1963).

Craig, G. Y., *The Geology of Scotland* (Oliver & Boyd, Edinburgh, 1965).

Dorn, P., *Geologie von Mitteleuropa*, 2nd ed. (Schweizerbart, Stuttgart, 1960).

Evans, J. W., and C. J. Stubblefield, *Handbook of the Geology of Great Britain* (Murby, London, 1929).

Faber, F. J., *Geologie von Nederland*, 3rd ed. (Naeff, The Hague, 1949).

Heim, A., *Geologie der Schweiz*, 2 vols. (Tauchnitz, Leipzig, 1919, 1922).

Holtedahl, Olaf, *Norges Geologie*, 3 vol. (Aschehoug, Oslo, 1953).

———, ed., "Geology of Norway," *Norges Geol. Unders.*, No. 208 (1960).

Johnson, M. R. W., and F. H. Stewart, eds., *The British Caledonides* (Oliver & Boyd, Edinburgh, 1963).

Knetsch, G., *Geologie von Deutschland* (Enke, Stuttgart, 1963).

Launay, L. de, *Géologie de la France* (Librairie Armand, Paris, 1921).

Magnusson, N. H., E. Granlund, and G. Lundquist, *Sveriges Geologie*, 2nd ed. (Norstedt & Söner, Stockholm, 1949).

Magnusson, N. H., *et al.*, "Description to Accompany the Map of the Pre-Quaternary Rocks of Sweden," *Sveriges Geol. Unders.*, No. 16 (1960).

Oncescu, N., *Geologia Republicii Populare Romine*, 2nd ed. (Editura Tehnica, Bucharest, 1959).

Rayner, D. H., *The Stratigraphy of the British Isles* (Cambridge Univ. Press, London, 1967).

Rutten, M. G., *The Geology of Western Europe* (Elsevier, Amsterdam, 1969).

Schaffer, F. X., *Geologie von Osterreich*, 2nd ed. (Deuticke, Vienna, 1951).

Svoboda, J., *et al.*, *Regional Geology of Czechoslovakia*, 2 vols. (Geol. Surv. Czechoslovakia, Prague, 1966).

Trümpy, R., "Paleotectonic Evolution of the Central and Western Alps," *Geol. Soc. Am. Bull.*, **71** (1960).

Wills, L. J., *A Paleogeographical Atlas of the British Isles and Adjacent Parts of Europe* (Blackie, London, 1951).

Africa

Cahen, L., *Géologie du Congo Belge* (Vaillant-Carmanne, Liège, 1954).

Du Toit, A. L., *The Geology of South Africa*, 3rd ed. (Hafner, New York, 1954).

Furon, R., *Geology of Africa*, transl. of 2nd ed. of *Géologie de L'Afrique*, 1960 (Oliver & Boyd, Edinburgh, 1963).

———, *Géologie de l'Afrique*, 3rd ed. (Payot, Paris, 1968).

Freitas, A. J. de, *A Geologia e o Desenvolvimenta Económico e Social de Moçambique* (Imprensa Nacional de Moçambique, Lourenço Marques, Mozambique, 1959).

Geology of the Arabian Peninsula. U.S. Geol. Surv., Prof. Pap. 560 F. Geukens, "Yemen," no. 560-B (1966); J. E. G. W. Greenwood, and D. Bleackley, "Aden Protectorate," no. 560-C (1967); R. W. Powers, L. F. Ramirez, C. D. Redmond, and E. L. Elberg, Jr., "Sedimentary Geology of Saudi Arabia," no. 560-D (1966); R. P. Willis, "Bahrain," no. 560-E (1967); D. I. Milton, "Kuwait," no. 560-F (1967); K. M. Al Naqib, "Southwestern Iraq," no. 560-G (1967); Z. R. Beydoun, "Eastern Aden Protectorate and part of Dhulfar," no. 560-H (1966).

Haughton, S. H., *Stratigraphic History of Africa South of the Sahara* (Oliver & Boyd, Edinburgh, 1963).

———, *Geological History of Southern Africa* (Geol. Soc. South Africa, 1969).

Krenkel, E., *Geologie Afrikas*, 2 vols. (Gebrüder Borntraeger, Berlin, 1928, 1934); 2nd ed. (Akademische Verlagsgesellshcaft, Leipzig, 1957).

Reed, F. R. C., *The Geology of the British Empire*, 2nd ed. Arnold, London, 1949).

Said, R., *The Geology of Egypt* (Elsevier, Amsterdam, 1962).

Middle East

Ball, M. W., and O. Ball, "Oil Prospects of Israel," *Bull. Am. Assoc. Petrol. Geol.*, **37** (1953).

Bender, F., *Geologie von Jordanien* (Gebrüder Borntraeger, Berlin, 1968).

Flugel, H., "Die Entwicklung des vorderasiatischen Paläozoikums," *Geotektonische Forschungen*, **18** (1964).

Furon, R., *Géologie du Plateau Iranien* (Mem. du Muséum National d'Histoire Naturelle, Paris, 1941).

Stöcklin, J., "Structural History and Tectonics of Iran: A Review," *Bull. Am. Assoc. Petrol. Geol.*, **52** (1968).

———, *Salt Deposits of the Middle East* (Geol. Soc. Am., Special Paper 88, 1968).

Wolfart, R., *Geologie von Syrien und dem Libanon* (Gebrüder Borntraeger, Berlin, 1967).

Asia

Anon., *Reconnaissance Geology of Part of West Pakistan* (A Colombo Plan cooperative Project. Hunting Surv. Corp., Toronto, 1961).

Borooah, K., *Elements of Indian Stratigraphy* (Dattsons, Nagpur, India, 1962).

Ch'ang, Ta, *The Geology of China* (Peking, 1959, in Chinese; transl. by U.S. Dept. of Commerce, Office of Technical Services, publ. JPRS-19209, 1963).

Endo, R., *et al.*, *Geology and Mineral Resources of the Far East* (Univ. Tokyo Press, Tokyo, 1967).

Fuchs, G., "Zum Bau des Himalayas," *Österr. Akad. Wiss.*, Denksch. **113** (1967).

Gansser, A., *Geology of the Himalayas* (Wiley, New York, 1964).

Huang, T. K., "Die geotektonischen Elemente im Aufbau Chinas," *Geologie*, **9**, No. 7 (1960).

Chi-ch'ing, Huang, *Basic Features of the Tectonic Structure of China (Preliminary Conclusions)*, *Intern. Geol. Rev.*, **5**, No. 3 (1963).

Krishnan, M. S., *Geology of India and Burma* (Madras Law Journal Office, India, 1949).

Lee, J. S., *The Geology of China* (Murby, London, 1939).

Leuchs, Kurt, *Geologie von Asien*, 2 vols. (Gebrüder Borntraeger, Berlin, 1935, 1937).

Liu. H. Y., *Palaeogeographic Maps of China* (Scientific Press, Peking, 1959); in Chinese.

Nalivkin, D. V., *The Geology of the U.S.S.R., A Short Outline,* transl. ed. J. E. Richey (Pergamon Press, New York, 1960).

Obrutschev, W. A., *Geologie von Sibirien* (Gebrüder Borntraeger, Berlin, 1926).

Pascoe, E. H., "A Manual of the Geology of India and Burma," 3rd ed., 3 vols. *Geol. Surv. India,* **1,** 1950; **2,** 1959; **3,** 1963).

Suslov, S. P., *Physical Geography of Asiatic Russia* (W. H. Freeman and Company, San Francisco, 1961).

Wadia, D. N., *Geology of India,* 3rd ed. (St. Martin's Press, New York, 1953).

Southeast Asia and Indonesia

Audley-Charles, M. G., "The Geology of Portuguese Timor," *Mem. Geol. Soc. London,* No. 4 (1968).

Bemmelen, R. W. van, *The Geology of Indonesia* (Government Printing Office, The Hague, 1949).

Chhibber, H. L., *The Geology of Burma* (Macmillan, London, 1934).

Liechti, P., F. W. Roe, and N. S. Haile, *The Geology of Sarawak, Brunei and the Western Part of North Borneo* [Borneo (British) Geol. Surv. Dept., **1, 2** (1960)].

Scrivenor, J. B., *The Geology of Malaya* (Macmillan, London, 1931).

Umbgrove, J. H. F., "Geological History of the East Indies," *Bull. Am. Assoc. Petrol. Geol.,* **22** (1938).

Australia and New Zealand

Brown, D. A., K. S. W. Campbell, and K. A. W. Crook, *The Geological Evolution of Australia and New Zealand* (Pergamon Press, Oxford, 1968).

Cotton, C. A., *Earth Beneath: An Introduction to Geology for Readers in New Zealand* (Whitcombe & Tombs, Christchurch, New Zealand, 1945).

David, T. W. E. (ed. and supp. by W. K. Browne), *The Geology of the Commonwealth of Australia* (Arnold, London, 1950).

Glaessner, M. F. and L. W. Parkin, "The Geology of South Australia," *J. Geol. Soc. Australia,* **5,** Pt. 2 (1958).

Hill, D., and A. K. Denmead, eds., "The Geology of Queensland," *J. Geol. Soc. Australia,* **7** (1960).

McWhae, J. R. H., *et al.,* "The Stratigraphy of Western Australia," *J. Geol. Soc. Australia,* **4,** Pt. 2 (1958).

Spry, A., and M. R. Banks, "The Geology of Tasmania," *J. Geol. Soc. Australia,* **9,** Pt. 2 (1962).

Japan and the Philippines

Corby, G. W., *et al., Geology and Oil Possibilities of the Philippines* (Philippine Dept. of Agriculture and Natural Resources, Tech. Bull. 21, Manila, 1951).

Geological Survey of Japan, *Geology and Mineral Resources of Japan* (Geol. Surv. Japan, Kawasaki-shi, 1956).

Minato, M., M. Gorai, M. Hunahashi, eds., *The Geologic Developments of the Japanese Islands* (Tsukiji Shokan, Tokyo, 1965).

Takai, F., T. Matsumoto, R. Toriyama, eds., *Geology of Japan* (Univ. of California Press, Berkeley, 1963).

Antarctica

Adie, R. J., ed., *Antarctic Geology* (Wiley, New York, 1964).

Cailleux, A., *Geologie de l'Antarctique* (Soc. Ed. Enseignement Superieur, Paris, 1963).

Fairbridge, R. W., "The Geology of Antarctica," *The Antarctic Today* (New Zealand Antarctic Society, Mid-Century Survey, 1952).

Ford, A. B., "Review of Antarctic Geology," *Intern. Geophys. Bull., Nat. Acad. Sci.,* No. 82 (1964).

Hadley, J. B., ed., *Geology and Paleontology of the Antarctic* (Am. Geophysical Union, Publ. 1299, 1965).

Hamilton, W., "Tectonics of Antarctica," *Am. Assoc. Petrol. Geol.,* Mem. 2 (1963).

Harrington, H. J., "Geology and Morphology of Antarctica," *Monographie Biologicae;* **15,** *Biogeography and Ecology in Antarctica* (Junk, The Hague, 1965).

Taylor, G., "Antarctica," *Regionale Geologie der Erde,* **1,** Sec. 8 (Akademische Verlagsgesellschaft, Leipzig, 1940).

Arctic

Harland, W. B., "An Outline Structural History of Spitsbergen," *Geology of the Arctic* (Univ. of Toronto Press, Toronto, 1961).

Orvin, A. K., "Outline of the Geological History of Spitzbergen," *Skrifter om Svalbard og Ishavet*, No. 78 (1940).

Raasch, G. O., ed., *Geology of the Arctic: International Symposium on Arctic Geology*, 2 vols. (Univ. of Toronto Press, Toronto, 1961).

Oceans

Ericson, D. B., and G. Wollin, *The Deep and the Past* (Knopf, New York, 1964).

Heezen, B. C., M. Tharp, and M. Ewing, *The Floors of the Oceans, I. The North Atlantic* (Geol. Soc. Am., Special Paper 65, 1959).

Hill, M. N., ed., *The Sea*, 3 vols. (Wiley, New York, 1966).

Menard, H. W., *Marine Geology of the Pacific* (McGraw-Hill, New York, 1964).

Sears, M., ed., *Progress in Oceanography; 4, The Quaternary History of the Ocean Basins* (Pergamon Press, New York, 1967).

Shephard, F. P., *Submarine Geology*, 2nd ed. (Harper & Row, New York, 1963).

Geologic History by Systems

Arkell, W. J., *Jurassic Geology of the World* (Oliver & Boyd, Edinburgh, 1956).

Charlesworth, J. K., *The Quaternary Era*, 2 vols. (Arnold, London, 1957).

Erben, H. K., ed., *Internationale Arbeitstagung über die Silur-Devon-Grenze und über die Stratigraphie des Silurs and Devons* (Schweizerbart, Stuttgart, 1964).

Hölder, H., *Jura, Handbuch der Stratigraphischen Geologie* (Enke, Stuttgart, 1964).

Lotze, F., and K. Schmidt, eds., *Präkambrium, Handbuch der Stratigraphischen Geologie; Part 1: Northern Hemisphere; Part 2: Southern Hemisphere* (Enke, Stuttgart, 1966, 1968).

Papp, A., and E. Thenius, *Tertiär, Handbuch der Stratigraphischen Geologie*, 2 parts (Enke, Stuttgart, 1959).

Rankama, K., ed., *The Quaternary: 1, Introduction, Quaternary of Sweden, Finland, and Norway* (1965); *2, France, British Isles, Netherlands, and Germany* (1967) (Wiley, New York).

———, ed., *The Precambrian: 1, Introduction, Precambrian of Denmark, Norway, Sweden, and Finland* (1964); *2, Spitsbergen and Bijornoya, British Isles, Canadian Shield, and Greenland* (1966); *3, The Congo, Madagasgar, Seychelles Archipelago, Ceylon, and India* (1968); *4, The United States of America and Mexico* (1969) (Wiley, New York).

Rogers, J., ed., *El Sistema Cámbrico, su Paleogeografia y el Problema de su Base; Part 1: Europa, Africa, Asia; Part 2: Australia, América* (20th Congreso Geológico International, Mexico, 1956).

Sherlock, R. L., *The Permo-Triassic Formations, A World Review* (Hutchinson Scientific and Technical, London, 1948).

Wolstedt, P., *1, Die allgemeinen Erscheinungen des Eiszeitalters* (1961); *2, Europa, Vorderasien, Nordafrika* (1958); *3, Afrika, Asien, Australien, Amerika* (1965) (Enke, Stuttgart).

Zeuner, F. E., *The Pleistocene Period*, 2nd ed. (Hutchinson Scientific and Technical, London, 1959).

5

THE PALEOZOIC ERA
I. North America

What a difference the presence of fossils makes in our interpretation of the history of the earth! Beginning with the Paleozoic Era, fossils become a part of the rock record, and by using them we are able to establish more refined correlations. We can now talk in more specific terms about the space and time relations of rock units and of the continental framework.

During the long Precambrian eras there were numerous cycles of diastrophism, erosion, and deposition. The geographic and temporal relations of these cycles, as we have just seen, are not well known. Precambrian rocks underlay most of the continent, the Canadian shield acting as a stable block, the rest of the continent going through a much more mobile history of subsidence and of both epeirogenic and orogenic movements. In some regions of the world, we

find, at or near the end of the Precambrian, evidence of largescale diastrophism. In other areas, such as the Appalachian and Cordilleran regions of the United States, eastern Australia, and China to name a few, the late Precambrian is marked by the deposition of thick sediments. The close of the Precambrian in these areas is not marked by any significant diastrophism. For the most part Cambrian geosynclines are situated in these areas of late Precambrian geosynclinal activity.

The framework of North America during the initial phases of Paleozoic time was discussed briefly in Chapter 3 and was illustrated in Figure 4-2 (p. 88). The main features are the stable central continental plate and the peripheral geosynclines. Vast platforms lie between the exposed Precambrian rocks of the shield and the inner edges of the geosynclines. The details of any portion of these main divisions of the continental framework are extremely complex; for our purposes, however, we shall focus our attention on the geosynclinal evolution of the continent.

One difficulty in working with early Paleozoic rocks arises from the fact that they have undergone many orogenic phases within mobile geosynclines where we also find extensive plutonic intrusions. We therefore encounter, in such geosynclines, vast areas in which gneiss, schist, and other metamorphic rocks predominate. Since such rocks are generally poor places to look for fossils, their ages are often difficult to determine. The eastern part of the Appalachian mountain system and certain areas along the western coast contain many such rocks. In the early days of geology in North America it was common to consider many or most of these metamorphic and plutonic terranes as of Precambrian age. Intensive investigation has clearly demonstrated, however, that many are actually Paleozoic or Mesozoic.

As we study the main depositional and organic events in the Phanerozoic history of the continents, we must keep in mind the changing main features of each continent's tectonic framework. Changes in the framework that are due to orogeny are always reflected immediately by changes in depositional and geosynclinal patterns, and these changes are merely incidents in a long, slow development. The physical and biological history of any geological system inevitably develops out of that of the preceding system. If we keep in mind that we are dealing with a continuous development, the relations of the main events in the Phanerozoic history of the earth will take on more meaning and become much easier to master.

The disposition of the main tectonic elements of the North American continent for the first three periods of the Paleozoic are shown by Figure 5-1.

5-1 THE CAMBRIAN PERIOD

The Early Cambrian seas were confined to the peripheral geosynclines. The vast central region of the continent was land. The miogeosynclines of the western and eastern sides of the continent were the sites of extensive deposition of quartz sand derived in part from the stable central region and, since the Middle Cambrian, from areas on the seaward side of the troughs. In the eugeosyncline of eastern North America we find Lower Cambrian shales, graywackes, and conglomerates. In the eugeosyncline of western North America we now recognize several Cambrian formations. One of the formations in north-central Nevada consists of approximately 5,000 feet of dark, thin-bedded chert, shale, greenstone, and pillow lava. Another formation in the same general region, of Late Cambrian age, consists of siliceous shale, chert, limestone, and feldspathic sandstone. Zircons from this sandstone have yielded an age of 680 m.y. That the feldspar in these sandstones is unweathered suggests the possibility of orogenic activity, probably to the north-west or west. In British Columbia the westerly outcrops of Cambrian strata contain varying amounts of volcanic rocks.

Middle Cambrian seas encroached slightly on the margins of the continental plate, but the deposits are largely confined to the geosynclines. The rocks of this age in the miogeosynclines are mostly carbonates, but there is some shale, and the quartz-sand facies persist in the near-shore regions. One of the shale units that crops out in the Rocky Mountains of British Columbia (the Burgess Shale Member of the Stephan Formation) is particularly famous for its large fauna of Middle Cambrian organisms, most of which are unknown elsewhere (Fig. 5-2). We shall discuss this fauna in Chapter 7. In the eastern eugeosyncline the deposition of shales, graywackes, and conglomerates continued.

For the first two-thirds of the Cambrian

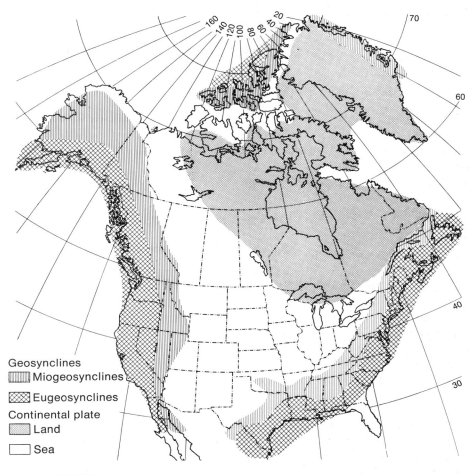

FIGURE 5-1

Paleotectonic map of North America for the Cambrian, Ordovician, and Silurian. (After A. J. Eardley, *Structural Geology of North America*, Harper & Brothers, New York, 1951.)

Period (which is believed to have lasted about 100 m.y.) geosynclinal seas were confined to the peripheral geosynclines. In Late Cambrian time they spread inward and inundated much of the stable interior of the continent. In the miogeosynclines, carbonate deposition continued; the eugeosynclines are represented mainly by clastic facies, as in the earlier epochs of the Cambrian. On the continental plate the first transgressive facies are quartz sands, which grade upward into carbonate facies. The near-shore deposits of the continental sea consisted predominantly of a quartz-sand facies.

The oldest fossiliferous strata in southern Oklahoma are of Late Cambrian age. These consist of a basal sandstone formation 450 feet thick that is overlain by 1,400 feet of limestone. Evidence now available strongly indicates that this Upper Cambrian deposit was preceded by extensive sedimentary and igneous accumulations of Early and Middle Cambrian age. The Upper Cambrian fossiliferous formations are underlain by rhyolites and granites that have been dated at 525 ± 25 m.y. These formations are underlain by basalts and gabbros that yield an age of about 535 m.y. Both dates suggest a Middle Cambrian age. The next oldest group of formations comprises meta-graywacke, argillite, quartzite, and bedded chert. This group, which is probably 15,000 feet thick, represents eugeosynclinal deposition that occurred probably in the Early Cambrian and perhaps partly in the Late Precambrian. There thus appears to have been an active

FIGURE 5-2

Collecting fossils in the Burgess Shale (Middle Cambrian) on Mount Wapta, British Columbia. (From the Smithsonian Institution, U.S. National Museum.) A few of the remarkable fossils obtainable from this formation were illustrated in Fig. 2-5; a reconstruction of the sea bottom on which the shale accumulated, with its fauna, is shown in Fig. 7-3.

geosynclinal belt in southern Oklahoma—a belt about 300 miles long and 100 miles wide—during late Precambrian and Early and Middle Cambrian time. The fossiliferous Upper Cambrian formations, which consist of platform facies, rest unconformably on these older Cambrian geosynclinal formations.

Known exposures of the Cambrian System in the Arctic islands of Canada are limited to small areas on Devon, Ellesmere, and Victoria islands. These formations are mainly limestones and dolomites, but there are some shale and sandstone beds. Our knowledge of the Cambrian in this area is still very limited.

At this point let us recapitulate the time-space relations of these Cambrian facies. Early and Middle Cambrian seas in North America were confined to the geosynclines. The facies in the miogeosynclines are mainly quartz sandstones in the Early Cambrian and carbonates in the Middle Cambrian. The eugeosyclines accumulated great thicknesses of shales, graywackes, conglomerates, and volcanic rocks. Late Cambrian seas spread across the United States and encircled the Canadian shield. The facies developed round the shield were quartz sand and carbonates. Restored cross sections of the Cambrian miogeosynclines are shown in Figure 5-3. Note the great difference in thickness between the Cambrian deposits of the geosynclines and those of the continental plate.

Cambrian rocks have been recognized in the eastern eugeosyncline only between Boston and Newfoundland. It is presumed that they continue south along the eastern seaboard, but they are completely masked by the metamorphic and granitic terranes of that area. Volcanic deposits, though rare in this Cambrian eugeosyncline, are found in several areas between Virginia and Newfoundland, where they occur in great thicknesses that underlie the fossiliferous Cambrian strata conformably and unconformably. They are considered late Precambrian or earliest Cambrian in age. What is noteworthy about them is their marginal geographic position and their early deposition.

5-2 THE ORDOVICIAN PERIOD

The Cambrian pattern of geosynclinal and continental seas and the general facies relations continue into Ordovician time. The Late Cambrian carbonate facies of the Appalachian miogeosyncline continue upward through the Early Ordovician, developing thick formations of limestones and dolomites. Across the continental plate of the central United States Early Ordovician formations are predominantly limestones and dolomites, and in the western miogeosyncline several thousands of feet of carbonate rocks record Early Ordovician time. The fossils of these carbonate facies are brachiopods, bryozoans, nautiloids, gastropods, corals, and algae, all of which have calcareous exoskeletons. The eugeosynclines, in contrast, contain volcanic rocks, and their predominant sedimentary rocks are dark, fine-grained clastics bearing an abundance of graptolites; the typical bottom-dwelling forms, such as brachiopods, corals, and mollusks, are generally missing. Early Ordovician deposits, in short, have two quite distinct facies: (1) a carbonate facies with abundant benthonic organisms, named the *shelly facies;* (2) a dark shale facies with planktonic graptolites, named the *graptolitic facies.*

The general paleogeographic picture of North America presented in Figure 3-16 (p. 80), representing the end of Early Ordovician time, brings to a close the first depositional phase of Paleozoic history. In Middle Ordovician time this picture begins to change, and we note significant regional variations. To facilitate our study, we shall discuss the Ordovician record region by region.

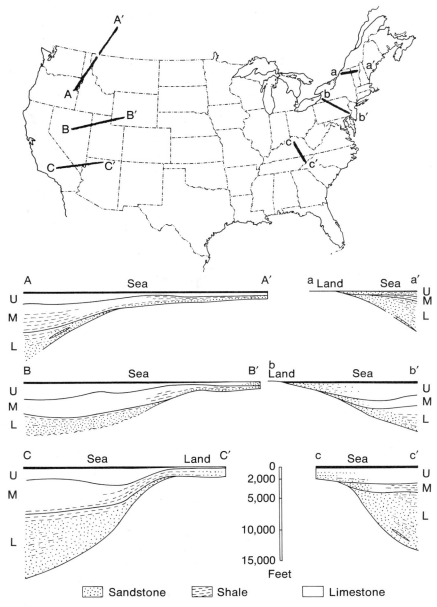

FIGURE 5-3

Restored cross sections of Cambrian miogeosynclines bordering the continental plate. (After M. Kay, *Geol. Soc. Am.*, Mem. 48, 1951.)

FIGURE 5-4

Thin-bedded limestones and shales of Early Ordovician age at Green Point, western Newfoundland. (Photograph by H. B. Whittington.)

The Eastern Geosyncline

We have already noted that in the Early Ordovician deposits of the eastern geosyncline we find a carbonate shelly facies in the miogeosyncline and an argillaceous, volcanic, graptolitic facies in the eugeosyncline. In many areas along the North Atlantic coast we find fossil-bearing rocks associated with volcanics: in western New Hampshire, for instance, the more than 15,000 feet of Ordovician argillites and graywackes include great thicknesses of volcanic rocks in the form of flows and fragmental material, and in central and eastern Newfoundland there are, in some places (around Notre Dame Bay, for example), more than 20,000 feet of Ordovician sedi-ments and volcanics. These volcanic-clastic facies are characteristic of all Ordovician time.

The Lower and lower Middle Ordovician facies of the miogeosyncline are part of the persisting carbonate facies begun in Late Cambrian time (Fig. 5-4). In the midpart of the Middle Ordovician Series, however, the carbonate strata grade upward into fine clastics that clearly show derivation from the east. The record shows that the whole facies development in the miogeosyncline became more and more clastic throughout the remainder of Ordovician time. What was going on? Since we know that the source of these clastic sediments was to the east, we attribute them to orogenesis in the adjoining eugeosyncline which increased

126

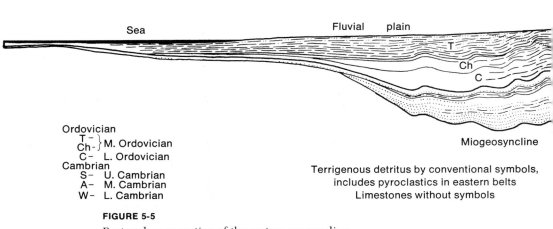

West   Eastern New York

Sea   Fluvial   plain

Ordovician
    T -
    Ch- } M. Ordovician
    C-   L. Ordovician
Cambrian   Miogeosyncline
    S-   U. Cambrian
    A-   M. Cambrian   Terrigenous detritus by conventional symbols,
    W-   L. Cambrian   includes pyroclastics in eastern belts
                                          Limestones without symbols

FIGURE 5-5

Restored cross section of the eastern geosyncline
from Maine to New York in Middle Ordovician time.
(After M. Kay, *Geol. Soc. Am.*, Mem. 48, 1951.)

the area and volume of land and thus formed the source of the clastic sediments carried westward into the miogeosyncline. The tectonic activity that formed this land mass, named Vermontia, which lay east of Pennsylvania and New York, began in Middle Ordovician time and continued almost until the close of the Ordovician. Figure 5-5 illustrates the relations of the various facies from Cambrian to Middle Ordovician and from the continental plate eastward to Maine at the beginning of the orogenesis in the western part of the eugeosyncline. The orogenic forces that caused this change continued and became stronger through the remainder of Ordovician time. At the close of the period this orogenic phase, named the Taconic orogeny, culminated in extensive thrusting of rocks deposited in the eugeosyncline onto the folded strata of the miogeosyncline. One of the principal remnants of this thrust mass is the Taconic Mountains of eastern New York, western Vermont, and Massachusetts, which consist of graptolitic shales (presumably eugeosynclinal facies), from Cambrian to Middle Ordovician in age, lying as an exotic block completely surrounded

by carbonates (miogeosynclinal facies) of the same age. The present relations of the Taconic Mountains to the carbonate facies of the miogeosyncline are shown in Figure 5-6, A. If the thrust blocks were pulled apart to their original position, the picture would look somewhat like Figure 5-6, B.

The area of the eugeosyncline that was undergoing orogenesis shed great volumes of detrital sediment into the adjoining miogeosyncline. The texture of these sediments tended to become progressively coarser from Middle to Late Ordovician time, probably reflecting increasing orogenic activity in the eugeosyncline. From the isopach[*] map of the Upper Ordovician Series (Fig. 5-7) we see that the subsiding area into which these sediments were deposited was semicircular in plan and that the center of subsidence was in southern New York and Pennsylvania. The sediments were deposited in a vast delta, named the Queenston delta, which formed on the western shores of the highlands of the eugeosyncline. A

[*] A line on a map drawn through points of equal thickness of a designated unit is an isopach; an isopach map thus shows the distribution of various thicknesses.

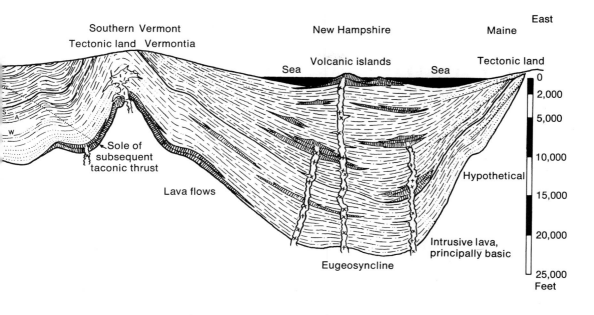

Southern Vermont
Tectonic land Vermontia

New Hampshire

East

Maine

Volcanic islands

Tectonic land

Sea

Sea

0

2,000

5,000

10,000

15,000

20,000

25,000
Feet

Sole of
subsequent
taconic thrust

Lava flows

Hypothetical

Intrusive lava,
principally basic

Eugeosyncline

section of them is illustrated in Figure 5-8, A.

In the central and southern Appalachians the eugeosyncline consists of metamorphic and plutonic rocks that are not yet well understood. It is generally assumed that at least some of them are early Paleozoic in age and are similar in kind and origin to the eugeosynclinal strata of the northern Appalachians and the maritime provinces. The early Paleozoic sediments of the Valley and Ridge province, which includes the miogeosyncline, are very similar to those of the northern Appalachians. After the Early Cambrian, which was, we have seen, a time of deposition of quartz sand derived only partly from the interior of the continent, deposits up to Middle Ordovician time were predominantly carbonates, aggregating more than 10,000 feet of strata. The first expressions of orogenesis are found in clastic formations of Middle Ordovician age, and the source of these clastics was to the east. Upper Ordovician sandstones and shales testify to continued orogeny in the east (presumably in the adjoining eugeosyncline).

In structurally complicated areas like the

Appalachians the reconstruction of geologic history is very difficult. Later orogenies, with their numerous thrust faults and varied igneous activity, have broken the once continuous sedimentary units into many isolated segments. Since those units are now complexes of intertonguing and gradational facies, many not as fossiliferous as we would like, correlation is not easy. Such difficulties open the way to several interpretations. Figures 5-9 and 5-10, for example, are independent attempts to show diagrammatically the facies relations of Ordovician rocks in the Knoxville region of eastern Tennessee. Figure 5-9 was constructed from a correlation based mainly on lithologic criteria, Figure 5-10 from a correlation based on fossils. To the historical geologist no manmade puzzle could be as complex or as fascinating as such problems.

The Southern Geosyncline

The southern margin of the continental plate, in southern Oklahoma, was geosynclinal during the Early and Middle Cambrian. This was a time of accumulation of

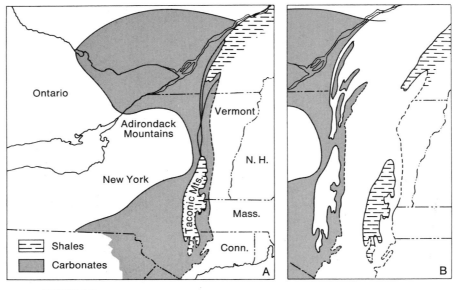

FIGURE 5-6

Distribution of Lower Ordovician facies in New York and adjacent states: (A) the present distribution, (B) before the Taconic orogeny. (After M. Kay, *Bull. Geol. Soc. Am.*, 1942.)

approximately 20,000 feet of sedimentary and igneous rocks derived from southerly sources and various kinds of intrusive igneous rocks. The Upper Cambrian formations are the same quartz-sand and carbonate facies that developed all over the central platform of the United States. In the Ordovician, however, geosynclinal subsidence and deposition began in two distinct facies.

In the Ouachita Mountains of Oklahoma and Arkansas and in the Marathon region of western Texas are found several thousand feet of graptolite-bearing shales, thin limestones, sandstones, and bedded chert. These strata may represent a facies transitional from the miogeosyncline to the eugeosyncline. Some of the graptolitic shales (now argillites) in southeastern Oklahoma contain pebbles of basalt and granite in conglomerate interbeds. In the Wichita mountain system of southwestern Oklahoma and in the Panhandle of Texas are found about 10,000 feet of Upper Cambrian and Ordovician carbonate formations of miogeosyn-

clinal aspect. Northward from this subsiding belt the formations grow much thinner as they encroach on the interior platform.

The Western Geosyncline

The broad pattern of subsidence established in the Cambrian from the margin of the continental plate westward to the margin of the continent continues throughout the Paleozoic. The miogeosynclinal deposits, during Ordovician time, consisted of quartz sand and carbonates. The thickness of the individual formations and of the whole Ordovician section varies considerably from place to place — a variation that suggests uneven subsidence.

West of the miogeosyncline, well-dated sequences of Ordovician rocks are not common. In central Nevada we find exposed, in thrust sheets, more than 20,000 feet of Lower, Middle, and Upper Ordovician graptolitic argillites and graywackes with volcanics. In central Idaho and eastern Washington we find a similar sequence, without volcanics, nearly 10,000 feet thick. In California, definitely dated Ordovician graptolitic rocks have been recognized in the central Sierra Nevada. The clastic sediments of these sequences were derived from the west. Presumably the sediments were provided by rejuvenation of the same general region as had supplied the Cambrian sediments, but neither its precise location nor the tectonics responsible for its relief are known.

Thick Ordovician sections with volcanics and sediments are also present in southeastern Alaska and adjacent regions of Canada where a widespread unconformity separates Upper Ordovician from Middle Ordovician and older strata. In addition, granitic rocks in southeastern Alaska were

FIGURE 5-7

Isopach map of Upper Ordovician formations in Pennsylvania and adjacent states. The cross sections of Fig. 5-8 are drawn along line A–A'. (After M. Kay, *Geol. Soc. Am.*, Mem. 48, 1951.)

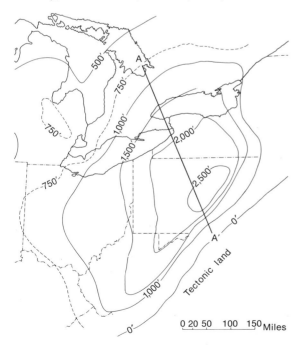

probably emplaced in late Ordovician time. These two features suggest orogenic activity.

The Northern Geosyncline

This geosyncline, named the Franklinian, and containing several thousand feet of sediments, lies across the northern periphery of North America from northern Greenland to Prince Patrick Island, a distance of more than 1,400 miles. Figure 5-11 shows its main structural elements. Within the belts of folded strata (Fig. 5-12) some Paleozoic formations are several thousand feet thick, but to the south they are thin and overlap the Canadian shield (Fig. 5-13).

Ordovician formations are widespread in the Arctic lowlands and plateaus and in the belts of folded Paleozoic rocks. The Ordovician rocks of the Arctic lowland and plateaus have a shelly fauna and consist mainly of carbonate facies. Within the belt of folded strata called the Innuitian region, however, the shelly strata become partly or entirely graptolitic. This change began in Early Ordovician time in the western part of the Parry Islands belt and in late Middle Ordovician time in the Cornwallis and Central Ellesmere belts. On northern Ellesmere and Axel Heiberg Islands graywackes, conglomerates, sandstones, and basic volcanics occur, indicating a eugeosynclinal belt.

An active zone of subsidence was present along the eastern coast of Greenland, where, in late Precambrian, Cambrian, and Ordovician time, more than 50,000 feet of carbonates and fine-to-coarse clastics of miogeosynclinal aspect accumulated (Fig. 5-14).

The Continental Interior

The vast continental interior contains a rich record of Ordovician history in thin, widespread formations. Quartz sand, shale, and carbonate facies predominate. All these

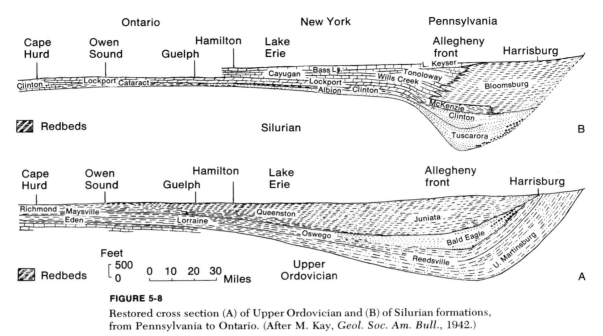

FIGURE 5-8

Restored cross section (A) of Upper Ordovician and (B) of Silurian formations, from Pennsylvania to Ontario. (After M. Kay, *Geol. Soc. Am. Bull.*, 1942.)

deposits testify to deposition in shallow, well-aerated seas, in which life was abundant.

Lower Ordovician formations are almost entirely carbonates. A pronounced erosional surface on the top of these testifies to a retreat of the seas and a period of erosion before deposition was resumed. Over more than 700,000 square miles of the north-central states the Lower Ordovician carbonates are disconformably overlain by a most remarkable, clean, quartz-sand deposit (St. Peter Sandstone), which attains a thickness of 300 feet in places. This is the most conspicuous clastic facies of Ordovician age in the continental interior (Fig. 5-15). The source of the sediments must have been some reworked sediments of the Canadian shield, for along the periphery of the St. Peter Sandstone—that is, to the south and east—are carbonate facies.

The Upper Ordovician formations from the Mississippi Valley east to the eastern miogeosyncline are largely shale. The detrital sediments of this facies are probably the peripheral deposits of the Queenston delta to the east. West of the Mississippi River we

encounter another remarkable Ordovician formation. In a wide belt across Arctic Canada and through the Rocky Mountains to northern Mexico is a dolomite facies that is generally some 300 feet thick. Its rich fauna is remarkably similar throughout its extent and is of Late Ordovician age (Fig. 5-16).

Subsidence throughout the central platform was not uniform. Beginning in the Ordovician, and increasing in number and distinctness as Paleozoic history unfolded, basins formed, and arches stood firm and resisted sinking. We shall take up some of these features later (§ 5-5).

The tectonic and depositional patterns established during the Ordovician laid the basis for Silurian history.

5-3 THE SILURIAN PERIOD

The Eastern Geosyncline

The highlands formed in the Taconic orogeny were the dominant tectonic element inherited from the Ordovician. At one place or another orogenic movements contined in the eugeosyncline through the Silurian.

Throughout much of the miogeosyncline, Middle and Upper Silurian strata consist of coarse-to-fine clastics, including some red-beds, derived from the eroding highland to the east. The relations of some of these clastic facies across Pennsylvania are shown in Figure 5-7, B. (Compare this with Fig. 5-7, A, a similar cross section for Upper Ordovician formations.) On the island of Anticosti in the Gulf of St. Lawrence we find Upper Ordovician carbonates conformably over-lain by Silurian carbonates; obviously the detrital materials from the tectonic lands of the eugeosyncline did not reach this area.

The eugeosyncline was an active belt of volcanism and subsidence throughout the Silurian. In Newfoundland, in the maritime provinces, and in parts of the New England states there are great thicknesses of clastic sediments with volcanics. Here, as in the miogeosyncline, the Silurian strata usually lie with angular unconformity on the Ordovician. In central Newfoundland, Silurian and older Paleozoic rocks were strongly

FIGURE 5-9

Diagram by John Rodgers interpreting relations in the Ordovician rocks of the Knoxville region, eastern Tennessee. The line of section is indicated by the line N–S in the inset map. The main basis of correlation was the lithologic similarity of the formational units. (From *Geol. Soc. Am. Bull.*, 1954.)

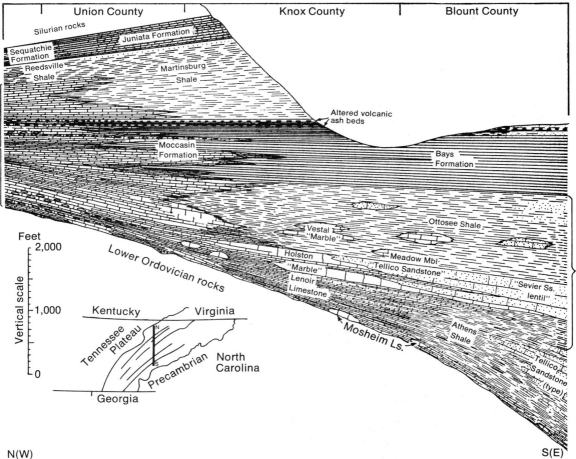

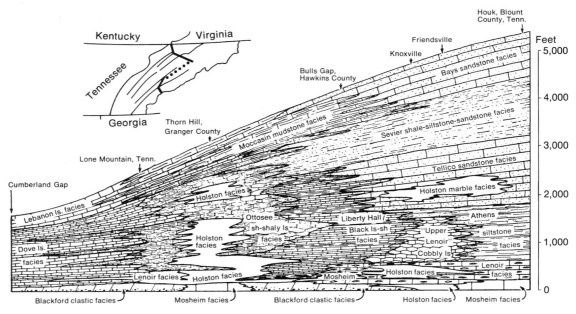

FIGURE 5-10

Diagram by Byron Cooper interpreting relations in the Ordovician rocks of the Knoxville region, eastern Tennessee. The line of section is indicated in the inset map. Note that it is in two parts separated by a considerable offset along the strike. The main basis of correlation was the fossil faunas of the various formations. (From *Geol. Soc. Am. Bull.*, 1954.)

deformed and subjected to granitic intrusion. This phase, apparently, did not affect any other part of the eastern geosyncline.

The Southern Geosyncline

The general character and distribution of Silurian facies in this region are very similar to those of the Ordovician. No significant change in the depositional pattern took place until after the mid-Paleozoic.

The Western Geosyncline

The Silurian deposits follow much the same patterns as the Ordovician. The miogeosyncline is characterized by carbonate and quartz sand formations; such are the types of Silurian rocks in central Alaska, from the Seward Peninsula east to the border, and

the same, one or two thousand feet thick and mostly carbonates, are found in Nevada, Utah, and Idaho.

The Silurian record for the eugeosyncline is very sparse. The best area in which to study it is southeastern Alaska, where we find more than 10,000 feet of carbonates, argillites, volcanic flows, graywackes, and conglomerates. In southern Alaska Silurian fossils have been found in limestone lenses in a quartzite that contains lava pebbles and lies on lava beds. In California there is a volcanic formation that is thought, on very indirect evidence, to be Silurian in age. In central Nevada the Silurian is represented by 4,000 feet of graptolitic shales, sandstone, arkose, chert, and siliceous pyroclastics in the eugeosynclinal facies, and by about 1,000 feet of carbonaceous limestone in the miogeosynclinal facies, with which it was telescoped during orogenesis in the Devonian.

The Northern Geosyncline

Here again we find the Silurian deposits very much like those of the underlying Ordovician. Within the Parry Islands fold belt, for example, there are about 2,000 feet of calcareous and dolomitic mudstones and shales, overlain by 3,000 feet of argillaceous and calcareous fine-grained sandstone, all with graptolites. In northern Greenland the Silurian consists of dark limestones and shales and some conglomerate and coarse-sandstone beds. Southward, on the Canadian shield, the Silurian, like the Ordovician, is represented by thin carbonate facies.

The Continental Interior

The interior of the continent was occupied by a vast, shallow sea that was very rich in organisms. The Early and Middle Silurian

FIGURE 5-11

Structural-stratigraphic elements of the Arctic islands of Canada. (Geol. Surv. Canada.)

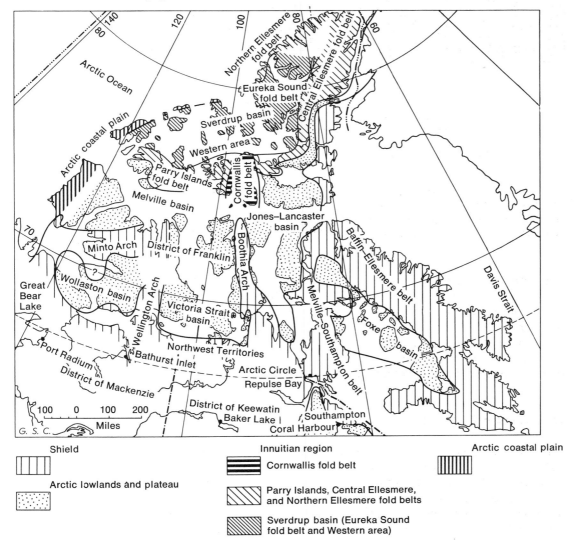

FIGURE 5-12
Western side of Melville Island (Canrobert Hills), in the Arctic archipelago of Canada, showing folded Ordovician, Silurian, and Devonian rocks of the Parry Islands fold belt. The folding of these strata took place in the interval between the Late Devonian and the Pennsylvanian. (Photograph by the Royal Canadian Air Force.)

were times of carbonate deposition throughout this area; in the Middle Silurian, particularly, numerous **organic reefs,** or **bioherms,** developed. These wave-resistant structures, formed by colonial plants and animals, are more or less circular, as much as three miles in diameter, and nearly 100 feet thick. The upper Mississippi Valley has many Middle Silurian organic reefs.

The circular pattern of some of these Silurian reefs was responsible for the deposition in Late Silurian time of evaporites: anhydrite, gypsum, and salt. These reefs occur in Michigan, Ohio, West Virginia, Pennsylvania, and New York (Fig. 5-17). The reef banks were cut by numerous chan-

nels through which sea water flowed. High rates of evaporation in the basins enclosed by the reefs led to the accumulation of gypsum, anhydrite, and salt. These deposits now occupy the central areas of the basins; most intertongue laterally into carbonate formations. A classic locality for the study of Silurian formations of the platform is Niagara Falls (Fig. 5-18).

5-4 THE DEVONIAN PERIOD

As we approach the Devonian Period, keep in mind that we have already discussed, though very briefly, nearly 200 million

FIGURE 5-13

Southeastern coast of Devon Island, in the Arctic archipelago of Canada, showing flat Paleozoic strata lying (in the left foreground) nonconformably on Precambrian gneiss. (Photograph by the Royal Canadian Air Force.)

years of Paleozoic history. The basic depositional patterns shown by the Cambrian formations persist—except for the Taconic orogeny—through the Ordovician and Silurian periods. The Taconic orogeny had a profound effect on the nature and distribution of sediments in the adjoining regions. In western North America orogenic episodes undoubtedly were part of the history of that eugeosynclinal area. Nevertheless, here the evidence is indirect, and it is difficult to pinpoint where these events took place and when. The shrinking of the Late Silurian sea produced great unconformities, over which the Devonian sea transgressed, generally onto Middle Silurian formations. The tectonic activity that commenced during the Devonian greatly altered the continental framework that had endured since the Cambrian and profoundly affected the whole depositional pattern for the remainder of the Paleozoic Era. We are at the halfway mark in Paleozoic history: at this point we see the first steps in the development of the modern North American continent. The tectonic framework of Devonian North America is illustrated by Figure 5-19.

The Eastern Geosyncline

The distribution of Lower Devonian facies in the eastern geosyncline follows a pattern already familiar to us. In the miogeosyncline carbonates and quartz sandstones prevail; in the eugeosyncline we encounter a greater variety of rock types. In Newfoundland, for instance, all Lower Devonian sediments appear to be nonmarine, many bearing fossil land plants. Many of these clastic formations contain volcanic and granite pebbles and are interbedded with volcanics. In the

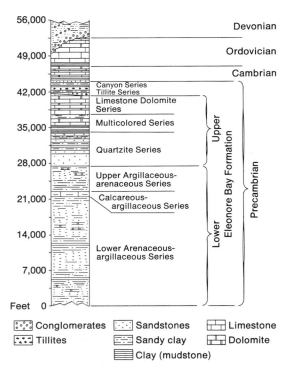

FIGURE 5-14

Stratigraphic column of late Precambrian and early Paleozoic formations in the East Greenland geosyncline. (After Fränkl, 1956.)

FIGURE 5-15

Generalized, reconstructed, north-south section of the Cambrian and Ordovician formations of the Mississippi valley. (After G. A. Thiel, *Geol. Soc. Am. Bull.,* 1935.)

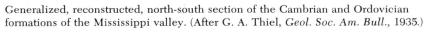
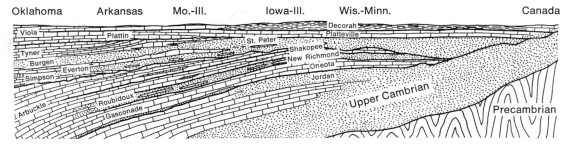

maritime provinces and the New England states several localities expose Lower Devonian formations, many containing volcanics and most of them very thick.

Beginning in early Middle Devonian time and continuing until the end of that period, or perhaps even later, a very severe orogeny, accompanied by plutonic intrusions of granite, disturbed the northern part of the eugeosyncline. The few Late Devonian rocks in this region are nonmarine clastics deposited in structural basins. The result of this Acadian orogeny, as it is called, was to consolidate the eugeosyncline. From that time to the present this area has not acted as a geosyncline. Later Paleozoic nonmarine deposits accumulated in isolated structural basins, but there was nothing like the pre-mid-Devonian geosynclinal deposits.

The establishment of a large, rigid tectonic land on the margin of the continent produced profound changes in the facies developed in the adjoining miogeosyncline. As the easterly tectonic lands rose, great quantities of terrigenous sediments were carried westward into the miogeosyncline — a pattern familiar from our study of Ordovician history. These Middle and Upper Devonian clastics were deposited in a broad, complex delta, called the Catskill delta, built out from the highlands. An isopach map of these deposits is shown in Figure 5-20.

The Southern Geosyncline

Facies relations here remain the same as they were in the Ordovician and Silurian. Carbonates predominate in the miogeosyncline, siliceous sediments (novaculites) in

FIGURE 5-16

Paleozoic formations exposed in Bighorn Canyon, Bighorn Mountains, Montana. The Gallatin Limestone and the Gros Ventre Formation are Middle and Upper Cambrian; the Bighorn Dolomite is Upper Ordovician; the Three Forks Shale and the Jefferson Limestone are Middle and Upper Devonian; the Madison Limestone is Lower Mississippian; the Amsden Formation is Upper Mississippian and Lower Pennsylvanian. (Photograph by G. E. Prichard, U.S. Geol. Surv.)

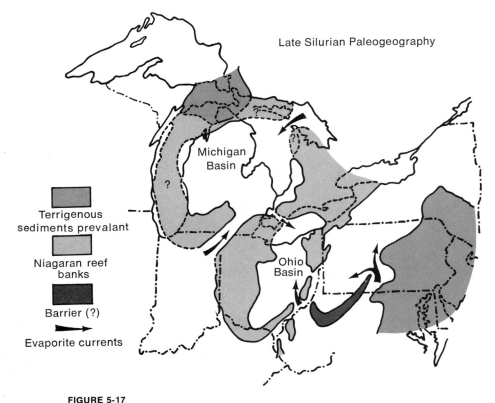

Late Silurian Paleogeography

Michigan
Basin

Ohio
Basin

Terrigenous
sediments prevalant

Niagaran reef
banks

Barrier (?)

Evaporite currents

FIGURE 5-17

Paleogeography of the Great Lakes region during the Late Silurian.
(After H. L. Alling and L. I. Briggs, *Bull. Am. Assoc. Petrol. Geol.*, 1961.)

the eugeosyncline. These Devonian formations are only about 1,000 feet thick. Note how all the early Paleozoic systems are thinner here than in the western and eastern geosynclines but generally thicker than on the interior platform.

The Western Geosyncline

Evidence of tectonic activity in the western geosyncline during the early Paleozoic is indirect, consisting mainly of clastic units within the eugeosyncline. Although the clastic sediments are clearly of westerly origin, we cannot be sure of the nature and location of the area that produced them. In

the latest Devonian we see clear evidence of orogenic activity along part of the western geosyncline.

The pre-orogenic Devonian formations of the western geosyncline are of the same pattern as those of the earlier Paleozoic systems (Fig. 5-21). That is, within the miogeosynclinal region carbonates and shale are dominant. An interesting variant is seen in Alberta. There on the inner margin of the miogeosyncline and on the adjacent platform, great thicknesses of limestones and evaporites accumulated. The evaporites were followed by extensive organic reefs in the Late Devonian. The eugeosyncline is characterized by graywackes, shales, volcanic rocks, and some limestones. Such

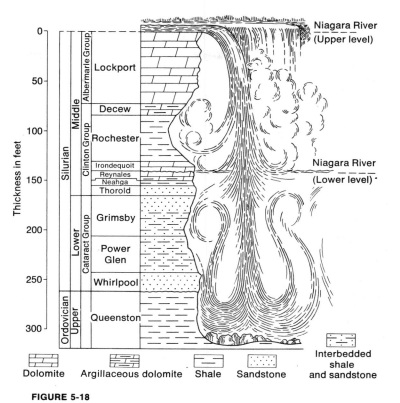

FIGURE 5-18

Stratigraphic section at Niagara Falls. (Geol. Surv. Canada, 1957.)

suites are known in western Nevada, northern California, and southeastern Alaska, to mention but a few.

In the latest Devonian an important orogeny, called the Antler orogeny, began in a belt extending from west-central Nevada northeasterly to central Idaho, approximately on the border between the eugeosyncline and the miogeosyncline. The main result of this orogeny was that Devonian and older eugeosynclinal rocks were thrust eastward over miogeosynclinal rocks, a telescoping of approximately 80 miles. After the thrusting there began slow uplift with two troughs developed on the flanks of the orogenic belt (Fig. 5-21). Volcanic and clastic deposits accumulated in the eugeosynclinal environment to the west, and clastics were deposited to the east. The trough to

the west has been named the Pumpernickel-Havallah trough, and that to the east the Chainman-Diamond Peak trough. At first the rate of uplift was slow, and only fine clastics were carried into the flanking troughs. As time went on, the rate of uplift increased, and coarser clastics were carried into the flanking troughs. This orogeny continued in effect into Early Pennsylvanian time and profoundly altered the depositional pattern in the western geosyncline of the United States during the Late Paleozoic. As significant effects of the Antler orogeny are to be seen in post-Devonian formations, we will defer further discussion of this problem until § 5-5.

There is some evidence for orogenic activity at this time in the western geosyncline of Canada and Alaska. This is especially well marked in northwestern British Columbia, in adjacent regions of the Yukon, and in northern Alaska.

The Northern Geosyncline

Lower and Middle Devonian formations in the Franklinian geosyncline are marine limestone, dolomite, shale, and siltstone. Similar facies are present on the adjoining platform, especially on Banks and Victoria islands. The Upper Devonian is represented by a widespread development of nonmarine clastic formations, which contain some coal beds. For instance, in the southwestern part of the Central Ellesmere belt, adjoining the Jones-Lancaster Basin, there are 1,900–3,800 feet of marine Middle Devonian limestone, dolomite, and calcareous shale, overlain by 1,700–2,900 feet of marine Middle Devonian limestone, sandy limestone,

FIGURE 5-19

Paleotectonic map of North America for the Devonian. (After A. J. Eardley, *Structural Geology of North America*, Harper & Brothers, New York, 1951.)

Geosynclines

Miogeosyncline

Eugeosyncline

Continental plate

Land

Sea

Stabler areas, surrounded, when not covered, by sea

sandy shale, and sandstone. The topmost Upper Devonian formation of this conformable sequence is more than 10,000 feet thick and is largely made up of nonmarine sandstone and shale and thin seams of bituminous coal. The similarity of this sequence and facies to the Devonian rocks of the northern Appalachian region is striking. These Upper Devonian clastics result from orogenic activity in the geosyncline to the north.

The deformed strata of the Cornwallis fold belt range in age from Ordovician to Late Silurian or Early Devonian (Fig. 5-11). They are overlain by flat or gently dipping strata ranging in age from latest Silurian to Pennsylvanian. Folds in this belt trend from northerly to northwesterly, at right angles to the fold trends of the Parry Islands belt and Central Ellesmere belt.

A somewhat similar but more complicated series of events is recorded in the East Greenland geosyncline. We have seen (§ 5-2) that this area contains fairly thick sediments of late Precambrian through Ordovician ages with a more sporadic distribution of Silurian formations (Fig. 5-14). Orogenic activity commenced in Silurian time and reached a climax at the end of that period. This orogeny was accompanied by granitic intrusions. As the activity continued into the Devonian, the earlier geosyncline was raised into lands of variable relief; in Middle and Late Devonian time more than 20,000 feet of red and gray clastic sediments accumulated in troughs and basins. These sediments are all nonmarine and contain important vertebrate fossils. They are of the same age and facies as the Old Red Sandstone of England. Figure 5-22 shows the present relations of the early Paleozoic formations in the fjord district of eastern Greenland.

The Continental Interior

Lower Devonian deposits are missing from many parts of the interior platform, where we commonly find Middle Devonian forma-

tions overlying Middle Silurian. More diverse topography and facies began to develop on the craton. In the Ordovician, as we have seen, distinct basins and arches had begun to form; in the Silurian they increased in number and development (the Cincinnati arch and the Michigan basin are examples); and the increase continued in the Devonian (see Fig. 5-19). The arches were not necessarily land, but they subsided much less than the surrounding regions. At times they were definite land barriers, surrounded by shallow seas.

The earliest Devonian deposits over much of the platform are carbonates; by Middle Devonian time shales began to appear. In Late Devonian all the formations from the Mississippi Valley eastward to the miogeosyncline were black shales derived from the Catskill delta. In the Michigan basin more than 4,000 feet of Devonian sediments accumulated. In most other regions the formations are less than 1,000 feet thick.

The most widespread Paleozoic formations of western Canada are of Devonian

FIGURE 5-20

Isopach map of Middle and Upper Devonian deposits in Pennsylvania and adjacent states. (After M. Kay, *Geol. Soc. Am.*, Mem. 48, 1951.)

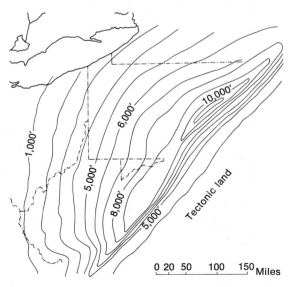

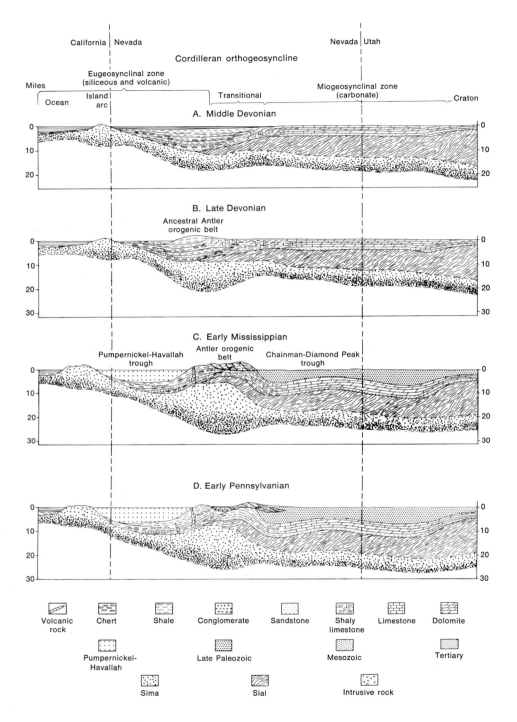

FIGURE 5-21

Schematic cross sections showing the evolution of the California-Utah region from Middle Devonian to Early Pennsylvanian time. (From R. J. Roberts, *UMR J.*, 1968.)

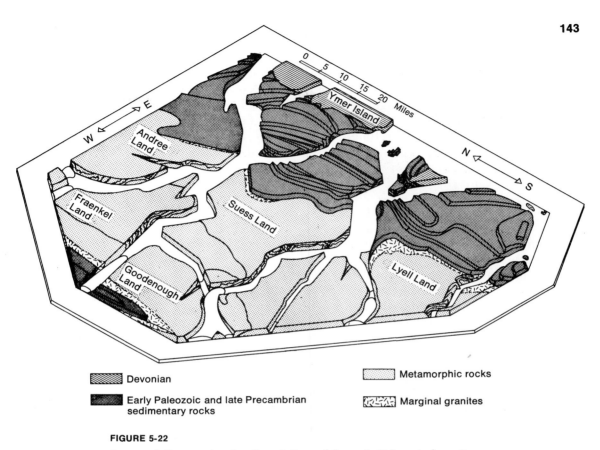

Devonian

Early Paleozoic and late Precambrian
sedimentary rocks

Metamorphic rocks

Marginal granites

FIGURE 5-22

Structural diagram showing the relations of the early Paleozoic formations
in the fjord district of eastern Greenland. (After Wegmann, 1935.)

age. Extending from the continental arch
(Fig. 5-19) to the Arctic, they form a great
wedge that thins toward the Canadian shield
and thickens toward the western miogeosyncline (of Middle and Late Devonian
age). The dominant facies are carbonates,
but there are some shales. In northern and
eastern Alberta extensive evaporites, chiefly
salt, are thought to be of Middle Devonian
age. In southern Alberta and western Montana another widespread evaporite deposit,
chiefly anhydrite, is of Late Devonian age.

Organic reefs are extensively developed
in the Canadian Devonian formations. The
recent discovery of petroleum in some
reefs has stimulated intensive exploration
and study, adding significantly to our knowledge of Devonian stratigraphy and of carbonate facies in general.

5-5 THE CARBONIFEROUS PERIOD

In North America the Carboniferous is
divided into two distinct periods—the Mississippian and the Pennsylvanian, which
are approximately equivalent to the Lower
and Upper Carboniferous of Europe. Because the distinction between Mississippian and Pennsylvanian is only made in
North America we here treat the Carboniferous as a unit, as is generally done the world
over.

Devonian tectonic developments had set
the stage for Early Mississippian deposition. Devonian orogenic movements in the
Franklinian, East Greenland, western and
eastern geosynclines had established tectonic lands that completely dominated sedimentation in adjoining basins. Basins and

arches continued to be important features of the platform; some of them were accentuated, in fact, and new ones were developed, during the Carboniferous. Figure 5-23 illustrates the tectonic framework of North America during Pennsylvanian time.

The Eastern Geosyncline

The eastern eugeosyncline no longer existed. With the Acadian orogeny and the accompanying plutonic intrusions, it—or at least its northern half—had been strongly folded and had become a rigid part of the continental framework. Isopach studies of the Upper Devonian Catskill delta deposits and the Early Mississippian rocks indicate a lessening influence of highlands toward the southwest. The complex of granites and metamorphic rocks in the southern part of the eugeosyncline is still only vaguely known. We assume that this area's history is similar to that of the northern reaches of the eugeosyncline. Radiometric age measurements indicate that the Devonian was a time of profound plutonism and metamorphism in the Piedmont belt. The movements within the southern part of the eugeosyncline are only known indirectly from the time of influx of easterly derived clastics into the adjoining miogeosyncline.

In New England, in the maritime provinces, and Newfoundland, distinct basins formed in the former eugeosyncline, in which enormous thicknesses of clastic sediments, including some volcanic materials, accumulated. In eastern Canada and Newfoundland, during part of Mississippian time, some basins were invaded by the sea, which deposited limestones, evaporites, and redbeds. In general, the Carboniferous formations of this area are clastics derived from adjacent highlands. A restored section of the sediments in New Brunswick and Nova Scotia is shown in Figure 5-24. Note the alignment of deeply subsiding basins adjacent to and between the axial areas that

were being uplifted. Similar but broader basins of Pennsylvanian age accumulated thick sequences of coarse clastics in eastern Massachusetts and Rhode Island.

In the adjoining miogeosyncline subsidence and deposition continued as in the Late Devonian. Early Mississippian deposits in West Virginia and Pennsylvania are coarse clastics derived from the Devonian tectonic lands to the east. South of this region of coarse clastics the Mississippian is represented mainly by carbonate facies. By Middle Mississippian time the area of coarse clastics was reduced to parts of the state of Pennsylvania and was surrounded by shale and carbonate facies. The Mississippian sequence is much thicker in the southern than in the northern portion of the miogeosyncline. In Late Mississippian time we again encounter great volumes of clastic sediments, coarsest in the east and becoming finer westward, where they eventually intertongue with carbonate formations.

The Pennsylvanian rocks of the miogeosyncline are the "coal measures" of North America, almost entirely nonmarine, with fine-to-coarse clastics and a few thin limestone beds. The whole sequence is about 2,500 feet thick in Pennsylvania, but in Alabama, even though only the lowest division of the Pennsylvanian is preserved there, it reaches, in some places, a thickness of 9,000 feet. Figure 5-25 is a restored cross section of the Mississippian and Pennsylvanian rocks of the Birmingham district of central Alabama. Notice the great thickening of the section, and the increase in clastic content, to the southeast. At the time these clastic sediments were accumulating, numerous and vast coal swamps periodically occupied great areas of the geosyncline. The fauna and flora of the period and the factors involved in coal formation will be discussed in Chapter 7.

We have already remarked that the sequence of events in the southern half of the geosyncline is obscure. Extensive granitic intrusions and metamorphism have masked the record, yet there are some clues. We note that beginning in Late Mississippian a thick wedge of clastics builds up through the remainder of the Carboniferous. The sediments of these wedges came from the east and southeast, and we therefore assume that one or more orogenic phases took place in the adjoining eugeosyncline in Late Mississippian and Pennsylvanian time.

The Southern Geosyncline

The southern geosyncline was also highly mobile during Carboniferous time. In this geosyncline we find a remarkable development of graywackes and shales, nearly 20,000 feet thick and from Middle Mississippian to Early Pennsylvanian in age, which must have been due to orogeny in

FIGURE 5-23

Paleotectonic map of North America for the Pennsylvanian. (After A. J. Eardley, *Structural Geology of North America*, Harper & Brothers, New York, 1951.)

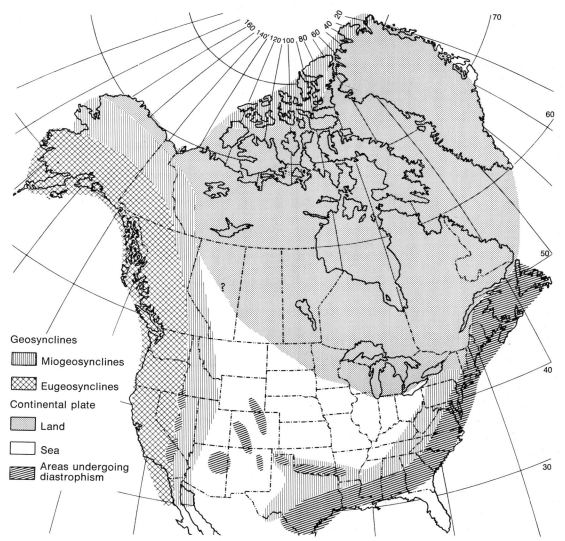

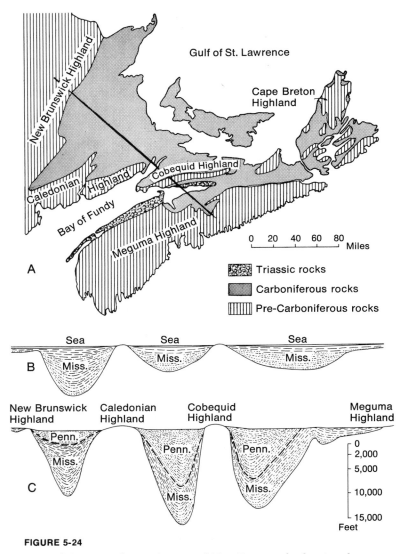

FIGURE 5-24

A. Geologic map of Nova Scotia and New Brunswick, showing the distribution of Carboniferous rocks and pre-Carboniferous highlands.

B. Restored section at the close of the Mississippian along the line marked on the map.

C. Restored section at the close of the Pennsylvanian along the line marked on the map.

(After M. Kay, *Geol. Soc. Am.*, Mem. 48, 1951.)

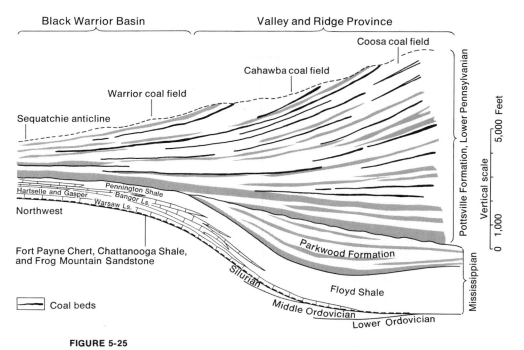

FIGURE 5-25

Restored cross section of the Mississippian and Pennsylvanian rocks of the Birmingham district of central Alabama, showing the thickening of the clastic rocks to the southeast. (After P. B. King, *The Tectonics of Middle North America,* Princeton Univ. Press, 1951.)

the eugeosyncline (Fig. 5-26). Later Pennsylvanian formations are likewise clastics derived from the south. The miogeosyncline received shale in Mississippian time and, during earliest Pennsylvanian, a great influx of sandstones and shales from the eugeosyncline. This was followed by orogeny in the miogeosyncline itself, forming a series of linear highlands (the Wichita Mountains) across southern Oklahoma and the Panhandle of Texas. Active erosion of these highlands supplied abundant sediments to the adjoining subsiding areas; the flanks of the highlands are overlapped by successive Pennsylvanian beds derived from the uplifts. If stripped of their flanking deposits, some of the highlands would stand today as mountains nearly two miles high. Late in Pennsylvanian time another strong orogenic phase affected the whole area. Post-orogenic deposits are Late Penn-

sylvanian nonmarine clastic sediments.

It is not possible to date precisely the last orogeny that affected the eugeosyncline; but, since this belt was thrust northward over the miogeosyncline in the Late Carboniferous, this thrust must post-date the Late Pennsylvanian orogeny of that belt. Thus a very late Pennsylvanian date for the thrusting of the eugeosyncline (Ouachita belt) northward over the miogeosyncline (Wichita belt) is probable.

The effects of the Carboniferous orogenies in the southern geosyncline can also be seen in the Marathon region of western Texas, where we find Early Permian rocks resting with great angular unconformity on Pennsylvanian and older strata. The orogenic forces that folded these strata began early in the Pennsylvanian. They are reflected in a thick accumulation of coarse-to-fine sediments in the geosyncline. The conglomerates consist of boulders or fragments of rocks from the geosyncline. Since deposition was going on during the Pennsylvanian in one place or another in the geosyncline, a good record of the orogenic history is preserved. Figure 5-27 illustrates the probable sequence of events. The orogeny that affected this area during the Pennsylvanian is named the Marathon orogeny.

In Early Pennsylvanian time the geosyncline accumulated a great thickness of sandstones and shales, which thicken and become coarser to the southeast and contain minute fragments of granitic and metamorphic rocks (Fig. 5-27, A). Early in Late Pennsylvanian time overthrusting commenced, and the associated deposits of this age consist largely of conglomerate beds

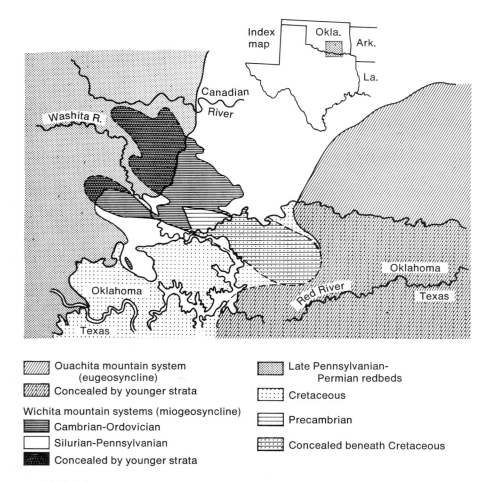

Ouachita mountain system
(eugeosyncline)

Concealed by younger strata

Wichita mountain systems (miogeosyncline)

Cambrian-Ordovician

Silurian-Pennsylvanian

Concealed by younger strata

Late Pennsylvanian-
Permian redbeds

Cretaceous

Precambrian

Concealed beneath Cretaceous

FIGURE 5-26

Geologic sketch map of southern Oklahoma, showing the relations of the Ouachita Mountains to the eastern part of the Wichita Mountains. The former are composed of eugeosynclinal sediments that were thrust northward, in the Late Pennsylvanian, over miogeosynclinal strata of the latter. (After King, 1951.)

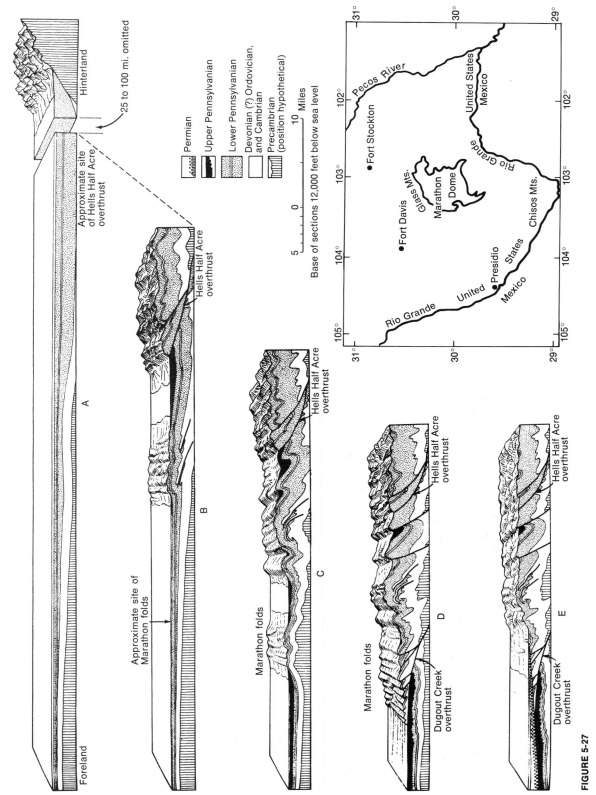

Permian

Upper Pennsylvanian

Lower Pennsylvanian

Devonian (?) Ordovician, and Cambrian

Precambrian (position hypothetical)

Base of sections 12,000 feet below sea level

5   0   10 Miles

Foreland

Hinterland

25 to 100 mi. omitted

Approximate site of Hells Half Acre overthrust

A

Approximate site of Marathon folds

B

Hells Half Acre overthrust

Marathon folds

C

Hells Half Acre overthrust

Marathon folds

D

Hells Half Acre overthrust

Dugout Creek overthrust

E

Hells Half Acre overthrust

Dugout Creek overthrust

Pecos River

Fort Stockton

Fort Davis

Presidio

Glass Mts.

Marathon Dome

Chisos Mts.

United States

Mexico

Rio Grande

United States

Mexico

Rio Grande

FIGURE 5-27

Hypothetical block diagrams illustrating the history of the Marathon orogeny. (After P. B. King, *U.S. Geol. Surv., Prof. Paper* 187, 1938.)

derived from uplifted areas in the geosyncline (Fig. 5-27, B). This phase was followed by further overthrusting and by folding (Fig. 5-27, C, D). Before Permian time the northern part of the folded and thrusted belt was leveled by erosion, and we therefore find Permian deposits resting on a profound angular unconformity (Fig. 5-27, E).

The Western Geosyncline

Because of the Antler orogeny the depositional pattern of Carboniferous rocks in western United States differs greatly from that of the pre-Upper Devonian. This orogeny, which began in the Late Devonian, continued in force into the Pennsylvanian. A diagram of the inferred sequence of events

before and during the orogeny in north-central Nevada is illustrated in Figure 5-21. The expression of this orogeny during the Late Devonian was primarily uplift along the orogenic belt. By earliest Mississippian time emergence had reached a point where a gravity thrust or glide plate broke away and moved eastward into the foredeep. The mass was 4 to 8 miles in thickness and made up mostly of siliceous and volcanic and transitional assemblage rocks; the frontal part reached a point 90 to 100 miles east or southeast of its point of origin. Although the plate was internally deformed during uplift and movement by folding and imbrication, the deformation seems to have been local, and the frontal part contains rocks that are transitional in facies or more nearly transitional than those in the middle

FIGURE 5-28

Angular unconformity (exposed along Highway 40, in western Carlin Canyon, Humboldt River, northeastern Nevada) between the nearly vertical Tonka Formation (Upper Mississippian) and the gently dipping Strathearn Formation (Upper Pennsylvanian), which forms the top of the hogback.
(Photograph by Marshall Kay.)

and rear. The distribution of facies in the plate, therefore, roughly parallels their original distribution in the geosyncline.

Concomitant with gliding, the emergent area was being eroded and clastics were being shed both to the east and west. These orogenic clastics of Mississippian and Pennsylvanian age ultimately overlapped the toe of the glide plate (Fig. 5-21). Westward, the Mississippian and Lower Pennsylvanian rocks wedged out, and Upper Pennsylvanian rocks lapped onto the central part of the orogenic belt (Fig. 5-28). By Permian time, except for local islands and peninsulas, most of the orogenic belt was eroded nearly to sea level and was overlapped by limestone of shallow-water facies.

The Carboniferous rocks of the central and eastern parts of the miogeosyncline consist almost entirely of orthoquartzites and carbonates, the amount of clastic material being greater in the Pennsylvanian formations. These Carboniferous formations are typical of geosynclinal accumulations, in that they are very thick. The enormous thickness of the miogeosynclinal section in the southern Wasatch and the Oquirrh mountains contrasts markedly with that of the platform section to the north and east in the Uinta Mountains (Fig. 5-29).

North of the Yukon River in Alaska the Mississippian is represented by argillaceous sandstone, shale, limestone, and bedded chert. Pennsylvanian formations are apparently absent, suggesting uplift and erosion during that time.

In the eugeosyncline the Carboniferous was a time of great accumulation of volcanics and clastic sediments. For instance, in the Klamath Mountains of northern California the Mississippian is represented by about 8,100 feet of shales, sandstones, lavas, and tuff beds. In the Sierra Nevada a thick, metamorphosed complex of sediments and volcanics is known to be, at least in part, of Carboniferous age. Carboniferous strata have also been recognized in various regions of Oregon and Washington; clastic sediments and volcanics are the predominant rocks, but fossiliferous limestones and calcareous sandstones are also present in some places. Several Pennsylvanian formations are thought to be nonmarine because they yield plant fossils. In southern British Columbia Carboniferous complexes of sediments and volcanics are known but are not yet well understood.

The Northern Geosyncline

Carboniferous strata occur along the southern, eastern, and northern periphery of the Sverdrup Basin, in the Innuitian region (Fig. 5-11), where they lie unconformably above Upper Devonian and older formations. They are also found, in widespread outliers, unconformably above metamorphic terranes of the Northern Ellesmere belt, and they include nonmarine beds that lie unconformably over lower Paleozoic strata in the Cornwallis belt. A widespread post-Devonian and pre-Middle-Pennsylvanian orogeny caused the folding in the Parry Islands fold belt (Fig. 5-11) and produced extensive deformation in the Central Ellesmere fold belt and also, apparently, in the Northern Ellesmere fold belt.

Limestone, commonly with chert, is the most widespread Pennsylvanian facies, but in the western part of the Parry Islands belt sandstone occurs almost to the exclusion of limestone.

In the East Greenland geosyncline the Devonian nonmarine clastic deposition continued into Carboniferous time.

The Continental Interior

The great mobility of the peripheral geosynclines, especially those to the east and south, profoundly affected the depositional facies of the interior platform. On the platform itself the arches and basins moved a

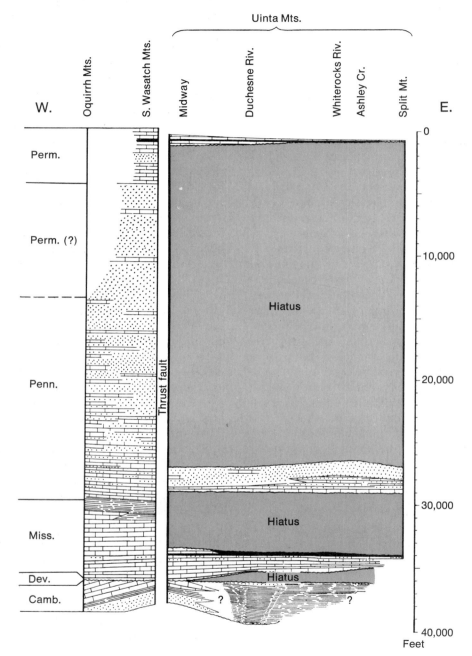

FIGURE 5-29

Composite diagram comparing stratigraphic sections along the southern flank of the Uinta Mountains with a section in the southern Wasatch and the Oquirrh Mountains, Utah. (From Baker, Huddle, and Kinney, *Bull. Am. Assoc. Petrol. Geol.*, 1949.)

good deal, and some new ones formed. These movements, aside from influencing the facies pattern, produced a complex pattern of unconformities.

In the earliest Mississippian time black shales were deposited in the area from the eastern miogeosyncline to the Mississippi Valley. These fine-grained clastics were derived from the eastern highlands that had risen in Late Devonian time and continued into Mississippian. By Middle Mississippian time great quantities of clastic materials no longer entered the central platform, and carbonate deposition prevailed. Many carbonate formations are made up almost entirely of crinoidal remains. On the periphery of the platform, toward the eastern miogeosyncline and the southern geosyncline, these carbonates intertongue with clastics.

In Late Mississippian time large quantities of clastics, reflecting uplift in the geosynclines, again entered the platform. In Kentucky, Illinois, and Indiana the Late Mississippian deposition was a complex alternation of sandstone and limestone formations, all rather thin but widespread. The clastic formations of the upper region of the Mississippi Valley retain clues to their sources. Detailed analysis of the grain size, composition, and maturity of the sand grains together with cross-bedding directions permits the dispersal centers to be pinpointed. The sediments of the Upper Cambrian formations were derived from Precambrian formations in the southern Canadian shield. The clastics of the Ordovician and Devonian came from erosion of Upper Cambrian formations. The Carboniferous sediments, however, were derived mainly from the Appalachian region, with only minor contributions from the Precambrian area to the north. Figure 5-30 summarizes the dispersal centers and source relations of the Phanerozoic clastics of the upper Mississippi Valley.

The western regions of the platform were the site of generally thin, principally carbonate deposition. In western North Dakota and eastern Montana a broad basin subsided twice as fast as the surrounding shelf regions. Carbonates are the main facies throughout this area except in the basin, where, in Middle Mississippian time, evaporites accumulated. In these northern states Upper Mississippian rocks, consisting of shales and sandstones, are largely confined to the Montana-Dakota basin.

Toward the end of Mississippian time the sea retreated from most of the interior platform, and a period of erosion followed. Figure 5-31 is a paleogeologic map of the platform prior to the Pennsylvanian transgression. Study this map carefully, for it shows how the basins and arches affected the platform. Some exposures of older Paleozoic rocks were emergent throughout Mississippian time, and others had been covered only by thin deposits of Mississippian sediments. Such rocks, obviously, had long resisted subsidence. The basins, of course, subsided and received Late Mississippian deposits. Most of these high and low areas, as we have seen, have long histories, some dating back to the Ordovician.

Early Pennsylvanian rocks are confined to the peripheral geosynclines. It was not until early Middle Pennsylvanian time that the sea again spread across the interior of the United States. The paleogeology over which it spread is shown in Figure 5-31. Its deposits reflected the great mobility of the tectonic lands to the south and east. In some regions the earliest deposits were on Late Mississippian, in others on earlier Paleozoic or even Precambrian, formations. In the states near the eastern miogeosyncline, Pennsylvanian rocks are largely nonmarine sandstone, shale, and coal. Farther west, in the Mississippi Valley, about half of the section is made up of marine shales and limestones. As these platform deposits approach the peripheral geosynclines, they thicken greatly.

One of the most remarkable aspects of Pennsylvanian formations in the middle and eastern interior states is the succession of cyclical deposits. Each series of beds associated with a single sedimentary cycle is called a **cyclothem.** The ideal cyclothem, as illustrated in Figure 5-32 has quartzose or subgraywacke sand at the base (member 1), commonly lying unconformably upon the uppermost member of the previous cyclothem. The basal sandstone grades upward into sandy shale and clay shale (member 2) and freshwater limestone nodules (member 3) above. This is succeeded by gray-to-drab underclay (member 4) and coal (member 5). Immediately above the coal is a thin marine shale (member 6), succeeded by dense or argillaceous limestone with marine fossils (member 7). Black, laminated shale (member 8), with an aberrant fauna composed of much-flattened forms, lies above the dense limestone and is succeeded by clean, fossiliferous, fragmental or foraminiferal limestone (member 9). The uppermost member is a bluish-gray marine shale, which becomes more silty near the top; plant fossils may occur in its upper portion. Though the ideal cyclothem is subject to numerous variations, which depend on the tectonic activity of the source of the clastic sediments and on the amount and rate of subsidence in the depositional area, many cyclothems can be reocgnized over wide areas covering one or more states.

The ideal cyclothem is the one commonly found in Illinois. Eastward, toward the miogeosyncline, the percentage of marine rocks decreases greatly, and here a cyclothem consists largely of nonmarine strata. Westward, the cyclothems have more marine strata and are quite different from those of Illinois or of the eastern states.

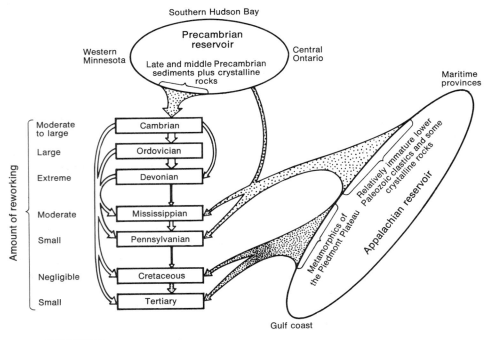

FIGURE 5-30

Dispersal centers and source relations of the Phanerozoic clastics for the upper Mississippi Valley and adjacent regions. (After P. E. Potter and W. A. Pryor, *Geol. Soc. Am. Bull.*, 1961.)

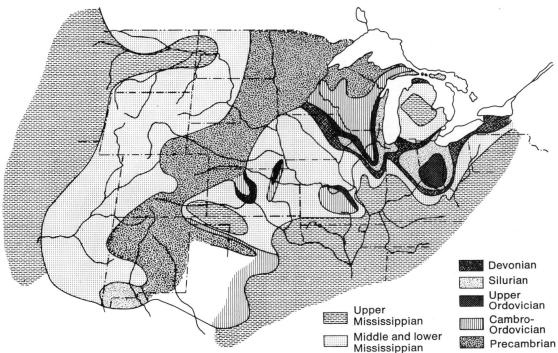

FIGURE 5-31

Paleogeologic map of the platform of the United States at the end of Mississippian time. (After A. I. Levorsen, *Bull. Am. Assoc. Petrol. Geol.*, 1933.)

Legend:
- Devonian
- Silurian
- Upper Ordovician
- Cambro-Ordovician
- Precambrian
- Upper Mississippian
- Middle and lower Mississippian

These differences in the lithologic composition of Pennsylvanian cyclothems are illustrated in Figure 5-33. Note the prevalence of marine strata in Kansas and—reflecting the adjoining tectonic land—the completely nonmarine sequence in West Virginia. The Illinois sequence is of an intermediate type.

On the western part of the platform Pennsylvanian deposition was of a quite different pattern. The dominant facies is quartz sandstone, but there are some carbonate formations (Fig. 5-16). In our discussion of the southern geosyncline we noted that a chain of mountains was formed along the southern border of Oklahoma and across the Panhandle of Texas. A similar group of mountains was formed in northern New Mexico and western Colorado and had a great effect on the facies developed in the adjoining basins. The depositional basins associated with these ancestral Rockies have thick, coarse, clastic deposits, which become finer-grained and thinner in all directions away from their source.

The North Dakota—Montana (Williston) basin continued to subside more than the surrounding regions. Carbonates and evaporites, as well as fine clastics, were deposited within the basin, but clastics prevailed in the surrounding area.

Another extremely interesting basin, about 200 miles long and 100 miles wide, developed in southeastern Utah and southwestern Colorado in the shadow of the ancestral Rockies and accumulated more than 10,000 feet of Pennsylvanian sediments. The Lower Pennsylvanian clastics and carbonates are followed by a very thick development of evaporites, and the Middle and Upper Pennsylvanian facies are clastics and carbonates. The Upper Pennsylvanian clastic facies indicate a marked uplift in an adjoining region of the ancestral Rockies. The salt deposits of this basin are the thickest salt beds of Pennsylvanian age in the United States.

5-6 THE PERMIAN PERIOD

As we approach the end of the first major phase of Phanerozoic history, we can begin to see that the tectonic framework of North America is a gradually evolving complex rather than the result of a series of catastrophes. That framework, as it was during Permian time, is illustrated by Figure 5-34. On that framework the Mesozoic history of the continent will be built.

The Eastern and Southern Geosynclines

The long Paleozoic history of deposition and subsidence in the eastern geosyncline came to an end in early Permian time. The only Permian rocks of this area, found in southeastern Ohio, northwestern West Virginia, and Pennsylvania and representing the same type of facies as the underlying Pennsylvanian strata, are clastic sediments of early Permian age. The main event in the Permian history of the eastern geosyncline was not depositional but orogenic: the whole mobile belt was transformed into a rigid part of the continental framework, and the folding of the present Appalachian Mountains was completed. The part of the crust affected by that process was shortened by about 200 miles. Through most of the Carboniferous the eugeosyncline was undergoing orogenesis at one place or another. This orogenic episode continued into the early Permian and is called the Alleghanian orogeny.

The deformation of the eastern geosyncline is best studied in the regions south of New England, for farther north it is difficult

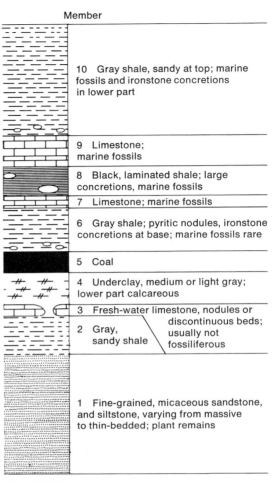

Member

10 Gray shale, sandy at top; marine fossils and ironstone concretions in lower part

9 Limestone; marine fossils

8 Black, laminated shale; large concretions, marine fossils

7 Limestone; marine fossils

6 Gray shale; pyritic nodules, ironstone concretions at base; marine fossils rare

5 Coal

4 Underclay, medium or light gray; lower part calcareous

3 Fresh-water limestone, nodules or discontinuous beds; usually not fossiliferous

2 Gray, sandy shale

1 Fine-grained, micaceous sandstone, and siltstone, varying from massive to thin-bedded; plant remains

FIGURE 5-32

The ideal cyclothem. (After *Stratigraphy and Sedimentation*, by W. C. Krumbein and L. L. Sloss. W. H. Freeman and Company. Copyright © 1963.)

to separate the effects of the Taconic and Acadian orogenies from those of the Alleghanian orogeny. In the central and southern parts of the Appalachian Mountains we find extensive folds, faults, and thrusts, most of them formed during the Alleghanian orogeny. The types of structures are related in a general way to the geosynclinal nature of the region, as can be seen in Figure 5-35, which shows the major structural systems of the eastern margin of the United States.

The Appalachian plateau province consists of flat or gently folded upper Paleozoic strata of the foreland of the eastern miogeosyncline. The folded and thrust-faulted province contains the Appalachian Mountains proper, consisting of numerous parallel or subparallel valleys and ridges. The rocks of this province consist of miogeosynclinal deposits that range from Cambrian through Pennsylvanian in age. The structures of the area are anticlines and synclines, some overturned to the northwest. The Blue Ridge province consists of highly folded Cambrian formations and Precambrian metamorphics and igneous rocks that are thrust-faulted toward the adjacent folded and faulted province. The rocks of this province are not only older than the westerly adjacent province but also more or less metamorphosed. The outer zone of

this old deformed belt is the Piedmont province, consisting mostly of schist, gneiss, and granite, and is probably partly Precambrian and partly Paleozoic in age. The rocks are metamorphosed sediments and volcanics of the eugeosyncline. Within this belt are some down-faulted blocks of Upper Triassic sediments, which will be discussed in Chapter 8. The relations of these various structural provinces are shown in Figure 5-36. As you study these cross-sections and the map of Figure 5-35, try to visualize the great complexities that we encounter when we try to decipher the history of such an area.

With the Alleghanian orogeny the eastern geosyncline became a rigid part of the continental framework, and remained so through the rest of geologic time.

FIGURE 5-33

Idealized cross section showing the variation in the lithologic composition of Pennsylvanian cyclothems from Kansas, where the strata are largely marine, to West Virginia, where they are entirely nonmarine. (From H. R. Wanless, 1950.)

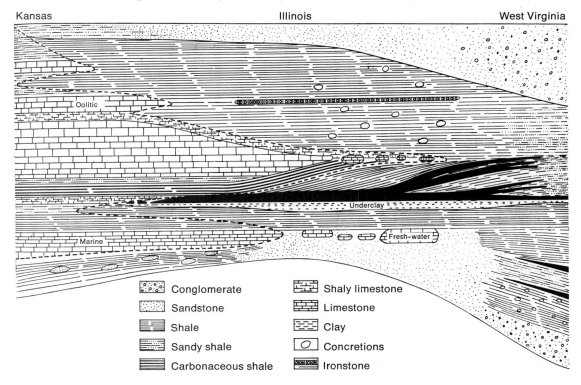

Kansas   Illinois   West Virginia

Oolitic
Underclay
Fresh-water
Marine

Conglomerate
Sandstone
Shale
Sandy shale
Carbonaceous shale
Shaly limestone
Limestone
Clay
Concretions
Ironstone

FIGURE 5-34

Paleotectonic map of North America for the Permian.
(After A. J. Eardley, *Structural Geology of North America*, Harper & Brothers, New York, 1951.)

The Western Geosyncline

Here the Permian formations are much more extensive and better understood than the older Paleozoic systems. Much of the eugeosyncline was a volcanic archipelago within which great thicknesses of volcanic flows, ash beds, and sedimentary rocks accumulated. Such deposits are to be found in California, western Nevada, Oregon, Washington, western Idaho, British Columbia, and southern Alaska.

Orogenic activity again characterizes north-central Nevada. The western flank of the Antler orogenic belt underwent orogenesis in the late Permian. This orogenic phase (Sonoma orogeny) first resulted in uplift of the area to the west of the Antler orogenic belt, and this was followed by a great thrust (Golconda thrust) onto the western area of the Antler orogenic belt. Orogenic activity most probably is part of the Permian history of the Pacific coastal regions, but the evidence for orogenesis is incomplete.

The miogeosyncline consists largely of quartz sandstone and carbonate formations. The Phosphoria Formation, centered in eastern Idaho and extending into western Wyoming, southwestern Montana, and northern Utah, is notable for its high content of phosphate minerals. Along the eastern border of the miogeosyncline and on the edge of the interior platform the Permian marine formations thin greatly and interbed with redbed formations. Much of the clastic sediment deposited in the miogeosyncline was derived from the ancestral Rockies, formed during the Pennsylvanian Period.

The Northern Geosyncline

The late Paleozoic history of this region is still poorly known. Several localities, from Melville Island eastward to Pearyland (northern Greenland), have yielded marine Permian fossils in clastic and carbonate formations. Some associated nonmarine strata are also thought to be of Permian age (Fig. 5-11). Near the northern tip of Axel Heiberg Island basalt flows are present in a Permian sequence. At one place these volcanic rocks are nearly 5,000 feet thick. For the present not much more can be said. Other Permian rocks do occur in this geosyncline, but their relations are still very vague.

Isolated Permian rocks along the coastal region of eastern Greenland seem clearly to be marginal marine deposits.

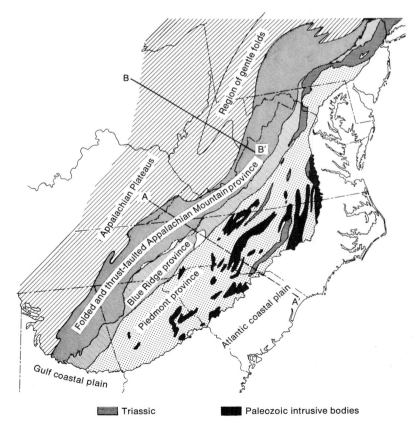

FIGURE 5-35

Structural provinces of the Appalachian Mountains and adjoining regions. The lines *AA'* and *BB'* indicate the cross sections of Fig. 5-36. (After Geol. Soc. Am., *Tectonic Map of United States*, 1944.)

Triassic   Paleozoic intrusive bodies

The Continental Interior

The most striking feature of the Permian deposits of the continental interior is the great extent of the redbed and evaporite formations. (Before we go on with our discussion, study again the map of Fig. 5-34, compare it with the similar maps for the earlier Paleozoic periods, and review the depositional and orogenic history of the eastern and southern geosynclines.)

Lower Permian marine rocks are found in a rather narrow zone from Nebraska south through western Texas. East of the Mississippi River the Permian is represented by clastic deposits only along the western margin of the miogeosyncline in West Virginia, Ohio, and Pennsylvania. These deposits, of early Permian age, closely resemble the underlying Pennsylvanian

rocks. Part of this sequence is marine; the seaway must have been a short-lived eastern arm of the mid-continent marine embayment.

The early Permian sea extending from western Texas to Nebraska was a typical, shallow epicontinental sea with a limy and muddy bottom, rich in marine life. Coarse clastic materials were contributed from the orogenic highlands formed during the Pennsylvanian in Colorado, the Panhandle of Texas, and southern Oklahoma. By the end of early Permian time most of these highlands were completely buried in their sediments. The deposits along the eastern margin of the interior sea have long since been eroded away, but along the western margin, in Colorado and Wyoming, the marine deposits grade into redbed deposits, largely of continental origin.

FIGURE 5-36

Diagrammatic cross sections showing the structural relations of the Appalachian Mountains. Study these cross sections along with the map of Fig. 5-35, where the lines of section are indicated. Section *AA'* extends across the Appalachian system from Kentucky to South Carolina. Section *BB'* extends from northwestern Ohio to Virginia and is based largely on the results of deep drilling. The vertical scale is exaggerated. The Allegheny synclinorium is a geologic province with gentle anticlines and synclines exposing mainly Pennsylvanian strata lying west of the folded true Appalachians. (After P. B. King, *Bull. Am. Assoc. Petrol. Geol.*, 1950.)

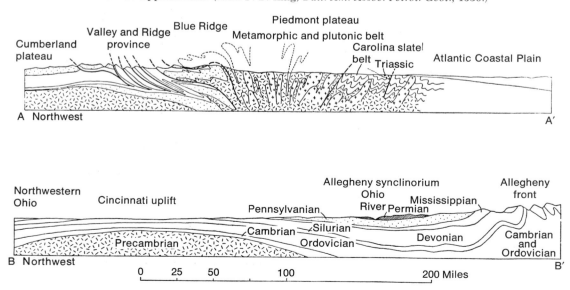

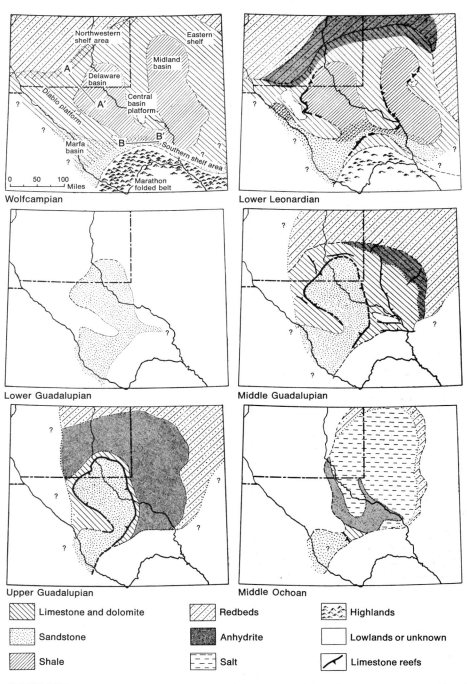

Wolfcampian

Lower Leonardian

Lower Guadalupian

Middle Guadalupian

Upper Guadalupian

Middle Ochoan

Limestone and dolomite   Redbeds   Highlands

Sandstone   Anhydrite   Lowlands or unknown

Shale   Salt   Limestone reefs

FIGURE 5-37

Paleogeographic maps of western Texas during Permian time. The time horizon for each map is indicated on the cross sections of Fig. 5-38. The Wolfcampian, Leonardian, Guadalupian, and Ochoan are (from bottom to top) the series of the Permian recognized in this area. (After P. B. King, *U.S. Geol. Surv.*, Prof. Paper 215, 1948.)

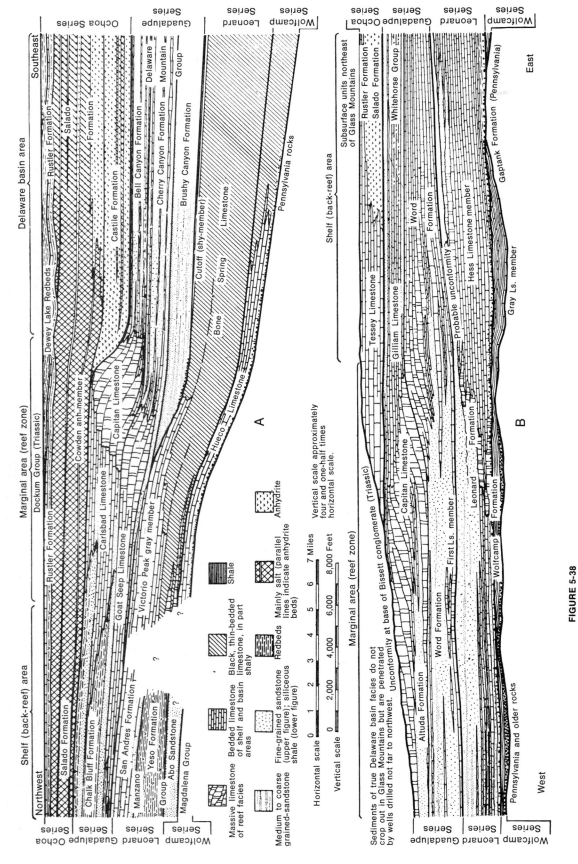

FIGURE 5-38

Diagrammatic cross sections of Permian rocks in western Texas. The locations of the sections are indicated in Fig. 5-37. (After P. B. King, *Bull. Am. Assoc. Petrol. Geol.,* 1942.)

FIGURE 5-39

Aerial view of the southern end of the Guadalupe Mountains, western Texas, showing middle Permian formations. Check the named units in the cross section of Fig. 5-38. (Photograph by Robert Muldrow III of Kargl Aerial Surveys, Inc.)

Capitan Ls.

Cherry Canyon Fm.

Bushy Canyon Fm.

Bone Springs Ls.

In early Permian time, as this sea began to retreat southward, redbed deposits and evaporites followed it. In southeastern Nebraska, for example, we find lower Permian marine facies followed by a middle Permian redbed facies. As we move southward, we find the marine-redbed transition coming at higher and higher horizons. The last stand of this regressive Permian sea was in western Texas, where we find the most nearly complete sequence of marine Permian rocks in North America; here it was not until late Permian time that redbed and evaporite facies completely replaced the normal marine facies.

The Permian strata of western Texas have been studied by a great many geologists. Few areas with such diversity of facies and fossils are better known or lend themselves so well to the study of fundamental problems of facies, paleoecology, and stratigraphy in general. For these reasons we shall delve into the Permian history of this region in some detail.

In § 5-5, in the discussion of the southern geosyncline during the Pennsylvanian, we pointed out that in western Texas the Late Pennsylvanian Marathon orogeny had folded and thrusted the earlier Paleozoic formations. Aside from the folded belt, which was a highland region, platforms with basins between them were established in the foreland. Thus, in early Permian time, western Texas had a folded mountain chain to the south, two prominent platforms, and adjoining basins surrounded by extensive shelf areas; these structural features are shown in Figure 5-37. Throughout the Permian history of this region the platforms remained firm and the basins subsided. The maps of Figure 5-37 summarize the distribution of facies during six distinct stages of Permian history. Study the cross section of Figure 5-38 along with these maps. Early Permian deposits on the platform and on the adjoining shelves consist of carbonate formations, which grade outward into redbed facies to the northwest. In the Delaware and Midland Basins dark shales prevail. Along the margins of the Marathon folded belt sandstone facies were developed. In early middle Permian time numerous narrow reef tracts developed along portions of the platform margins facing the basins. The platform sediments are limestones grading away into redbeds and anhydrite facies. The basin sediments are dark shales and limestones.

Later in middle Permian time deposition was confined entirely to the Delaware basin, where great quantities of fine sandstone and sandy limestone accumulated. Presumably the surrounding areas were land undergoing erosion. This phase was followed, in middle Guadalupian time, by an expansion of the marine waters beyond the Delaware basin. The margins of the platforms surrounding the basin became the site of a phenomenal reef development, behind which back-reef carbonate facies are found grading into redbed-anhydrite facies. This general facies pattern continued into upper Guadalupian time, when the reef rimmed the entire basin on the margin of the surrounding platform (Fig. 5-39). The basin sediments are dark, fine-grained sandstones and sandy limestones; the platform rocks are light-colored lagoonal carbonates grading into anhydrite-redbed facies. The final depositional phase in late Permian time was one of redbeds and evaporites.

From the study of these Permian rocks we can come to several interesting conclusions: The topography of the surrounding areas was subdued, and the extensive development of evaporites suggests that the climate was generally warm and dry. The water was much deeper in the basins than on the platforms; a maximum of 1,800 feet is estimated for the depth in the Delaware basin, but on the shelves it was probably not more than a few tens of feet. The predominance of dark organic sediments in the basins suggests reducing conditions and stagnant bottoms. The reefs formed along the rim of the Delaware basin are made up mainly of calcareous algae, sponges, bryozoans, and hydrocorallines, but other invertebrate groups, of course, added their skeletons to the calcareous deposits. The association of the reefs and the adjoining deep basin suggests that an upwelling of stagnant deep water may have provided a particularly rich source of nutrient salts to the phytoplankton and the reef-forming algae.

Suggested Readings

See the list at the end of Chapter 4.

6

THE PALEOZOIC ERA

II. The World Outside North America

The history of the Paleozoic is generally better known for North America than for any other continent. In Europe, however, where geology began, and where Mesozoic and Cenozoic strata are widespread, fossiliferous, and little disturbed, the Paleozoic presents many difficult problems. We saw, in Chapter 1, that geologists of the early nineteenth century were perplexed by the Transition class of rocks; we now know that this class includes the early Paleozoic systems, which, in most areas, have been involved in one or more orogenic phases and are, therefore, highly disturbed. The outcrops of these Paleozoic systems, in much of Europe, are like windows piercing a blanket of Mesozoic and Cenozoic strata. The discontinuity of the Paleozoic outcrops and the superposition of later geologic events, both depositional and orogenic, make the interpretation and synthesis of Paleozoic history difficult, especially for the earlier systems. We have enough clues, however, to organize the main series of events into a reasonably comprehensible evolution.

We saw, in Chapter 5, that the main theme in the physical development of Paleozoic North America was the interplay between the stable shield, including the surrounding shelves, and the peripheral geosynclines. This interplay between less mobile and more mobile areas is likewise recorded in the other continental areas. The map in Figure 3-19 (p. 83) shows the Precambrian shields and the mobile belts. In South America a prominent mobile belt borders the western periphery of the continent, and broad shields occupy the central and eastern parts. The vast mass of Eurasia is dominated by two shields on the north, the Baltic and the Angaran, and two on the south, the Ethiopian and the Indian. Between these shields were once mobile belts of diverse ages. The southern mobile belt, encompassing the great mountain chains beginning with the Alps and extending eastward through the Himalayas, is primarily of Cenozoic age. The mobile belts to the north were deformed in the Paleozoic; these include the mobile belts north of the Alps in Europe, the Ural Mountains, and those around the Angaran shield. The mobile belts of eastern Australia also evolved entirely in the Paleozoic. In the East Indies there is hardly any record of Paleozoic history.

For simplicity we will develop the main events of Paleozoic history separately in each of these continental areas, focusing our attention on the evolution and interrelations of mobile and immobile belts. First we consider Europe, where historical geology began, including all the area between the Baltic shield on the north and the Ethiopian shield on the south, from the Atlantic Ocean on the west to the Ural Mountains on the east.

6-1 EUROPE

The Early Paleozoic

Paleozoic rocks crop out in many parts of Europe, most extensively around the Baltic shield—along the folded belt of Norway and on the southern and eastern margins of the shield (Fig. 4-7, p. 95). In most of Scotland, western England, Wales, and Ireland Paleozoic rocks are exposed. Across continental Europe, highly disturbed Paleozoic rocks crop out in massifs, the remnants of Paleozoic mountains, now completely surrounded by Mesozoic and Cenozoic rocks. Paleozoic formations in geosynclinal and shelf facies are found fringing the Ethiopian

shield in North Africa. Finally, the Ural Mountains consist entirely of Paleozoic formations.

The Cambrian rocks along the southern margin of the Gulf of Finland (in Estonia) make up a remarkable series of flat sandstones with an intercalated blue-clay formation, the whole sequence being some 600 feet thick. Their fossils, though not abundant, are enough to date the beds and to identify several hiatuses in the section. Farther west important Cambrian exposures are found in the Oslo region of Norway and in southern Sweden, but here the rocks are mostly shales, but include some limestone beds and a basal sandstone unit. In Norway these strata conformably overlie upper Precambrian arkose, but to the east the Cambrian rests on older peneplaned Precambrian formations (Fig. 6-1). The Scandinavian Mountains of Norway are made up mainly of metamorphic and igneous rocks, from Precambrian to Silurian in age. The relations and facies are poorly known, but we do know that thick Cambrian deposits are present.

We look to Wales as the standard of the lower Paleozoic systems, since the type areas for the Cambrian, Ordovician, and Silurian are there. In northern Wales (in the Harlech dome) more than 12,000 feet of Cambrian conglomerates, sandstones, and shales are exposed. To the east, in the Welsh border country, incomplete Cambrian sections are only about 1,000 feet thick. Many shale beds are rich in fossil trilobites. In the Scottish Highlands, a thin basal quartzite, resting uncomformably on the Precambrian, and a dolomite formation yield Lower Cambrian fossils of a facies very different from that of Wales.

The major massifs of Europe have undergone at least two orogenic phases, and it is therefore not surprising that their oldest formations, of early Paleozoic age, are generally poorly known. Most of their Cambrian facies are fine-to-coarse clastics, of which, in

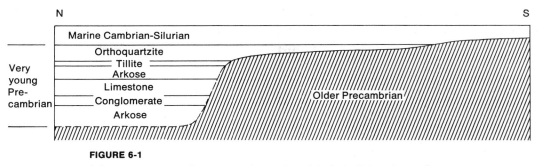

FIGURE 6-1

Schematic section of pre-orogenic stratigraphical conditions in southern Norway. (After O. Holtedahl, *Geol. Soc. London, Quart. J.*, 1952.)

some places (in the Polish massif, for example), more than 6,000 feet were deposited. A classic area of Cambrian rocks is found in the Bohemian massif, where highly fossiliferous shale and sandstone strata nearly 5,000 feet thick are preserved; in the Upper Cambrian, volcanic rocks are intercalated. In the Sudeten Mountains, in the northern part of the Bohemian massif, the Lower Cambrian is represented by 1,500 feet of limestone consisting mainly of the shells of an early reef-building, sponge-like organism, *Archaeocyathus*.

Lower Paleozoic rocks are few in southern Europe. This region was the center of intense depositional and diastrophic activity during the Mesozoic and Cenozoic.

Cambrian formations are known in southeastern France, in Spain, and in Sardinia. The predominant rock is shale, but immediately above the Precambrian there is usually a sandstone formation, and archaeocyathid limestones are not uncommon. There are also Cambrian formations in the Atlas Mountains of Morocco and southward into the Sahara Desert. The sequences are thin sandstone beds unconformably overlying the Precambrian rocks of the Ethiopian shield. To the north the basal sandstones are overlain by much thicker archaeocyathid limestones and shale. The thickest sequences, consisting largely of shales, are in northern Morocco. Thus these

North African Cambrian strata include a well-developed sequence of shelf sediments deposited on and along the margin of the old shield and thickening to geosynclinal proportions in the north (Fig. 6-2).

Cambrian rocks have not been recognized in the parts of Europe adjoining the central and eastern Mediterranean Sea. Likewise much of Russia west of the Ural Mountains has no Cambrian deposits and presumably was not invaded by Cambrian seas.

The Cambrian record permits certain generalizations on the tectonic framework of Europe at that time. The most conspicuous feature is the existence of a geosyncline extending northeast from Wales through Norway. The Swedish and Estonian deposits are clearly shelf or platform facies lying on the margins of the shield and merging westward with the geosyncline. This Caledonian geosyncline, as it is called, received sediment from the late Precambrian through the Silurian. Geosynclinal conditions existed elsewhere in Europe, but the data are too sparse for generalization. The Moroccan Cambrian deposits clearly indicate that a geosyncline in North Africa transgressed onto the Ethiopian shield, where shelf facies accumulated. The generalized land-sea relations of the Cambrian, Ordovician, and Silurian are shown in Figure 6-3. For North America, Cambrian facies, though incomplete, permit the

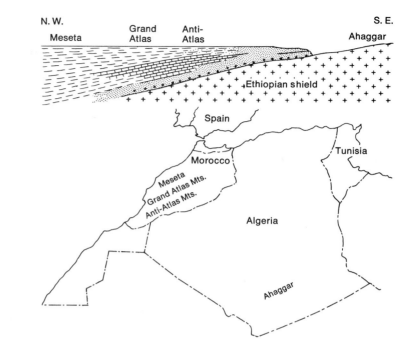

FIGURE 6-2

Schematic diagram of Cambrian facies in Morocco and Algeria. (After M. Gignoux, 1955.)

recognition of the broad features of the continental tectonic framework, but the Ordovician furnishes a more comprehensive picture. The same situation is found in Europe.

The Ordovician Period

Wales, at the southwestern end of the Caledonian geosyncline, is the type area of the Ordovician system. Whereas the Cambrian deposits of Wales are fairly homogeneous, the Ordovician varies greatly in both lithology and thickness. The initial Ordovician deposits are generally sandstones and conglomerates, resting unconformably on various divisions of the Cambrian, showing that the Cambrian strata had been uplifted and eroded before the Ordovician. As in North America, Ordovician sedimentary rocks occur in two distinct facies: (1) the graptolitic facies, mainly of shale and graywacke, with numerous graptolites; (2) the shelly facies, of sandstone and limestone, bearing

a rich benthonic fauna with calcareous exoskeletons. Lava flows, agglomerates,* and ash beds are very extensive (Figs. 6-4 and 6-5). Unconformities, diastems, and great lateral changes in thickness also characterize the Ordovician in this belt. The sedimentary record indicates a geosyncline in an active tectonic state and including, at various times and places, volcanic islands and tectonic lands.

Similar Ordovician rocks in Scotland most strikingly illustrate the complex facies in this geosyncline. Near the western coastal town of Girvan the Upper Ordovician consists of 4,000 feet of conglomerate, graywacke, shale, and limestone; but at Moffat, 55 miles to the east, this segment of time is represented by only 100 feet of graptolitic shales and mudstones (Figs. 6-6 and 6-7). The Girvan sequence is interpreted as a littoral and shallow-water shelly facies, the

---

* A pyroclastic rock containing a predominance of rounded or subangular fragments greater than 32 mm in diameter.

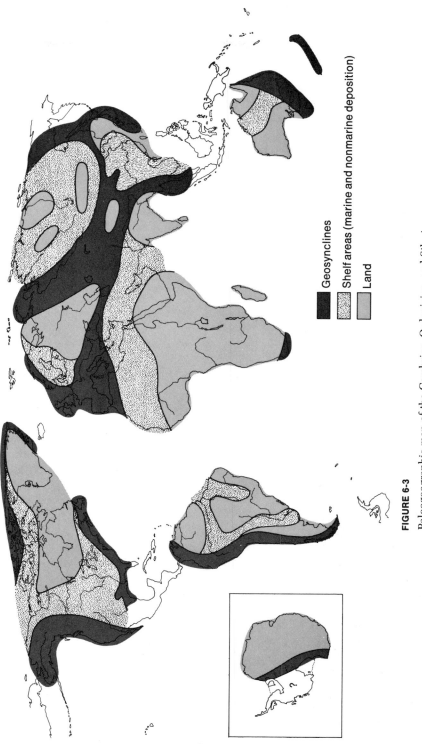

■ Geosynclines

▨ Shelf areas (marine and nonmarine deposition)

▧ Land

FIGURE 6-3

Paleogeographic map of the Cambrian, Ordovician, and Silurian. (After many sources but especially Brinkmann, 1954.)

Moffat sequence—an unusually good example of a condensed section (see p. 174)—as a deep-water facies whose sediments accumulated extremely slowly.

In western Norway the Ordovician is represented by very similar rock types; in the Trondheim district, for example, more than 20,000 feet of fine-to-coarse clastics and volcanic rocks crop out. This Welsh-Norwegian belt is clearly a eugeosyncline. As we proceed eastward, the facies change considerably. Around Oslo the Ordovician is represented by about 1,000 feet of very fossiliferous shales and limestones. The same facies, but thinner, appears in outcrops across southern Sweden and on into Estonia (Fig. 6-8).

The Ordovician in the massifs of central Europe presents the same problems as the Cambrian. In the northern massifs, shale and sandstone, rich in graptolites and conformable on the Cambrian, predominate. In the massifs across central Europe, from Brittany to the Bohemian massif, the Ordovician is generally unconformable on the Cambrian. That of the Bohemian massif is a richly fossiliferous sequence of graptolitic shales and trilobitic sandy beds.

In the Mediterranean region Ordovician rocks are not common, but fossiliferous shales and sandstones are known in southeastern France and in the Carnic Alps. In Sardinia graptolitic shale rests with angular unconformity on the folded Cambrian. In

FIGURE 6-4

Distribution of Ordovician volcanic rocks in the Caledonian geosyncline. Dots indicate pillow lavas. (After O. Holtedahl, *Geol. Soc. London, Quart. J.*, 1952.)

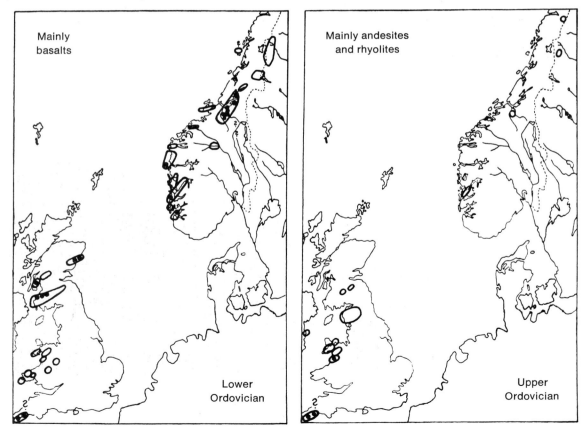

Mainly basalts

Lower Ordovician

Mainly andesites and rhyolites

Upper Ordovician

western North Africa the Ordovician is represented by thick shale in the northern zones of the Atlas Mountains, but southward the facies change to sandstone and thin out near the Ethiopian shield. Orodovician and Silurian glacial deposits crop out in a broad belt from southern Morocco to Lake Tchad. As some of these deposits rest on striated pavements, they are probably products of continental glaciation.

East of the Baltic shield and covering most of European Russia (that is, to the Urals) is the vast Russian platform, which has acted as a stable shelf during much of Phanerozoic history. Throughout the early Paleozoic most of this platform was land, and along its eastern margin, now occupied by the Ural Mountains, there was a very

FIGURE 6-5

Pillow structures in lava flows of Ordovician age exposed in Ayrshire, southwestern Scotland. (Courtesy of Her Majesty's Geol. Surv., British Crown copyright.)

FIGURE 6-6

Cliff of Ordovician Benan Conglomerate near Girvan, Scotland (see Fig. 6-7). (Courtesy of Her Majesty's Geol. Surv., British Crown copyright.)

active geosyncline, the Uralian. The broad outcrop patterns of the late Paleozoic and Mesozoic rocks around Moscow and eastward reflect the platform character of that region. The Ural Mountain belt has narrow, discontinuous, complex outcrops with much igneous material, reflecting its orogenic history. East of the Urals young Cenozoic rocks bury the ancient formations. This geosyncline is unusual because it lies within a continent and separates two great shields, the Baltic to the west and the Angaran to the east. The oldest definitely dated rocks are of Ordovician age. Underlying these are several thousand feet of unfossiliferous strata that may be at least partly of Cambrian age. Beginning in the Ordovician, which produced fossiliferous strata, and continuing through the mid-Carboniferous, the Uralian geosyncline consisted of a western miogeosyncline and an eastern eugeosyncline. During the Ordovician the miogeosyncline accumulated about 2,000 feet of shale, limestone, and sandstone, which gradually thin westward until they pinch out and end on the Russian platform. The eugeosyncline accumulated sandstones, conglomerates, limestones, and volcanics.

The extensive volcanism of Ordovician time and the advent of the Uralian geosyncline are evidence of the great tectonic instability of the belts surrounding the Baltic shield. Silurian history, as we shall now see, follows closely the pattern established in the Ordovician.

The Silurian Period

The tectonic and depositional facies of the Ordovician persisted into Silurian time with one important difference: volcanism ceased except in small areas. In the center of the Caledonian geosyncline the Ordovician and Silurian rocks are conformable and form a continuous series. In the marginal areas, however, and in local belts within the geosyncline, incomplete sections, overlapping,

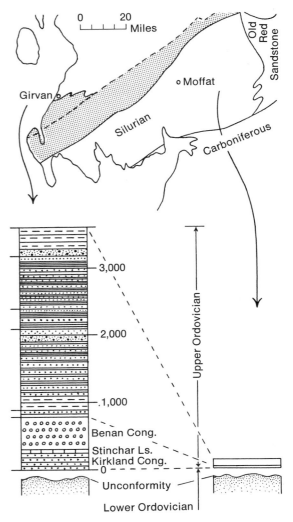

FIGURE 6-7

Stratigraphic sections of Upper Ordovician rocks in the southern uplands of Scotland. (After Wells and Kirkaldy, *Outline of Historical Geology*, Murby, London, 1956.)

and unconformities are common. The type area of the Silurian is the Welsh borderland.

The graptolitic and shelly facies of the Ordovician persist into the lower Silurian, but not into the Middle Silurian. The geosynclines are characterized by generally thick shale, mudstone, and some graywacke and conglomerate strata. Limestone, common in the shelly facies, includes fine reef facies. Local unconformities and striking changes of facies show that the Caledonian geosyncline was very active tectonically through much of the Silurian.

The Silurian has been recognized in some of the metamorphic complexes of the Scandinavian Mountains, and presumably a thick geosyncline, somewhat like that in Wales, developed there. In the Oslo region the Silurian record is much thinner and consists chiefly of shale. Farther east, along the Gulf of Finland, the Silurian is represented predominantly by carbonates, with splendid development of reefs.

Deep bore holes in northern Poland penetrate 6,000 feet of Silurian graptolitic mudstone, probably recording an elongated basin whose bottom environment was deficient in oxygen. In the coastal waters on the northern margin of this basin, shelly faunas flourished on a well-ventilated bottom. The rich coral reefs of Gotland, in the Baltic Sea, are probably part of this coastal facies.

Silurian formations in the various massifs of Europe are principally graptolitic shale, which, in many places, grades upward into carbonate facies. The depositional pattern

FIGURE 6-8

Upper Cambrian and Lower Ordovician limestone formations, in Vestergotland, southwestern Sweden. (Photograph by H. B. Whittington.)

in the Uralian geosyncline is similar to that of the Ordovician: the miogeosyncline has much limestone, and the eugeosyncline was volcanically active.

The Caledonian Orogeny

The early Paleozoic depositional phase—during the Cambrian, Ordovician, and Silurian Periods—culminated in a major orogeny that affected the main geosynclines. The orogeny is recorded by structural relations such as those shown in Figure 6-9. The Caledonian geosyncline was transformed into a rigid part of the continental framework; folding also affected the Uralian geosyncline and a belt across northern Europe.

Orogenic activity had been more or less continuous at one place or another in the Caledonian geosyncline from Late Cambrian to Middle Devonian time (Fig. 6-10). All of these events make up the Caledonian orogeny. In the final phases the increased tempo of orogenic and plutonic activity folded and thrust the Caledonian geosyncline into a vast mountain system, ending geosynclinal deposition.

The dominant structural trends established by the Caledonian orogeny are northeast-southwest; similar folded belts, in a different direction, are recognized in the Ardennes and Sudeten massifs. Spitsbergen and Bear Island contain extensive outcrops of early Paleozoic sediments that were severely folded, faulted, intruded, and metamorphosed in the Caledonian orogeny. This rock series, the Hecla Hoek Formation, may be partly Precambrian in age. A general lack of fossils makes correlation difficult. The Barents Sea shelf is thought to be a submerged extension of the Baltic shield. The boundary between Spitsbergen, including the Barents shelf, and the deep Norwegian sea is a direct continuation of the boundary off the Norwegian coast (see Fig. 6-11). The trend of the Caledonian mountain chain indicates the presence of a large, continuous, and stable region to the east.

The Caledonian geosyncline was little affected by later depositional or orogenic phases, but the belt across northern Europe became the site of active geosynclinal deposition in the later Paleozoic. The effect of Silurian orogenic activity in the Uralian geosyncline was minor compared with that in the Caledonian.

The Devonian Period

The Caledonian orogeny created a vast land area that included the Caledonian geosyncline and the Baltic shield (Fig. 6-12). It was upon this land area that the famous Old Red Sandstone was deposited. The Old Red Sandstone consists chiefly of continental deposits of sandstone, shale, conglomerate, marl, and, locally, volcanics, laid down in numerous, more or less isolated basins within the Caledonian Mountains. The highlands of this ancient mountain chain were the active sources of the sediments that accumulated in the adjoining basins. In Scotland the accumulated thickness of Old Red is 40,000 feet, but much of this volcanic. The most important faunas of the Old Red are fossil fish, which are discussed in the next chapter. The faunas and sediments strongly suggest that the Old Red continent had a semi-arid and tropical climate.

The Old Red Sandstone is best known in Ireland, Wales, England, and Scotland. It is also found in patches in Norway and widely along the Finnish-Russian border. South of the Old Red continent lay Devonian seas extending to the Ethiopian shield. The Russian platform received a blanket of terrestrial sediments, of the Old Red facies, derived from the Baltic shield. Only in Late Devonian time did the sea spread from the Uralian and the central European geosynclinal regions to inundate the Russian platform (Fig. 6-13).

FIGURE 6-9

Unconformable contact of upper Old Red Sandstone beds on vertical Silurian rocks
at Siccar Point, Berwickshire, Scotland. It was from a study of this locality that
James Hutton, in 1788, first realized the meaning of unconformity.
(Courtesy of Her Majesty's Geol. Surv., British Crown copyright.)

The Devonian System was established by Murchison's and Sedgwick's studies in Devonshire, England; but the intense structural complications of that region have made the Devonian of the Rhine-Meuse region of Germany and Belgium a preferable standard. The northern shore of the Early Devonian sea extended across southern England and through the northern massifs of Europe. Along this belt transgressive and regressive facies produced an intertonguing of marine and continental Old Red facies. The Devonian in this belt generally rests unconformably on older Paleozoic systems. The marine sediments consist mostly of shale and sandstone, locally as thick as 15,000 feet.

The Devonian of southern Europe and the Mediterranean region is generally conformable on the underlying Silurian. The facies here, in contrast to the clastics of the north, is largely calcareous, with thick limestone and shale. Most outcrops are isolated, small, and commonly metamorphosed, but are nevertheless more widespread than those of any of the earlier Paleozoic systems. In the Sahara and Libyan deserts, Devonian sandstones of near-shore facies accumulated on the northern edge of the Ethiopian shield.

Three very distinct facies, each with its own unique fauna, are recognized between the Femnoscandian and Ethiopian shields. The Rhenish facies, a clastic facies containing little limestone; the Hercynian facies, a limestone facies with little or no clastic rocks except shales; and an intermediate facies representing a mixture of the first two are distributed as shown in Figure 6-12. The Rhenish facies is generally close to land and the Hercynian facies offshore. The map in Figure 6-12 shows late Lower Devonian, but comparable relationships prevailed through most of the Devonian.

The Devonian facies in the Uralian geosyncline are much like those of the earlier

FIGURE 6-10

The distribution of Caledonian folding in the British Isles. Broken lines indicate uncertainty in either time or in area. Each of the five groups of folding covers movements during an appreciable geological time; all the folding in any one group is not necessarily closely associated. (After W. S. McKerrow, *Proc. Nat. Acad. Sci.*, 48, 1962.)

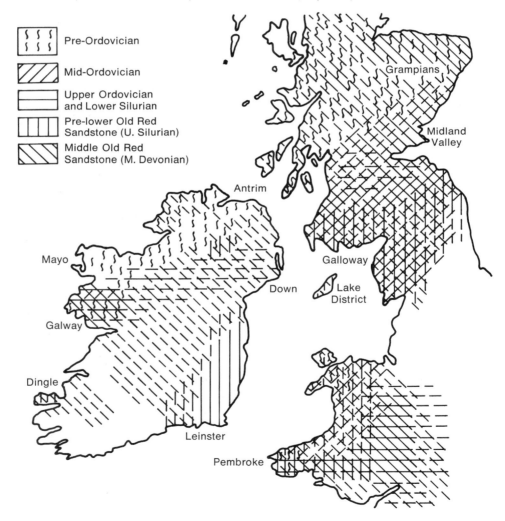

Pre-Ordovician

Mid-Ordovician

Upper Ordovician and Lower Silurian

Pre-lower Old Red Sandstone (U. Silurian)

Middle Old Red Sandstone (M. Devonian)

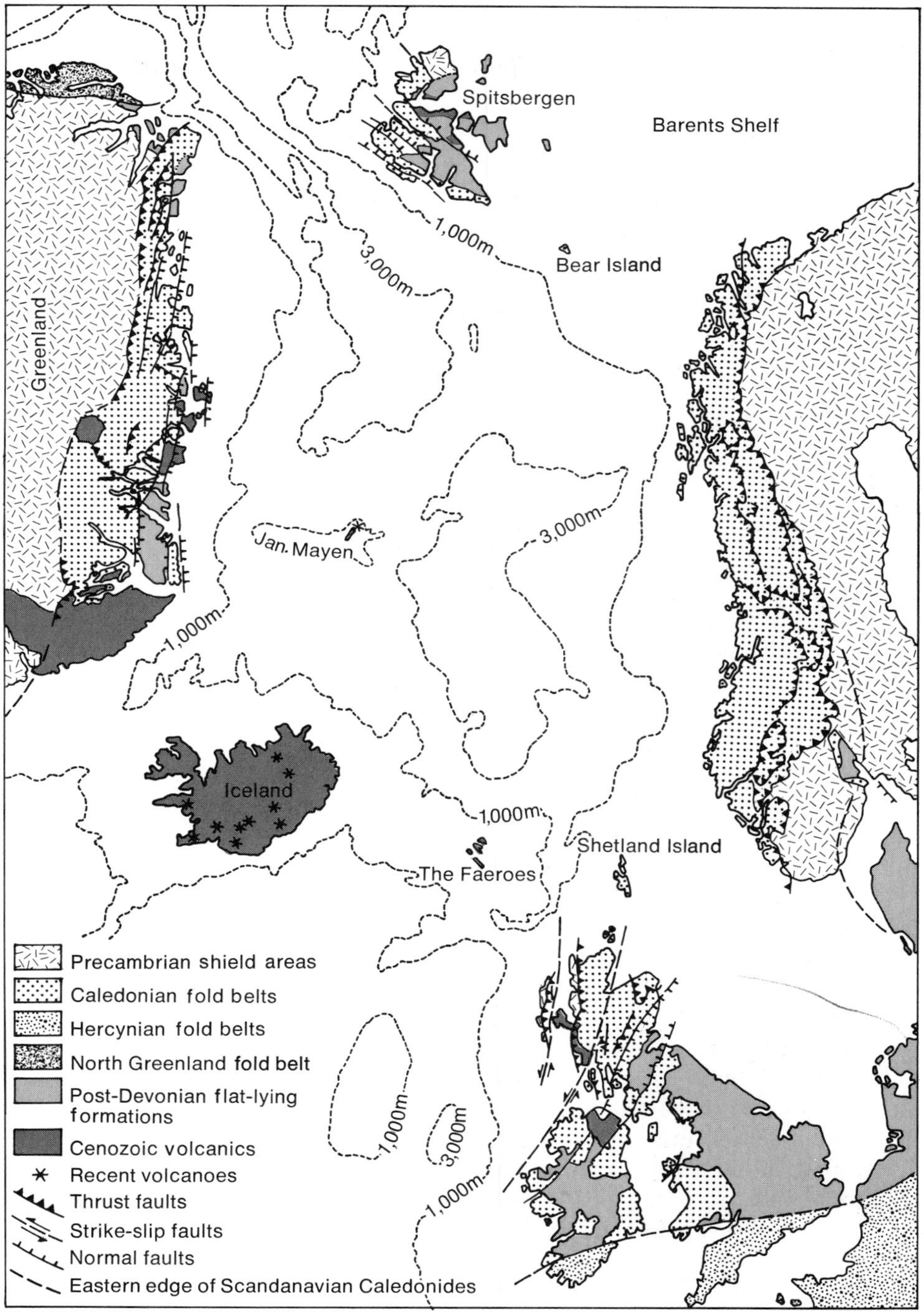

Spitsbergen
Barents Shelf
1,000m
3,000m
Bear Island
Greenland
Jan. Mayen
3,000m
1,000m
Iceland
1,000m
Shetland Island
The Faeroes
1,000m
3,000m
1,000m
Precambrian shield areas
Caledonian fold belts
Hercynian fold belts
North Greenland fold belt
Post-Devonian flat-lying formations
Cenozoic volcanics
Recent volcanoes
Thrust faults
Strike-slip faults
Normal faults
Eastern edge of Scandanavian Caledonides

FIGURE 6-11

Tectonic map of north Atlantic region.
(After J. Haller, *Med. om Grønland*, **171.5**, 1969.)

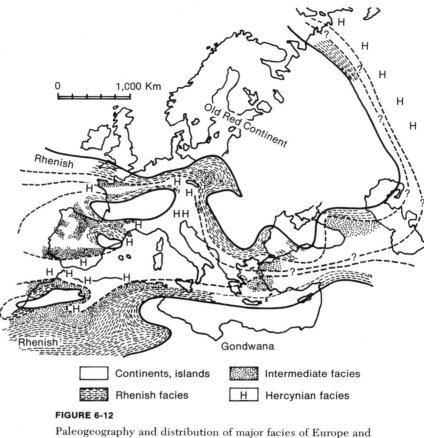

Continents, islands   **Intermediate facies**
Rhenish facies   **H Hercynian facies**

FIGURE 6-12

Paleogeography and distribution of major facies of Europe and
North Africa during the late Early Devonian. (After H. K.
Erben, *Proc. Ussher Soc.*, 1964.)

Paleozoic systems, but there was greater
tectonic activity in the eugeosyncline,
where the strata consist of 10,000–20,000
feet of sandstone, shale, radiolarian chert,
lava, tuff, and agglomerate. In the Late
Devonian, when orogenic activity was
greatest, there was extensive emergence in
the miogeosyncline, and numerous separate
basins developed. The miogeosyncline has
mainly carbonates.

Massive Lower and Middle Devonian
formations are restricted to the geosyncline.
Westward, on the Russian platform, deep
wells (in Moscow) have encountered nearly
2,000 feet of Old Red Sandstone, probably
of Middle Devonian age, resting on crystalline rocks. The Late Devonian sea, spreading far west of the Uralian miogeosyncline,
inundated the Russian platform until it
reached the Baltic states and connected
with the sea of central Europe. The facies
relations of the Upper Devonian rocks from
the Baltic shield to the Uralian geosyncline
are diagrammatically illustrated in Figure
6-13.

The Carboniferous Period

The tectonic and depositional events of the
Devonian of Europe were, in a sense, transitional between the Caledonian orogeny and

a late Paleozoic one, the Hercynian (Fig. 6-14). The Carboniferous was the great coal-forming period. There is a difference in systematic nomenclature between North America and the rest of the world. In Europe the Carboniferous embraces all the rocks between the Devonian and the Permian. The principal subdivisions are the Lower Carboniferous, mainly marine, and the Upper Carboniferous, which includes the coal beds and is largely nonmarine. In North America we recognize a Mississippian System, roughly equivalent to the Lower Carboniferous, and a Pennsylvanian System, roughly equivalent to the Upper Carboniferous.

The Early Carboniferous sea, with some significant exceptions, covered the same area as the Late Devonian sea, inundating much of the Old Red area of the British Isles and the vast geosyncline that covered Europe, except for numerous islands south of the relic of the Old Red continent.

The many facies deposited testify to extremely varied conditions. Central England and Ireland were sites of mainly carbonate deposition, much developed in reef facies. Toward the highlands of southern Scotland, coarse clastics and even minor coal beds crop out, and various volcanic rocks are common. Southern Ireland, Wales, and England have geosynclinal shale-and-sand facies. A cross section of these Lower Carboniferous facies from southwestern England to Scotland is shown in Figure 6-15. The shale-sand facies, derived from uplifts in the geosyncline, is widespread across Europe and is adjoined on the north, along the fringe of the Old Red continent, by carbonate shelf facies (Fig. 6-16). Marine shales and carbonates in southern Europe indicate an open sea.

The minor orogenic movements that occurred during the Early Carboniferous are reflected in the complex facies developed at that time. A major movement, which occurred at the close of the epoch, folded and consolidated much of Europe south of the relic Old Red continent (Fig. 6-17). This was the first phase in the formation of the ancient Hercynian Mountains of Europe. It was around these Hercynian Mountains that the Upper Carboniferous coal basins of Europe were formed. Between the northern edge of the Hercynian Mountains and the land mass covering Wales, southern England, and Belgium (Wales-Brabant Island), a deeply subsiding basin formed, and there the principal coal fields of Europe are now found (Fig. 6-18). A similar basin extended

FIGURE 6-13

Schematic cross section of Devonian facies from the Baltic shield across the Russian platform to the Uralian geosyncline. (After M. Gignoux, 1955.)

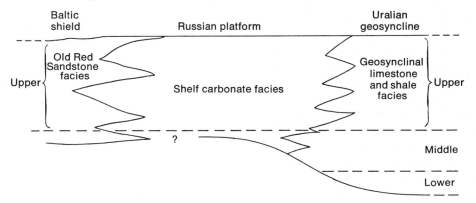

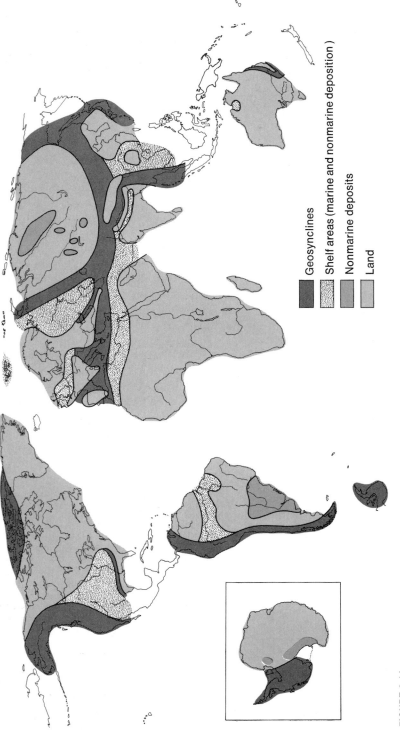

FIGURE 6-14

Paleogeographic map of the Upper Carboniferous.
(After many sources but especially Brinkmann, 1954.)

Geosynclines

Shelf areas (marine and nonmarine deposition)

Nonmarine deposits

Land

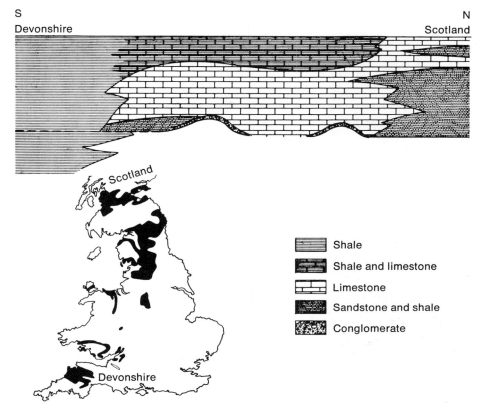

FIGURE 6-15

Schematic cross section and outcrop map of Lower Carboniferous facies in England, Wales, and Scotland. (After M. Gignoux, 1955.)

across Ireland and central England, between the highlands of Scotland and Wales-Brabant Island. Only the lower stages of the Upper Carboniferous still show occasional marine invasions between clastic terrestrial deposits. The later ages produced only nonmarine delta and swamp deposits of shale, sandstone, and, frequently, coal. Within the Hercynian Mountains numerous isolated basins accumulated nonmarine clastic facies that included some coal beds.

Tectonic instability marked all of Late Carboniferous time throughout Europe: minor pulsations and orogenies occurred in various places, and toward the end of the epoch they became more intense. They are all part of the Hercynian orogeny. The later Carboniferous movements strongly folded and faulted most of the remaining depositional basins and also affected some of the areas folded in earlier orogenies. The trend lines of the Hercynian folded belts within the massifs are indicated in Figure 6-19. The massifs are merely relics of the Hercynian Mountains, which today form windows in the cover of Mesozoic and Cenozoic formations. We are approaching the last stage in the formation of Paleozoic Europe.

The minor orogenic phases in the Uralian eugeosyncline produced emergent areas where Lower Carboniferous coal beds accumulated; these were later overlain by marine carbonate facies. Volcanic activity

184

FIGURE 6-16

Escarpment of Lower Carboniferous limestone formations near Llangollen, northern Wales. (Courtesy of Her Majesty's Geol. Surv., British Crown copyright.)

was still evident in the Early Carboniferous. By mid-Carboniferous time the eugeosyncline was uplifted, and the final orogenic phase had begun. The next marine deposits to be found in this area are Cretaceous. The Uralian miogeosyncline, except for thin coal beds in the lowest Carboniferous, is the site of mainly carbonate facies. Across the Russian platform extensive carbonate facies conformable with the underlying Devonian are found. The region around Moscow is especially important for the study of these strata.

The northern coast of Africa, from Morocco to Libya, was widely inundated by Carboniferous seas. The pattern and distribution of facies, because of a series of Hercynian orogenies, are highly complex. The Hercynian orogenic phases created massifs in parts of the North African geosyncline—the Moroccan Meseta, the Hauts Plateaux, and the Kabylie massif—that remained rigid during the Mesozoic and Cenozoic (Fig. 13-9).

The Permian Period

The culmination of Carboniferous depositional and tectonic history found the geography of Europe profoundly changed (Fig. 6-20). The geosynclinal seas, by then, were confined to the Mediterranean region and were part of a great seaway, the Tethyan geosyncline, that continued eastward through the Middle East to the Himalayas. The Tethys was the main geosyncline of Mesozoic Eurasia. Europe became a desert and remained so for much of Permian and Triassic time.

The earliest Permian rocks are in the regions of the latest Carboniferous deposits; they are conformable and in some places even transitional. The later deposits, however, are widely unconformable across folded Carboniferous rocks. The Lower Permian consists of continental deposits, mainly red, laid down in broad basins by temporary streams or by wind. Volcanoes were active in some places (for example, in Scotland and Devonshire). In Late Permian time an arm of the Atlantic extended across northern Europe south of Norway and Sweden—in Germany and Poland—and also entered east-central England. The initial deposits of this sea—thin clastic formations—were followed by carbonates (mainly dolomites) with peculiar faunas, rich in individuals but poor in number of species and genera. Similar faunas exist today in the Caspian Sea, where the water is super-saline, and in the Baltic, where it is brackish. This inland sea, the Zechstein Sea, opened to the Atlantic only between England and Norway. Deposition of the carbonate facies was followed by that of several hundred feet of evaporites, including the famous German potassium salts. The overlying Triassic deposits are chiefly evaporites, indicating that arid conditions persisted on the continent.

In the Uralian miogeosyncline carbonate deposition continued from the Late Carboniferous into the Permian with no significant breaks. Along its eastern edge coarse clastics tongue out into the carbonate facies. The eugeosyncline was emergent; as it underwent orogeny in the mid-Carboniferous,

FIGURE 6-17

Zigzag folding of Lower Carboniferous sandstones and shales, Millook Haven, southwestern England (Cornwall). (Courtesy of Her Majesty's Geol. Surv., British Crown copyright.)

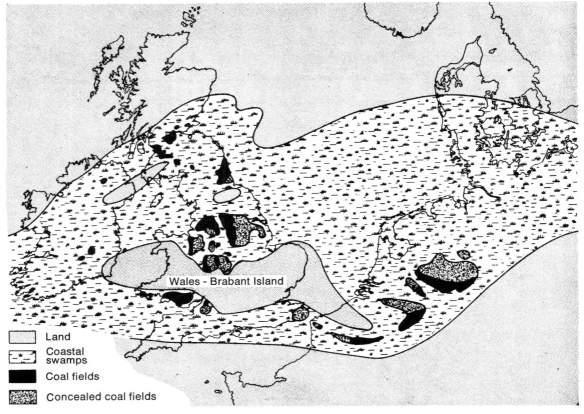

Wales - Brabant Island

Land

Coastal swamps

Coal fields

Concealed coal fields

FIGURE 6-18

Paleogeography of Late Carboniferous northern Europe, showing the location of the principal coal fields. (After Wills, 1951.)

some of its coarse clastics were transported westward into the adjoining miogeosyncline. The carbonate facies of the geosyncline continued westward over the Russian platform to the Moscow region. In Late Permian time the increased influx of clastics from the east, reflecting continued orogeny, displaced the carbonate facies in a gradual change to terrestrial deposition of redbeds that contain thick beds of salt, gypsum, and anhydrite (Fig. 6-21). These are approximately equivalent in age to similar facies in other parts of Europe and in England.

With this Permian orogeny the Uralian geosyncline came to its end. The emergence of this mobile belt united Europe and Asia to form the great Eurasian continent.

South of the Hercynian Mountains, geosynclinal conditions persisted in the great Tethyan Sea. In northern Italy mixed marine and nonmarine sequences, commonly including volcanics, were deposited on the margins of the Tethys. The westernmost deposits with marine Tethyan faunas are in Sicily and Tunisia; farther west in Algeria, Morocco, and the Pyrenees, the Permian is represented by terrestrial redbeds.

6-2 ASIA

Although less is known about the Paleozoic history of Asia than about that of any other continent except Antarctica, enough is known to put the broad depositional and tectonic patterns into a satisfactory framework.

Figure 3-19 (p. 83) shows that Asia is made up of two major shields, the Indian in the south and the Angaran in the north. Between the two is a vast complex of folded Phanerozoic strata and rather large blocks of Precambrian rocks (some not shown on the map). The folded rocks immediately north of the Indian shield are Cenozoic mountain chains; those immediately next to the Angaran shield are Paleozoic mountain chains, except along the east and northeast, where they are Mesozoic. The folded chains of extreme eastern Siberia and Japan are largely Cenozoic. Some of these folded belts have undergone more than one orogeny.

Stratigraphic data are much too incomplete for a general synthesis of the whole region period by period; therefore, the major geological areas are treated separately. The first of these comprises the Angaran shield and the immediately surrounding geosynclines.

The Angaran Shield

This great shield differs from most others in the world in having a very extensive cover of lower Paleozoic formations. Precambrian rocks crop out, in fact, in three main areas separated from one another by vast stretches of nearly horizontal early Paleozoic rocks. Figure 4-17 (p. 108) summarizes the main geological features of the Angaran shield. Surrounding the shield itself, with its flat early Paleozoic rocks, are folded geosynclinal belts of Paleozoic and Mesozoic age.

FIGURE 6-19

Paleozoic massifs of Europe, showing the trend lines of Hercynian folding. The parallel straight lines indicate the area of later, Alpine folding; the blank area between the massifs is occupied by virtually flat Mesozoic and Cenozoic formations. (A) Armorican. (B) Ardennes-Rhine. (C) Harz. (D) Bohemian. (E) Vosges. (F) Black Forest. (G) Massif Central. (H) Spanish Meseta.

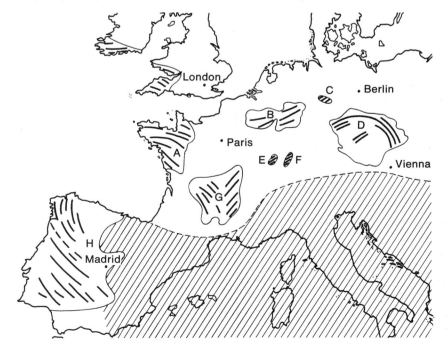

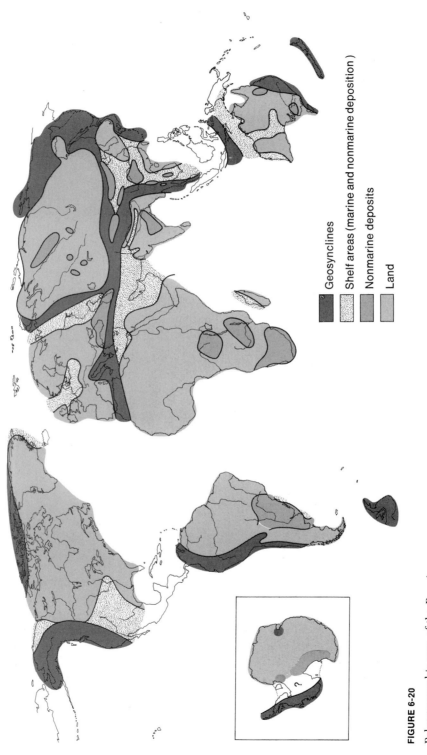

Geosynclines

Shelf areas (marine and nonmarine deposition)

Nonmarine deposits

Land

FIGURE 6-20

Paleogeographic map of the Permian.
(After many sources but especially Brinkmann, 1954.)

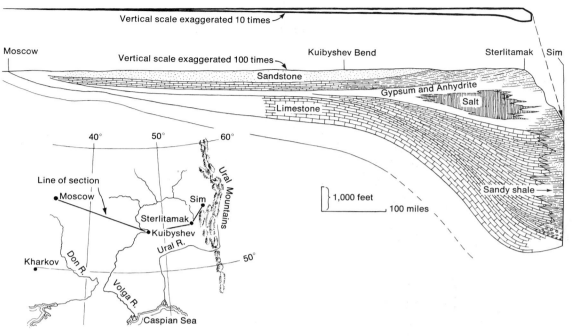

FIGURE 6-21

Diagrammatic cross section showing facies relations of Permian formations from Moscow across the Russian platform to the Ural Mountains. Note the significance of the vertical exaggeration. (After C. O. Dunbar.)

The Paleozoic rocks on the shield—of Cambrian, Ordovician, and Silurian age—consist mainly of shales, limestones, and dolomites. Near the borders of the shield, most of these formations thicken appreciably, but on the shield itself each measures only a few hundred feet. On the shield there are, however, several basin-like areas that received thousands of feet of early Paleozoic sediments. In the Paleozoic folded belt on the southwestern edge of the shield (Fig. 4-17) the early Paleozoic rocks are metamorphosed and contain much extrusive and intrusive igneous rock. Caledonian fiolding occurred in the area to the southwest of the shield, in the geosyncline on the northwest of the shield, and along the eastern margin of the shield. In this eastern area later folding has almost entirely masked the early phase. The folded belt to the southwest of the shield became a

unified, welded mass during this orogeny and has remained largely unaltered since that time.

With the termination of the Caledonian orogeny, Devonian seas occupied the geosynclines but did not encroach upon the shield, which henceforth was emergent and received only terrestrial basin deposits. Marine conditions prevailed in the geosynclines well into the Early Carboniferous; but then began a long series of orogenic phases (Hercynian) with much terrestrial and coal deposition. The whole eastern and southeastern margins of the shield were folded at this time (and much of this region was folded again in the Mesozoic). By the end of the Paleozoic all of Europe and Asia down to the Tethyan geosyncline, with the exception of easternmost Siberia, was a rigid unit. No further geosynclinal episodes have altered this huge land mass, for the shallow marginal seas and terrestrial basins of Mesozoic and Cenozoic time added only superficial deposits.

China and Mongolia

China and Mongolia constitute a very ancient continental mass whose geosynclinal history is largely Paleozoic. The basic structural elements, illustrated in Figure 6-22, are geosynclines, platform areas, and massifs. The eastern and western parts of China belong to two different structural systems. Western China is the more mobile and is characterized by linear geosynclinal regions separated by stable massifs. These geosynclines were folded and consolidated during the Paleozoic. Eastern China, on the other hand, is a vast platform area covered by marine Paleozoic and nonmarine Mesozoic and Cenozoic deposits. Northern China and Mongolia was geosynclinal in the Paleozoic and now makes up the Tien Shan-Mongolian fold belt.

The platform areas of China differ from that of the central United States in being more mobile, as is evidenced by the generally thicker sequence of sedimentary rocks. The Chinese platforms underwent true orogenies with folding and faulting and much igneous activity, generally during more than one cycle. Orogenies are almost completely unknown in most platforms. Because of the above differences Chinese geologists have introduced the term "paraplatform" for these regions.

Most of the Chinese platforms consist of late Precambrian through Paleozoic sequences several thousand feet thick resting on highly deformed Precambrian strata. The South China platform differs in that the sedimentary mantle is Devonian through Permian in age, resting on (1) deformed late Precambrian-Silurian strata and (2) deformed earlier Precambrian formations. The scant available data suggest that Tibet is a platform area like that of South China, with a mantle of Carboniferous-Mesozoic sedimentary rocks resting on deformed earlier Paleozoic and Precambrian.

The Upper Precambrian of China—the Sinian System—is quartzite and limestone, in places as much as 30,000 feet thick. It is largely confined to the major platforms of eastern China. Cambrian deposits were laid down in virtually the same regions, in many places conformable on the Upper Precambrian (Fig. 6-23, A). Thus a broad Cambrian seaway, in which limestone and shale were deposited, extended from Korea to South China. The paleontological and stratigraphic record indicates a series of transgressions and regressions of the Cambrian seas. The Tien Shan-Mongolian geosyncline began to develop in the Cambrian and in Mongolia contains much volcanic material.

Lower and Middle Ordovician deposits were laid down in eastern China in the same seaway. In the north these Ordovician formations are mainly limestone and dolomite; in the south there is some graptolitic shale. In the Kun-lun and Nan-shan geosynclinal regions the thick Ordovician includes extensive volcanics. These geosynclinal regions became well-marked troughs during the Cambrian. The paleogeographic map of Figure 6-23, A is for earliest Cambrian time, prior to the origin of these geosynclines. The Ordovician is represented in the Tien Shan-Mongolian geosyncline by thick graywacke formations. In the Late Ordovician the seas occupying much of north China and Korea retreated southward (Fig. 6-23, B), leaving emergent masses. In these areas subsidence did not begin again until the mid-Carboniferous, when coal measures and interbedded marine sediments were laid down. These are followed by some marine Permian and continental Triassic formations.

The paleogeography and facies of the Silurian resemble those of the Upper Ordovician (Fig. 6-23, C). The northern platform areas were emergent, and seas occupied much of the southern platform and the geosyncline of western China and Mongolia.

The absence of lowest Devonian strata suggests that in Late Silurian time the seas completely regressed from south China. The first Devonian deposits here are red clastic formations of late Early Devonian age, followed in many areas by marine shale and limestone. The South China platform, however, underwent orogeny toward the close of the Silurian. The Devonian deposits here are an interbedded mixture of marine and nonmarine facies (Fig. 6-23, D). Devonian marine sedimentary and volcanic rocks are very thick in the Tien Shan-Mongolian geosyncline.

The Carboniferous and Permian were times of much crustal movement throughout China (Fig. 6-23, E, F). Major folding, commonly accompanied by granitic intrusions, took place in the northeastern platform region, the Tien Shan-Mongolian geosyncline, the Kun-lun geosyncline, the Nan-shan geosyncline, and Tibet. The exceedingly complex sequences of orogenic and depositional events differ in each area, resulting in highly diverse Carboniferous and Permian facies. In general, marine conditions prevailed in the platform areas of south China. In the more marginal regions, however, continental basins were extensive, and in many of them coal deposits were formed. Toward the close of the Paleozoic a marked regression of seas from China and Mongolia took place. Late Permian marine deposits are known only in South China; elsewhere in China and

FIGURE 6-22

Major tectonic units of China. (After Huang Chi-ch'ing, 1963.)

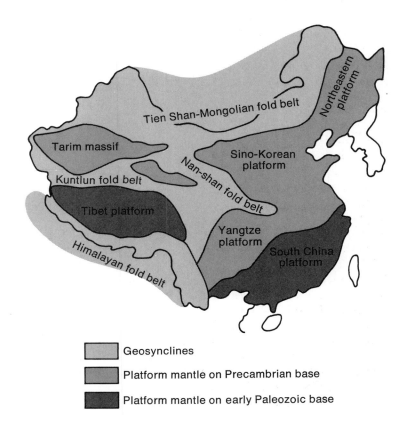

Geosynclines

Platform mantle on Precambrian base

Platform mantle on early Paleozoic base

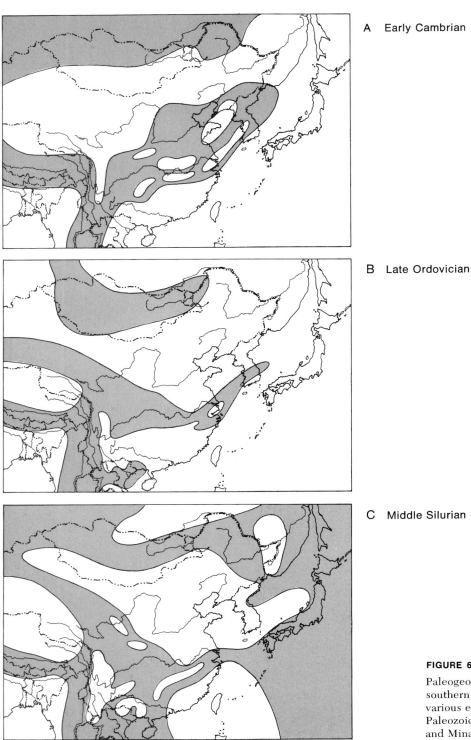

A Early Cambrian

B Late Ordovician

C Middle Silurian

FIGURE 6-23

Paleogeographic maps of
southern Asia during
various epochs of the
Paleozoic. (After Liu, 1959,
and Minato *et al.*, 1965.)

D Late Devonian

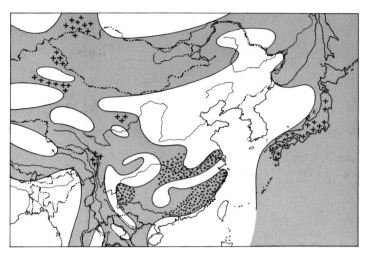

E Late Carboniferous

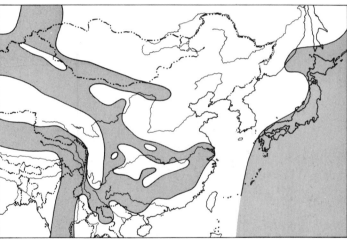

F Middle Permian

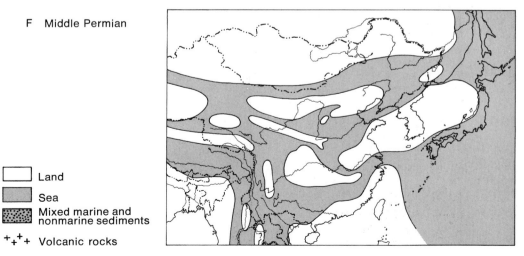

☐ Land

▨ Sea

▨ Mixed marine and
nonmarine sediments

+₊+₊ Volcanic rocks

Mongolia any late Permian deposits are continental. At the close of the Paleozoic the Sino-Korean, Yangtze, and South China platforms became emergent but did not undergo orogeny.

Japan

The geologic history of Japan contrasts markedly with that of the adjoining Asiatic regions. There are apparently no Precambrian or Lower Paleozoic strata. Silurian fossiliferous formations are the oldest rocks definitely dated. The depositional record from Silurian to Recent is predominantly marine, but marine deposition was interrupted by many orogenic cycles, resulting in extremely complicated geology. Volcanism has been important since the mid-Paleozoic.

The main tectonic provinces of Japan are shown in Figure 6-24. A primary fault belt, called the Fossa Magna, crosses central Honshu and separates northeast from southwest Japan. The latter is in turn divided into outer and inner zones by a median dislocation (tectonic) line trending parallel to the island arc.

The inner zone of southwest Japan exposes widespread, intensely folded Paleozoic formations; most Mesozoic formations are restricted to isolated basins. The whole region has received much granitic intrusion and has undergone widespread regional metamorphism. The outer zone is divided by two conspicuous strike-fault lines into three units of quite different geologic character. The innermost is a metamorphic terrane derived mainly from Paleozoic sediments and volcanics. The central is mainly nonmetamorphic Carboniferous and Permian with some Silurian and Mesozoic formations. The outer unit is composed of folded Mesozoic and Lower Cenozoic formations with a few Cenozoic granitic intrusions.

Northeast Japan exposes mostly Cenozoic rocks; the older rocks are in isolated massifs (Fig. 6-24). The Paleozoic rocks generally differ in facies from those of southwest Japan. The structural trends of the massifs of northeast Japan differ from each other and from those of southwest Japan. These massifs are east of the extension of the general northeasterly trend of southwest Japan, suggesting that northeast Japan has moved eastward relative to southwest Japan.

Hokkaido is situated at the junctions of three island arcs. Its southwesterly peninsula is an extension of Honshsu and belongs to the arc of Japan proper. The central part is on the fold trends of Sakhalin Island, and the fold trends of the eastern part are those of the Kuril Islands to the northeast.

The oldest strata in Japan dated by fossils are of Silurian age, known mainly from the outer zone of southwest Japan on Kyushu and Shikoku Islands, but there are also some in the Kitakami massif of northeast Japan. The facies are primarily limestone and shale with minor amounts of volcanic rocks. The maximum exposed thickness of these Silurian strata is about 1,000 feet, but because of faulting no base has been found. Fewer Devonian fossil localities are known than for the Silurian, but the Devonian outcrops are much more extensive. Devonian facies are true geosynclinal limestones and clastic rocks with abundant volcanics (Fig. 6-25, A). Some marine Devonian formations have yielded fragments of land plants. Carboniferous strata lie unconformably upon the Devonian. The lower to mid-Lower Carboniferous is generally characterized by great thicknesses of clastic sediments and volcanics, with little limestone; the upper part consists mainly of carbonate facies plus minor amounts of volcanic rocks. The two divisions of the Carboniferous are separated by an orogeny with extensive folding, faulting, and metamorphism. The Upper Carboniferous record on the inner side of Japan consists predominantly of limestone-shale facies, and there is little evidence of volcanism; the outer side, in contrast, was the site of much volcanism, and the sedimentary record is highly varied (Fig. 6-25, B).

Orogeny affected at least parts of the Japanese geosyncline in the late Carboniferous. In most areas Permian strata lie unconformably on Carboniferous and older formations. Permian formations made up of highly varied facies and having extremely complex structure cover about 80 percent of the total Paleozoic area. Limestone, clastic sediments, and volcanics are extensive (Fig. 6-25, C, D). There are widespread conglomerates, including granite boulders at several horizons. Uplift progressed intermittently on the inner margin of the geosyncline and within it. In the inner belt of Southwest Japan and in Northeast Japan, orogenic pulses began in the mid-Permian and reached a climax in the mid-Triassic. This profoundly altered the basic structural pattern upon which Mesozoic history was developed. Geosynclinal conditions thereafter prevailed only along the Pacific margin of Japan. Although the inner region entered a post-orogenic phase, it remained the site of much basinal, but non-geosynclinal, deposition.

FIGURE 6-24

Tectonic provinces of Japan. (Geol. Soc. Japan, 1960.)

Southern Asia

The vast mountain systems that run from the Middle East into the Himalayas and thence swing south through Burma mark the site of the great Tethyan geosyncline (Fig. 6-26). The mountain systems are Cenozoic, but earlier minor orogenic phases had occurred there, each phase being accompanied by igneous intrusions that mask the geological structure and complicate the picture enormously. Inevitably, the older the geologic period, the more incomplete the record. Knowledge of the early Paleozoic history is sketchy, but more is known of the later Paleozoic.

The eastern half of the Himalayas is geologically little known, but some mapping has been done in the western half. The difficulties of access and of working in the high terrain make progress very slow. Four structural stratigraphic zones make up the Himalayas. The southernmost, the foothills zone, consists entirely of Cenozoic formations. Immediately to the north lies the sub-Himalayan zone, which includes strata ranging in age from the Precambrian to the Cenozoic, generally of peninsular foreland facies, largely unfossiliferous. The structure is highly complicated with many overthrusts and nappes. The next northerly zone, the central Himalayan, has rocks like those of the sub-Himalayan zone but also much gneiss and granite. The most northerly zone, the trans-Himalayan or Tibetan zone, has a complete sequence of marine, fossiliferous, geosynclinal strata ranging in age from the Cambrian to Eocene.

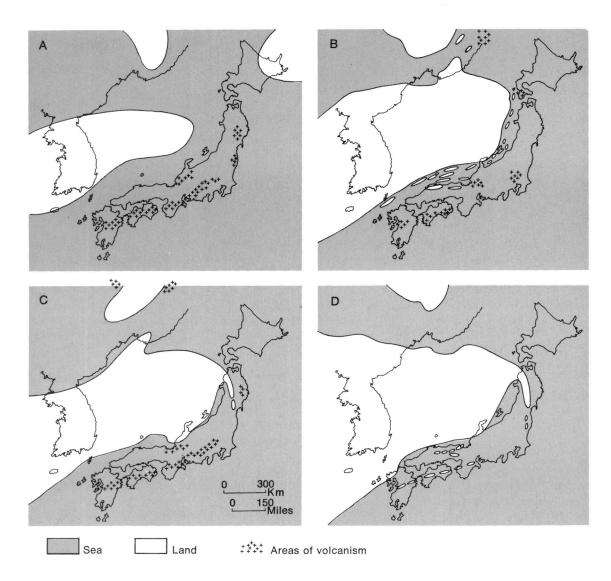

FIGURE 6-25

Paleogeographic map of Japanese Islands and adjoining regions for (A) the Early Devonian to mid-Carboniferous, (B) the mid-part of the Late Carboniferous, (C) the early Permian, and (D) the late Permian. (After Minato *et al.*, 1965.)

The Paleozoic formations of the Himalayas are principally known from a few classic areas, which are indicated in Figure 6-27. It is generally assumed that the sedimentary pattern in the geosyncline between the Indus and Brahmaputra rivers is more or less uniform, and that the few known localities are representative of the whole belt. Marine fossiliferous Paleozoic strata (except for a small patch of Lower Permian

rocks) are absent from peninsular India, and from the southern Himalayas.

Cambrian strata are known from three areas, the Salt Range, Spiti, and Kashmir. Those of the Salt Range consist of about 3,000 feet of fossiliferous sandstones, shales, and dolomites. The lower half of the sequence is made up of red marls, salt, and gypsum, possibly of late Precambrian age. Similar salt deposits in the same stratigraphic position are widespread in Iran, where the evidence suggests a late Precambrian age.

The classic area for Himalayan stratigraphy is the Spiti Valley, where a remarkably complete sequence of Paleozoic and Mesozoic formations crops out (Fig. 6-28). Spiti lies in the trans-Himalayan, or Tibetan, zone; the Cambrian is represented by approximately 5,000 feet of quartzites, slates, and dolomites. In Kashmir the Cambrian is represented by fossiliferous arenaceous shales and thin-bedded limestones.

Ordovician and Silurian formations, mainly of clastic and carbonate facies, are known in Spiti, Kashmir, and Burma but not in the Salt Range or in Hazara, west of Kashmir (Fig. 6-27). These systems are particularly well developed in Burma, where both shelly and graptolitic facies are found. The graptolitic facies is not known in the Himalayas.

The available data suggest that rather uniform depositional conditions prevailed in the Himalayan region from the late Precambrian through the Ordovician (Fig. 6-29, A). Carbonates predominate in Nepal but fine-grained clastic facies predominate to the northwest and south. Orogeny affected the Himalayan region during the early half of the Silurian, producing a linear tectonic ridge that separated the Tethyan geosyncline from a long, linear basin between the ridge and the Indian shield wherein accumulated mainly continental sediments (Fig. 6-29, B). Devonian formations are of many different facies. An unfossiliferous quartzite that overlies the Silurian and underlies fossiliferous Lower Carboniferous in Spiti and Kashmir is generally considered Devonian. Fossiliferous Devonian limestone is known in Chitral on the Afghan border (Fig. 6-27). In Burma the Devonian is the basal part of the Plateau Limestone formation, which ranges up to early Permian and is about 3,000 feet thick.

The late Paleozoic depositional patterns resemble those established in the Devonian following the orogenic episode (Fig. 6-29, C): thick continental sediments in the lower Himalayan basin southwest of the tectonic ridge and marine sediments in the Tethyan realm. Peninsular India, which had been dormant since the Precambrian, became an area of deposition, beginning with tillite, in the early Permian. The peninsular deposits belong to peculiar facies that are found throughout most of the southern hemisphere and are known as the Gondwana formations (see Chapter 11).

From the Spiti region to Nepal the Upper Carboniferous and Permian are represented by coarse clastic facies at the base and by fossiliferous shale beds above; near Mount Everest these upper Paleozoic systems are very thick carbonate and sandstone rocks. The top of Mount Everest is composed of massive sandy Carboniferous limestone. In Kashmir a volcanic episode commenced that continued well into Triassic time, and in places as much as 7,000 feet of lavas and agglomerates accumulated. Overlying the volcanic rocks and also interbedded with them are slaty rocks containing plants peculiar to the Gondwana facies of peninsular India. Tongues of Gondwana facies are also found in the sub-Himalayan zone in western Kashmir, in Hazara, and in the eastern Himalayas. In general, however, the late Paleozoic is represented in the sub-Himalayan zone by carbonate and shale facies, for the most part unfossiliferous.

South of the Himalayas, in the Salt Range, the Cambrian is overlain by late Paleozoic formations, beginning with a glacial boulder bed 200 feet thick, immediately overlain by marine sandstones and shales. These are overlain by one of the most nearly complete marine Permian sequences in the world, the Zaluch Group of middle and late Permian age, about 800 feet thick, which consists of very fossiliferous limestones and calcareous sandstones.

The Middle East

The mountain systems between the Himalayas and Turkey form the bridge between the Mediterranean and the Himalayas

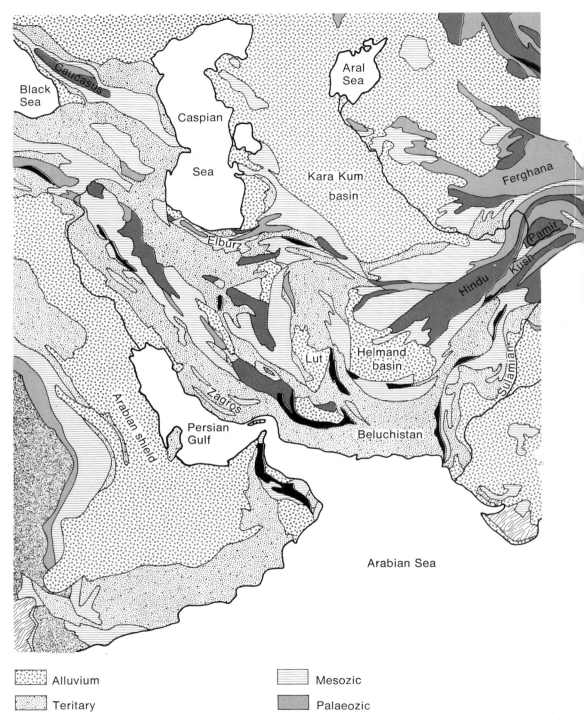

Black
Sea

Caucasus

Caspian

Sea

Aral
Sea

Kara Kum

basin

Ferghana

Elburz

Pamir

Hindu

Kush

Lut

Helmand
basin

Zagros

Arabian shield

Persian
Gulf

Beluchistan

Suliaman

Arabian Sea

Alluvium

Teritary

Mesozic

Palaeozic

FIGURE 6-26

Generalized geologic map of southern Asia. (After Gansser, 1964.)

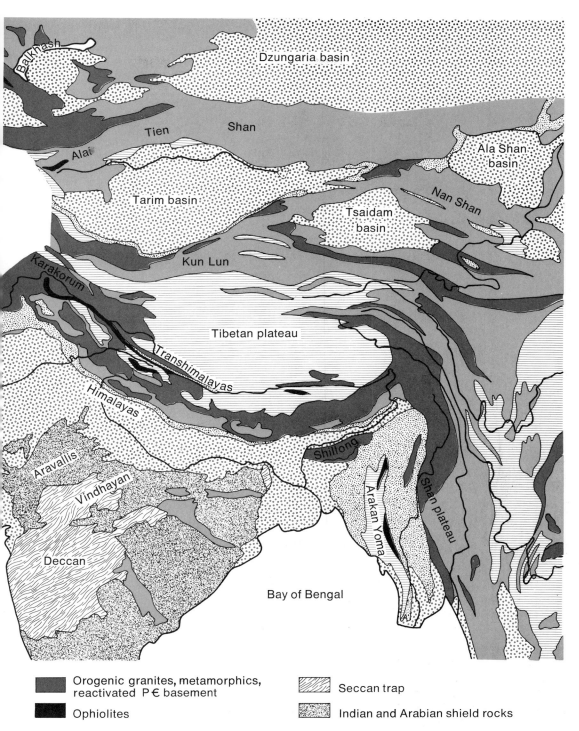

Orogenic granites, metamorphics, reactivated P€ basement

Ophiolites

Seccan trap

Indian and Arabian shield rocks

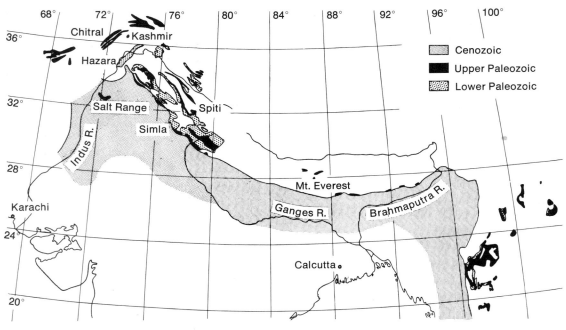

FIGURE 6-27

Outcrops of Paleozoic formations in Pakistan, India, and Burma. (After *Geological Map of India*, 1940.)

(Fig. 6-26). This area includes, in Mesozoic and Cenozoic formations, the world's largest known reserves of petroleum. The geology of Iran is better known than that of the rest of the Middle East, so our discussion largely focuses on this country.

Iran has two great mountain ranges, the Elburz in the north and the Zagros-Makran in the south, framing a vaguely defined "Median Mass" in central Iran, which some have thought to be a shield. The "Median Mass" has been considered by other investigators as a deformed eugeosynclinal belt. Intensive geological investigations in central and north Iran have recently shown that for much of geologic time central and north Iran was a vast mobile platform area and neither shield-like nor geosynclinal as had previously been thought.

The Paleozoic formations particularly support this picture. The shields, platform areas, and geosynclines of this area are shown in Figure 6-30. The Paleozoic geosyncline occupied the area of northwest-

ern Turkey, the Caucasus Mountains, and southern U.S.S.R. east of the Caspian Sea. To the south were the Ethiopian and Indian shields. The vast intervening area was a mobile platform throughout the Paleozoic. In Iran the Paleozoic sequences show many large gaps in the record and with minor exceptions no angular unconformities, thus reflecting widespread epeirogenic movements but essentially no orogenic movements. Sediments laid down on the periphery of the Ethiopian shield in southern Israel, Egypt, and Central Saudi Arabia are intertonguing sequences of marine and nonmarine facies, generally thin and with many hiatuses. Figure 6-31 illustrates the transgressive and regressive aspects of the Phanerozoic history of the region from Saudi Arabia to Israel.

Although the Paleozoic history of the Middle East is becoming better known, only a most generalized synthesis is yet possible. Precambrian basement rocks have been recognized widely in central, northern, and eastern Iran, but no synthesis of their history is yet possible. These Precambrian strata are overlain unconformably by a thick (up to 6,000 feet) sequence of dolomite, limestone, sandstone, and colored shale all of apparent platform facies. These strata change laterally into salt, gypsum, and redbeds in central and southern Iran. Equivalent facies are present in the Salt Range of West Pakistan. In Iran these salt deposits are now exposed in many salt domes in the Persian Gulf and southern Iran, some of which form mountains of salt, as much as 6,000 feet high (Fig. 6-32). The

salt began moving in the Late Cretaceous but was most active in the Late Cenozoic.

These dolomite and salt strata are unfossiliferous; because they are overlain by fossiliferous Cambrian strata and overlie with angular unconformity older Precambrian formations, they are thought to be late Precambrian in age. In Iran this group of rocks is referred to as the Infracambrian. Above the evaporites occur a remarkably widespread and persistent purple to red cross-bedded sandstone, the Lalun Sandstone. A similar sandstone overlies the salt in the Salt Range of West Pakistan and near

FIGURE 6-28

Paleozoic and Mesozoic beds behind Muth in the Spiti region of the Himalayas.
(A) Cambrian slates and quartzite. (B) Lower Silurian conglomerate. (C) Red Lower Silurian quartzite. (D) Silurian limestone. (E) Carboniferous Muth quartzite.
(F) Permian Productus Shale. (G) Triassic shale and limestone. (A) and (B) are separated by a fault. (Photograph by H. H. Hayden, Geol. Surv. India.)

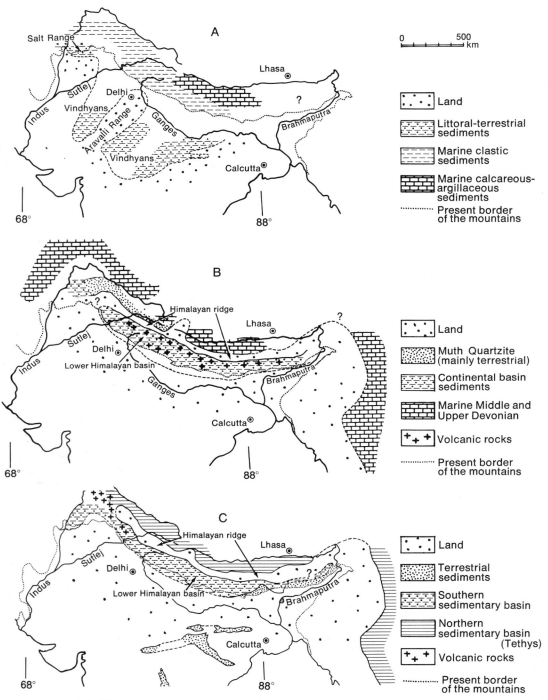

FIGURE 6-29

Paleogeographic maps of the Himalayas for (A) late Precambrian-Ordovician,
(B) Upper Silurian-Devonian, and (C) Permian. (After Fuchs, 1967.)

Mardin in southeast Turkey. The Lalun Sandstone is overlain by fossiliferous limestones and shales of Middle and Upper Cambrian age. A schematic paleogeographic map of the Middle East for the Cambrian is shown in Figure 6-33, A.

Ordovician and Silurian formations are not well represented in the mobile platform area of the Middle East. Where present they are sandstone, shale, and limestone only a few hundred feet thick. Much of northern Iran and southeast Turkey was emergent during most of these two periods. In the geosynclinal area very thick accumulations of sediments and volcanics are recorded. Active epeirogenic movements constantly changed the pattern of land and sea relations. During the Devonian in the geosynclinal region very thick graywacke and volcanic facies are common (Fig. 6-33, B). On the mobile platform the early Devonian is mainly nonmarine clastic facies, and the Middle and Upper Devonian, where present, are marine. In northern Saudi Arabia the Lower Devonian is represented by a marine clastic facies overlain by nonmarine Devonian sediments. The area of the Elburz Mountains and adjoining regions, which had been emergent since the Early Ordovician, were again inundated by Devonian seas.

Paleogeographic conditions remained quite similar in Early Carboniferous time, with the main changes in Iraq and adjoining regions (Fig. 6-33, C). The Carboniferous formations of Iran are generally thin limestone and shale with some sandstone units, mainly of Early Carboniferous age. The widespread Permian carbonates of the Middle East rest on a conspicuous unconformity, which in Iran is due mainly to the almost complete absence of Upper Carboniferous strata (Fig. 6-33, D). The Permian of Iran includes strata of very late Permian age, especially in the northwestern part of the country. Through most of the world, seas regressed off the continents during the Late Permian, so that strata of latest Permian age are rare and accordingly of great interest.

6-3 THE SOUTHWESTERN PACIFIC

The nucleus of the southwestern Pacific is the large Australian shield, which includes slightly more than the western half of Australia (see Fig. 4-15, p. 105). The island chains between Indonesia and New Zealand were extremely active tectonic and depositional regions in the Mesozoic and Cenozoic, and their Paleozoic records are largely masked by the later events.

Australia

Australia, however, is an ancient continent whose geosynclinal history culminated at the end of the Paleozoic (Fig. 4-15). Note

FIGURE 6-30

Geologic provinces of the Middle East during the Paleozoic. (After Wolfart, Erdöl, and Kohle, 1967.)

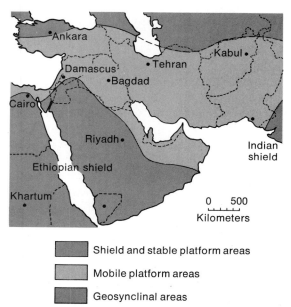

Shield and stable platform areas

Mobile platform areas

Geosynclinal areas

the vast exposure of Precambrian rocks. The only folded belts are in the east; they consist of sedimentary and volcanic rocks deposited in the Tasman geosyncline, which is entirely Paleozoic in age. Mesozoic and Cenozoic formations are flat plates overlying the folded Paleozoic and Precambrian formations. These plates of younger sediments isolate a western belt of Paleozoic folded strata from an eastern belt, making paleogeographical interpretation, in many instances, uncertain but challenging.

Australia has a vast development of upper Precambrian rocks, which in some places grade into the overlying Lower Cambrian formations, the boundary between the two often being arbitrarily chosen because of thick, unfossiliferous conformable sequences beneath the first fossils (Fig. 6-34).

The Cambrian history of Australia is incompletely known. In the Tasman geosynclinal area of eastern Australia, events of this period are very difficult to interpret (Fig. 6-35, A). Fossiliferous Cambrian strata are known mainly from Victoria and Tasmania, where they consist of very thick graywacke and volcanics. Orogenic movements (the Tyennan orogeny) greatly disturbed the geosynclinal region of Tasmania during the Middle and Late Cambrian, but no such movements took place in Victoria. In extreme eastern Australia no fossiliferous Cambrian strata have been identified. Nevertheless, thick metamorphosed sequences of graywacke and volcanics, which locally unconformably underlie the Ordovician, are interpreted as probably Cambrain. West of the southern part of the Tasman geosyncline was the very prominent Adelaide geosyncline wherein accumulated 45,000 feet of late Precambrian orthoquartzites, limestones, and other strata (Fig. 6-34). Subsidence in this belt continued into the Lower and Middle Cambrian with the deposition of orthoquartzite and thick limestone largely made up of the sponge-like *Archaeocyathus*. The basin was uplifted and folded in the mid-Cambrian.

Cambrian seas progressively transgressed

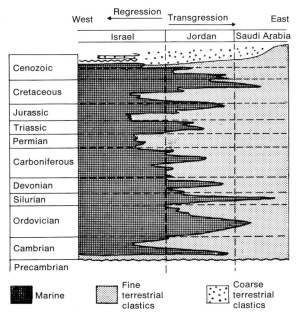

FIGURE 6-31

Diagrammatic sketch illustrating the intertonguing of marine and nonmarine facies between Saudi Arabia and Israel. (After Ball and Ball, 1953.)

across central and northwestern Australia. The Lower Cambrian seas extended only to central Australia, but at the same time thick lava flows spread across northern and western Australia, followed by a widespread Middle Cambrian transgression. The greatly variable thickness of the Cambrian formations indicates highly uneven subsidence in the different parts of this vast platform region. The Cambrian formations are mainly limestone, dolomite, and sandstone. The many erosional disconformities within the Cambrian reflect a very complex pattern of transgressions and regressions.

As with many other geosynclines, the Cambrian record is very incomplete, but the Ordovician record is significantly better. A broad shelf sea occupied a belt across central and western Australia (Fig. 6-35, B) in the Early Ordovician, becoming increasingly smaller through Middle and Late Ordovician time. The sediments of this sea were primarily sandstone and limestone, many containing an abundant and diverse shelly fauna. Two other small marginal marine basins appeared on the west coast. Within the Tasman geosyncline graywackes and volcanics predominated, and graptolites were abundant. The sediments deposited there contain one of the most complete Ordovician sequences of these fossils in the world. The wide exposures of these fossiliferous rocks allow the recognition of several north-south troughs separated by geanticlinal ridges, mainly submerged, especially in the southern Tasman geosyncline. Thick sediments and volcanics accumulated in the troughs and thinner ones on the intervening ridges. The Ordovician was a period of relative tectonic stability for Australia, though there was much epeirogenic movement, reflected in the transgressive and regressive patterns on the central platform area. Orogeny along the southwestern margin of the geosyncline produced tight folds and shifted the shoreline eastward. Another very local orogeny (the Benambran) affected eastern Victoria and adjoining regions of New South Wales in the Late Ordovician, continuing on into the Early Silurian.

The Late Ordovician orogenic phase altered the general pattern of the Tasman geosyncline, primarily by shifting and more sharply delineating the troughs and arches (Fig. 6-35, C). The Silurian deposits, like those of the Ordovician, are mainly graywacke and volcanics. The volcanics are found mainly in the regions most affected by the Late Ordovician orogeny. In central Australia terrestrial sandstones accumulated during the Silurian. Along the western coast of Australia only a single marginal trough existed, in which thin wedges of sandstone and limestone accumulated. Toward the close of the Silurian, the Tasman geosyncline underwent another orogenic phase (the Bowning orogeny) that had its greatest effects in eastern Victoria and New South Wales. This phase continued into the Devonian and was accompanied by extensive granitic intrusions.

The Devonian was a period of dramatic change in the Tasman geosyncline. The Late Silurian orogeny considerably altered the pattern of troughs and arches, at least in the southern part, though the Early and Middle Devonian sediments were similar to those of the Silurian. In extreme eastern Australia extremely thick unfossiliferous graywacke-volcanic strata prevail; only because they are interspersed with a few beds of coralline limestone is it possible to

FIGURE 6-32
Gakum Salt Dome (Cambrian), which has pierced Cenozoic formations in southwestern Iran. (Photograph by J. V. Harrison.)

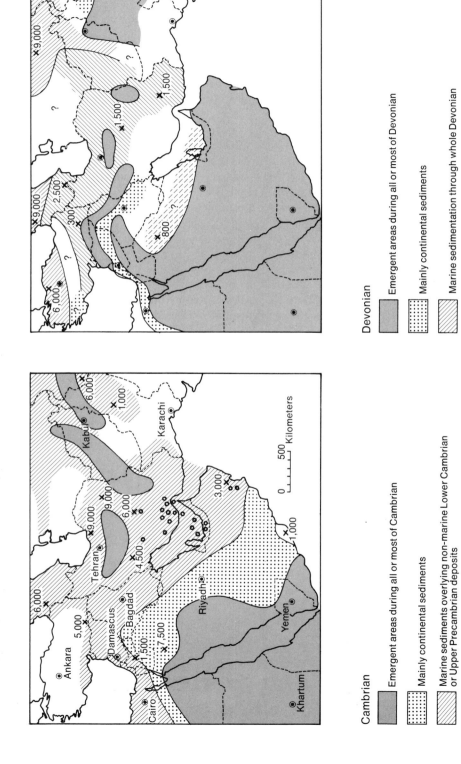

Cambrian

■ Emergent areas during all or most of Cambrian

▨ Mainly continental sediments

▧ Marine sediments overlying non-marine Lower Cambrian or Upper Precambrian deposits

○ Principle areas of late Precambrian or early Cambrian salt deposits

✕ Thickness in feet

Devonian

■ Emergent areas during all or most of Devonian

▨ Mainly continental sediments

▧ Marine sedimentation through whole Devonian

▧ Marine Devonian overlying non-marine Devonian deposits

▧ Marine Lower Devonian overlain by non-marine Devonian deposits

✕ Thickness in feet

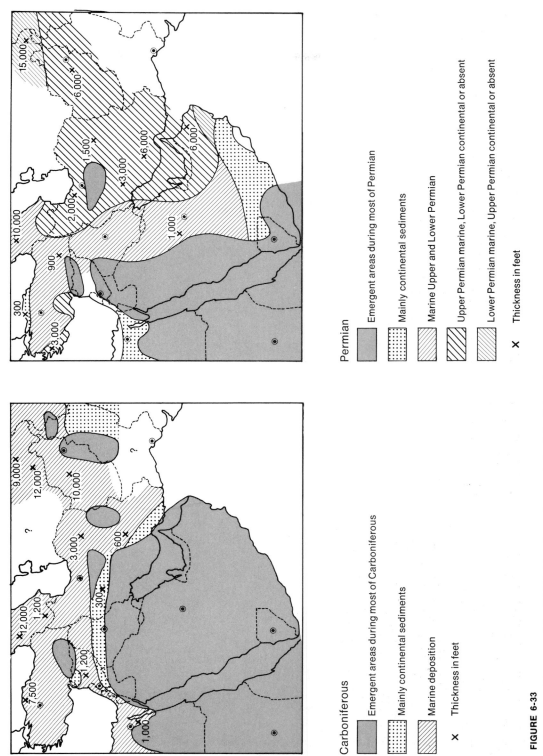

Carboniferous

▨ Emergent areas during most of Carboniferous

⬚ Mainly continental sediments

▨ Marine deposition

✕ Thickness in feet

Permian

▨ Emergent areas during most of Permian

⬚ Mainly continental sediments

▨ Marine Upper and Lower Permian

▨ Upper Permian marine, Lower Permian continental or absent

▨ Lower Permian marine, Upper Permian continental or absent

✕ Thickness in feet

FIGURE 6-33

Paleogeographic maps of the Middle East for various periods of the Paleozoic. (After Wolfart, Erdöl, and Kohle, 1967.)

date these beds. In late Middle Devonian time the inner half of the Tasman geosyncline underwent a most intense orogeny (the Tabberabberan) that was accompanied by plutonic intrusions and resulted in the uplift of this region. This orogenic movement brought marine deposition to a close in the inner belt of the Tasman geosyncline. Consequently, the Late Devonian is represented predominantly by stream and lake deposits (Fig. 6-35, D). In easternmost Australia, where the mid-Devonian orogeny had little effect, geosynclinal conditions continued through the Late Devonian, with the accumulation of thick graywacke and volcanics.

The Devonian history of Western Australia was comparatively uneventful except

FIGURE 6-34

Natural basin of late Precambrian (Pound Quartzite) resting on shale and quartzite (valley and low hills in foreground: Wilpena Pound, Flinders Range, South Australia). The Pound Quartzite was previously considered as Cambrian, but in recent years it has yielded a most remarkable fossil assemblage of soft-bodied animals and algae, many of unknown affinity, that clearly indicates a late Precambrian age. (Photograph by the South Australian Government Tourist Bureau.)

for the deposition of highly fossiliferous sediments in marginal marine basins. Along the northern margin of the Canning basin (Fig. 6-35, D) a remarkable development of reef complexes extends for 180 miles. These reef complexes are composed of four distinct facies: the reef, back-reef, fore-reef, and inter-reef facies. The reef and back-reef facies together make up limestone platforms that once stood tens to hundreds of feet above the floor of the surrounding inter-reef basins. These platforms are similar to present-day platforms of the Bahama Banks and associated limestone bank areas of Florida, Cuba, and the Gulf of Mexico. The reef facies forms narrow, discontinuous rims around the platforms, which are flanked by fore-reef talus deposits that have steep depositional dips and interfinger with the surrounding inter-reef deposits. The reefs are atolls, fringing reefs, and barrier reefs, and the platforms cover areas ranging from a few acres to hundreds of square miles. The maximum thickness of the reef complexes may exceed 3,000 feet.

The reef complexes developed around islands and promontories in the basement of Precambrian rocks, except in one area where they rest on Ordovician strata that formed the Devonian mainland shore. They generally trend northwesterly, parallel to the regional strike of the Precambrian rocks.

These reef complexes are particularly well developed in the Napier Range at Windjana Gorge (Fig. 6-36). The reef, fore-reef, and back-reef facies are well displayed, but the inter-reef facies lies under the Quaternary alluvium in the upper left of the photograph. The low land area in the upper right of the photograph is Precambrian schist.

The pattern of geosynclines, marginal marine basins, and terrestrial basins established in the Late Devonian continued through Carboniferous time (Fig. 6-35, E). In the Early Carboniferous the geosynclinal region of eastern Australia was entirely marine, but in the Late Carboniferous the seas receded from the southern

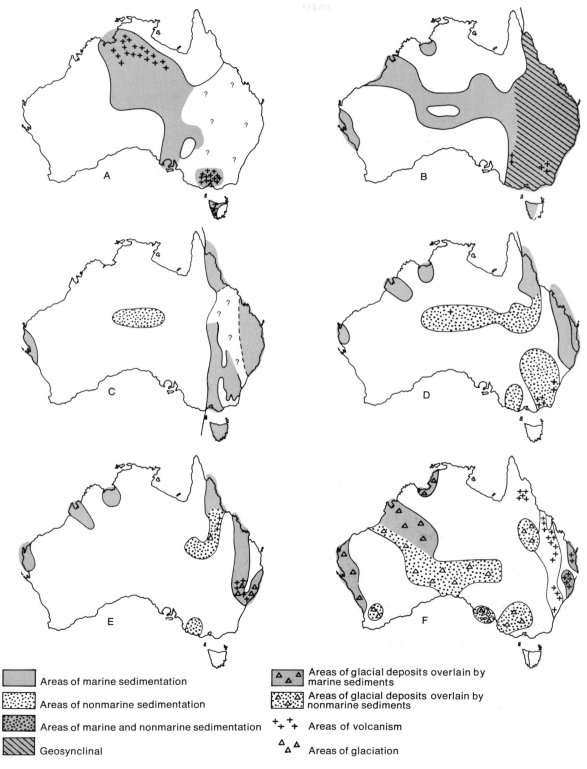

FIGURE 6-35

Paleogeographic maps of Australia for various periods of the Paleozoic: (A) Cambrian, (B) Ordovician, (C) Silurian, (D) Late Devonian, (E) Early Carboniferous, and (F) Permian. (After Brown *et al.*, 1968.)

Areas of marine sedimentation

Areas of nonmarine sedimentation

Areas of marine and nonmarine sedimentation

Geosynclinal

Areas of glacial deposits overlain by marine sediments

Areas of glacial deposits overlain by nonmarine sediments

Areas of volcanism

Areas of glaciation

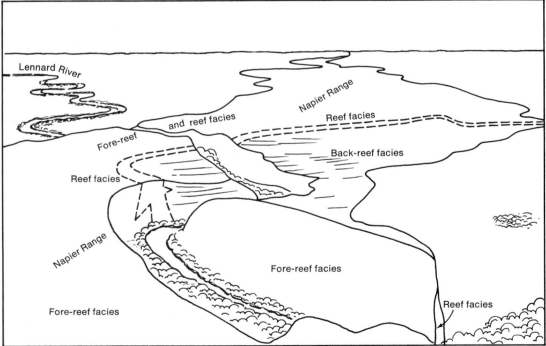

FIGURE 6-36

Aerial view looking northwest along Napier Range at Windjana Gorge, Canning Basin, Western Australia. (Photo by Department of Lands and Surveys, Perth, Western Australia.)

part, and thick terrestrial deposits accumulated along with volcanics. Glacial deposits are intercalated in the upper half of the Carboniferous sequence. Upland areas on the margin of the trough appear to have been sites of ice accumulation. On the inner side of the Tasman geosyncline, some terrestrial basins that formed during the Late Devonian continued to receive sediments in the Carboniferous but on a smaller scale. Western Australia continued to exhibit a pattern of isolated marginal basins, much as it did in the Devonian. Orogeny with many plutonic intrusions affected one place or another in eastern Australia throughout the Carboniferous (the Kanimblan orogeny), but in general this orogeny affected the inner areas of the Tasman geosyncline much more than the outer areas.

The Permian was a climactic period in the geological evolution of Australia (Fig. 6-35, F), during which orogenic episodes finally consolidated the Tasman geosyncline, the last true geosyncline in the Australian region. Permian strata display greater variety and wider distribution than those of any other period of the Paleozoic. Aside from the typical shelf, platform, and geosynclinal facies, glacial deposits and significant coal deposits were widely developed in eastern Australia.

The Permian geosynclines in eastern Australia followed the pattern of those that existed during the Carboniferous; on the eastern coastal region extremely thick marine sandstones, shales, and volcanics accumulated in marginal troughs. A low arch separated these troughs from a much larger trough system to the west, wherein accumulated a mixture of marine and nonmarine facies plus coal and glacial deposits. This variety of facies indicated that within the trough active geanticlinal ridges were rising and being eroded. A late Permian orogeny (the Hunter-Bowen), accompanied by extensive igneous intrusion, folded and consolidated the whole geosynclinal region.

Glaciation was widespread during the Permian of Australia. During the early Permian, ice sheets occupied much of the southern half of Australia except for the Tasman geosyncline in the southeast. The ice moved from south to north, suggesting ice accumulation in the regions now covered by sea off southern Australia. These glacial deposits and their associated faunas and floras resemble others that occur widely in the southern hemisphere (see Chapter 11).

Indonesia

Paleozoic strata are sparse in Indonesia, and no worthwhile synthesis can be made. The Permian is exceptional mainly because of a fantastically rich fossiliferous limestone on Timor. These rich faunas, monographed in many volumes, occur in shallow-water limestone associated with volcanic rocks that obviously had their source in some other region. Another Permian formation, attributed to tectonic activity somewhere outside of Timor, consists of slightly metamorphosed sandstone and shale that, because of the absence of a benthonic fauna, may be deep-water facies. Finally, there are two widespread Permian formations of clastic facies that include much angular to subangular quartz and very fresh plagioclase grains, suggesting they are first-cycle deposits, derived from acid igneous rocks. The Permian glacial, fluvio-glacial, and marine glacial deposits of northwest Australia are the most probable source of these detrital deposits. Furthermore, these formations show clear evidence of large-scale slumping from south to north. Paleogeographical reconstruction of this area for the Permian is particularly difficult in that no outcrops are known in situ of the formations whose sediments have obviously been transported. A suggested paleogeographical reconstruction for Permian time is shown on Figure 6-37. The in-situ clastic formations clearly have their source in northwest Australia. The direction of transport and slumping of these clastic formations indicates a progressive deepening of the Permian sea northward from Australia. It thus seems logical to align the deep-water facies in a zone beyond Timor and to place the original site of the carbonate platform further northwest. The deep-water facies probably marks the axial area of a geosyncline. Movement of the deep-water sediments and the limestone facies southeast onto Timor was part of an important Miocene orogeny, but the mechanism of this movement remains a mystery. The Triassic formations of Timor have similar relations, and we shall discuss this problem again in § 9-5.

New Zealand

The geological history of New Zealand can be divided into three main depositional phases, each forming a well-defined natural unit without major unconformities. These are (1) a lower division consisting of lower Paleozoic rocks, (2) a middle division of upper Paleozoic, Triassic and Jurassic rocks, and (3) an upper division of Upper Cretaceous and Cenozoic formations (Fig. 6-38).

The lower-division strata are confined to the western part of South Island, where very thick complexes of metamorphosed volcanic and clastic rocks are perhaps of Precambrian age, but whose relations to the lower Paleozoic are unclear. Fossiliferous Cambrian and Ordovician strata are present in the Nelson region of northwest South Island and in Fiordland of southwest South Island. Cambrian formations in the Nelson region consist of conglomerate, graywacke, and volcanic strata some 25,000 feet thick: one formation has yielded Middle Cambrian fossils. Ordovician formations in the same region consist of graptolitic shale, graywacke, marble, and volcanics reported to be 40,000 feet thick! Rocks of Silurian age are unknown in New Zealand.

Lower Devonian sandstone and coral limestone in the Nelson region record the end of the first phase in the Phanerozoic evolution of New Zealand. The second phase began in the Late Carboniferous. The long time span (Middle Devonian to Late Carboniferous) between phases is not represented by any rock record, but sometime between the two phases the geosynclinal complex underwent orogeny and plutonic intrusion. In the Late Carboniferous a new geosyncline was established to the east of the axial area of the earlier geosyncline. This new geosyncline, the New Zealand geosyncline, dominates the second phase in the Phanerozoic history of New Zealand. Deposition in this geosyncline was rapid, and enormous thicknesses of graywackes and volcanics of Permian, Triassic, and Jurassic age accumulated. Further history of the New Zealand geosyncline is discussed in Chapter 9.

6-4 ANTARCTICA

Few Phanerozoic strata of Antarctica are fossiliferous, making age assignments difficult. During the past decade the region between the Weddell and Ross seas and adjacent to the Transantarctic Mountains (Fig. 4-19) was identified as the site of Paleozoic geosynclines.

The Transantarctic Mountains are composed mainly of metamorphic and granitic rocks, in some of which archaeocyathids of Early and Middle Cambrian age have been found. Although much more exploration remains to be done, the rocks that contain the fossils appear to form a belt that trends across the continent, broadly co-extensive with the Transantarctic Mountains. Before metamorphism the rocks that now form the mountains presumably consisted of thick strata of graywacke, shale, and limestone. After metamorphism, in late Cambrian, Ordovician, or Silurian time, they were folded and intruded by granitic batholiths.

The Antarctic Peninsula consists of meta-

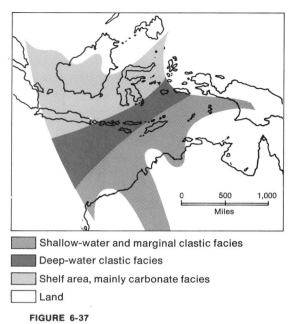

Shallow-water and marginal clastic facies

Deep-water clastic facies

Shelf area, mainly carbonate facies

Land

FIGURE 6-37

Suggested relationships of major depositional
zones in the Indonesian region for the Permian.
(After M. G. Audley-Charles, *Paleogeography,
Palaeoclimatology, and Palaeoecology*, Fig. 3,
Elsevier, Amsterdam, 1965.)

morphic and igneous rocks of late Paleozoic
and Mesozoic age, an extension of the
Andean orogenic belt. The region between
the Antarctic Peninsula and the Transantarctic Mountains is interpreted as a complex
Paleozoic geosynclinal belt folded and intruded at several times between the Devonian and Permian or possibly early Mesozoic. The evidence for this, however, is
still tenuous because few fossils have been
found and few radiometric age determinations are available, but, briefly, here is
what is known. Structures on opposite sides
of the Ross Sea strike toward each other.
The rocks on the East Antarctic side of the
Ross Sea are thick sequences of metamorphosed graywacke and shale, intruded by
granitic plutons that have yielded radiometric dates of about 350 m.y. (Late Devonian or Early Carboniferous). The Late
Cambrian or Ordovician orogenic complexes that make up the Transantarctic

Mountains are unconformably overlain by
Lower Devonian marine and nonmarine
clastic sequences that apparently are marginal geosynclinal deposits; these are overlain by plant-bearing, nonmarine Upper
Devonian strata.

The mountainous tops of islands project
through the ice cap of interior West Antarctica between the Ross and Weddell seas
and the Antarctic Peninsula. The pre-Terti-
ary rocks of these islands are low-grade
metamorphic rocks that are intruded in
some regions by granitic batholiths. Available data suggest that the complexes are
disconnected from each other and from the
terranes of East Antarctica. Radiometric
dating gave an age of 280 m.y. for a gneiss in
this region (Late Carboniferous or Early
Permian and an age of 199 m.y. (Middle
Triassic) for a granite. Finally, the granitic
rocks of the Antarctic Peninsula, central

FIGURE 6-38

Geologic sketch map of New Zealand.
(Adapted from H. W. Wellman, 1952.)

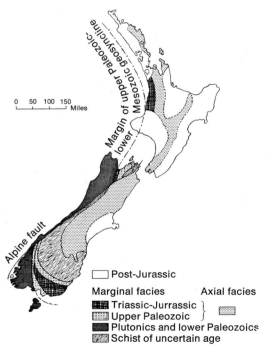

Post-Jurassic

Marginal facies   Axial facies

Triassic-Jurrassic ⎫
Upper Paleozoic   ⎬
Plutonics and lower Paleozoics
Schist of uncertain age

West Antarctica, and the Transantarctic Mountains form distinct petrologic groups. Even though the identification of this region as a mid- and late-Paleozoic orogenic belt is still somewhat tenuous, the available data make this reasonable.

The rest of the Paleozoic record in Antarctica consists of deposits of glacial tillites and nonmarine sandstones like those on all southern continents (see Chapter 11).

6-5 SOUTH AMERICA

South America differs from the continents already considered in having a great geosyncline only along its western edge, the central and eastern parts being dominated by Precambrian shields (Fig. 4-6). Throughout the Andean geosyncline the Paleozoic rocks, especially the earliest deposits, are poorly known, for intense orogenic and plutonic activity in the Mesozoic has obscured much of the depositional history. East of the Andes a wide belt of Cenozoic deposits, built up from sediments derived from the rising mountains, contains Paleozoic rocks that are deeply buried except in isolated uplifted structures. The Paleozoic rocks of central and eastern South America are largely confined to troughs, one along the present course of the Amazon River and one in a belt across eastern and southern Brazil, the latter extending westward through Paraguay, northern Argentina, and Bolivia. These troughs accumulated fairly thick sequences of sediment, much of it nonmarine. Extensive marine invasion of this shield area was largely confined to the Paleozoic. During the Mesozoic and Cenozoic, except in several basins along the eastern coast, the depositional record was written in terrestrial basin deposits.

It is obvious that little is known in detail of the Paleozoic of South America, only the broad outlines can be discerned. In the Andean region, early Paleozoic rocks are generally very thick, and the fossil evidence

suggests that most periods are represented. In contrast, the stratigraphy and paleogeography of the early Paleozoic in central and eastern South America are quite incomplete.

The paleogeographic evolution of South America in the early Paleozoic is interpreted as follows. In the Cambrian, geosynclinal seas occupied the western margin of the continent, and there was a shallow basin in the Amazon region (Fig. 6-39, A). Nonmarine basins existed in parts of Brazil. The Ordovician seas were confined to the Andean geosynclinal belt; during the Early Silurian a sea covered Ecuador and Peru but apparently did not occupy the Andean regions either to the north or south (Fig. 6-39, B). During this time the geosynclinal area of Ecuador and Peru was connected to the Amazon basin and to another trough extending across Bolivia, Paraguay, and into eastern Argentina. By Middle Silurian time the seas had retreated from the extra-Andean region and were confined to Ecuador, Peru, Bolivia, and Chile. The absence of Upper Silurian strata in all of South America suggests retreat of the seas and gentle uplift of the Andean region.

Exceedingly thick sequences of shale and sandstone of Devonian age are found in the Andes of Peru, Bolivia, and western Argentina, where Devonian rocks more than 1,000 feet thick include some of nonmarine origin. The Lower Devonian is particularly widespread, extending far east of the geosyncline, but the Middle and Upper Devonian are largely confined to the foothill and Andean regions (Fig. 6-39, C). The east-west basin south of Buenos Aires accumulated 3,500 feet of coarse-to-fine sandstones. The Falkland Islands were also an active site of deposition in the Devonian; there 10,000 feet of clastics rest on Precambrian rocks.

In the Amazonian trough 1,500 feet of marine Devonian sandstone, shale, and limestone were laid down. The broad basin that crossed eastern Brazil south of the

Amazon River was also occupied by seas, wherein sandstone and shale accumulated. During the Middle Devonian, glaciers appeared in the extreme northeast of the central Brazilian shield, and discharged into the Amazonian trough and adjacent basins, where glacial-marine conglomerates are interbedded in dominantly marine sequences. The massifs of western Argentina were most probably covered by ice caps during the Early Devonian. Glacial-marine conglomerates are present in Lower Devonian formations to the west and also on the Falkland Islands.

Sometime during the Late Devonian the Andean geosynclinal troughs were compressed and uplifted above sea level. Folding was extremely intense in the south but moderate to weak north of Bolivia.

With the advent of Carboniferous time great changes in the depositional pattern took place over much of South America. A series of orogenies (Hercynian) affected various parts of the Andean geosyncline. Deposition continued along the same patterns in the Amazonian trough, but south of there the remainder of the Paleozoic is marked mainly by continental deposits that include several horizons of extensive glacial deposits. These deposits, the Santa Catharina "System", are correlated with the Gondwana complex recognized in peninsular India. They are discussed further in Chapter 11.

Lower Carboniferous formations are scarce in South America, the continent appears to have been emergent during that time (Fig. 6-39, D). Marine formations are known from Colombia and Chile in the Andean geosynclinal belt, but in Peru and Bolivia only continental facies are known. No deposition appears to have taken place in the Amazonian trough and only a little in the linear basin to the southeast. In the mountain ranges that resulted from the Late Devonian orogeny in western Argentina, Alpine glaciers shed tillites and glacial-marine conglomerates onto the surrounding regions. Similar glacial deposits are known from southern Bolivia and northern Argentina. Strong tectonic activity took place in western Argentina toward the close of the Early Carboniferous but is not recognized elsewhere.

Marine geosynclinal conditions returned to the Andean region in the Late Carboniferous (Fig. 6-39, E). A great variety of clastic and limestone formations are known from Venezuela south to Chile (Fig. 6-40). Although the Amazon trough no longer opened to the Atlantic, it did open to the Andean geosynclinal sea to the west, whose successive transgressions and regressions gave rise to an alternating sequence of limestone, clastic, and evaporitic facies.

Through the first half of the Permian (Fig. 6-39, G) marine conditions persisted in the Andean geosynclinal belt, where extensive limestone formations are known. Regression of the seas and orogeny affected the Andean geosyncline in the late Permian, particularly in Peru, where continental redbeds and abundant volcanic deposits are present (Fig. 6-39, H). In the lower part of these continental redbeds are some thin marine limestones, suggesting that during the initial tectonic phase rows of volcanic islands were surrounded by sea. Continued ejection and deposition of volcanic materials forced the seas to the west. In northern Chile granitic batholiths have yielded lead-alpha ages of 265 ± 30 m.y. (Permian).

6-6 375 MILLION YEARS— A RECAPITULATION

Though it lasted some 375 million years, the Paleozoic Era was short in comparison with the Precambrian, which had lasted some 3 billion. The Paleozoic, however, is the first segment of the earth's development that can be outlined comprehensively, for correlation by means of fossils permits comparatively fine time-space relations to be

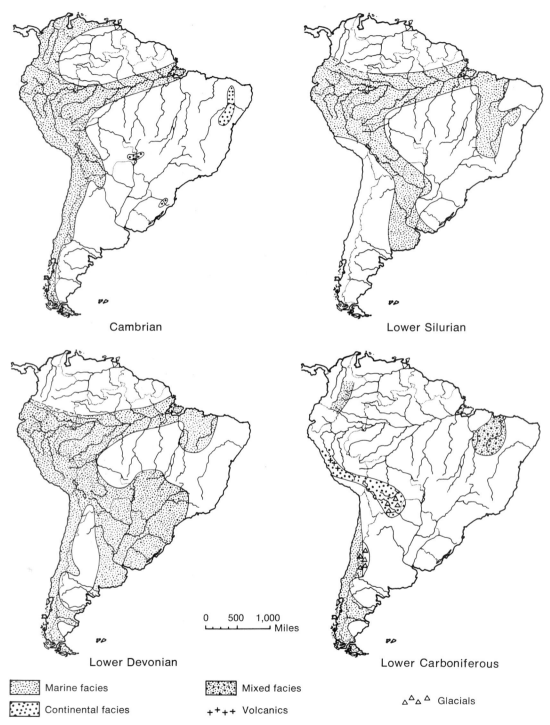

Cambrian

Lower Silurian

Lower Devonian

0   500   1,000
|__|__|__|_____| Miles

Lower Carboniferous

Marine facies   Mixed facies

Continental facies   + + + Volcanics   △ △ △ Glacials

FIGURE 6-39

Paleogeographic maps of South America for various periods of the Paleozoic. (After Harrington, 1962.)

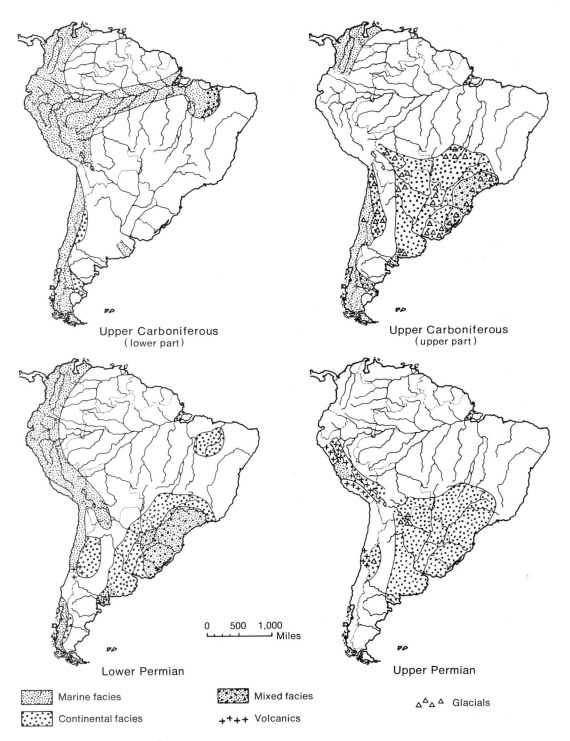

Upper Carboniferous
(lower part)

Upper Carboniferous
(upper part)

Lower Permian

Upper Permian

0   500   1,000
Miles

Marine facies

Mixed facies

Continental facies

Volcanics

Glacials

FIGURE 6-40

Upper Paleozoic formation exposed at Ambo village in the Andes of central Peru:
P, Permian red sandstone, conglomerates, and shales; *PL*, lower Permian or Upper
Carboniferous limestone and sandstone; *Ms*, Lower Carboniferous tuff, sandstone,
and shale. (Photograph by the author.)

established. Paleozoic rocks, moreover, have a larger outcrop area per million years of duration than the Precambrian rocks and therefore present more physical data for study.

During the Paleozoic Era striking changes took place in the distribution of lands and seas and in the tectonic frameworks of the continents. The early Paleozoic seas were very widespread (Fig. 6-3) and occupied many continental margins around and between the stable shields. As the era came to a close, there was a gradual decrease in the extent of the geosynclinal and continental seas, and this was accompanied by a consolidation of geosynclines through orogenies.

The pattern of widespread geosynclinal and continental seas began to change noticeably in mid-Paleozoic time. Orogenies occurred in the eastern eugeosyncline of North America in the Ordovician, the Silurian, and the Devonian. This belt, in fact, or at least the northern part, became a firm part of the continental framework with the advent of the Devonian Acadian orogeny, which brought geosynclinal deposition to a

close in this area. Orogenies of Late Silurian and Devonian age affected the Arctic regions of North America, particularly Alaska and eastern Greenland. The Caledonian geosyncline of Wales, Norway, and Spitsbergen, after an active depositional history, became consolidated in the Caledonian orogeny. The first phase of this orogeny occurred in the Late Cambrian and the last phases in the Devonian. Orogenies of Caledonian age also consolidated parts of the geosyncline at the southwestern edge of the Angaran shield. The areas affected by these Paleozoic orogenies are indicated by Figure 6-41. Where the orogenies were particularly intense, they brought geosynclinal conditions to a close. The adjoining shields were enlarged. Thus the Baltic shield was enlarged by the addition of the consolidated Caledonian geosyncline. The same applies to the geosyncline southwest of the Angaran shield. These changes in the tectonic frameworks of the continents and in the geography of lands and seas prepared the continents for the geosynclinal patterns of the Devonian Period. Despite the transgressive and

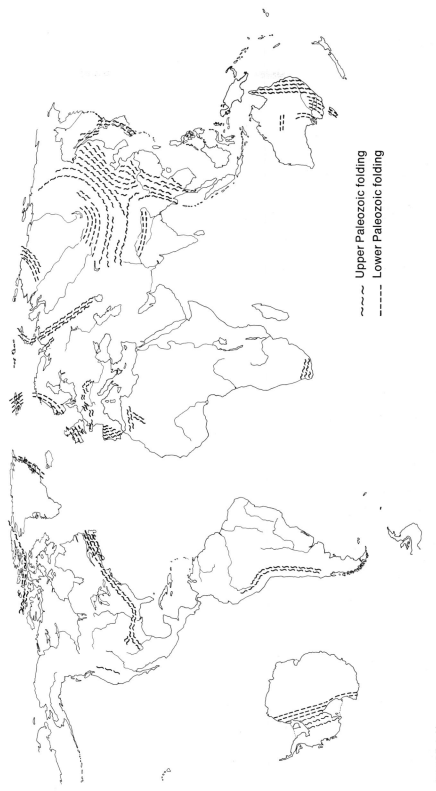

~~~ Upper Paleozoic folding

- - - - Lower Paleozoic folding

FIGURE 6-41

Distribution of folded strata formed during the Caledonian orogenies (dotted lines) and the Hercynian orogenies (dashed lines). (After J. H. F. Umbgrove, *The Pulse of the Earth*, Nijhoff, The Hague, 1947.)

regressive movements of the sea, Devonian seas are nearly as extensive as those of the earlier Paleozoic — except, of course, in the belts severely affected by orogeny.

By Carboniferous time a new series of orogenies had begun, and it continued until the end of the Paleozoic. The areas affected are indicated by Figure 6-41. Note how, in Europe, the folded belts were aligned south of the Old Red Sandstone continent; how the Uralian geosyncline, lying between the Angaran and Baltic shields, was converted into a solid continental mass; how the whole peripheral belt of the Angaran shield was involved in severe folding. In the last area, however, only south and northwest of the shield did this orogeny bring the geosynclinal phase to a close; the eastern belt adjacent to the shield underwent further deposition and folding in the Mesozoic. Australia had been formed by the close of this late Paleozoic orogenic phase; the great Tasman geosyncline, which lay along the east-ern edge of the Australian shield, was completely consolidated by numerous distinct orogenies. The remaining geologic history of Australia is recorded mainly in terrestrial basin deposits and indications of minor shallow-sea incursions.

By Permian time most of the intense orogenic activity was over. In Eurasia the actively subsiding depositional belt coincided with the Alpine-Himalayan trend. This was the great Tethyan geosyncline, which was to play a most important role in the Mesozoic and Cenozoic. In North America the eastern and southern geosynclines came to the end of their depositional histories, and the whole tectonic framework of the continent was radically altered. The new arrangement set the pattern for the Mesozoic. In South America the continental seas covering the shields came to an end; terrestrial basin deposits predominated by late Paleozoic time, and geosynclinal conditions persisted in the western mobile belt.

Suggested Readings

See the list at the end of Chapter 4.

PALEOZOIC LIFE

The earliest substantial record of ancient life is found in the Paleozoic System. Even in the first epoch of the Paleozoic Era (the Cambrian) representatives of most of the major phyla of invertebrate animals had appeared. This first documented phase in the evolution of life lasted some 375 million years, in the course of which the evolutionary radiation of the invertebrate phyla began; the first vertebrates appeared in the form of primitive, armored, jawless fish; and, finally, the reptiles evolved. During this era the land became populated by plants, and the first forests and the great coal-forming swamps appeared. The era came to a close with the extinction of many major invertebrate stocks, but this extinc-

tion had little effect on the land vertebrates or the plants.

During the Cambrian, Ordovician, and Silurian Periods marine invertebrates, the dominant forms of life, underwent an extremely diverse evolutionary radiation. The first section of this chapter will focus upon these early marine populations. We will then follow the evolutionary history of vertebrates from their appearance in the Ordovician, in the form of jawless fish, to the first reptiles, which appeared in the late Paleozoic. Next we take up the appearance, evolution, and spread of plants. The first land plants were Late Silurian, but not until the Devonian does there appear a really abundant land-plant fossil record, which reaches its climax in the Late Carboniferous. Then we will return to the marine environment and investigate the evolutionary progress and population structures of life in the mid-Paleozoic seas, Finally, we take up the end of this era and the problem of extinction.

7-1 EARLY PALEOZOIC MARINE LIFE

The earliest Cambrian sea must have been a strange sight, with its rather small fauna consisting of primitive coelenterates, brachiopods, sponges, gastropods, and arthropods. By the Late Silurian marine populations had become extremely diverse both in variety of organisms and in number of individuals.

A Middle Cambrian sea floor in central Bohemia is illustrated in Figure 7-1, and one in British Columbia (during the deposition of the Burgess Shale) in Figure 7-3. The first striking feature is the small number of different kinds of organisms shown in Figure 7-1, where the population consists only of trilobites, sponges, jellyfish, and seaweeds. The British Columbian scene includes many animals not normally preserved in the fossil record.

Paleontologically, the most significant

FIGURE 7-1

Reconstruction of a Middle Cambrian sea floor in Bohemia. The fauna includes siliceous sponges (the upright cones), jellyfish, and two genera of trilobites (*Paradoxides*, the large form, and *Ellipsocephalus*, the small form). (Drawing by Z. Burian under the supervision of Prof. J. Augusta.)

organisms are the trilobites—arthropods that were extremely abundant and diverse in the early Paleozoic seas. The fossil record indicates that they were most diverse early in their history, soon after their first appearance, and thereafter steadily declined until they died out at the close of the Paleozoic. Cambrian chronology is based mainly on the sequence and distribution of trilobites. The large trilobite in Figure 7-1 (*Paradoxides*) is an extremely important Middle Cambrian guide fossil in Europe and eastern North America. Most trilobites were small, but some grew to lengths of more than a foot. Since, as a group, they are extinct, their ecology and mode of life are difficult to reconstruct. Their great diversity of form (Fig. 7-2) and the great variation in details of their anatomy surely reflect a wide range of adaptations in the marine environment: some forms wandered about the sea bottom, others were fairly active swimmers, and still others probably burrowed.

The trilobites are the best-known form of Cambrian life, but we do know something about other animals of the same period. The restoration of the Burgess Shale sea (Fig. 7-3) includes a few of them. Most are arthropods, but there are also jellyfish, holothurians, and annelid worms. Compare this figure with the photographs of the fossils of the same animals in Figure 2-5 (p. 28). The colonies of large tubes are the sponge-like *Archaeocyathus*. These organisms were exclusively Cambrian in age and had a world-wide distribution. They lived in large numbers on calcareous sea bottoms, forming "gardens" attached to rocks or shells (sessile benthos) but not building topographically prominent reefs. Apparently they lived in narrow belts parallel to the coasts of the shallow seas. Such was the appearance of a Middle Cambrian sea floor.

Figure 7-4 is a reconstruction of a Silurian sea floor in Bohemia. The whole fauna has changed; many new groups of animals have appeared; nautiloids, corals, brachiopods, pelecypods, gastropods, stalked echinoderms, and trilobites are prominent.

The large, handsome nautiloids were among the largest animals living during this period. The first nautiloids, found in Upper Cambrian formations, had small, slightly curved, conical shells. In the Ordovician the group underwent a tremendous evolutionary radiation, more than a third of all known genera being of Ordovician age. The diversity of the group is shown by differences in the shell: its shape varied from straight to tightly coiled, and variations in internal structure affected its strength and buoyancy. Throughout the Ordovician, nautiloids abounded and were world-wide in distribution. They began to decline thereafter and now only one genus, *Nautilus*, remains.

This history—a great early radiation, then a gradual but steady decline—is very similar to the history of the trilobites. The group's radiation into new ecologic niches is easily understood, but its decline and extinction are more difficult to understand. A partial explanation lies in the rise of predators and of closely related but more efficient organisms that competed for the same ecologic niches. The trilobites were probably a succulent dish for the nautiloids and some of the fishes, and their decline may be related to the rise of those two groups. The nautiloids, in turn, may have declined as a result of the rise of the ammonoids after the mid-Paleozoic. Some problems of extinction are discussed at the end of this chapter.

Other conspicuous members of the Silurian fauna depicted in Figure 7-4 are the corals. The large circular masses are colonies of tabulate corals of the genus *Favosites* (see also Fig. 7-5), and there are also two types of solitary cup corals belonging to the Rugosa. Both classes of corals, first appearing in the Ordovician, are confined to the Paleozoic. The Silurian was particularly

FIGURE 7-2

Diversity of form among the trilobites.

A. *Lonchodomas*, Ordovician.
B. *Proetus*, Lower Devonian.
C. *Pemphigaspis*, Cambrian.
D. *Staurocephalus*, Silurian.

E. *Radiaspis*, Devonian.
F. *Condylopyge*, Middle Cambrian.
G. *Triarthrus*, Ordovician.
H. *Olenellus*, Lower Cambrian.
I. *Deiphon*, Silurian.
J. *Megalaspis*, Ordovician.

K. *Cyclopyge*, Ordovician.
L. *Asaphus*, Ordovician.
M. *Illaenus*, Ordovician.
N. *Illaenus*, Ordovician.
O. *Harpes*, Silurian.
P. *Trimerus*, Silurian.
Q. *Ceratarges*, Devonian.

(After L. Størmer, 1944.)

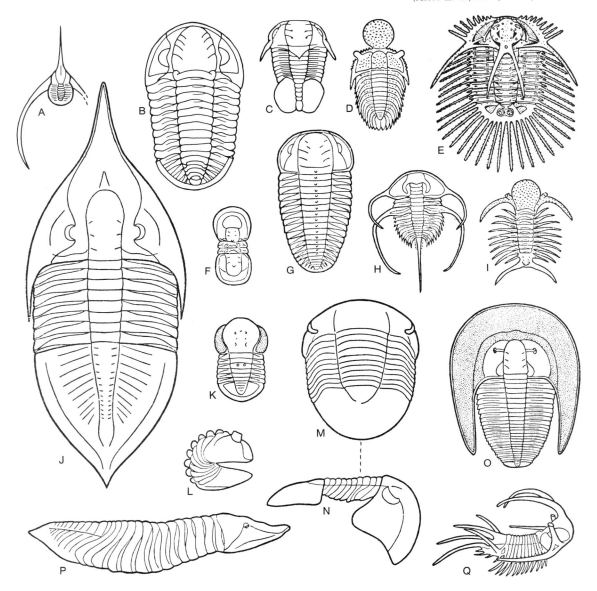

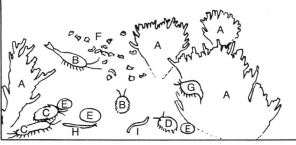

FIGURE 7-3

Reconstruction of a Middle Cambrian sea floor in western North America.
The fauna represented is that of the Burgess Shale of British Columbia.
See Fig. 2-5 for the fossil forms.

A. *Archaeocyathus*, a primitive sponge-like organism.
B. *Sidneyta*, an arachnid.
C. *Ogygopsis*, a trilobite.
D. *Neolenus*, a trilobite.
E. Jellyfish.
F. *Marrella*, an extinct arthropod of uncertain affinities.
G. *Hymenocaris*, an extinct arthropod of uncertain affinities.
H. An annelid worm.
I. A holothurian (echinoderm).

(Chicago Museum of Natural History.)

favorable for the corals, and they built extensive, widely distributed reefs. Paleozoic reefs are largely made up of algal and algoid organisms (mostly of uncertain affinities), with some tabulate corals and stromatoporoids. Rugose corals were not important as reef-builders. In Chapter 5 it was noted that reef development was particularly widespread during the Middle Silurian in the upper Mississippi Valley.

A reconstruction of part of such a reef is shown in Figure 7-5. Besides the large colonies of tabulate corals, the fauna includes stalked echinoderms (cystoids), nautiloids, trilobites, brachiopods, and gastropods. The calcareous remains of these organisms make up the Middle Silurian carbonate formations of the upper Mississippi Valley.

A few of the more common Paleozoic

FIGURE 7-4

Reconstruction of a Silurian sea bottom in central Bohemia.

A. *Orthoceras*, a nautiloid cephalopod; color patterns based on actual specimens.
B. *Cyrtoceras*, another nautiloid.
C. *Favosites*, a tabulate coral.
D. *Omphyma*, a rugose coral.
E. *Xylodes*, another rugose coral.

F. *Cheirurus*, a trilobite.
G. *Aulacopleura*, another trilobite.
H. *Conchidium*, a brachiopod.
I. *Glasia*, another brachiopod.
J. *Scyphocrinus*, a crinoid.
K. Gastropods.

(Drawing by Z. Burian under the supervision of Prof. J. Augusta.)

FIGURE 7-5

Middle Silurian coral-reef community, Illinois.

A. *Halysites*, the chain coral.
B. *Syringopora*, the tube coral.
C. *Favosites*, the honeycomb coral.
D. Cystoid echinoderms.
E. *Phragmoceras*, a nautiloid cephalopod.
F. Another nautiloid.

G. Another nautiloid.
H. *Pentamerus*, a brachiopod.
I. *Actinurus*, a trilobite.
J. *Isotelus*, another trilobite.
K. *Leptaena*, a brachiopod.
L. Two solitary corals.

(Chicago Museum of Natural History.)

rugose and tabulate corals are illustrated in Figure 7-6. Compare this with Figures 7-4 and 7-5.

Another important Silurian group consists of the brachiopods. These small, attached organisms constitute one of the most abundant and widely distributed of Paleozoic phyla. Cambrian brachiopods were chiefly small, inarticulate forms. In the Late Cambrian or Early Ordovician the ar-ticulate brachiopods began a very wide-ranging radiation into the main evolutionary lines that continued through the Paleozoic. A few common Paleozoic brachiopods are shown in Figure 7-7.

Environments favorable for brachiopods during the Paleozoic were generally also favorable for bryozoans. The remains of these delicate colonial animals often make up a large part of some deposits. The

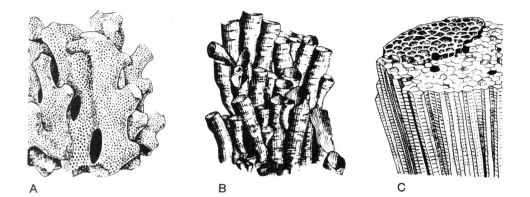

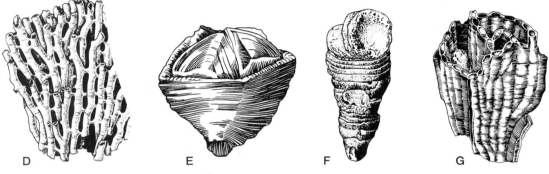

FIGURE 7-6

Representative Paleozoic corals.

A. *Thamnopora,* Silurian-Permian.
B. *Palaeophyllum,* Ordovician.
C. *Favosites,* Ordovician-Devonian.
D. *Syringopora,* Silurian-Pennsylvanian.

E. *Goniophyllum,* Silurian.
F. *Araeopoma,* Silurian.
G. *Halysites,* Ordovician-Silurian.

A, C, D, and G are tabulate corals; B, E, and F are rugose corals. Compare the habitat reconstructions of Figs. 7-4 and 7-5. (From *Coelenterata,* Vol. F of *Treatise on Invertebrate Paleontology,* Geol. Soc. Am. and Univ. of Kansas Press, 1956.)

slab of rock illustrated in Figure 7-8 shows a typical bryozoan-brachiopod fossil assemblage of the Silurian. Four genera of bryozoans and four of brachiopods are recognizable.

Very important rock-builders of the Paleozoic seas were the stalked echinoderms, which had a long stalk, a cup-like body, and feathery arms. Those illustrated in Figure 7-5 are cystoids, and those in Figure 7-4 are crinoids.

Many distinct groups of stalked echinoderms (Fig. 7-9) were more primitive than the blastoids and crinoids and, unlike them, became extinct in the mid-Paleozoic. None of these groups was very abundant. Their decline may be related to the evolutionary radiation of the crinoids, which began in the Ordovician. The high point in the evolution of the crinoids came in the mid-Paleozoic. Further comments on this group are deferred until § 7-11.

All the marine faunas so far discussed belong to the so-called shelly facies, being composed largely of benthonic organisms with calcareous shells, such as brachiopods, bryozoans, corals, and nautiloids; trilobites and associated arthropods had shells of the organic horn-like material called chitin.

The fauna of the graptolitic facies was quite distinct; as the name implies, graptolites are the commonest animals in this facies, benthonic animals with calcareous shells being rare or absent. Graptolites were colonial organisms that secreted an external skeleton of chitinous material in the form of cups or tubes distributed along one or more long branches that diverged from a central point of attachment. Graptolites are most often found in fine black shales; in coarser rocks the delicate

chitinous shells are seldom preserved. Most fossil graptolites look like little more than pencil marks on a piece of shale, but complete three-dimensional specimens have been recovered by acid treatment of calcareous rocks. The study and preparation of these fossils are enormously complex.

The two major groups of graptolites, the dendroids and the graptoloids, differed both in structure and in mode of life. Dendroids were mainly bottom-living creatures attached to hard objects on the sea floor. That dendroids were associated with brachiopods, bryozoans, and nautiloids indicates that their environment was shallow,

FIGURE 7-7

Representative Paleozoic brachiopods. *Discina* (A) and *Lingula* (B) are inarticulate; *Strophomena* (C), *Spirifer* (D), *Atrypa* (E), and *Productus* (F) are articulate. (From T. Davidson, 1858–1863.)

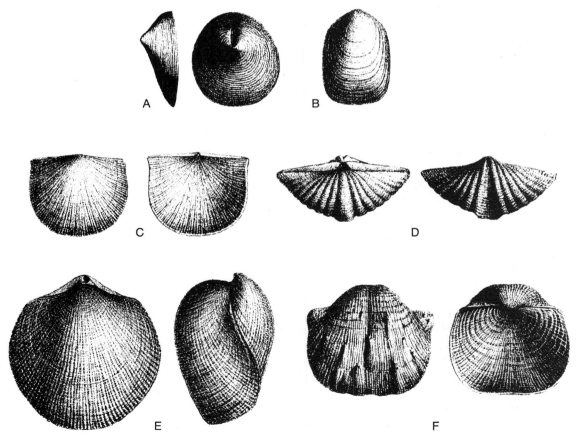

FIGURE 7-8

Fossil bryozoans and brachiopods of the Rochester shale (Silurian), from the Niagara gorge, New York. (Specimen 35,852, U.S. National Museum; photograph by H. B. Whittington.)

well-oxygenated water. Where they were abundant the sea floor must have looked like a miniature forest of tough seaweed and bushy dendroid graptolites. One type of dendroid, however, did not live on the sea bottom but floated on the surface, attached by a threadlike process either to a bit of seaweed or to its own float. The best-known and most significant of this group of dendroids is *Dictyonema flabelliforme*, which lived throughout the world during the earliest Ordovician (Fig. 7-10). This is a guide fossil *par excellence;* its cosmopolitan distribution is in great contrast to that of the sessil forms. The dendroids range from the Upper Cambrian to the Lower Carboniferous and hence, except for *Dictyonema* and related types, are not very good guide fossils.

The second group of graptolites, the graptoloids, were all floating organisms, attached either to seaweed or to their own floats. Graptoloids were very widely dis-

tributed, many being truly cosmopolitan. They lived only from the Early Ordovician to the Early Devonian but underwent an extremely rapid evolution. The graptolites, perhaps more than any other group, illustrate an evolutionary pattern in which changes took place along a few guiding trends and affected nearly all the stocks. The early graptoloids had many branches in a colony; later, however, the number of branches became fewer, until, in the Silurian, all forms had only one. Early Ordovician graptoloids with numerous branches are illustrated in Figure 7-11, A, E. A very stable phase of two branches, having the appearance of a tuning fork, was reached by the Middle and Late Ordovician (C). Along with the reduction in the number of branches there was a change in the orientation of the branches in relation to the supporting thread. The early forms had the branches suspended downward from the thread. The orientation of the branches

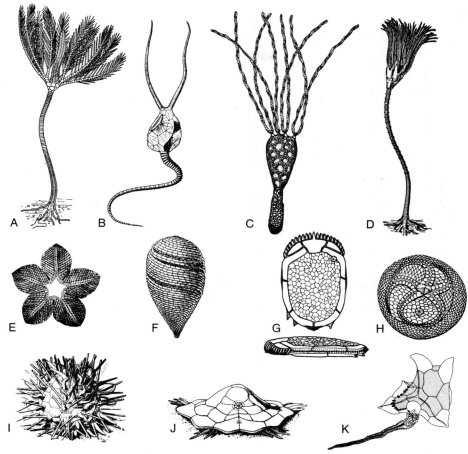

FIGURE 7-9

Representative echinoderms of the Paleozoic.

A. The crinoid (*Botryocrinus*) has a regular arrangement of calcareous plates and numerous food-gathering arms, generally branched.

B. The cystoid (*Pleurocystites*) has numerous irregularly arranged plates with pores and a long stem that was not used for attachment to the substrate.

C. The eocrinoid (*Gogia*) has irregularly arranged calcareous plates with pores along their margins and slender food-gathering appendages.

D. The blastoid (*Pyramiblastus*) has plates arranged in three circlets, complicated internal structures and numerous food-gathering appendages.

E. The starfish (*Villebrunaster*) has five radial arms; this genus of Lower Ordovician age is not too unlike modern forms.

F. The helicoplacoid (*Helicoplacus*) has a spirally pleated skeleton with food grooves.

G. The ctenocystoid (*Ctenocystis*) has a flattened, bilateral skeleton with a grill-like mouth apparatus at one end. Upper drawing is dorsal view, lower drawing is a side view.

H. The edrioasteroid (*Carneyella*) has a round, flattened body with five curved, unbranched food areas on its upper surface.

I. The echinoid (*Bothriocidaris*) has regular arrangement of calcareous plates bearing numerous spines and fleshy food-gathering tube feet.

J. The ophiocistoid (*Volchovia*) has a flattened, saucer-like skeleton with plated appendages on the under side.

K. The stylophora (*Ceratocystis*) has a tail-like appendage and a flattened skeleton with irregular plates. This form probably lived lying horizontal on the sea floor.

Part A from Bather, 1910; B from Paul, 1968; C, D, and K from Vol. S of *Treatise on Invertebrate Paleontology*, Geol. Soc. Am. and Univ. of Kansas Press, 1967; F, H, and J from Vol. U, 1966; E from Durham, 1964; G from Robison and Sprinkle, 1969; I from Durham, 1966.)

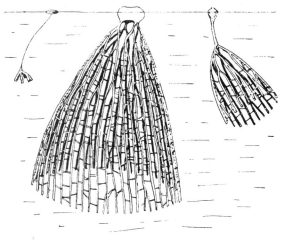

FIGURE 7-10

Reconstruction of the probable habit of life of
Dictyonema. (From Størmer, 1933.)

gradually changed, decreasing the angle between the thread and the branch. A common Middle Ordovician form with the branches roughly perpendicular to the thread is shown in B. Finally, by the Late Ordovician, the two branches were vertical, back to back (D, I). In the Silurian, one of these branches had been lost entirely, and a row of cup or tube-like living chambers had formed along only one side of the branch (F, G). Such forms are called monograptids. Another major evolutionary trend affected the shape of the living chamber of the individual animal of the colony. This trend is especially characteristic of the Silurian monograptids.

For the Ordovician and Silurian probably no other group of fossils excel the graptoloids as zonal markers; because of their characteristic evolutionary pattern along a few dominant trends, in fact, we can determine age, at least approximately, solely by the morphology, without resorting to specific identification.

The eurypterids, spider-like arthropods, were especially abundant in the Silurian and Devonian (Fig. 7-12). Some grew to be ten feet long. The long, segmented body

was covered with thin chitin, and on the ventral side of the head were six pairs of appendages, variously modified for walking, swimming, defense, and grasping prey. The environment of the eurypterids is not known in detail; they were certainly aquatic, but we do not know whether they preferred marine, fresh, or brackish water. The distribution of their fossils suggests, at least, that they did not live in water as saline as the seawater of today. Whatever their environment, they were among the largest animals of the period, and it seems likely that many eurypterids were fierce predators.

FIGURE 7-11

Representative graptoloids.

A. *Dichograptus*, Lower Ordovician.
B. *Leptograptus*, Middle and Upper Ordovician.
C. *Didymograptus*, Lower Ordovician.
D. *Diplograptus*, Middle Ordovician and Lower Silurian.
E. *Goniograptus*, Lower Ordovician.
F. *Monograptus*, Silurian.
G. *Monograptus*, Silurian.
H. *Paraplectograptus*, Middle Silurian.
I. *Hallograptus*, Ordovician.

(From *Graptolithina*, Vol. V of *Treatise on Invertebrate Paleontology*, Geol. Soc. Am. and Univ. of Kansas Press, 1955.)

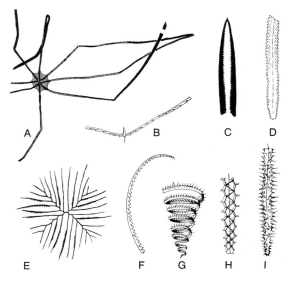

7-2 THE FIRST VERTEBRATES

As far as the direct paleontological record is concerned, the early Paleozoic seas were the domain of invertebrates, for not until the Upper Silurian do we find any considerable fauna of fossil fish, and not until the Devonian do we find fossil fish diverse, abundant, and widespread. The earliest unquestioned remains of fossil vertebrates are from an Ordovician formation in Russia and one in Colorado (Harding Formation) but these remains consist of only small bony scales. The Late Silurian fish were jawless; they belonged to the class Agnatha, the most primitive of all vertebrate classes. In the Late Silurian and the Devonian the Agna-

tha were represented by a diverse group of jawless fish, commonly known as ostracoderms, which, except for a few types, became extinct at the close of the Devonian. The modern lamprey is a surviving member of this class. The first fossil lampreys were recently discovered in Middle Pennsylvanian strata of northeastern Illinois. The fossils are flattened impressions in concretions that show the eyes, gill basket, liver, intestinal tract, and body outline preserved as dark stains.

One of the best-known fossil ostracoderms is *Hemicyclaspis*, illustrated in Figure 7-13. It was about a foot long, and was covered with a heavy bony armor. The head consisted of a single piece of bone, rounded in

FIGURE 7-12

Representative eurypterids of New York State during the Late Silurian.
A. *Pterygotus*.   B. *Stylonurus*.   C. *Hughmilleria*.   D. *Carcinosoma*.
(Chicago Museum of Natural History.)

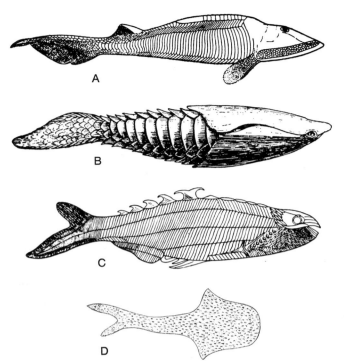

FIGURE 7-13

Ostracoderms.

A. *Hemicyclaspis*. (From Romer, 1946.)
B. *Anglaspis*. (From Romer, 1946.)
C. *Birkenia*. (From Stetson, 1928.)
D. *Lanarkia*. (From Traquair, 1899.)

front and tapering from the body to the anterior margin. The top of the head had openings for a pair of eyes and a nostril (Fig. 7-14). Between the eyes was the small pineal opening, whose function may have been that of a light-receptor. Immediately behind the eyes and along the lateral dorsal surface of the head were depressed areas covered with polygonal bones. These may also have been sensory fields of some sort, but this is very uncertain. The ventral side of the head was flat and covered by small plates. The mouth was round and anterior. Along the sides of the ventral surface of the head were gill openings, ten on each side. The body of *Hemicyclaspis* merged directly with the head and tapered back to a long tail. The body was roughly triangular, flat on the ventral side and arched on the dorsal side. The body armor consisted of vertically arranged, slender bony plates. In the middle of the back was a dorsal fin. Fin-like structures were also present on each side of the rear end of the head. The anatomical features of this particular os-

tracoderm—the flattened body outline (especially the flat ventral region), the ventral position of the mouth, and the dorsal eyes—suggest that it lived on the bottom and sucked up particles from the mud.

A very different type of ostracoderm was *Anglaspis* (Fig. 7-13, B), which had a tubular, heavily armored body without median or paired fins. The group to which *Anglaspis* belonged lacked a dorsal nostril, was wider between the eyes, and lacked bone cells in the armor.

The most fish-like ostracoderm was *Birkenia*, which had a compressed, deep body and an anterior mouth like a transverse slit rather than a simple round hole (Fig. 7-13, C). Instead of massive bony armor on the head and throat it had a complex pattern of small scales. The body was covered with narrow, elongated plates in several rows. *Birkenia* had no fins but had rows of flattened spines, on the dorsal and ventral parts of the body, and a small spine about where a pectoral fin is normally found. The tail was somewhat like that of *Anglaspis* in that the lower part was longer than the upper part. This tail pattern tends to drive the animal upward, whereas a longer upper part, as in *Hemicyclaspis*, tends to drive the animal downward. The body shape, the anterior position of the mouth, and the type of tail suggest that *Birkenia* swam near the surface of the water feeding off plankton.

The fourth ostracoderm group is illustrated by *Lanarkia* (Fig. 7-13, D). Fossils of *Lanarkia* are generally not much more than flattened impressions showing the body outline. *Lanarkia* had no armor like that of the other groups, but the body was covered with scales.

The four early fish illustrated in Figure 7-13 are typical of the four ostracoderm orders. It is readily seen that there was great divergence in form and structure within the class Agnatha, and within each of the orders there was great variability. The fossil record demonstrates that they were fairly successful during the Late Silurian and Devonian and then became nearly extinct. Probably one of the main reasons for their extinction was the rise of primitive jawed fish in the Devonian. The Agnatha, being jawless, were too restricted in their adaptive possibilities to compete successfully with animals that invaded the same ecological niches.

Although no known ostracoderm type has qualified as ancestral, scientists agree that the primitive jawed fish evolved from ostracoderms. This situation may not seem strange on closer inspection. Clearly, the ostracoderm types discussed here were all rather specialized. The occurrences of isolated bony plates in the Ordovician of Colorado and Russia (probably Upper Ordovician) carry the history of the ostracoderms back a whole geologic period, the gap between this Ordovician record and the beginning of fairly abundant fossils in the Late Silurian is fifty million years. We know, therefore, that the faunas of the Late Silurian and Devonian represent the final rather than the beginning phases of the ostracoderms' evolutionary history. Why should there be this tremendous gap in the record?

Professor A. S. Romer of Harvard University has offered the following explanation, which appears very plausible. The evidence of the sedimentary formations containing ostracoderms, and the structures of the fish themselves, strongly suggest that they lived in fresh-water streams and lakes or in estuaries. The answer to our question lies in the virtual absence of nonmarine facies from the record left us of the early Paleozoic. Unquestionably such deposits were formed, but they were subsequently eroded. Not until the middle and late Paleozoic orogenic phases were tectonic conditions favorable to the preservation of nonmarine facies.

One other point of interest about the first fish: Why should they have been armored? Their mobility must have been greatly decreased by their bony coating. Presumably the armor, as in all animals, had one function—protection. Since these fish were jawless and no other vertebrates were present to feed on them, the predators must have been invertebrates. Eurypterids had the size, the mouths, and the claws to cope with the early fish. These creatures dwindled into insignificance once the ostracoderms were on the decline, at about the same time during the Devonian that the primitive jawed fish evolved and became established. Most jawed fish were more agile swimmers than the ostracoderms and thus better able to escape the predatory eurypterids.

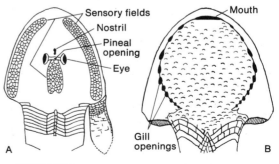

FIGURE 7-14

Upper (A) and lower (B) surface of the head of the ostracoderm *Hemicyclaspis*. (After Romer, 1946.)

7-3 PRIMITIVE JAWED FISH— THE PLACODERMS

The evolutionary history of the vertebrates is punctuated by tremendously significant morphological changes that immediately shifted the whole adaptive range of the group. The appearance of jaws is one of these significant morphological innovations. The adaptive range of the jawless fish was very restricted; but, once an effective jaw developed, the whole evolutionary future of the vertebrates became assured.

The ostracoderms had gills along the sides of the skull; in *Hemicyclaspis* there were as many as ten on each side (Fig. 7-14). The gills were supported by cartilaginous or bony processes called gill arches. Jaws evolved from one of these; the first two arches disappeared, and probably the third became modified into jaws. In the first jawed fish, the placoderms, the upper jaw was attached to the skull only by ligaments, and immediately behind the jaw was a full, normal gill slit (Fig. 7-15, B). In the higher fishes the upper jaw became anchored to the skull through firm attachment of the upper part of the first arch behind the jaw (the hyoid arch), close to the ear capsule and to the point of jaw articulation. This transformation crowded out the gill slit that had been immediately behind the jaw; the slit became a vestigial structure consisting only of a small hole, called the spiracle, near the ear.

The placoderms evolved during the Late Silurian and Devonian into a highly variable group, many of which became unquestioned masters of their environment. They died out by the close of the Paleozoic—the only class of vertebrates without living descendants. A few of the better-known placoderms are illustrated in Figure 7-16.

Lunaspis (Fig. 7-16, B) is a placoderm typical of the early Devonian marine habitat, known only from Europe. The head and body were protected by a well-ossified shield, which on the body bore large spines,

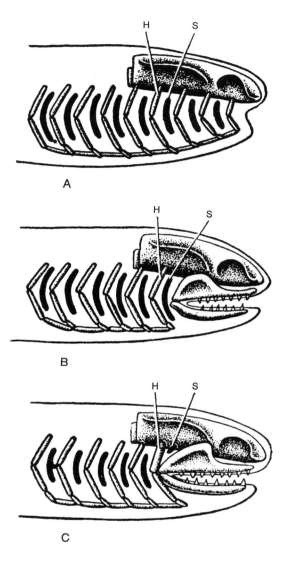

FIGURE 7-15

Evolution of the jaws in the vertebrates: (A) The primitive condition, in which the jaws had not developed, the head having only gill slits and gill arches. (B) In the placoderms the jaws were formed from the forward gill arch (mandibular arch) but were quite separate from the brain case and were attached to it only by ligaments; the gill slit between the jaw and the hyoid arch was unspecialized. (C) In the higher fishes the upper part of the hyoid arch (the hyomandibular) is attached to the brain case near the ear capsule, and the other end is connected with the point of jaw articulation, which thus becomes firmly braced to the skull. (After A. S. Romer, 1946.)

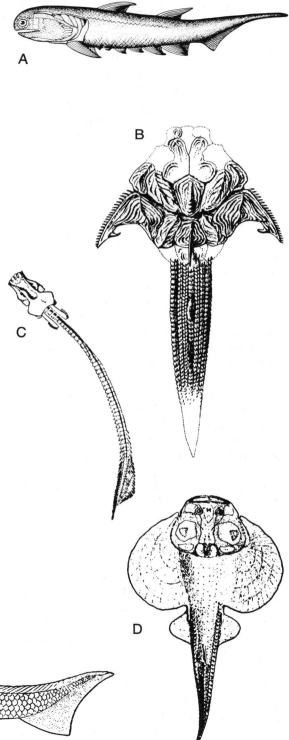

and the posterior part was protected by large spines. Another placoderm, *Gemuendina* (Fig. 7-16, D), looked like a modern rayfish with greatly enlarged pectoral fins. The body was very flat and covered by small plates. The mouth was anterior, the eyes and nostrils dorsal. Probably it lived on the bottom, much as do the present-day skates and rays.

A strange group of placoderms is represented by *Pterichthyodes* (*Pterichthys*) (Fig. 7-16, E). The head and the anterior part of the body were heavily armored; the posterior and tail had heavy scales. The mouth was ventral, and the jaw rather weak. Most unusual were the bony, articulated pectoral appendages. It is hard to see how these appendages could have been effective propulsive organs, but they may have served as steering devices and have been useful for life on the bottom.

The most spectacular and largest of all the placoderms was the group exemplified by *Dinichthys* (Fig. 7-17), which underwent a widespread evolution in the Devonian but became extinct at the close of that period. It evolved the giants of that time: *Dinichthys*, for instance, attained a length of thirty feet, and was the unquestioned

FIGURE 7-16

Primitive jawed fish. (A) *Climatius*. (B) *Lunaspis*. (C) *Palaeospondylus*. (D) *Gemuendina*. (E) *Pterichthyodes* (*Pterichthys*). Those represented by B, D, and E are typical placoderms. The group represented by A is considered by some authorities to be a primitive bony fish; that represented by C is unique, and its placement in the placoderms a matter of convenience. (From A. S. Romer, 1946.)

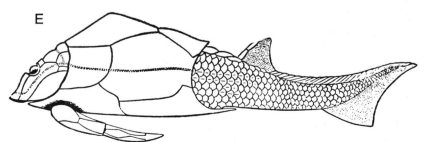

238

FIGURE 7-17

The largest of the Devonian placoderms, *Dinichthys*, pursuing a typical Devonian shark, *Cladoselache*. (Drawing by Z. Burian under the supervision of Prof. J. Augusta.)

master of the Late Devonian seas. *Dinichthys* and its relatives were bony fishes with good armor, but the armor was confined to the head and the thorax. Since the head's armor and the thoracic armor were quite distinct and were joined by well-formed ball-and-socket joints, the head was freely movable. There were no true teeth, but the edges of the jaws were sharply serrated.

Two very strange Devonian fish have generally been considered placoderms but doubtfully so. The first was a strange creature called *Palaeospondylus* (Fig. 7-16, C), known from one quarry in Middle Devonian strata in Scotland. This very small fish has no dermal armor but does have ossified internal structures, differing so from other Devonian fishes that its placement among the placoderms is a matter of convenience.

The second of these strange fishes can be illustrated by the genus *Climatius* (Fig. 7-16, A), which was generally only a few inches long with a shark-like body and a long, slender tail. The middle of the body, dorsal and ventral, was well supplied with fins supported by stout spines. Besides the

median fins there were ventral paired pectoral (shoulder) and pelvic fins. Unusual were the five pairs of fins between the pectoral and pelvic pairs; higher vertebrate fishes have only two pairs of paired fins. The body covering consisted of small diamond-shaped scales. *Climatius* lived in fresh-water streams and lakes and was fairly common from the Late Silurian to the close of the Paleozoic, when it became extinct. Some students think *Climatius* belongs to the bony fishes, the Osteichthyes, its taxonomic placement is a matter of controversy, merely emphasizing that much remains to be learned about primitive vertebrates.

This roster of placoderms, presenting a form typical of each recognized order, demonstrates the great diversity of the primitive jawed fish, but we do not find among them the ancestors of the higher fishes. Presumably the ancestral form is a still-undiscovered placoderm, probably of Silurian age.

The higher fishes include two distinct and well-defined groups, which probably evolved independently from a placoderm ancestor. These are the sharks (class Chondrichthyes) and the bony fishes (class Osteichtyes). In both classes the jaw is firmly attached to the brain case and is much more efficient than the placoderm jaw (Fig. 7-15). There seems no question but that the rise of the higher fishes was primarily responsible for the demise of the ostracoderms and placoderms.

7-4 THE SHARKS

The sharks, though never really abundant, have been and still are successful fish, persisting from the Devonian into Recent time. The cartilaginous skeleton was previously thought to be primitive, but it is now generally assumed that the cartilaginous skeleton was a secondary development and that the bony skeletons of the ostracoderms and placoderms were truly primitive. The only hard parts of the shark's skeleton are the teeth and certain spines, and generally these are the only parts found as fossils.

A well-known and typical Devonian shark is *Cladoselache* (Fig. 7-17), which in outward appearance resembles many modern sharks. This shark represents a persisting primitive type, from which the other major groups of sharks evolved.

7-5 THE BONY FISHES

The most successful vertebrate animals, and the best adapted to life in water, are the bony fishes. Their evolutionary success and diversity is shown by the number of species, of which there are more than all other vertebrate groups put together. Soon after evolving from some placoderm ancestor the bony fish diverged into two main evolutionary lines, the one including the familiar ray-finned fish (subclass Actinopterygii), and the other including the air-breathing fish (subclass Choanichthyes). The first subclass is more diverse and abundant than the second. The success of these fish arises partly from their swimming efficiency, which is the result of better body streamlining and well-developed fins.

The evolutionary history of the bony fishes is complex and, for our purposes, really a sideshow to the evolution of vertebrate life. The ray-finned fish were extremely successful and important in the marine environment, but they did not give rise to any other vertebrate types; the air-breathing fish are much more important in the evolutionary picture.

The Choanichthyes comprise two principal orders, the dipnoans, or lungfishes, and the crossopterygians, or lobe-finned fishes. Throughout its history the latter order was never very abundant or diverse, but it is of major evolutionary significance because it included the ancestors of the first land animals.

At present there are three genera of lungfishes, one in Australia, one in Africa, and one in South America (Fig. 7-18). The Australian genus (*Epiceratodus*) is able to survive in stagnant pools by coming to the surface and breathing air, but it cannot live out of water. The other genera can survive for months out of water by burying themselves in mud. They breathe by leaving openings through the mud to the outer air. The lungfishes' ability to breathe air convinced many early students that they were surely the ancestors of land vertebrates. The modern genera, however, are highly specialized and somewhat degenerate forms. Tracing the history of the lungfishes back in the fossil record, we find that in the Devonian they were highly specialized—too much so to have played a role in the evolution of land-living vertebrates.

Among the crossopterygians, however, we find a fish, the typical and well-known Devonian *Osteolepis* (Fig. 7-19), whose anatomy was perfect as a base for the evolution of land-living vertebrates. Superficially it was like any typical fish, but it differed from the bony fishes in having two dorsal fins and a very different sort of scales, and its pair of lobed fins contained bones that closely resembled those of the limbs of a typical quadruped (Fig. 7-20). The bone pattern of the skull and the jaws resembles that of other bony fishes and of primitive quadrupeds. Detailed study of the crossopterygian braincase shows that the brain cavity, blood vessels, nerve channels, and internal ear are highly similar to those of amphibians.

Typical crossopterygians became extinct in the Permian, but one aberrant branch, the coelacanths, survived much longer. The crossopterygians were fresh-water fish, but the coelacanths were marine, and their structure modified with the change in environment. For many years it was believed that the coelacanths became extinct in the Cretaceous. A really noteworthy discovery was that of living coelacanths from the deep waters off the Comoro Islands, between Madagascar and East Africa (Fig. 7-21). The gap in the fossil record is presumably due to the possibility that the coelacanths have long been confined to ocean depths from which sediments able to preserve them as fossils are seldom recovered.

7-6 THE FIRST AMPHIBIANS

Danish explorations in eastern Greenland uncovered, in 1932, Late Devonian fish and the oldest known amphibians. The amphibians, called icthyostegids, are very primitive and clearly related to the crossopterygians in many anatomical features. The skeleton and a sketched restoration are shown in Figure 7-22. These creatures were not yet well adapted to land life and must still have spent most of their time in the water. The backbone, little advanced over that of a crossopterygian, was not efficiently designed for supporting the animal out of water. Except for the four short legs the animal was very fish-like with its torpedo shape and finned tail. The jaws were lined with conical teeth whose labyrinthine structure was very similar to that of the crossopterygians (Fig. 7-23). These ancient quadrupeds are known as labyrinthodont amphibians.

Why should four-legged amphibians have evolved? When they were living in the same environment as the crossopterygians and their diets were probably the same (fish), there was little food for them on land. During the Caledonian orogeny, in the Devonian, there was a period of aridity; as pools and streams dried up, animals that could live for long periods out of water had a distinct advantage. This explanation is attractive, but another suggestion put forth by Professor A. S. Romer, in view of all the data available, is even more so. Perhaps amphibians evolved, not to escape from water, but to stay in water. If many regions were

FIGURE 7-18

The three living genera of lungfishes.

A. *Protopterus*, the African mudfish.
B. *Lepidosiren*, the South American lungfish.
C. *Epiceratodus*, the Australian lungfish; this is the most like the fossil lungfish.

(From J. R. Norman, *A History of Fishes*, Benn, London, 1931.)

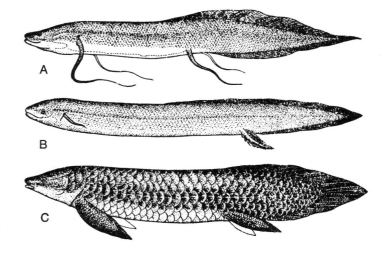

FIGURE 7-19

Osteolepis, a typical Devonian crossopterygian. (From Romer, *Proc. Am. Phil. Soc.*, 1956.)

excessively dry during the late Devonian, fish caught in a drying pool were doomed. The crossopterygians, however, if their lobed fins were slightly modified and strenghtened, could, as a pool dried up, go across country looking for another pool and, of course, a supply of food. Which explanation is closer to the truth is still an open question, but one thing is sure: this chapter of the evolution of life has not yet been written in final form.

The amphibians underwent an interesting adaptive radiation from an ichthyostegid ancestor. The development toward more efficiency of movement on land continued and culminated in a common Permian amphibian known as *Eryops* (Fig. 7-24). This creature grew to about six feet in length, was very low, but had very stout legs, and the head was large. There seems no question but that *Eryops* could take care of himself; his size and massive build enabled him to cope with any other land animal of the time. *Eryops* probably lived somewhat like the alligator, along the banks of streams and pools, in and out of the water.

From *Eryops* evolved, in the Triassic, a group of amphibians that became almost entirely adapted to life in the water. They had very small, weak legs, greatly flattened bodies, and very large and extremely flat heads. They made up the fossilized graveyard illustrated in Figure 2-3 (p. 26).

The labyrinthodont amphibians became extinct at the close of the Triassic. The modern amphibians—toads, frogs, and salamanders—came from stocks that originated in the late Paleozoic. Another group of labyrinthodont amphibians, the embolomeres, were at least close to the group from which the reptiles evolved. Without going into anatomical details we may say that the embolomeres, as illustrated by *Archeria* (Fig. 7-25), though retaining many amphibian characteristics, also showed relationship to the reptiles.

7-7 THE FIRST REPTILES

In § 7-3 we mentioned the appearance of jaws in the placoderms, a very significant breakthrough in evolution. A second such step was the appearance of amphibians, transitional to land-living animals. Another major step was the development of the **amniote egg** (a large, yolked, shelled egg with characteristic embryonic membranes). Amphibians, though they could live and move about on land, were never completely free of the water, for they had to return to it to reproduce. With the development of the amniote egg by the reptiles, a new mode of reproduction was introduced. Land animals no longer were tied to water for reproduction, a change that greatly enlarged their evolutionary and adaptive potential. The improved method of reproduction, more than anything else, accounted for the reptiles' evolutionary advantage over the amphibians.

The transformation from amphibians to

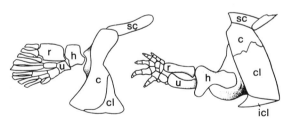

FIGURE 7-20

The fish-amphibian transition: the shoulder girdle of a Devonian crossopterygian (left) compared with that of a primitive quadruped (right).

sc, supracleithrum   *h*, humerus
c, cleithrum   *r*, radius
cl, clavicle   *u*, ulna
icl, interclavicle

(From Swinton, British Museum, Natural History.)

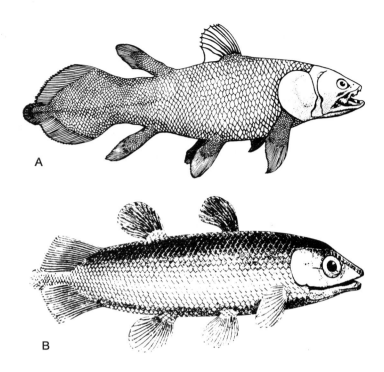

FIGURE 7-21

The surviving coelacanth, *Latimeria* (A), and a fossil Mesozoic coelacanth crossopterygian (B). (From Romer, *Proc. Am. Phil. Soc.*, 1956.)

A

B

reptiles took place in the early Carboniferous. One early reptile is represented by *Hylonomus*, collected from hollow tree stumps in the Coal Measures of Nova Scotia (Fig. 7-26). *Hylonomus* belonged to the cotylosaurs, sometimes called the stem reptiles because all other reptile groups evolved from them.

Reptiles living today can easily be distinguished from amphibians by means of soft parts and skeletal features. But fossilization generally obliterates many critical features. It is almost impossible to distinguish primitive reptiles from some of the earliest amphibians because their skeletons are so similar. The true test is the type of egg the animal laid; but few extinct forms can be distinguished thus. A classic example of the difficulty encountered in distinguishing amphibian from reptile is *Seymouria* (Fig. 7-27), from the Early Permian of Texas. *Seymouria* has been variously classified as an amphibian or a reptile. The discovery of a fossil egg in the same beds with *Seymouria* skeletons has been thought by some to favor the assignment as a reptile. But the

same strata contain several true reptiles. On the basis of evidence obtained primarily from other forms related to *Seymouria*, the consensus today is that it was an amphibian. Thus the controversy emphasizes the transitional nature of skeletal changes from amphibian to reptile. Soon after their appearance in the Carboniferous the cotylosaurs underwent elaborate evolutionary radiation that established nearly all the major rep-

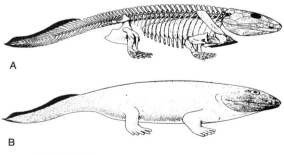

A

B

FIGURE 7-22

Skeleton (A) of a Devonian ichthyostegid amphibian and a sketch (B) of its probable appearance in life. (From Jarvik, Am. Assoc. Adv. Sci., *Scientific Monthly*, 1955.)

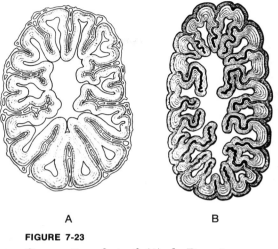

A   B

FIGURE 7-23

Cross section of a tooth (A) of a Devonian
crossopterygian (*Eusthenopteron*) and (B) of
a labyrinthodont amphibian (*Benthosuchus*).
(From Bystrow, 1938, 1939.)

tilian groups. In the Permian there was
already a very diverse and abundant reptile
fauna, but it was not until the Mesozoic that
the reptiles reached the climax of their
development.

In contrast to marine invertebrate faunas,
land animals did not undergo a severe
crisis of extinction at the close of the Paleozoic; on the contrary, reptiles and amphibians continued to evolve with no significant
breaks or drastic changes.

7-8   FROM FISH TO REPTILES—
A RECAPITULATION

A diagrammatic summary in the form of an
evolutionary tree, with sketches of all the
main animals discussed so far in this chapter, is shown in Figure 7-28.

FIGURE 7-24

Restoration of *Eryops*, a large, clumsy, Permian
amphibian. (From Romer, *Proc. Am. Phil. Soc.*, 1956.)

The oldest evidence we have of vertebrates consists of isolated bony plates and
scales from Ordovician strata in Colorado
and Russia. Not until Upper Silurian, however, do fossil vertebrates become significant in the paleontological record. In these
strata and in the succeeding Devonian are
found a large variety of armored fishes that
had no jaws. These are the ostracoderms,
belonging to the class Agnatha, the most
primitive of all vertebrates. The modern
lamprey and hagfish are members of the
Agnatha. Although their variety indicates
adaptation to several different modes of
life, the lack of jaws and the bony armor
limited the ostracoderms' adaptability.
These early fish seem to have lived in fresh
water.

From an unknown ancestry within the
ostracoderms evolved a group of primitive
fish with jaws, the placoderms, known from
Upper Silurian and Devonian formations.
The evolution of the placoderms produced
a large number of varied types, some marine, among which was the giant *Dinichthys,*
the largest vertebrate of its day. With the
close of the Devonian most ostracoderms
and placoderms became extinct, doubtless
because of the appearance of higher fishes
with much more efficient jaws.

The higher fishes (class Osteichthyes),
soon split into two main evolutionary
lines. One is that of the ray-finned fishes,
which continues to the present and whose
evolutionary success and diversity are
shown by the fact that there are more
genera of modern ray-finned fish than of
all other vertebrates. The other line includes two groups of air-breathing fish,
the lungfish, or dipnoans, and the crossopterygians. These two groups were closely
related but diverged early in their history
when the lung-fish became particularly
specialized for breathing air. The crossopterygians are of special importance
because they had anatomical structures
from which it was possible for land-living
animals to evolve. The key feature was the
lobed fins, which had bony supports.

In the Late Devonian the first amphibians, represented by the icthyostegids of eastern Greenland, evolved from the crossopterygians. Their origin is clearly shown by such anatomical features as the structure of their teeth and limbs. Soon after the appearance of the icthyostegids the amphibians rapidly evolved many different types. *Eryops* represents the full development of one of these lines into an animal thoroughly adjusted to a life spent mostly on land. The big evolutionary phase of the amphibians came to an end in the Triassic, and since then the group has been represented by the frogs, toads, and salamanders. One early amphibian radiation resulted in the embolomeres, represented by *Archeria*. This branch is significant in that it shows characteristics transitional to the reptiles.

In the Carboniferous, the first reptiles evolved and soon underwent an evolutionary radiation, establishing most of the reptile orders that dominated the world during the Mesozoic. The significant innovation in the reptile was the changed mode of reproduction, the amniote egg, which released land animals from their intimate tie with an aqueous environment and opened the way for an adaptive radiation that had been impossible for the amphibians.

7-9 THE APPEARANCE OF LAND PLANTS

The land surface of the early Paleozoic world must have been a strange sight, for there was no plant cover. The first evidence we have of land plants are specimens from Upper Silurian rocks in Bohemia. Lower Devonian strata from several places in Europe have yielded fossil floras, and the fossil record of land plants improves greatly from Lower Devonian upward. The earliest land floras were undoubtedly derived from marine forms, but just when and how this transformation took place is still a mystery because of the sparse fossil floral record.

As one would expect, the Early Devonian plants were the most primitive of the vascular plants — that is, of plants having specialized tissues for the conduction of food and water. A reconstruction of a coastal marsh of this time is illustrated in Figure 7-29. Nearly all the plants in this habitat group belong to the Psilophytales, the oldest and simplest group of the vascular plants. They had no proper root system; instead, the subterranean part of the stem served as the root. The stems were naked or bore small leaves, and the spore-bearing organs were at the end of the main shoot or its branches. Even though

FIGURE 7-25

Restoration of *Archeria*, a late Paleozoic embolomere amphibian. (From Romer, *Proc. Am. Phil. Soc.*, 1956.)

FIGURE 7-26

Hylonomus, one of the early cotylosaur reptiles. (Reproduced with permission from R. L. Carroll, *J. Linn. Soc. Zool.*, **45**, 1964.)

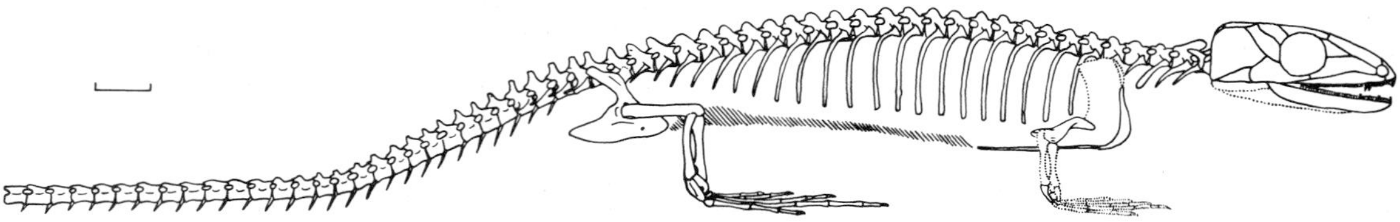

this group was highly primitive in its structures, it evolved considerably diverse forms, four of which are illustrated in the figure.

One of the commonest and best-known Early Devonian plants is *Psilophyton*, which grew to a height of about three feet and inhabited wet marshy areas along streams and swamps. The slender stems rarely exceeded one centimeter in diameter. The spore-bearing organs were borne on the tips of bifurcated branchlets. *Psilophyton* produced no real leaves, but the longer stems were covered with short, rigid, pointed spines. The subterranean part of the plant was a horizontal stem, from which the vertical, above-ground portions grew.

Right in front of *Psilophyton*, in Figure 7-29, are specimens of *Sciadophyton*, which look like rosettes of slender leaves on the ground. Other forms of the Psilophytales (*Zosterophyllum* and *Taeniocrada*) lived in the water at all times; they look much like algae but are true vascular plants.

The most abundant plant of those shown in Figure 7-29 is *Protolepidodendron*, which belongs to a slightly more advanced group of plants that became particularly widespread and diverse in the Carboniferous. This plant had branched stems that bore small, closely set, linear leaves. The horizontal stem had a weak root system.

As we can see from Figure 7-29, the Early Devonian flora consisted of small plants of primitive structure that were generally confined to coastal marshes or other well-watered environments. Once plants became established on land, however, they underwent a rapid evolutionary development. By the Middle Devonian the number and diversity of plant types had increased greatly, and the first real forests had been established. A reconstruction of a Middle Devonian landscape is shown in Figure 7-30. The plants (Psilophytales) that had dominated the Early Devonian lands were now a minor element of the flora; for the most part they occupied the marshy areas (*Rhynia, Asteroxylon*) or formed leafless tree-like forms (*Pseudosporochnus*). The other elements of this flora are lycopods, scouring rushes, and ancient ferns, all groups that played dominant roles during the later Paleozoic. The large tree with the fern-like leaves is *Aneurophyton*, and the slender leafless tree is *Archaeosigillaria*. These genera are particularly well known from Gilboa, New York, where several successive horizons in a Middle Devonian formation have yielded numerous stumps, still in upright positions. The geological setting of these fossil trees clearly shows

FIGURE 7-27

The difficulty of distinguishing between amphibians and early reptiles is well illustrated by this form from the Permian of Texas, called *Seymouria*. It was long considered to be a reptile, thus the eggs in the picture; the consensus now is that it is an amphibian. (American Museum of Natural History.)

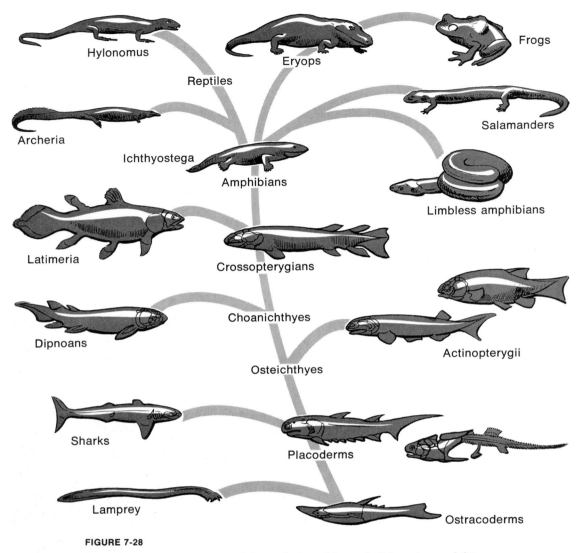

FIGURE 7-28

Diagrammatic representation of the evolution of the early fishes, the amphibians, and the earliest reptiles. (Data from A. S. Romer.)

that each horizon represents a forest that became inundated by an advancing sea. The stumps are all broken off at five and a half feet or less. They had bulbous bases as much as three feet in diameter. The trees (*Aneurophyton*) must have had a height of twenty feet or more. Because of their imperfect preservation their precise botanical affinities remain unknown.

7-10 THE CARBONIFEROUS FLORAS

The evolution of plants during the Paleozoic reached its highest point in the Pennsylvanian (Late Carboniferous), the great coal-forming period of the earth's history. Let us first investigate the nature of the principal elements of this flora and then discuss the formation of coal.

Figure 7-31 illustrates some of the principal plants of the Carboniferous. The tall, jointed, reed-like plants in the outer circle are *Calamites*, closely related to modern scouring rushes or horsetails. These usually grew fifteen to twenty feet high but some giant forms attained heights of forty feet. They were tall, slender, tapering plants with a thick pith core and a circle of branches at each joint. The smaller branches bore leaves also arranged in a circle. The *Calamites*

bore cones that in some species were a foot long and an inch or more in diameter. These plants are among the commonest in Carboniferous formations. Their general structure and geological setting suggest that they lived in swampy places, forming dense jungles somewhat like the canebrakes of the South. They were already well established in the Devonian and are represented in the flora of Figure 7-30, but not until the Carboniferous did they evolve such large, diverse, and abundant forms.

The large tree in the right foreground of

FIGURE 7-29

Early Devonian landscape showing some of the first land plants. (A) *Zosterophyllum*. (B) *Psilophyton*. (C) *Taeniocrada*. (D) *Protolepidodendron*. (E) *Sciadophyton*. (Drawing by A. Burian under the supervision of Prof. J. Augusta.)

FIGURE 7-30

Middle Devonian landscape showing the first forest that covered the earth.

A. *Aneurophyton.*	E. *Protolepidodendron.*	I. *Pseudosporochnus.*
B. *Archaeosigillaria.*	F. *Cladoxylon.*	J. *Asteroxylon.*
C. *Calamophyton.*	G. *Hyenia.*	K. *Protopteridium.*
D. *Duisbergia.*	H. *Barrandeina.*	L. *Rhynia.*

(Drawing by Z. Burian under the supervision of Prof. J. Augusta.)

Figure 7-31 is a *Lepidodendron,* or scale tree, which grew to impressive size. A trunk of this tree that was uncovered in an English coal mine was found to measure 114 feet up to its first branches and the crown of branches to extend for at least another twenty feet. The living tree must have been more than 135 feet tall. Trunks four feet in diameter were common. Branching did not begin until near the top. The branches bore thickly set awl-shaped leaves, set in a close spiral; when the leaves fell,

they left diamond-shaped scars arranged in oblique rows round the stem. The branches also bore cones at their tips. Cones twenty inches long and two inches in diameter have been recorded. The underground system is rather peculiar; it does not seem to consist of true roots. The so-called roots of the tree, massive and dichotomously branched, and known as stigmariae (Fig. 7-32), are extremely common fossils in coal formations.

The other large tree in Figure 7-31, in the

FIGURE 7-31

Composite group of leading Carboniferous plants: in the foreground at the right, *Lepidodendron;* at the left, *Sigillaria;* in the right center middle ground, a tree fern; in the left center, *Cordaites;* at the extreme right and left, *Calamites.* (From Chamberlain and Salisbury, *Geology* (Vol. 2 of *Earth History*), 1907, by permission of Henry Holt, New York.)

left foreground, is *Sigillaria,* another widely distributed tree, generally found with *Lepidodendron.* It grew to a hundred feet or more. The sword-shaped leaves were nearly three feet long and concentrated in a cluster at the top. When the leaves fell, they left scars, as in *Lepidodendron,* but the scars were six-sided and arranged in vertical rows. The cones, rather than being at the tip of the smaller branches, were arranged in clusters round the trunk, just below the crown of leaves. The underground system of *Sigillaria* was identical with that of *Lepidodendron* and indistinguishable from it.

Cordaites, the tree in the central background of Figure 7-31, is another very common and widely distributed tree of the Carboniferous. It was a tall, slender tree that grew to a height of nearly a hundred feet with a trunk two feet in diameter. The trunk had a central core of pith surrounded by woody tissue. The branches were generally confined to the upper part of the trunk and were clothed with a lush crown of leaves. The leaves varied in size and shape, ranging from narrow, strap-shaped, sharp-pointed leaves nearly three feet long to spatulate leaves fifteen inches wide. Some trees had long, narrow, grass-like leaves.

The great diversity and abundance of fern-like leaves in the Carboniferous floras led early workers to believe that ferns were among the predominant plant groups of the period, but it was later discovered that many of the fern-like leaves belonged to plants that bore seeds rather than spores. Even so, the ferns were significant in the Carboniferous floras. The tree ferns were especially impressive in their size, many growing seventy feet high with trunks two feet in diameter (Fig. 7-31). Large, spreading, compound leaves were concentrated at the top of the tree.

7-11 THE FORMATION OF COAL

Coal is a combustible rock that originated by the accumulation and partial decomposition of vegetation. Conditions favorable for the formation of coal—lush and diversified vegetation and suitable climatic conditions—were particularly widespread during the Carboniferous. In general, a warm and humid climate, like the temperate and subtropical climates of today, were most suit-

able for the accumulation of coal-forming material. Accumulation of plant debris is controlled by climatic conditions; other important factors are the environments of deposition and the geological processes that led to the deposition and preservation of inorganic sediments. The most suitable environment for the accumulation of coal-forming peat (a soft, spongy, brownish deposit in which plant structures are easily recognized) is afforded by flat swamps. Extensive, poorly drained coastal areas in deltas and estuaries of sluggish rivers are ideal for the growth of plants and the accumulation of their remains under water, protected from rot. An impression of a forest swamp of the Carboniferous is shown in Figure 7-33. To identify the main plants in this scene, compare with Figure 7-31.

As plant debris accumulates, a slow depression of the earth's crust may permit a counterbalancing increase of its thickness. If, for any reason, this balance is disturbed, conditions change rapidly. A rise of the water level—say by a marine invasion—would kill all the growing plants and would probably be accompanied by an influx and deposition of inorganic sediments; a lower-

FIGURE 7-32

Grove of stigmariae, casts of the lower, rooted portions of *Lepidodendron* trees in Lower Carboniferous strata, Scotland. (Courtesy of Her Majesty's Geol. Surv., British Crown copyright.)

ing of the water level would expose the previously accumulated plant debris to erosive agents. The sequence of coal-bearing formations of Carboniferous age throughout the world reflects constantly changing conditions. We commonly find several horizons of coal separated by barren beds.

The transformation of peat into the various kinds of coal comes about through burial, the consequent rise in pressure and temperature causing chemical changes. The first products of this transformation are the brown coals, or lignites, which have a high water content and an appreciable quantity of volatiles. As pressure and temperature rise and time elapses, the lignites give off some of their volatiles and much of their water and become bituminous coals.

If the process goes on, it produces anthracite, the highest-ranking coal, which may contain as much as 95 percent carbon and less than 5 percent volatile material.

The Carboniferous was not the only coal-forming period in the earth's history. The principal occurrences of coal are shown in relation to geologic age in Figure 7-34. The oldest known coal deposits are the anthracite in Precambrian strata of Michigan; these deposits were formed from algae, as trees had not yet evolved. The next oldest, on Bear Island (near Spitsbergen) and in northern Russia, are Devonian. During the Carboniferous coal was formed in both the northern and the southern hemisphere—the extensive deposits that now contribute two-thirds of the world's coal supplies. In some regions,

FIGURE 7-33

Reconstruction of a Carboniferous coal-forming forest swamp. (Courtesy of Her Majesty's Geol. Surv., British Crown copyright.)

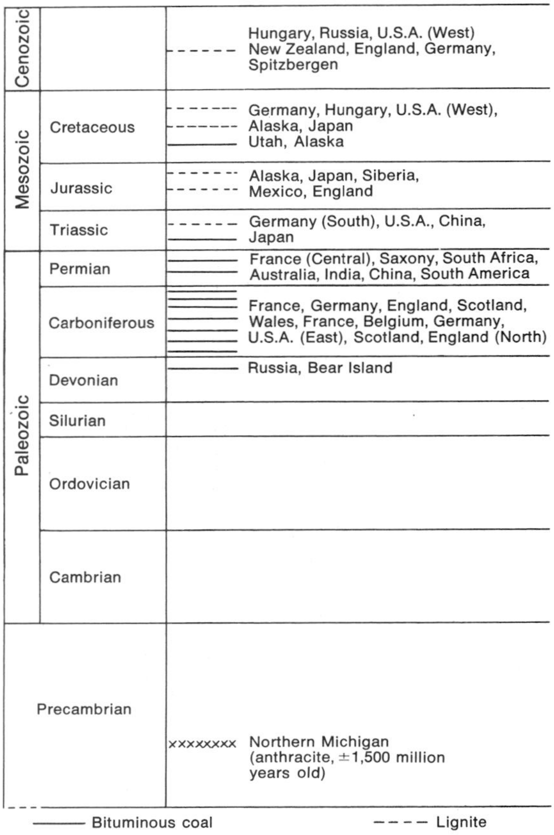

Cenozoic		-----	Hungary, Russia, U.S.A. (West) New Zealand, England, Germany, Spitzbergen
Mesozoic	Cretaceous	----- ------- _____	Germany, Hungary, U.S.A. (West), Alaska, Japan Utah, Alaska
	Jurassic	----- -----	Alaska, Japan, Siberia, Mexico, England
	Triassic	----- _____	Germany (South), U.S.A., China, Japan
Paleozoic	Permian	======	France (Central), Saxony, South Africa, Australia, India, China, South America
	Carboniferous	======	France, Germany, England, Scotland, Wales, France, Belgium, Germany, U.S.A. (East), Scotland, England (North)
	Devonian	_____	Russia, Bear Island
	Silurian		
	Ordovician		
	Cambrian		
	Precambrian	xxxxxxx	Northern Michigan (anthracite, ±1,500 million years old)

————— Bituminous coal   - - - - Lignite

FIGURE 7-34

Principal occurrences of coal in relation to age of formation. (After Adams, 1955.)

notably India, China, Manchuria, Australia, and South Africa, the rhythmic cycle of coal formation, which began during the Carboniferous Period, continued into the Permian.

Most Mesozoic coal deposits are in small scattered basins in central Europe, western North America, central Asia, and southeast Asia. These often equal in richness the Carboniferous coals but yield mostly low-ranking coals. Lignite is the characteristic coal of the Cenozoic, its principal occurrences being in central and southern Europe, Russia, western North America, Australia, and New Zealand.

7-12 MID-PALEOZOIC MARINE INVERTEBRATES

The composition of marine faunas changed considerably from the early Paleozoic to the mid-Paleozoic, by which time many phyla that had been differentiated into many major groups were reduced to only a few. The abundant, diverse, primitive echinodermal groups of the Cambrian and Ordovician were greatly reduced by the Devonian. With the extinction of these groups the blastoids and crinoids greatly expanded in evolutionary diversity, reaching their greatest development in the Early Carboniferous (Mississippian). Trilobites and nautiloids gradually declined after their great development in the Cambrian and Ordovician. One extremely important early Paleozoic group that became extinct before the mid-Paleozoic was the graptoloids. A few dendroids persisted into the Carboniferous.

The corals, brachiopods, bryozoans, pelecypods, gastropods, and foraminifers continued to evolve fairly rapidly in the second half of the Paleozoic. A reconstruction of a Devonian sea bottom is shown in Figure 7-35. As in the Silurian, conditions favorable for the growth of coral reefs were very widespread. Rugose and tabulate corals were still the main groups, and many genera that thrived during the Silurian persisted into the Devonian. Other members of the fauna shown in Figure 7-35 are brachiopods, nautiloids, trilobites, sponges, blastoids, crinoids, and gastropods.

The Devonian faunas were enriched by the first appearance of a new order of invertebrate animals, the ammonoids. These coiled mollusks looked like coiled nautiloids, but anatomically were quite different. Ammonoid fossils are particularly important for correlation in the middle and upper Paleozoic; in the Mesozoic they underwent a phenomenal evolution, producing numerous diverse types that were geographic-

FIGURE 7-35

Restoration of a Middle Devonian sea bottom in the central United States.
This habitat group contains crinoids, blastoids, corals, nautiloids, brachiopods,
trilobites, gastropods, and sponges. Can you recognize all these groups?
(Chicago Museum of Natural History.)

ally widespread. In Chapter 10 we will discuss in detail the ecology and morphology of ammonoids. Here we merely note that the junction of a shell partition (septum) with the shell wall forms a pattern (the suture), which varies from fairly simple to complex. Nautiloids generally have simpler sutures. Also the fleshy tube (the siphuncle) extending back from the body through the partitions is generally in an extreme ventral position in the ammonoids, but in the nautiloids it has various positions, most commonly central. An aberrant group of ammonoids that lived only during the Devonian, the clymenids, had the siphuncle in an extreme dorsal position. A representative group of Paleozoic ammonoids is illustrated in Figure 7-36. Most Paleozoic ammonoids were smooth and small, with simple sutures. Because they evolved rapidly, were widely

distributed, and can readily be identified, they are excellent zonal fossils. Ammonoids are of particular value because intercontinental correlations are largely based on their sequence and character.

Another group of late Paleozoic animals that are exceptionally valuable for purposes of correlation is the fusulinids, a family of foraminifers whose shell looks like, and is about as big as, a grain of wheat. The internal structures are extremely complicated (Fig. 7-37). The fusulinids were widely distributed geographically during the Pennsylvanian and Permian and underwent very rapid evolution. In North America at least ten faunal zones of fusulinids are recognized. Many Pennsylvanian and Permian formations consist almost entirely of the remains of these animals. Generally they occur in carbonate rocks to the exclusion of

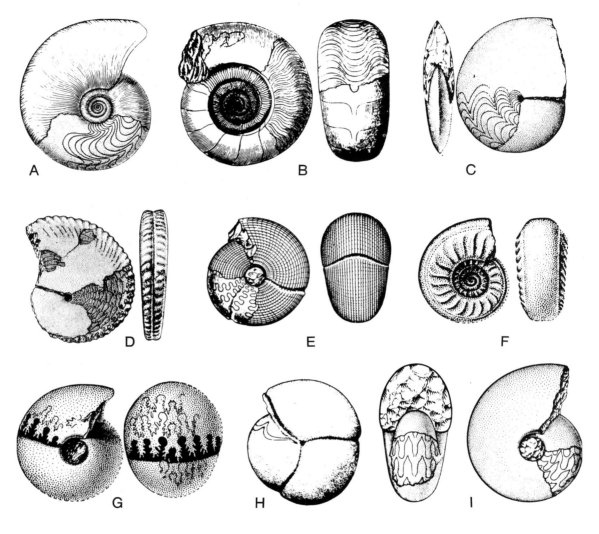

FIGURE 7-36

Representative Paleozoic ammonoids.

A. *Manticoceras*, Upper Devonian.
B. *Anarcestes* (*Anarcestes*), Lower and
Middle Devonian.
C. *Timanites*, Upper Devonian.
D. *Artinskia*, Permian.
E. *Adrianites*, Permian.

F. *Eumorphoceras*, Mississippian.
G. *Waagenoceras*, Permian.
H. *Epiwocklumaria*, a clymenid, Upper Devonian.
I. *Schistoceras*, Upper Pennsylvanian and
lower Permian.

(From Vol. L of *Treatise on Invertebrate Paleontology*, Geol. Soc. Am. and Univ. of Kansas Press, 1957.)

ammonoids. Because the two animal groups, ammonoids and fusulinids, evolved so rapidly and are so widespread and abundant, they are the most important zonal guides in late Paleozoic rocks.

Crinoids were particularly abundant and diverse during the Mississippian. Many formations of this age in the upper Mississippi Valley are composed mainly of the remains of these organisms. The sea bottom must have looked like a forest of waving slender trees with large delicate crowns (Fig. 7-38). Crinoids, like nearly all members of the phylum Echinodermata, were gregarious animals living in large, fairly dense populations where conditions were favorable, to the exclusion of almost all other animals.

7-13 THE END OF AN ERA

The history of life from the Cambrian to the Recent follows a pattern of increasing numbers and diversities of adaptive types; but there were sporadic crises in which the diversity was greatly reduced by extinc-

tions. The first major wave of extinction affecting many marine invertebrates came in the Permian, but it had little effect on land-living vertebrates and plants.

A survey of Early and Middle Permian seas offers no clue to the severe crisis that lay ahead for most of the invertebrate animals. A reconstruction of a typical Permian bottom fauna in the lower Middle Permian strata of the Glass Mountains of western Texas is shown in Figure 7-39. This includes several varieties of brachiopods, nautiloids, sponges, corals, and gastropods. Particularly striking are two brachiopods, the spiny *Aulosteges* and the cone-shaped *Richtofenia*, which is also illustrated in Figure 2-9 (p. 29). In the bottom of the picture are brachiopods that have adopted an oyster-like habit, reflected in their shape and mode of attachment. The nautiloids and sponges are equally diverse. Even a superficial inspection of Figure 7-39 shows this Permian fauna to be a varied one with many special adaptive types. Many of the animals belong to, or are closely related to genera that flourished in the Late Carboniferous;

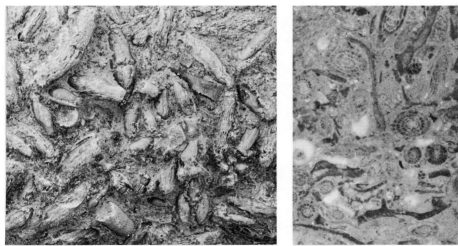

FIGURE 7-37

Upper Pennsylvanian fusulinids from western Texas; to the left is a weathered specimen, showing the external surfaces of many fusulinids; to the right is a polished section of the same slab, showing the internal structures of the fusulinids. (Photographs by Whittington.)

FIGURE 7-38
Reconstruction of a Mississippian sea bottom, in the central United States, populated
by crinoids and blastoids. (Chicago Museum of Natural History.)

others are specialized types that appeared during the Permian.

An interesting aspect of Permian marine life is the group of relic faunas that persisted in Indonesia and nowhere else. Dutch geological expeditions discovered on Timor a most remarkable Permian fauna containing, among other things, blastoids, certain crinoids, and corals. The blastoid, crinoid, and coral faunas are unique because of the many endemic forms. The Indonesian region was a haven for some of these groups long after they had become extinct in the rest of the world.

The question, "Why should apparently successful Paleozoic invertebrates become extinct by the close of the Permian?" has not been answered to everybody's satisfaction. A popular hypothesis suggests a catastrophe as an explanation, but this does not seem plausible. (See the last paragraph in this chapter.)

A survey of the main invertebrate groups of the Paleozoic shows highly variable rates of evolution. In Figure 7-40 the rate is measured by the number of new orders and classes that appear in each period. Only the best-known invertebrate groups were taken into account. The salient periods of evolutionary activity during the Paleozoic were obviously the Ordovician, Silurian, Mississippian, and Permian. The high points for the Silurian and Permian probably represent the culminations of trends begun in the preceding periods and reflect the increased numbers of lower categories (families and genera). The Ordovician and Mississippian highs, however, represent real peaks of evolutionary activity. Times of reduced activity and diversity were the Devonian and Pennsylvanian. Ordinal and class extinctions were concentrated at, but were not confined to, the end of the Paleozoic Era.

The periods of high and low evolutionary rates indicated by Figure 7-40 reflect the changing patterns of land and sea during the Paleozoic. The basic pattern of early Paleozoic seas, which widely inundated the continents, developed gradually from the earliest Cambrian and reached a maximum in the Ordovician. By the mid-Paleozoic many groups (nautiloids, trilobites, graptolites, etc.), which had earlier produced many diverse major divisions, underwent a wave of partial extinction: many became extinct, but others survived. This wave resulted in the low evolutionary rate of the Devonian. This may be called opportunistic evolution: the rise to the Ordovician peak reflects evolution toward the occupation of all, or nearly all, the environmental niches available. The Silurian presents a sort of anticlimax: instead of new

FIGURE 7-39

Reconstruction of a middle Permian sea bottom, Glass Mountains, western Texas.
Sponges: (A) *Heliospongia,* (B) *Girtyocoelia,* and (C) *Defordia.* Nautiloid cephalopods:
(D) *Cooperoceras,* (E) *Metacoceras,* and (F) *Stenopoceras.* Brachiopods:
(G) *Richtofenia,* (H) *Aulosteges,* (I) *Neospirifer,* (J) *Dialasma,* and (K) *Leptodus.*
(Photograph from the Smithsonian Institution.)

forms, the fauna consists of forms surviving from the Ordovician or elaboration on a lower taxonomic level. The low ebb of the Devonian perhaps reflects a decrease in environmental opportunities for the then existing stocks. The mid-Paleozoic orogenies (Caledonian, Taconic, Acadian) did restrict the seas, but we have no sure proof that at the close of the Devonian the seas dried up enough to cause the wholesale extinction that we know took place. In any case, higher rates of extinction in the Devonian, together with lower rates of differentiation in several groups, may have opened the way for the Mississippian eruption. Simultaneous expansion of many groups during the Mississippian suggests that far more opportunities for adaptation were available then than in either the Devonian or the Pennsylvanian. The favorable condition may have been brought about by reduction of competition through Devonian extinction, and it may have produced new habitats by means of vast environmental changes.

The Pennsylvanian depression may have been, in part, an effect of the Mississippian peak. Since the Pennsylvanian seas were populated largely by genera surviving from earlier periods, it appears that a condition of evolutionary equilibrium had been achieved in a majority of stocks. Perhaps available environmental niches were generally occupied to capacity. The balance continued into the Permian with only minor expansions in a few groups but was upset by environmental changes at the close of the Permian.

The most plausible explanation of the extinction of a large part of the Paleozoic fauna lies in the greatly reduced extent of continental seas at the end of the Paleozoic. The late Paleozoic orogenies led to enlargement of the stable shields and to expulsion of the seas from all the continents. Compare the paleogeographic map of the early Paleozoic (Fig. 6-3, p. 171) with that of the Per-

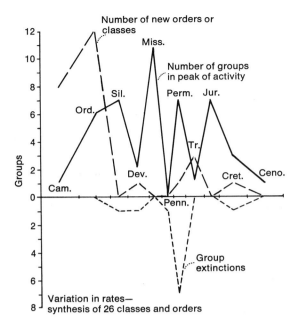

FIGURE 7-40

Synthetic graph showing times of highest rates of generic differentiation and times of first appearance and extinction of the major groups of invertebrates. (After Newell, 1952.)

mian (Fig. 6-20, p. 188). By the close of the Permian the Eurasian continent north of the Tethys was a large land area, geosynclinal conditions had ceased in Australia, and all of the eastern United States was land. A complete withdrawal of the seas from the continents seems most likely at the end of the Permian and might have been caused by a universal drop in sea level. Such a withdrawal of the seas would, of course, greatly restrict the area and variety of the environmental niches available to marine invertebrates and would greatly reduce the populations. It is well established that reduction of a natural population below a critical size, from whatever cause, commonly results in an excess of deaths over births. Reduction of the normal population density greatly diminishes the chance of reproduction in all bisexual animals, of course, but it has an even greater effect among aquatic forms, in which the chance union of gametes varies logarithmically with population density. Thus the final extinction of a species may be brought about by intrinsic factors after it has suffered severe reduction in number from extrinsic causes.

Some schools of thought have advanced the theory that the Permian wave of extinction may have resulted from some such catastrophe as a large dose of cosmic or solar radiation. If such a catastrophe had occurred, however, the effects would have been just the reverse of what the record indicates. Ultraviolet or cosmic rays intense enough to be lethal to terrestrial animals of ordinary tolerance would probably not affect aquatic organisms at depths of a few meters. Such rays would extinguish land animals and plants but would probably not have much effect on marine organisms. The paleontological record, however, shows that it was the marine invertebrates that underwent extinction while the land animals and plants changed but slightly.

SUGGESTED READINGS

See list at the end of Chapter 2.

Scientific American **Offprints**

831 Jacques Millot, *The Coelacanth* (December, 1955).

867 Norman D. Newell, *Crises in the History of Life* (February, 1963).

1125 Kjell Johansen, *Air-Breathing Fishes* (October, 1968).

THE MESOZOIC ERA

I. North America

The richly fossiliferous Mesozoic strata that crop out extensively in northwestern Europe, generally in simple structures, furnished the basis of modern historical geology. William Smith's studies of Jurassic rocks led to the discovery of the relation between fossils and rock strata, a relation that is the key to the correlation of rock formations. The concept of zones and the method of dating and correlating rocks by fossils were derived, early in the nineteenth century, from studies of Jurassic and Cretaceous rocks in the Old World.

When Phillips introduced the divisions called eras, stratigraphy and paleontology had hardly begun. Later, however, the eras were found to coincide with gross evolutionary features of at least parts of the animal kingdom. The whole aspect and composition of the marine populations of the Mesozoic Era are different from those of the Paleozoic and are, in general, more highly evolved. Many conspicuous inhabitants of the Paleozoic seas did not survive into the Mesozoic. Whole phyla, such as echinoderms, Sarcodina (foraminifers), bryozoans, coelenterates, and sponges, are unknown or sparse in the Lower Triassic record, and brachiopods and gastropods are not well represented; only pelecypods and ammonoids are abundant. Yet the wave of extinction that wiped out numerous orders, superfamilies, and classes of animals during the latter half of the Paleozoic left numerous ecologic niches free for occupation by the surviving stocks that gave rise to the Mesozoic faunas. At no other time in the Phanerozoic history of the earth did such a striking change in faunal composition take place as between the Paleozoic and the Mesozoic.

8-1 THE AMMONOIDS

The physical and biological history of the Mesozoic world is better known than that of the Paleozoic, possibly even better than that of the Cenozoic, partly because the outcrop area is greater than that of the Paleozoic and partly because the Mesozoic formations have been involved in fewer orogenies. The essential fact, however, is that fairly refined correlations that are nearly world-wide can be made. These correlations are possible because of the remarkable evolution of the ammonoids. No other invertebrate group underwent such a rapid evolution for so long a period, and thus no other equals the ammonoids as a tool for the student of evolution or earth history. These pelagic animals, adapted to a very wide ecologic range and a correspondingly wide geographic distribution, were the predominant inhabitants of the Mesozoic seas.

The ammonoids first appeared in the Devonian. Through the remainder of the Paleozoic they evolved along several distinct lines, increasing in structural complexity but not attaining anything near the range of diversity they were to show in the Mesozoic. About 200 genera are known from Paleozoic formations. Toward the end of the Paleozoic the ammonoids declined rapidly, and most main stocks became extinct. One small group, of rather simple, generalized morphology, survived into the Early Triassic. The extinction of the other, more abundant stocks left an open field for the surviving group. The Paleozoic-Mesozoic boundary is placed where the evolutionary resurgence and expansion of the surviving Paleozoic ammonoid stock took place. During the Triassic nearly 400 different genera evolved. Their morphological diversity is interpreted as due both to an increase in the number of ecologic niches available and to an increase in the number of genera evolving per unit of time.

Toward the end of the Triassic the ammonoids underwent another near extinction: practically all the diverse, well-developed stocks died out, and a solitary stock survived into the Jurassic. The Triassic-Jurassic boundary, like the Permian-Triassic, is determined by the resurgence of ammonoid evolution. This resurgence gave rise to an even greater evolutionary complexity: about 1,200 genera of ammonoids lived during the Jurassic and Cretaceous. No striking changes in evolutionary rate took place during that time, but at the end of the Cretaceous the ammonoids finally became extinct.

Because the Mesozoic Era and its systems, series, and stages can be distinguished by fossil records of the fluctuations of ammonoid evolution they are more clearly defined and distinct that any other units of the geologic column. Each series and stage is characterized by particular genera of ammonoids: six in the Triassic System, ten in the Jurassic, and thirteen in the Cretaceous. The type areas of these stages are in continental Europe or in England, but their equivalents can be recognized wherever fossil ammonoids are found.

8-2 THE MIDDLE ERA

The systems of the Mesozoic, following those of the Paleozoic and preceding those of the Cenozoic, were, as the name indicates, "middle," or intermediate, both in the evolution of flora and fauna and in the depositional and tectonic history of the earth's crust. Many of the broad structural features of the crust were molded by events of the Mesozoic, which in turn had developed on the Paleozoic framework.

The early Paleozoic geosynclines of North America were more or less symmetrically disposed around the North American continent. This arrangement began to alter with the Acadian orogeny in the eastern eugeosyncline. Continued tectonic activity in the eugeosyncline in Carboniferous time formed the source of the extensive Carboniferous deposits in the continental interior. During the Alleghanian orogeny the miogeosyncline became consolidated as an integral part of the continent. In the western half of the country, during the Paleozoic, conditions did not alter as radically as in the east, and geosynclinal conditions persisted into the Mesozoic. Thus, with the change from a balanced geosynclinal framework in the early Paleozoic to an unbalanced framework, with a geosyncline only in the west, at the end of the Paleozoic, the major tectonic and sedimentary framework for Mesozoic deposition was established. Mesozoic depositional history began with, and continued with, a pattern established during the Paleozoic.

The Mesozoic outcrops in North America are extremely variable. The largest by far are Cretaceous, and represent the time of maximum inundation by geosynclinal and epicontinental seas. The outcrops of Triassic and Jurassic strata are much smaller.

8-3 THE TRIASSIC PERIOD

With the close of the Alleghanian orogeny, the eastern geosyncline had become a land area undergoing erosion, as it remains to this day (Fig. 8-1). Throughout Early and Middle Triassic time, erosion greatly subdued the topography. Beginning in the Late Triassic, a series of linear fault troughs formed, and in them accumulated more than 20,000 feet of land-laid, clastic rocks, with some volcanic deposits. The rocks of this phase are called the Newark Group, and their present distribution is shown by Figure 8-2. Each subsiding trough had a large fault on one side or the other; some had faults on both sides. The rocks deposited in these troughs comprise a heterogeneous assortment of arkosic conglomerate and sandstone, shale, mudstone, and coal, with numerous basalt flows, sills, and dikes. The coarser textures reflect the faulted margins of the troughs, where thick boulder beds accumulated in fans. In the northern troughs the rocks are mostly red; in the southern red coloration is minor, and coal beds are present. The shale and mudstone commonly show mud cracks, rain pits, and reptilian footprints. The rocks of the Newark Group are alluvial fan, swamp, fluvial, and lacustrine deposits. Fossils are, in general, rare, but in some places dinosaur footprints and the preserved skeletons of fresh-water fish are abundant.

The lithology and structural relations of these Triassic rocks permit a fairly satisfactory reconstruction of the development of the different troughs. Figure 8-3 shows the development of the Triassic trough of central Connecticut. Cross section A shows the general character of the region at the end of the Alleghany orogeny. Cross section B is an interpretation of the area at the beginning of Late Triassic deposition. This was followed by movement along the great fault bounding the trough on the east; the uplifted lands east of this fault formed the source of sediments carried westward into the trough. Sedimentation was interrupted sporadically by great outpourings of basaltic lava across much of the trough. The displacement on the great fault was roughly equal to the thickness of the sediments accumulated—16,000 feet or more. In very late Triassic time, the whole area was gently uplifted and complexly faulted (Fig. 8-3, D). The disturbance as a whole is known as the Palisade orogeny, very different from the Paleozoic orogenies in this same belt.

The western geosyncline continued as an active site of deposition and tectonism. Its two main components were an outer belt, containing thick volcanic deposits, and an inner, nonvolcanic belt adjacent to the stabler continental interior, much as it had been since early Paleozoic time.

Unconformities separate Permian and Triassic deposits in some places, but in others deposition seems to have been continuous, and in still others minor hiatuses are recognized, mainly on the basis of paleontological data. The exact significance of many of these hiatuses is not known; they could represent withdrawals of the sea, followed by erosion before inundation again in the Triassic, or they could represent areas of nondeposition.

Early Triassic marine deposits are confined entirely to the western and circum-Arctic geosynclines and to a narrow faulted trough on the east coast of Greenland. Continental deposits, mostly redbeds, characterize the western platform regions. The arid conditions of the central-western United States, which began during Permian time and were marked by redbed formations including various evaporites, persisted into the Triassic. Throughout Wyoming, southern Montana, and northern Colorado vast alluvial plains accumulated red sand, silt, and clay. Various embayments opened out to the geosynclinal seaways to the west. The shoreline area between normal open

FIGURE 8-1

Paleotectonic map of North America for the Triassic. (Adapted partly from A. J. Eardley, *Structural Geology of North America*, Harper & Brothers, New York, 1951.)

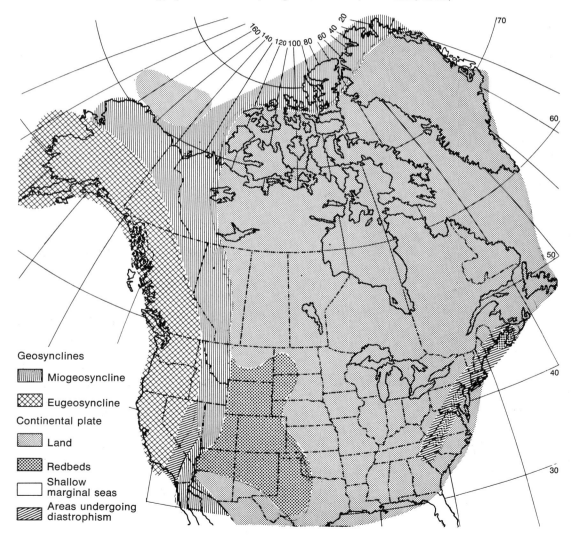

Geosynclines

▥ Miogeosyncline

▨ Eugeosyncline

Continental plate

▢ Land

▩ Redbeds

☐ Shallow marginal seas

▨ Areas undergoing diastrophism

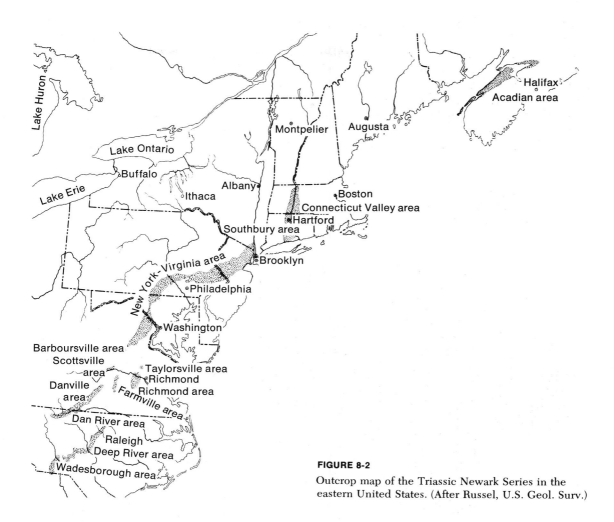

FIGURE 8-2

Outcrop map of the Triassic Newark Series in the eastern United States. (After Russel, U.S. Geol. Surv.)

sea and the area of predominantly continental deposition fluctuated constantly. Within and adjoining the continental redbeds were highlands that were undergoing erosion. Since the underlying Permian deposits are likewise red sandstone, siltstones, and shales deposited under very similar, if not identical, environmental conditions, the boundary between the two geological systems is difficult to place accurately. The total thickness of these Triassic continental deposits is about 1,000 feet.

The seaway of the western geosyncline probably covered much of the area west of the interior platform. Early Triassic deposition in the miogeosyncline is represented by shallow-water deposits consisting of siltstone, sandstone, and limestone. Eastward from the geosyncline these marine deposits intertongue with continental redbeds (Fig. 8-4). The maximum subsidence was in southeastern Idaho, where more than 5,000 feet of Lower Triassic sediments accumulated. Here a nearly complete sequence of Lower Triassic ammonoid zones is present, and the strata are conformable

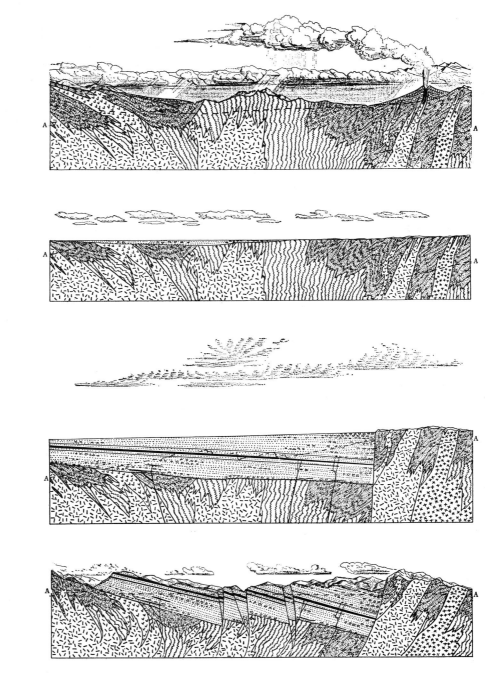

FIGURE 8-3

Diagrammatic cross sections illustrating four stages in the evolution of the Triassic trough of central Connecticut, the length of the section being about thirty miles: (A) complex structure and topography of the region after the Appalachian Revolution; (B) beginning of Triassic deposition on the greatly weathered, nearly peneplaned surface; (C) profile of the Triassic trough toward the end of its depositional history; (D) end of the Palisade orogeny, showing the complex of normal faults and the high-relief. (From Barrell, 1915.)

on the underlying Permian. Eastward, on the interior platform in Wyoming, the Triassic formations transgress and overlap upon the Permian formations.

In southern Utah, Arizona, and New Mexico continental deposition prevailed through most of Early Triassic time except toward the middle of the epoch when there were some marine transgressions. The continental deposits are red sandstones and shales; the marine deposits, mainly limestones.

By Middle Triassic time the seas had retreated westward from the miogeosyncline of the United States, but they persisted in Canada. No fossiliferous Middle Triassic formations are recognized in the former miogeosyncline or on the adjoining platform. The whole area was probably undergoing erosion with no significant deposition. The Lower Triassic formations of Utah and adjoining areas are disconformably overlain by continental deposits, some of which have yielded Upper Triassic reptiles and plants. This terrestrial depositional phase continued into the Jurassic. Since fossils are extremely scarce, precise dating is difficult, and placement of the Triassic-Jurassic boundary is therefore still uncertain. Figure 8-5 shows a diagrammatic stratigraphic column of the main formations that crop out near Zion National Park, Utah (Fig. 8-6). The lowest unit (Shinarump) is a thin conglomerate that covers tens of thousands of square miles in Utah and adjacent states. This conglomerate reflects uplift in the neighboring areas of Arizona, western Colorado, and central Idaho. The overlying (Chinle and Kayenta) formations of sandstone and shale represent chiefly fluvial and lacustrine deposition; they are interbedded with aeolian sand-dune accumulations (Wingate and Navajo). The Wingate Sandstone was probably derived from areas to the west, in Arizona and eastern Nevada. The sedimentary structures of the Navajo Sandstone indicate that the sands were deposited by winds blowing from the west and northwest, from source areas in western and

southern Nevada, Arizona, and California.

In the western eugeosyncline thick sediments and volcanics record an active deposition throughout the Triassic. Lower Triassic deposits are not well represented over the whole belt, but classic areas exist in southwestern Nevada and in the Inyo Range of southeastern California. In southwestern Nevada the whole Triassic sequence comprises about 30,000 feet of

FIGURE 8-4

Diagrammatic cross sections of Lower Triassic formations from the miogeosyncline onto the adjoining platform in the western United States.

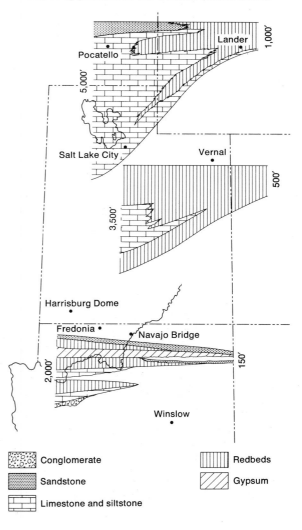

Conglomerate

Sandstone

Limestone and siltstone

Redbeds

Gypsum

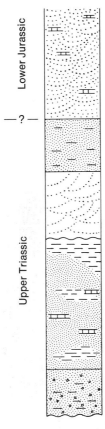

Navajo Sandstone

From tan to light gray; massive, cross-bedded limy sandstone, with few thin sandstone lenses. Principally aeolian, very slight ephemeral, lacustrine stages. 400'–800'

Kayenta Formation

Sandstone, red, argillaceous, cross-bedded in part; shale, red and green. Fluvial. 40'–300'

Wingate Sandstone

Massive, cross-bedded, from buff to brownish red; medium-grained sandstone; cliff-former, Aeolian. 300'–400'

Chinle Formation

Sandstone, variegated shale, limestone, and conglomerate; lenticular and intertonguing. Fluvial and lacustrine. 150'–850'

Shinarump Conglomerate

Cross-bedded, lenticular sandstone, conglomerate, variegated shale. Fluvial. 0'–275'

FIGURE 8-5

Diagrammatic columnar section of Upper Triassic and Lower Jurassic terrestrial deposits exposed in central Utah.

sedimentary and volcanic rocks in a complex array of facies. During Middle Triassic time alone more than 12,000 feet of tuffs, flows, and breccias accumulated. The Lower Triassic consists of shales, sandstones, conglomerates, and volcanics resting with angular unconformity on Paleozoic strata. The Upper Triassic strata are mainly shales and limestones. The sequence of formation and facies is illustrated in Figure 8-7. In the Inyo Range of southern California a shale-and-limestone formation contains Lower Triassic fossils identical to the forms known across Nevada to southeastern Idaho; the overlying Middle and Upper Triassic are represented by more than 5,000 feet of volcanic and sedimentary deposits.

In Oregon, western Washington, and southern Alaska, no Lower or Middle Triassic has been recognized, but the Upper Triassic record begins with a basal conglomerate resting unconformably on Permian or older strata. It seems highly probable that these parts of the eugeosyncline underwent orogeny from the Late Permian, when the area consisted of linear island belts, until the Late Triassic, when subsidence once again became widespread.

The stratigraphic record of Alaska can be divided into a southern eugeosyncline and a northern miogeosyncline—a relation clearly reflected in the Paleozoic and continuing into the Mesozoic. The eugeosyncline has yielded only Upper Triassic fossils, but an underlying volcanic sequence may be Lower or Middle Triassic, although it is generally considered Permian (Fig. 8-8). Northern Alaska has an exceptionally widespread limestone and clastic formation of Late Triassic age, and Middle and Lower Triassic fossils have been found in a few places on the Arctic slope.

Well-developed Triassic formations remain in the Canadian Arctic Archipelago, in Peary Land, and on the east coast of Greenland. The formations in the Sverdrup Basin of the Canadian Arctic consist primarily of shale, siltstone, and sandstone derived mainly from the south and east. Those of East Greenland accumulated in faulted troughs that opened to the north. The earliest Triassic rocks contain exceptionally rich ammonoid and fish faunas and are overlain by latest Triassic continental deposits.

8-4 THE JURASSIC PERIOD

The uppermost Triassic strata (those of Rhaetian age) are recognized only in southwestern Nevada, where they are conformable and continuous with strata containing the lowest Jurassic ammonoids. In no other place in western North America is the fossil record rich enough for us to recognize this time relation specifically. There seems no

FIGURE 8-6

Triassic and Jurassic nonmarine formations exposed along the eastern side of the valley of the Virgin River near Springdale, Zion National Park, Utah. The Triassic-Jurassic boundary is probably within the massive sandstone unit forming the top of the mountain. (Photograph by the National Park Service.)

question but that the seas withdrew from other parts of the area, but the pattern and extent of the withdrawal cannot yet be deciphered from the data available. In Oregon the Jurassic rests with considerable angular unconformity on Upper Triassic and Paleozoic strata.

The gradual changes in the continental framework and in the disposition of stable areas and geosynclines that began in the Triassic continued and were intensified during the Jurassic. Lower Jurassic marine strata are known in isolated places in south-western Nevada and in California, Oregon, British Columbia, Alaska, Canadian Arctic islands, and East Greenland. Their facies and geosynclinal relations are not unlike those of the Upper Triassic. Because few Jurassic deposits are recognized in eastern Nevada, it is thought that most of the area was part of a linear emergent peninsula extending northward through Idaho into British Columbia and known as the Mesocordilleran geanticline (Fig. 8-9). In the interior of the United States—that is, from the area of the former miogeosyncline eastward to the western part of the interior platform — desert conditions and sedimentation continued the patterns established in the Triassic.

In the Middle and Late Jurassic a complex series of marine formations was deposited in a shallow sea whose southern end occupied Montana, Wyoming, and parts of Idaho, Utah, and Colorado. This sea, which has been named the Sundance Sea, after one of the more widespread formations deposited in this area, extended from the Arctic through Alberta. Its transgression began in Early Jurassic time, but it did not reach the United States until the Middle Jurassic.

Western Nevada and California have extremely thick Jurassic formations, a large part of which are volcanic. Although a scarcity of fossils and a discontinuity of outcrops hamper stratigraphic analysis of these formations, the oldest of them rest conformably on Triassic sediments. The thickest of these occupy a north-south trough in western Nevada, where they are part of the continuous sequence already mentioned. Deposition was interrupted early in Jurassic time by an orogenic disturbance, in the course of which eastward-moving thrust faults formed.

More dramatic events took place in the Sierra Nevada of California. This range is composed chiefly of Mesozoic granitic rocks and metamorphosed Paleozoic and Mesozoic sedimentary and volcanic rocks. The granitic rocks are part of a discontinuous belt that extends from Baja California northward through the Peninsular Ranges and the Mojave Desert, through the Sierra Nevada, and into western Nevada. North of the Mojave Desert this is called the Sierra Nevada batholith, part of a belt of Mesozoic plutonic rocks that encircles the Pacific ocean (Fig. 8-10). The Sierra Nevada batholith is composed chiefly of granitic rocks but includes scattered smaller masses of darker and generally older plutonic rocks and remnants of metamorphic rocks.

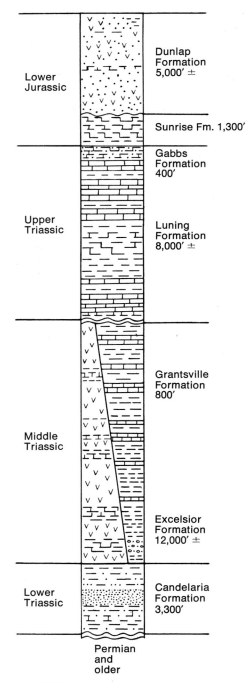

FIGURE 8-7

Columnar section of Triassic formations exposed in southwestern Nevada. (Data from Muller and Ferguson, 1939.)

The granitic rocks are in discrete masses or plutons, which generally are in sharp contact with one another or are separated by thin septa of metamorphic or mafic igneous rocks. Individual plutons range in outcrop area from less than 1 kilometer to more than 1,000 kilometers (Fig. 8-11). Isotopic age dates indicate that the pattern of intrusions of the plutons is highly complicated. The dates obtained from the central Sierra Nevada indicate that most of the granitic rocks along the east side of the batholith and in adjacent areas to the east are 170 to 210 million years old (Late Triassic or Early Jurassic); that plutons in lower Yosemite Valley and others intruded into the western metamorphic belt are 125 to 145 million years old (Late Jurassic); and that plutons along and just to the west of the range crest are 80 to 90 million years old (early Late Cretaceous). An interpretation of the history of this region is shown by the series of cross sections in Figure 8-12. Emplacement of the intrusions was accompanied by diastrophism. The major episode of disturbance came toward the close of the Jurassic and is known as the Nevadan orogeny. It is important to remember, however, that both earlier and later disturbances have been identified.

Subsiding troughs formed along the western margin of this orogenic belt and received during the last stages of the Jurassic enormous thicknesses of coarse-to-fine clastic sediments both from the rising Nevadides and from tectonic lands to the west. The sediments of this trough are arkoses, graywackes, radiolarian cherts, tuffs, agglomerates, and basalt flows. The

FIGURE 8-8

Lost Creek, Alaska Range, eastern Alaska. The dark Permian lava flows in the foreground are overlain by thick and thin beds of Upper Triassic limestone. The dark rocks of the mountain's top are shale and sandstone of Late Jurassic and Early Cretaceous age. (Photograph by F. H. Moffit, U.S. Geol. Surv.)

FIGURE 8-9

Paleotectonic map of North America for the Late Jurassic. (After A. J. Eardley, *Structural Geology of North America*, Harper & Brothers, New York, 1951.)

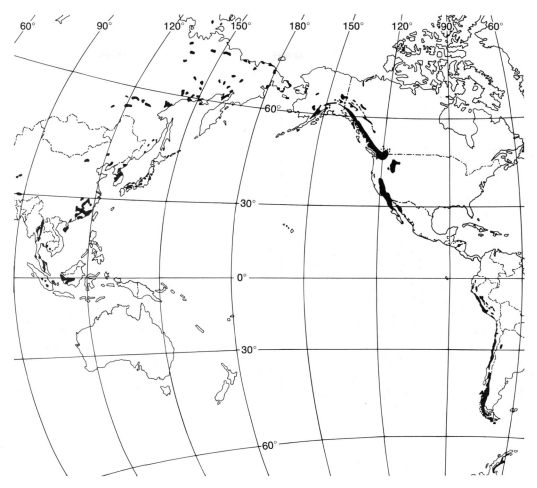

FIGURE 8-10

Distribution of batholiths of Mesozoic age in the circum-Pacific region. (From
P. C. Bateman and J. P. Eaton, *Science,* **158**, Dec. 15, 1967. Copyright 1967
by the American Association for the Advancement of Science.)

eugeosyncline north of California underwent similar depositional and orogenic
events.

One result of the Nevadan orogeny was to
enlarge and extend the area separating the
Sundance Sea from the Pacific eugeosyncline. Before the orogeny the Mesocordilleran geanticline was not an important
source of the sediments deposited in the adjoining basin to the east, but after the orog-

eny this relation changed conspicuously,
and during Late Jurassic and all of Cretaceous time it was one of the most important
sources of sediments in North America (see
§ 8.5).

During the early history of the Sundance
Sea, streams, mainly from the south and
east, contributed sand, silt, and clay to a
complex of facies. The maximum subsidence
was in southwestern Wyoming. Here many

FIGURE 8-11

Yosemite Valley, California. The bedrock of the region is massive granite, emplaced during and after the Nevadan orogeny. The present topography was formed during the Pleistocene. Note the beautiful U-shaped valleys. (Photograph by J. T. Boysen, U.S. Geol. Surv.)

shore lines were constantly changing because the rate of subsidence changed and because local uplifts influenced the kind of facies developed. Islands formed within the sea also influenced the character and distribution of sediments. Clastic rocks, at times associated with evaporites, form the main facies, and much of the near-shore and extremely shallow-water area was the site of redbed deposition. Only one significant limestone formation was formed.

The principal source of clastic sediments entering this inland sea changed to the west during Late Jurassic time, reflecting both increased instability in the source area and the Nevadan orogeny. Late in the Jurassic

the sea retreated northward, and 400–500 feet of varicolored clays and coarser clastic sediments—the Morrison Formation—one of the most spectacular formations in North America—was laid down in fluvial and lacustrine environments. The Morrison contains rich beds of fossils of the very large dinosaurs. Large, well-preserved dinosaurs are not common fossils, and their presence in the formation left its age in doubt, since it could be either Late Jurassic or Early Cretaceous. Eventually, a sequence of alternating marine and nonmarine facies in Tanzania yielded dinosaurs very similar to those of the Morrison Formation. The marine strata overlying the Tanzanian dinosaur

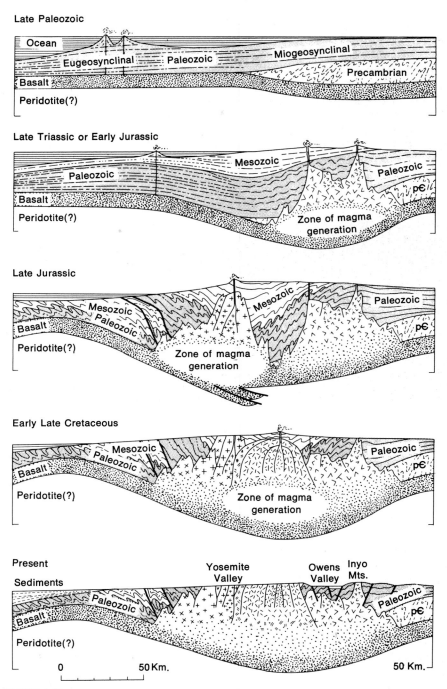

FIGURE 8-12

Cross sections illustrating (bottom) the existing crustal structure of the Sierra Nevada and (top four sections) a model for the evolution of that structure. (From P. C. Bateman and J. P. Eaton, *Science*, **158**, Dec. 15, 1967. Copyright 1967 by the American Association for the Advancement of Science.)

beds contain ammonoids that correlate with European ammonoids of known Late Jurassic age. The fossil mammals and ostracodes of the Morrison Formation resemble those of the Upper Jurassic in England. The Morrison Formation also contains some of the largest deposits of uranium ore known in the United States.

A third distinct area of Jurassic marine embayment introduces a striking new aspect to the continental framework. In northern Louisiana and southern Arkansas many oil drills have penetrated a thick sequence of Upper Jurassic rocks—a complex of facies with near-shore clastic sediments, generally red, and evaporites. Southward these facies pass into normal marine, fossiliferous shale and limestone. These formations do not crop out; they are known only in the subsurface and are overlain by Cretaceous formations (Fig. 8-13).

Farther westward, most of the eastern half of Mexico was occupied by a very extensive geosyncline, in which several thousand feet of shale and limestone were deposited. The fossils are closely related to those in Jurassic formations of northern South America and the Mediterranean region. They show no affinities to the Jurassic faunas of the Pacific coastal region of North America or of the Sundance Sea, thus substantiating the independence of the three main depositional areas in North America during Jurassic time.

The facies relations of the Jurassic of Mexico and northern Louisiana indicate open sea to the east and south and furnish the first evidence of a Gulf of Mexico. Data are too sparse to reveal the nature of this region before Jurassic time, but from the Jurassic on, the history is recorded in the sedimentary formations. For brief periods during the Late Jurassic a marine connection with the Pacific may have existed in west-central Mexico.

Jurassic rocks are probably present at depth along the east coast of the United States. Deep wells in southern Florida have penetrated layers of limestone, dolomite, and anhydrite that differ from overlying Lower Cretaceous rocks both in lithology and microfaunal content, suggesting a Jurassic age. A deep well on Cape Hatteras, North Carolina, bottomed in 1,400 feet of sedimentary rocks that may be Jurassic.

By the end of Jurassic time the framework of the continent had been fundamentally changed. At the beginning of the Mesozoic the geosynclinal and depositional patterns were merely a continuation of those of the late Paleozoic. By Middle Triassic time the western miogeosyncline of the United States had become an area of continental deposition and remained so into the Jurassic. Marine conditions returned to this region with the transgression of a Middle Jurassic sea from the Arctic Ocean along the east side of the Mesocordilleran geanticline. The Pacific coastal belt included many large embayments that were rapidly subsiding and receiving sediments from the surrounding highlands. The ancestral Gulf of Mexico came into existence, and its seaways encroached upon Louisiana, Texas, and Mexico. Deposition also began on the eastern coastal plain. The older geosynclines were destroyed; the new patterns established during the Jurassic were a preparation for the geosynclinal picture of the following Cretaceous Period.

8-5   THE CRETACEOUS PERIOD

During the Cretaceous the last large-scale inundation of the continent took place. More than half of the present area of North America was covered by Cretaceous seas. At one time or another during this period marine transgressions occupied the Atlantic coastal plain, all of Florida, the Gulf coast states, and the eastern half of Mexico. A very wide seaway extended north from Texas across the west-central part of the United States and Canada to the Arctic Ocean. Marginal embayments existed on

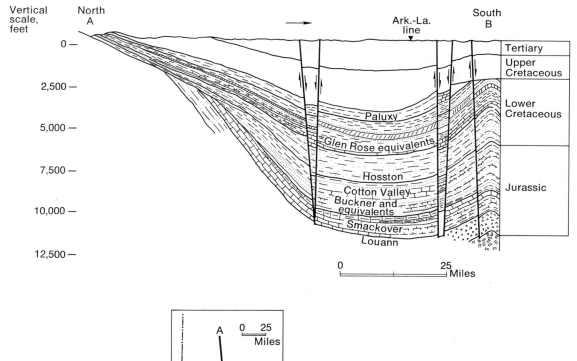

FIGURE 8-13

Cross section of Jurassic and Cretaceous formations in southwestern Arkansas and northwestern Louisiana. (After M. T. Halbouty, *Bull. Am. Assoc. Petrol. Geol.*, 1969.)

the Pacific coastal belt. This was a drastic change in geography from Paleozoic seaways, but the change was gradual, and the pattern of slow change can be followed in the physical history of the Triassic and Jurassic.

The development of the Atlantic coastal plain began during Late Jurassic time. After the late Paleozoic orogenies in the Appalachians, the only rocks deposited were those of the Upper Triassic Newark Group, which accumulated in linear faulted troughs. During the intervening time this area had been undergoing profound erosion, and by the Late Jurassic a low, flat plain had developed. Early in the Cretaceous the coastal plain began to subside while the Appalachian area to the west began to rise. It is generally believed that

peneplaned remnants can be recognized in the Appalachians and that these show that the area was raised as a low, broad arch, elongated parallel to the strike of the Paleozoic deformation. The various levels of the peneplains indicate that uplift took place in several stages. The sculpturing of the Appalachian Mountains we shall discuss in Chapter 12.

On the subsiding coastal plain clastic deposits of continental and marine origin accumulated to form a thick, wedge-shaped deposit dipping gently seaward. Cretaceous formations of the Atlantic coastal plain crop out in a narrow belt from New Jersey into Maryland and Virginia. There is another large outcrop in North and South Carolina. Northeastward of New Jersey the coastal plain Cretaceous is beneath the waters of the Atlantic, on the continental shelf. Well borings and small outcrops indicate that Cretaceous sediments underlie Long Island, Block Island, Martha's Vineyard, and Nantucket, as well as the Cape Cod peninsula. Dredging operations on Georges Banks (off the New England coast) have brought up fossiliferous Cretaceous and Cenozoic rocks like those farther southwest in the coastal plain. Fossiliferous Cretaceous rocks have also been dredged from off the coast of Nova Scotia.

The coastal-plain Cretaceous formations consist mainly of sand and clay sediments deposited by streams, in swamps, and in shallow off-shore marine environments. Differences in rate of subsidence and transgressions and regressions of the sea produced a complex intertonguing of marine and nonmarine formations. The wedge-shaped profile of these Cretaceous deposits is illustrated by Figure 8-14. The semi-consolidated bed shown in A and B is thought to consist of the lower half of the Cretaceous sequence. The unconsolidated zone above would thus include the later Cretaceous deposits and those of the Cenozoic. These profiles clearly show that the continental shelf has subsided about three miles since the beginning of the Cretaceous.

Study of the continental margin off eastern United States has greatly intensified in recent years. The surface upon which the wedge of Cretaceous and younger sediments rest has been found not to be a simple, seaward-dipping plane but to contain deep trenches (Fig. 8-15). Off the continental margin from New Jersey northward to Nova Scotia, a trench near the edge of the continental shelf is filled with 3 to 5 kilometers of sediment. Further seaward on the upper part of the continental shelf is another trench filled with 3 to 6.5 kilometers of sediment. The trenches are separated by a ridge over which the sediment

cover is only 1.6–2.5 kilometers. A composite cross section of the continental margin off New England is shown in Figure 8-16. The inferred ages indicated on the cross section are based on correlations with geology of the adjacent land areas. Most of the sediments are of Cretaceous age. To the south, off the coast of Florida, is a single deep trench filled with as much as 10 kilometers of sediments. In Figure 8-15 this southern trench is shown provisionally as joining the double trenches to the north.

The Florida peninsula was a shallow submarine bank; more than 10,000 feet of mostly carbonate rocks accumulated on it during the Cretaceous. Streams coming off the Appalachian highlands of northern Georgia and Alabama brought clastic sediments to Florida. Beyond the areas covered by the clastic sediments, marls and limestones were deposited. The transition from clastic to carbonate facies was, in Early Cretaceous time, in central Florida, but the transition gradually receded northward during the remainder of the Cretaceous (Fig. 8-17).

The Bahama Banks are one of the largest areas of contemporaneous limestone deposition. A well on Andros Island penetrated more than 14,500 feet of carbonates and some evaporites of Cretaceous and Cenozoic age. In fact, during much of Cretaceous time, nearly all the marginal shelf areas of the Gulf of Mexico—that is, from the Bahama Banks to eastern Mexico—were covered by shallow shelf seas studded by numerous reefs and had bottoms of carbonate mud. The predominantly carbonate deposition was altered only when an influx of terrigenous sediments masked the carbonate phase. The volume of Cretaceous rocks in Florida and southern Georgia has been estimated at 75,000 cubic miles.

The relation of the facies of a sedimentary basin to the source and amount of clastic material is well illustrated by the profile of Cretaceous formations extending from Georgia through Alabama and Mississippi into Tennessee (Fig. 8-18). Here the Cretaceous seas were wrapped around the southern edge of the old Appalachian Mountains. The earliest deposition was of nonmarine sandstone, but there is a gradual upward gradation into marine facies. At the two extremes of this bow-shaped outcrop that is, in Georgia and in Tennessee—clastic deposition prevailed throughout Cretaceous time, testifying to adjacent streams coming off the Appalachian highlands and building large deltas. From the centers of clastic deposition there is a gradation toward Alabama, first with intertonguing clay shales, which in turn pass into chalk deposits (Fig. 8-19). Thus the

FIGURE 8-14

Sections across the Atlantic coastal plain from the fall line to the edge of the continental shelf: (A) near Cape May, New Jersey; (B) in southern Virginia; (C) in east-central North Carolina. Sections A and B are based on seismic surveys, black dots indicating points determined by the surveys. Section C is based on drilling and is the most accurate in its stratigraphic boundaries. (After King, 1951.)

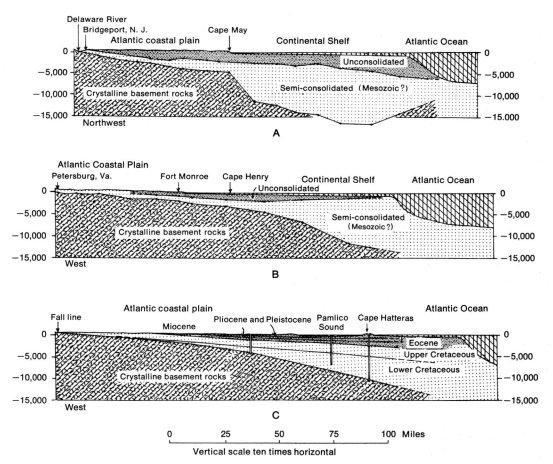

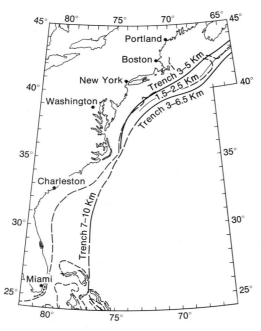

FIGURE 8-15

Axes of trenches off the Atlantic coast. Numbers indicate thickness of sedimentary fill atop basement rocks. (After K. O. Emery, 1966.)

shelf marls, clay formations, and extensive organic reefs. This complex of facies and their relations are reflected in the distribution of ammonoids in the Lower Cretaceous of Texas. Analysis of the sediment types and their distribution, of the whole fauna, and of the ammonoid genera has permitted recognition of four ecologically significant assemblages assigned to depth zones.

The Early Cretaceous transgressive sea spread across eastern Mexico and the central Gulf coast states, through Texas as far north as Kansas and southeastern Colorado. The overlying Upper Cretaceous formations are generally unconformable on the Lower Cretaceous, testifying to an extensive retreat of the seas gulfward. This regression was shortly followed, in early Late Cretaceous time, by a renewed transgression that resulted in one of the most extensive inundations of North America (Fig. 8-20).

It is worthwhile to review the three transgressive phases that affected this area. The first Mesozoic inundations of the Gulf coastal states came in the Late Jurassic and extended only as far as southern Arkansas. The second, in the Early Cretaceous, reached Kansas and southeastern Colorado. Each of these transgressions was followed by a regression. The third and greatest transgression was that of the Late Cretaceous, which eventually produced a wide seaway extending from the Gulf of Mexico to the Arctic Ocean.

The long land area that separated the depositional basins along the Pacific coast from the Sundance Sea became an even more important source of clastic sediments in the Cretaceous. When these features first developed during the Jurassic, the main sources of the terrigenous sediments of the Sundance Sea were to the northwest and east. Toward the end of Jurassic time, however, the main source became the land to the west and south of the Sundance Sea (part of the Mesocordilleran geanticline). The last phase of Jurassic history

distance of central Alabama from an active source of terrigenous sediments allowed the accumulation of fairly pure carbonate deposits. The facies and time relations of these formations are shown diagrammatically in Figure 8-18.

The Gulf coast states were transgressed upon to a limited extent in the Late Jurassic, but the sea had regressed by the end of the Jurassic. Early in the Cretaceous, the sea again encroached northward across a low peneplaned surface of marine Paleozoic and Jurassic formations in Texas and Louisiana. The initial deposits of this transgressive sea were sandstones, which were succeeded by marls and limestone deposits as the shoreline moved northward. Lower Cretaceous formations, well exposed in Texas, present a variety of facies, including near-shore sandstones, shallow-

was the extensive fluvial deposition represented by the Morrison Formation which, as noted above, covers much of Montana, Wyoming, and Colorado. Continental deposition continued into Early Cretaceous time.

Renewed uplift in the westerly source area supplied vast quantities of coarse-to-fine clastic sediments to streams flowing eastward. This increase in the supply of sediments was the direct result of orogeny (Sevier orogeny) in eastern Nevada and western Utah (Fig. 8-21). Within the orogenic belt uplift and thrust faulting that continued through most of Cretaceous time caused tens of miles of displacement. Great thicknesses of nonmarine clastic formations of earliest Cretaceous age were laid down in a belt from Colorado northward through British Columbia (Fig. 8-22). While this vast interior fluvial plain was receiving great quantities of sediments from the adjoining western highlands, the marine transgression of the Gulf coast states was well under way. Likewise, during Early Cretaceous

time, an arm of the Arctic Ocean began to transgress across the Yukon Territory into British Columbia. The transgressive sea from the south reached through Texas only as far as Kansas and southeastern Colorado toward the end of Early Cretaceous time. The sea from the north was extending itself to northern Colorado at the same time. The area between these two seaways, across central and southern Utah and Colorado, was undergoing erosion and nonmarine deposition. Early in Late Cretaceous time these two seaways joined and, in effect, separated the North American continent into two large islands.

Throughout Late Cretaceous time the main depositional facies in the Gulf coastal states and in the Mexican geosyncline was carbonate, as it had been in the Early Cretaceous. North of Texas, however, very little limestone accumulated, and the great bulk of the formations are composed of fine-to-coarse clastic sediments. The facies are generally coarse clastic sediments on the

FIGURE 8-16

Structural section inferred from seismic data for continental shelf, slope, and rise south of Massachusetts. The inferred ages are based on correlation with geology from adjacent land areas. (After K. O. Emery, 1966.)

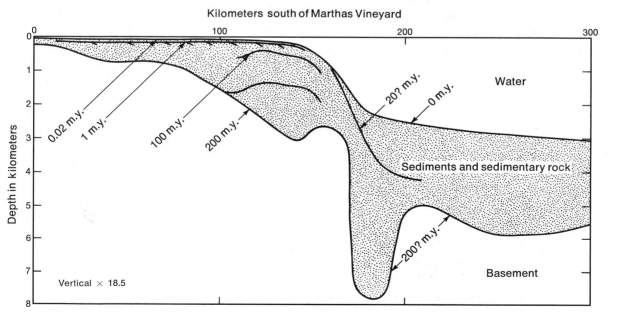

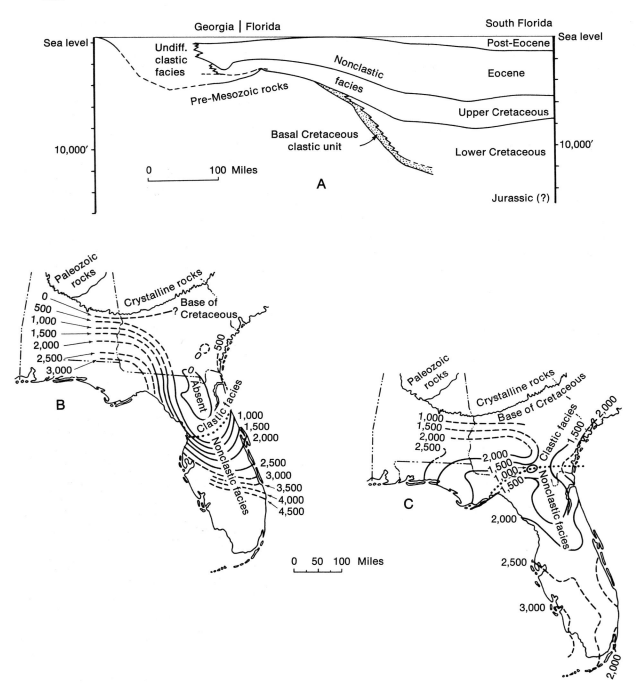

FIGURE 8-17

Cretaceous deposits of Georgia and Florida: (A) generalized cross section from the crystalline rocks of central Georgia to southern Florida; (B) isopach map of Lower Cretaceous formations; (C) isopach map of Upper Cretaceous formations. (After P. L. Applin, *Geol. Soc. Am. Bull.*, 1952.)

western side of this geosyncline and clay shales on the eastern side. The maximum subsidence was closer to the western edge of the geosyncline. The gross depositional pattern shows massive sandstone wedges along the western margin of the geosyncline; these grade eastward into shale (Fig. 8-23). The sand wedges consist of both marine and nonmarine facies. The sequence of facies reflects several transgressive and regressive movements of the Cretaceous seas. The constant shifting of the shore line resulted in the deposition of facies units that cut across time lines. The interrelations of transgressive and regressive facies and of planes of time equivalence are illustrated in Figure 8-24. The facies relations and distribution shown in this diagram are typical of the western edge of the Cretaceous geosyncline. The eastern half of the geosyncline was almost unaffected by these

influxes of coarser detrital sediments, and the whole sequence is made up of clay shale with a few thin limestone beds.

These general facies relations characterize the Cretaceous geosyncline from Texas to the Arctic Ocean. At about the middle of Late Cretaceous time the connection of this seaway with the Arctic Ocean was broken by the building of vast deltaic and alluvial plains across northeastern British Columbia and Alberta. With minor fluctuations the interior sea began to recede southward toward the United States. The earlier route north to the Arctic had become an area of continental deposition, mainly of coarse clastic sediments. The eastward spreading of deltaic and alluvial plains gradually extended southward through the Rocky Mountains and Great Plains toward Texas. Cretaceous depositional history thus ended with the accumulation of thick sequences

FIGURE 8-18

Diagrammatic cross section and outcrop map of Cretaceous deposits in the southeastern states. Formations are shown in their precise time relations, and the hiatuses are noted. (After W. H. Monroe, *Bull. Am. Assoc. Petrol. Geol.*, 1947.)

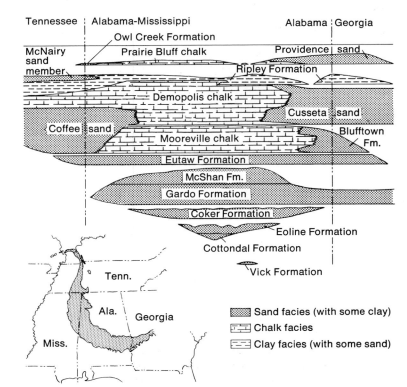

FIGURE 8-19

Typical exposure of the Cretaceous Selma Chalk, Jones Bluff, Tombigbee River,
Sumter County, Alabama. Compare the cross section of Fig. 8-19. (Photograph
by L. W. Stephenson, U.S. Geol. Surv.)

of sandstones, some coal-bearing. The sea
retreated southward toward the Gulf of
Mexico and by the end of Cretaceous time
had receded completely from the continent.

The implications of the depositional patterns of the geosyncline are noteworthy.
First, as shown in Figure 8-23, the geosyncline was asymmetrical, with the

greatest subsidence near the western margin and the least in the Great Plains of
Nebraska and Kansas. Secondly, the coarser
clastic sediments are along the western
margin, where a nearly continuous sequence of sandstones and conglomerates,
generally of continental origin, grades
eastward into fine-grained clastics (shales),

FIGURE 8-20

Paleotectonic map of North America for the Late Cretaceous. (After A. J. Eardley, *Structural Geology of North America*, Harper & Brothers, New York, 1951.)

generally of marine origin. The eastern (thin) edge of the geosyncline consists wholly of shales and thin limestone beds. It has been calculated that the Cretaceous sediments in the United States would make a blanket 1,750 feet thick over the whole country. The west-to-east gradation of coarse-to-fine sediments, as illustrated by Figure 8-23, clearly indicates that the source of most of this sediment was the Sevier orogenic belt to the west. The area of this adjoining land is something like 200,000 square miles. If a cubic mile of geosyncline sediment necessitated the erosion and transport of a cubic mile from the source area, an average thickness of 5 miles of sediments

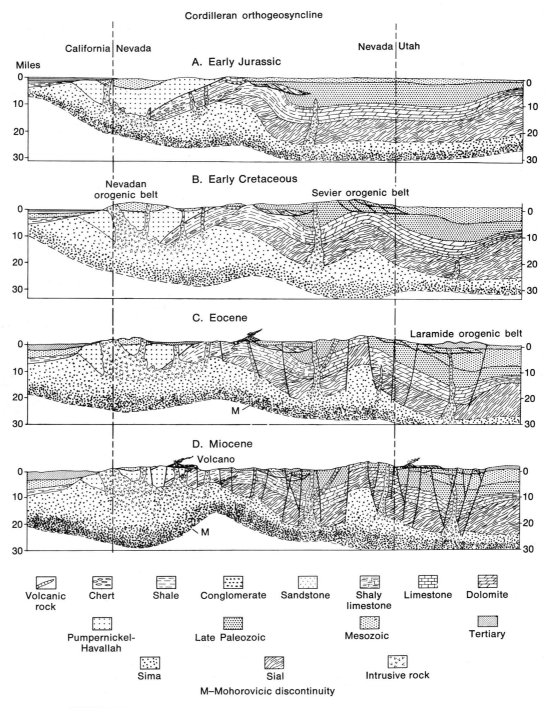

Cordilleran orthogeosyncline

California | Nevada

A. Early Jurassic

Nevada | Utah

Miles

Nevadan
orogenic belt

B. Early Cretaceous

Sevier orogenic belt

C. Eocene

Laramide orogenic belt

M

D. Miocene

Volcano

M

Volcanic rock · Chert · Shale · Conglomerate · Sandstone · Shaly limestone · Limestone · Dolomite

Pumpernickel-Havallah · Late Paleozoic · Mesozoic · Tertiary

Sima · Sial · Intrusive rock

M—Mohorovicic discontinuity

FIGURE 8-21

Diagrammatic cross sections showing the evolution of the California-Utah region from Early Jurassic to Miocene time. (From R. J. Roberts, *UMR J.*, 1968.)

FIGURE 8-22

Nonmarine coal-bearing sandstones and shales of Early Cretaceous age, Peace River Canyon, Rocky Mountain foothills of British Columbia. (Geol. Surv. Canada.)

would have been eroded from each square mile of source area during the Cretaceous. A generous allowance for volcanic rocks (mostly volcanic ash beds) in the geosynclinal suite still leaves an enormous thickness of rocks to be eroded in the source area (approximately a million cubic miles). The homogeneity of the coarse clastic facies along the western margin of the geosyncline throughout the Cretaceous is interrupted by several unconformities that disappear eastward, suggesting that the source area was mountainous and that there had been repeated tectonic movements, probably both epeirogenic and orogenic. Thus the tectonic and depositional relations here were quite similar to those in the Carboniferous geosyncline of the eastern United States.

Toward the end of the Cretaceous began a great orogenic cycle in a belt just east of the Sevier orogenic belt that lasted well into the Cenozoic and deformed a belt that included the Cretaceous geosyncline. Not since Precambrian time had there been such a major mountain-building epoch in North America or, indeed, in the world as a whole. Folding, faulting, and overthrusting on a gigantic scale formed the present Rocky Mountain system with its extensions northward into Alaska and southward into Mexico and the Andes of western South

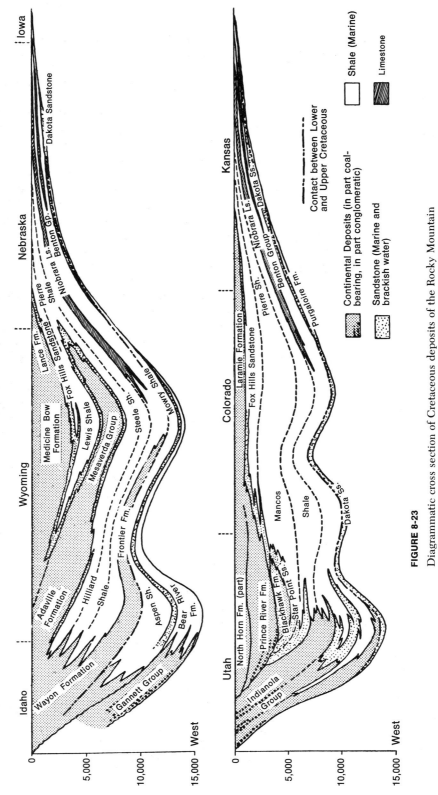

FIGURE 8-23

Diagrammatic cross section of Cretaceous deposits of the Rocky Mountain region. (After P. B. King, *Geol. Soc. Am.*, Special Paper 62, 1955.)

America. This phase of deformation was named, years ago, the Laramide orogeny, though it should be emphasized that it is actually a continuation of the earlier Cretaceous deformation referred to above as the Sevier orogeny. The intensity of deformation was greatest in the western margin of the geosyncline and less severe eastward. Along the western margin are found numerous large overthrust faults; in eastern Idaho and western Wyoming, in fact, a whole series of parallel thrust belts are found. Movement on some of these great thrusts actually began in the Jurassic. Most of the structure of the Rocky Mountains is the result of the Laramide orogeny; their present form, however, is the result of Cenozoic deposition and erosion (Fig. 8-25). In contrast to the Nevadan orogeny, which was accompanied and followed by great batholithic intrusions, only the Boulder batholith seems to have been emplaced during the Laramide orogeny. Most of the states west of the Great Plains had active volcanoes spreading tuff and lava from Jurassic to Pliocene.

The Pacific coastal belt of the United States and Canada and most of Alaska were sites of active deposition and differential movement during Cretaceous time. The types of basins, the sediments deposited, and the faunas are quite different from those of the interior sea. Numerous rapidly subsiding embayments and troughs characterized this coastal area—a geographic pattern first established in the early Mesozoic, for most of the troughs of Cretaceous deposition in California and Oregon were, in fact, inherited from Jurassic time.

The present outcrop of Cretaceous rocks in California includes a narrow belt along the western edge of the Great Valley and several areas in the Coast Ranges. The tectonic instability of this whole area is shown by the accumulation, in some places, of more than 50,000 feet of clastic sediments, including volcanics, and marked by numerous hiatuses, disconformities, and angular unconformities. Throughout the Cretaceous the source of the sediments appears to have been the tectonic lands to the west, east, and north. Early Cretaceous sediments are generally fine-grained clastics. Toward the latter part of Early Cretaceous time orogenic disturbances in parts of the Coast Ranges transformed these regions into active sources of coarse clastic sediments, which were transported to the adjoining subsiding basins. Similar local orogenic disturbances occurred in the Late Cretaceous. In most cases these orogenic disturbances produced striking angular unconformities

FIGURE 8-24

Diagram showing the relations of transgressive and regressive deposits and stages to planes of time equivalence, as found in the Cretaceous strata of the interior geosyncline: a, marine; b, near-shore sands; c, coastal deposits; d, flood-plain deposits; A–A', B–B', etc., planes of time equivalence. (From J. D. Sears, C. B. Hunt, and T. A. Hendricks, U.S. Geol. Surv., Prof. Paper 193F.)

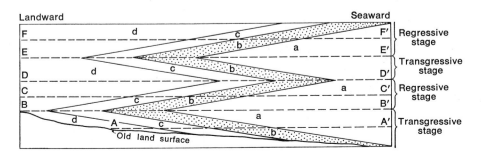

FIGURE 8-25

Oblique aerial view of the foothills of the Front Range of Colorado, near Morrison, Colorado. The prominent hogback in the center is the basal Cretaceous sandstone (Dakota Sandstone) overlain by a thick series of Cretaceous shales and sandstones. The face of the hogback formed by the Cretaceous sandstone is the Morrison Formation of Late Jurassic age. The highlands on the far left are composed of Precambrian schists; between these strata and the Morrison are nonmarine redbed formations of late Paleozoic and possibly Triassic age. (U.S. Geol. Surv., 1941.)

on the margins of the basins, but eastward only disconformities. The relations of these orogenies to the subsiding basins are shown in Figure 8-12.

The Cretaceous record of the western half of British Columbia consists of poorly known, isolated, thick sequences of both marine and continental clastic formations, including volcanics. The lithology and the general unfossiliferous character of most of the formations make correlation difficult. All that we can be sure of is that there are numerous formations of variable clastic lithologies, with associated volcanic members, and, commonly, coal beds.

The eugeosyncline in southern Alaska and the miogeosyncline in northern Alaska persisted into Cretaceous time. Since few Jurassic or Cretaceous formations are known between the Alaskan and Brooks ranges, this area was probably occupied by an extension of the Mesocordilleran geanticline.

Lower Cretaceous facies of the eugeosyncline consist of argillite, chert, graywacke, tuff, and basaltic beds. The miogeosynclinal facies consists of shale and sandstone, probably derived from land in the Brooks Range. Late in Early Cretaceous time a profound orogeny and extensive batholithic intrusions deformed and metamorphosed most of the eugeosyncline. This orogenic phase ended the basic geosynclinal pattern that had persisted since early in the Paleozoic, and Alaska became differentiated into recognizable positive and negative (subsiding) areas. Within the negative areas clastic sediments of marine continental origin accumulated.

In the Arctic islands of Canada the Cretaceous is recorded by alternations of marine and nonmarine clastic deposits. Minor volcanic activity occurred in the Early Cretaceous and at some time between the late Early and the middle Late Cretaceous.

These formations are especially widespread on Axel Heiberg Island.

Along the eastern Greenland coast varied facies record several transgressive and regressive movements of the sea and periods of erosion. A Late Jurassic orogeny was expressed in eastern Greenland by north-south faults. The initial Cretaceous deposits are coarse clastics deposited adjacent to the faulted regions. The marine fossils of these strata indicate a very early Cretaceous age. Overlying strata are of carbonate facies, red clastic facies, and shallow near-shore clastic facies, testifying to periodic transgressions and regressions, important gaps in the sedimentary record marking times of emergence and erosion.

SUGGESTED READINGS

See the list at the end of Chapter 4.

THE MESOZOIC ERA

The widespread diastrophism that occurred in many parts of the world during the latter half of the Paleozoic Era gradually altered the geography of most of the continents. By Late Permian time the world had more land area than in any earlier period of the Paleozoic. It was somewhat comparable to the earth today. Both times of great emergence have been periods of vast continental glaciation. What relation there may have been between these two conditions is discussed in Chapter 15.

Early Triassic geography was largely a continuation of Late Permian geography. In Permo-Carboniferous time, as we have seen, there began a gradual change from predominantly marine geosynclinal deposition to a greater proportion of terrestrial deposition. This shift in the types of sedi-

ments closely reflects the diastrophic history. By the end of the Paleozoic several of the main early Paleozoic geosynclines were rigid parts of continental frameworks.

Late Paleozoic marine formations are few, and most provide incomplete records. The record of the succeeding Early Triassic Epoch is also scanty. We do know, however, that Early Triassic seaways covered the Tethyan geosyncline from the Alpine-Mediterranean region eastward through the Middle East to the Himalayan and Indonesian regions. Broad shelf seas covered southeastern China, parts of Japan, parts of the eastern Siberian coast around Vladivostok, and the circum-Arctic area. The main continental areas of Eurasia, Africa, peninsular India, Australia, and South America were sites of continental deposition. The geosynclinal seas—except, of course, for the Tethys, which was between two continental masses—were marginal to the continents. The great preponderance of continental deposition over marine characterized most of the Triassic Period.

Beginning in the Jurassic, continental seas transgressed widely over Eurasia, the circum-Arctic region, the Andes region of South America, and the Indonesia–New Zealand region. Marginal transgressions also occurred in eastern Africa and western Australia. During the Cretaceous the Mesozoic seas attained their greatest size. All the continents were invaded at one time or another during the Cretaceous, and the seas left a very extensive sedimentary record. By the end of Cretaceous time a great regression resulted in renewed emergence of the continents. The paleogeography of the Mesozoic periods is presented in Figures 9-1, 9-4, and 9-9.

9-1 EUROPE

The Triassic Period

The arid continental physiography of northern Europe during Permian time continued into the Triassic. Here the Triassic record consists of two continental sequences separated by a marine complex—a relation responsible for the name of this system. The Germanic basin of north-central Europe extended from England across southern Scandinavia and, during at least part of the Triassic Period, into Poland. The predominant facies are red and varicolored sandstones and shales. Fossils are rare, even in the Middle Triassic marine complex.

In the Alpine and eastern Mediterranean regions the Triassic is represented by a nearly complete sequence of geosynclinal strata of complex facies relations. These strata contain abundant, well-preserved ammonoids, the main basis of Triassic correlation. Although the type area of the Triassic System is in Germany, where it is based on a sequence of Germanic facies, the Alpine region is now the standard of reference because of its complete marine development.

Germanic Facies. The extent of the Germanic facies coincides closely with that of the continental facies developed during Permian time. During the earliest Triassic the Germanic basin was traversed by aggrading streams and spotted by lakes. About a thousand feet of coarse-to-fine, generally red, clastic, nonmarine sediments accumulated.

Gradually marine invasion began, and lagoonal facies (some including evaporites), with a small marine and brackish-water fauna, were deposited. Most of the basin was marine during the Middle Triassic, and limestones, marls, marginal sandstone facies, and evaporites were deposited. In the last stages some coal-bearing continental sandstones were deposited; these testify to a change in the rate of subsidence and a change from normal marine to lagoonal conditions. The faunas of the Middle Triassic marine Germanic facies are peculiar. There were few species, but many animals of each kind. Most species were different from those living in the Alpine region at the same time.

The Upper Triassic deposits of the Germanic Basin greatly resemble those of the Lower Triassic. They also are lagoonal and continental. Variable continental facies, with evaporites, dolomites, and even some coal, characterize the last stages of deposition.

The Triassic formations of Great Britain include most of the New Red Sandstone. In a broad triangular belt across central England, with an apex in northern Ireland, continental deposition continued from Late Permian time (Fig. 9-2) into the Early Triassic. The surrounding highlands contributed fine-to-coarse clastic sediments. The Middle Triassic marine invasion of the Germanic basin did not reach England. The

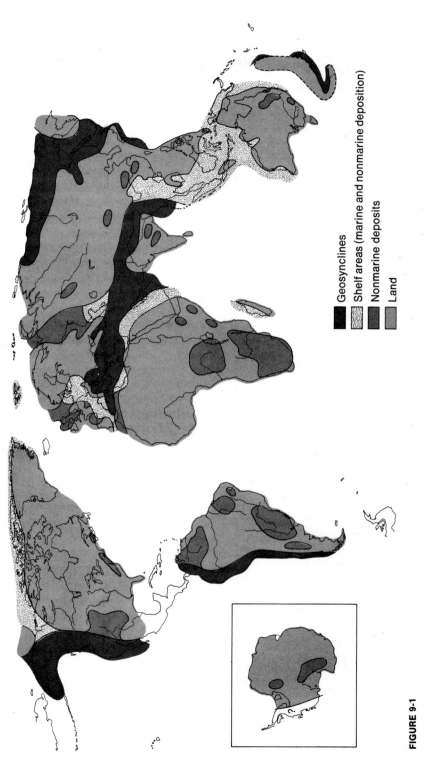

Geosynclines

Shelf areas (marine and nonmarine deposition)

Nonmarine deposits

Land

FIGURE 9-1

Paleogeographic map of the Triassic. (After many sources but especially Brinkmann, 1954.)

Upper Triassic consists mostly of red marls, with extensive deposits of gypsum and salt which cover a slightly larger area than that of the Lower Triassic. Like the Germanic basin, most of England apparently had an arid or semi-arid climate during Triassic time.

The Germanic facies were developed in the same sequence and general character in the Pyrenees, eastern and southern Spain, North Africa, the Balearic Islands, and Sardinia, though in eastern Spain and the western Mediterranean islands, the marine Middle Triassic fauna includes some Al-pine pelecypods and ammonoids mixed with Germanic types.

Thus much of Europe and North Africa was emergent during the Triassic, but contained several subsiding basins where aggrading streams, lakes, lagoons, and swamps accumulated a thick and varied sequence of clastic and chemical sediments. A single major phase of marine transgression interrupted the continental deposition, but the evaporite deposits, the redness of the sedimentary formations, and some sedimentary structures suggest that arid or semi-arid conditions prevailed over most of the area. The

FIGURE 9-2

Aeolian cross-bedding in the New Red Sandstone of southern Scotland. The interval between the Carboniferous and the Jurassic is represented in England by terrestrial desert and lacustrine facies. These particular beds are of Permian age but are similar to much of the Triassic of this region. (Courtesy of Her Majesty's Geol. Surv., British Crown copyright.)

rather homogeneous aspect of these Triassic deposits suggests similar environmental conditions over the whole region where the Germanic facies were developed.

Western Tethys–Alpine Facies. Few places have had a more complicated structural history than the Alps, with such richly diverse facies that make interpretation extremely difficult. The great orogenic movements that formed the Alps (Chap. 13) had not yet started in Triassic time.

The Middle Triassic seaway of the Germanic basin had few and probably only intermittent connections with the great Tethyan geosyncline to the south. One connecting channel was in central Czechoslovakia, and another was near the Rhone River in eastern France. The contrast between the facies and faunas of the Germanic and Alpine areas suggests a land barrier — the Vindelician arch — separating them. The Alps are composed of a complex of nappes — huge recumbent folds, overturned toward the north or northwest. Neither the geographic relations of the various facies nor their stratigraphic sequence is normal because of post-Triassic orogenies. A detailed paleogeography of the Alpine Triassic is beyond the scope of this book; merely a brief review of the types of facies represented is given here.

Triassic facies are developed in curving, roughly parallel bands from the French Alps to Austria. The most northern facies, near the Vindelician arch, are Germanic and consist of sandstone, dolomite, shale, and gypsum. South of these are deeper-water shale, limestone, and dolomite. In the southern Alps are reef dolomite and limestone deposited in shallower water than the earlier facies.

During the earliest Triassic much of the Alpine area was covered by a shallow sea with sandy and clayey bottoms. Life was neither specifically diverse nor particularly abundant. In the littoral zones plant debris was buried with the silts, muds, and sands.

In the Middle Triassic the number of facies increased. The principal lithologic type was limestone, deposited under many different environmental conditions and differing in bedding, composition, faunal content, and color. Reef complexes made up of algae are common; some limestone facies contain abundant cephalopods, but others contain mainly crinoids or pelecypods. Later in Middle Triassic time volcanoes supplied local tuff and agglomerate. Although a detailed paleogeography is nearly impossible, the numerous diverse facies must have required a wide variety of environments.

In Late Triassic time principally carbonate deposition continued, along with some volcanics. Many scattered algal and coral reefs formed large, structureless masses of limestone, between which accumulated well-bedded carbonate deposits (Fig. 9-3). In some places massive carbonate deposits continued through several Triassic ages, but stage boundaries are commonly uncertain because diagnostic fossils are few.

The sequence and zonation of the ammonoids in the Alpine area form the standard for correlation of marine Triassic strata.

The Jurassic Period

The paleogeography of the Jurassic differs greatly from that of the Triassic (Fig. 9-4).

European Shelf Seas. Shallow seas spreading from the Tethys and the Atlantic early in Jurassic time invaded much of England, France, Germany, and Denmark. The geography of these seas was influenced greatly by the massifs, the remnants of the old Hercynian Mountains, which acted as stable blocks (Fig. 6-19, p. 187). The areas between the massifs became subsiding basins and preserved a rich and complicated sedimentary history. The Jurassic history of this area is thus an interplay between the massifs and the basins in the transgressive

FIGURE 9-3
Upper Triassic dolomite reefs and associated marly beds exposed in the southern Tyrol. (From Kayser, 1893.)

and regressive movements of the sea. Along the margins of the basins, near-shore sediments are common and show many breaks in the stratigraphic record. In the central parts the history is more nearly complete. Some massifs did not remain islands throughout Jurassic time but were occasionally completely inundated. The shoreline patterns thus changed markedly with the various movements of sea level. Figure 9-5 shows the shorelines around the French central massif at three stages of the Jurassic and three of the Cretaceous.

As the islands were low, little coarse clastic sediment was deposited. The Jurassic formations thus consist mainly of clay shale, generally dark, and carbonate deposits (Fig. 9-6). Most formations contain abundant fossils.

The Early Jurassic shelf sea did not reach Poland or most of European Russia. Southern Russia was occupied by the Tethys (Fig. 9-4). In early Late Jurassic time a great transgression covered much of European Russia, connecting the Arctic Ocean with the Tethys. The eastern shore of this Late Jurassic sea lay west of the Urals; the northwestern shore was on the Baltic shield. The sea connected westward through Poland

and Germany with England and the North Sea. With the opening of this seaway from Russia across northern Europe new members were added to the European faunas. The connection was not long-lived, for the fossiliferous clays and carbonates preserving its record are succeeded by lagoonal and continental facies, indicating that the sea retreated from northern Europe. The presence of fossils of Russian Upper Jurassic affinities on one of the Lofoten Islands, off the northwestern coast of Norway, indicates that there may possibly have been migration of marine organisms through the Arctic region, north of the Baltic shield, between the Russian sea and the North Sea.

Toward the close of the Jurassic the shallow seas that had inundated most of Europe retreated northward to the Arctic and Atlantic Oceans and southward to the Tethys. This regressive phase is reflected in marked facies changes, which terminate in lagoonal and continental deposits. In the Jura Mountains of eastern France the Upper Jurassic sedimentary patterns record the facies changes. Through much of Late Jurassic time the French Jura was the site of extensive building of organic reefs, which were

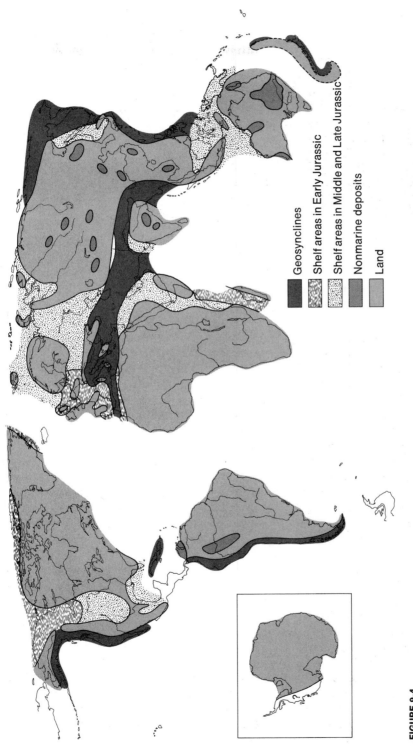

FIGURE 9-4

Paleogeographic map of the Jurassic. (After many sources but especially Brinkmann, 1954.)

Geosynclines

Shelf areas in Early Jurassic

Shelf areas in Middle and Late Jurassic

Nonmarine deposits

Land

aligned in bands parallel to the structural axis of the Jura Mountains. With successive younger stages of the Late Jurassic, these reef trends migrated southeastward (Fig. 9-7). In the latest Jurassic the reef zone was on the border of the Alpine sea, and behind the reef, toward Paris, lagoonal conditions prevailed amid low emergent lands.

Western Tethys–Alpine Region. The original Alpine geosynclinal belt included a number of longitudinal swells and basins. Long linear welts or geanticlinal ridges were separated by downwarps. The welts were earlier considered to be incipient anticlinal folds on the bottom of the geosyncline and were thought to represent the early stages of development of the great nappes. These folds were held to be asymmetric, with a steep external flank and a more gentle slope toward the south or east. The crest of each anticlinal fold was considered to be the site of accumulation of shallow-water sediments, and each intervening downwarp the site of deep-water sediments. Figure 9-8 presents in schematic form this interpretation of the Alpine geosyncline during the Jurassic.

The above interpretation of the Alpine geosyncline has until quite recently been accepted as dogma, but it now appears to be a gross oversimplification. Detailed research in different parts of the Alps has failed to prove the existence of embryonic folds. Many submarine or temporarily emergent welts were discovered, but they were not systematically asymmetric, did not show anticlinal structure, were not all persistent, and are not necessarily linked to the fronts of nappes. Many steep scarps were shown to face the wrong way—toward the south or, in the western Alps, toward the east. Extensive geologic studies in the Alps over the past two decades demonstrate that the theory of embryotectonics is not applicable for the Jurassic and early Cretaceous. Rather, there now emerge data to show that during these times the geosyncline was the site of extensive normal faulting, resulting in a complex pattern of uplifted and downwarped blocks. Facies patterns associated with these two main tectonic elements are very complex, and changes in thickness often very rapid. Conglomerates that accumulated in front of active faults are common. Faulting was highly variable but was especially strong during the first half of the Jurassic.

The Cretaceous Period

The final period of the Mesozoic is by far the most spectacular of that era and perhaps of all Phanerozoic time. It was a period of great transgressions (Fig. 9-9). The widespread Late Cretaceous transgression, already described for North America, also occurred in South America, Africa, Europe, and Australia. Many portions of these continents that had been largely or entirely

FIGURE 9-5

Paleogeographic map of the Massif Central of France, showing the shoreline during three stages of the Jurassic (J1, J2, J3) and three stages of the Cretaceous (C1, C2, C3). (After J. H. F. Umbgrove, *The Pulse of the Earth*, Nijhoff, The Hague, 1947.)

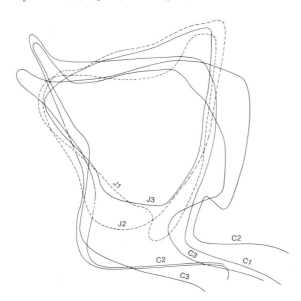

FIGURE 9-6

Alternating shale and limestone of Early Jurassic age (Blue Lias) exposed along the southern coast of England (Lyme Regis, Dorset). (Courtesy of Her Majesty's Geol. Surv., British Crown copyright.)

emergent before were covered by vast continental seas.

Chalk and limestone are characteristic of the Cretaceous shallow shelf seas adjoining the Tethys. Tectonic activity increased, especially during the latter half of the period; it was accompanied by widespread vulcanism and plutonism.

European Shelf Seas. The regressive facies of the uppermost Jurassic marked a retreat of the seas from Europe, both toward the north and southward to the Tethys. In earliest Cretaceous time many of the old massifs were joined, and much of north-central Europe, England, western France, and Spain became land. Continuous sequences of marine deposits dating from the Jurassic to the Cretaceous are found only in the Tethyan realm, and especially in southeastern France. Shale and limestone

of diverse facies and with abundant fossils record the transition (Fig. 9-10).

Early in the Cretaceous the Tethyan seas began to spread northward. Marine deposits occur in Yorkshire and Lincolnshire, northern Germany, Denmark, and Poland,

FIGURE 9-7

Diagrammatic reconstruction of Upper Jurassic facies across the Jura Mountains, illustrating the southward shift, in time, of the organic-reef facies. The broken lines are stage boundaries. (After M. Gignoux, 1955.)

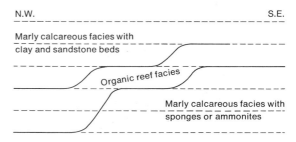

N.W.   S.E.

Marly calcareous facies with
clay and sandstone beds

Organic reef facies

Marly calcareous facies with
sponges or ammonites

extending from there to the Urals. The transgression began earlier in Yorkshire than in northern Germany. As in the Late Jurassic, the Early Cretaceous faunas of the northern basins were completely unlike those of the southern basin, which were allied to those of the Tethys. Not until near the end of Early Cretaceous time was there a mixture of boreal and Tethyan elements. In addition to the marine embayments in the northern province and the Tethys in the south, several basins receiving stream and swamp deposits, several thousands of feet thick, marked the European landscape. The best known of these are the Wealden beds of southeastern England and northwestern France, but similar facies are recognized on the northeastern coast of Spain and in northern Germany.

The age and facies relations of the English Wealden deposits to the Tethyan facies are illustrated in Figure 9-11, which shows the gradation from Tethyan deeper-water facies through shallow shelf-sea facies to the fluvial-deltaic Wealden beds (Fig. 9-12). The route of these early transgressive seaways was largely controlled by the old Hercynian massifs. By late Early Cretaceous time, most of these terrestrial basins were completely inundated, and the boreal and Tethyan seas were connected.

Much of the Lower Cretaceous consists of fine-grained clastic facies. In the margins of the Tethys richly fossiliferous carbonate facies are common; some are accumulations of corals, mollusks, foraminifers, hydrozoans, and bryozoans—an assemblage characteristic of the shelf seas adjoining the Tethys throughout the Cretaceous.

There was no significant regression in Early Cretaceous time, and by Late Cretaceous a really widespread transgression had begun. The most characteristic facies of the Upper Cretaceous in Europe is white chalk (Fig. 9-13). The initial facies of the Upper Cretaceous, where they transgress beyond the limits of the Early Cretaceous seas, are glauconitic sandstones. The intricate pattern of transgression and regression on the flanks of the old Hercynian massifs indicates widespread differential subsidence over the area (Fig. 9-5).

FIGURE 9-8

(A) Schematic map of paleogeographic units of the western Alps in the Jurassic. (B) Profile showing ridges and basins of the Jurassic Alpine geosyncline. (After M. Gignoux, 1955.)

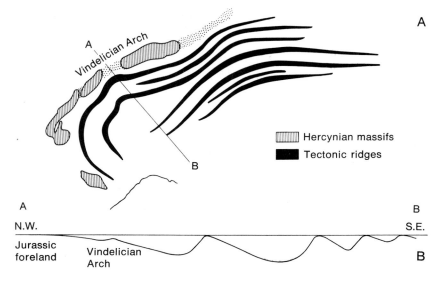

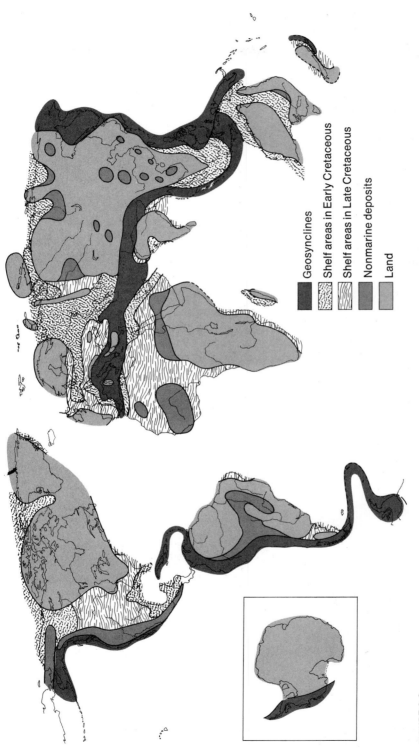

	Geosynclines
	Shelf areas in Early Cretaceous
	Shelf areas in Late Cretaceous
	Nonmarine deposits
	Land

FIGURE 9-9

Paleogeographic map of the Cretaceous. (After many sources but especially Matsumoto, 1967.)

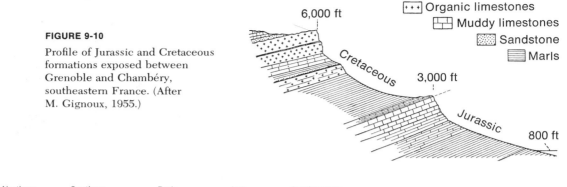

FIGURE 9-10

Profile of Jurassic and Cretaceous formations exposed between Grenoble and Chambéry, southeastern France. (After M. Gignoux, 1955.)

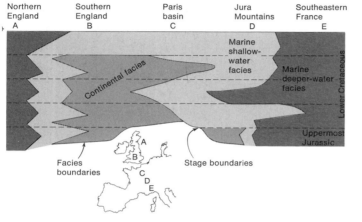

FIGURE 9-11

Diagrammatic reconstruction of Lower Cretaceous facies between northern England and southeastern France, showing the Early Cretaceous transgression and the establishment of open communication between the northern seas and the Tethys. (After Gignoux, 1955.)

FIGURE 9-12

Lower Cretaceous cross-bedded sandstones (Folkestone beds) of southeastern England. These strata represent the initial marine deposits of the transgressive Early Cretaceous sea that inundated the Weald delta. (Courtesy of Her Majesty's Geol. Surv., British Crown copyright.)

FIGURE 9-13

Cretaceous chalk cliffs of southeastern England. (Courtesy of Her Majesty's Geol. Surv., British Crown copyright.)

Western Tethys–Alpine Region. The geosynclinal pattern of the Late Jurassic continued into Cretaceous time in the Alpine region. The pattern consisted of positive and negative blocks separated by normal faults (Fig. 9-14). The position and duration of these tectonic units was ephemeral. In mid-Cretaceous the pattern changed from normal faulting to folding, establishing linear island festoons unrelated to the positive fault blocks of the Jurassic and early Cretaceous. Throughout much of the Alpine geosyncline, these anticlinal islands spilled into the adjoining deep basins a unique shale and sand facies called Flysch. Orogenic movements in the island areas and deposition of Flysch continued throughout the Upper Cretaceous. The schematic cross sections of Figure 9-14 clearly illustrate this change in tectonic pattern within the Alpine geosyncline during the Mesozoic.

The several orogenic phases of the Cenozoic that formed the present Alpine structure entailed enormous compressions, which caused large segments of the crust to override the frontal masses. A complex series of nappes was thus formed in the Alps. With the nappes forced up and away from their original position, and overriding masses that may have been of the same or later age, we can easily understand that the present geographic and structural relations of the facies sequences for any particular period are quite different from the original. A thorough knowledge of facies is necessary if we are to "unroll" the nappes and reconstruct the form and history of the ancient geosyncline.

This type of analysis can be well illustrated by Cretaceous formations cropping out of the western Swiss Alps, in the Helvetic nappes. The tectonic relations of the various segments of the Helvetic nappes is illustrated in Figure 9-15, A. The sequence of Cretaceous facies in each of the nappes is diagrammatically represented, in their present geographic position, in Figure 9-15, B. Here we see an unintelligible mixture of near-shore shallow-water organic limestones of various types and deeper-water marls with cephalopods. Reconstruction of these various segments of facies sequences in terms of the gross facies relations gives the cross section of Figure 9-15, C, where the near-shore shallow-water facies to the left can be seen to tongue out into the deeper-water marl facies to the right. Thus, by combining stratigraphic and tectonic data, geologists have been able to reconstruct this segment of the Alpine Cretaceous geosyncline as it was before the Cenozoic orogenic disturbances.

9-2 AFRICA

While the continental area to the north of the Tethys underwent a constant series of extensive transgressions and regressions, testifying to great instability, the foreland to the south of the Tethys (northern Africa) remained much more stable, with only very minor marine transgressions during the Tri-

assic. During the Jurassic the Mediterranean coast of Africa was marked by a series of embayments from Algeria to Egypt, none of which extended very far southward. The eastern coast, however, from Ethiopia and Somaliland to Mozambique, was widely inundated, as was western Madagascar. In the Cretaceous, even more extensive transgressions occurred, partly on the previous patterns and partly in new regions.

During the Cretaceous the Tethyan seas spread south, invading much of the northern tier of countries, the Late Cretaceous seas extending much farther south than the Early Cretaceous. In this sea fine muds and

FIGURE 9-14

Schematic cross sections showing evolution of the Alpine geosyncline during the Mesozoic and early Cenozoic. (After R. Trümpy, *Umschau*, **18**, 1965.)

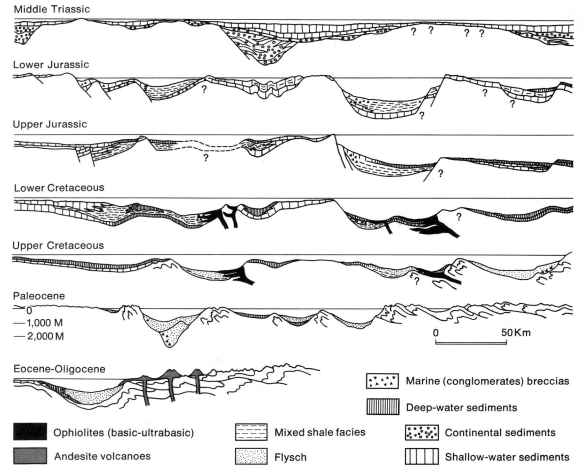

Middle Triassic

Lower Jurassic

Upper Jurassic

Lower Cretaceous

Upper Cretaceous

Paleocene
0
—1,000 M
—2,000 M

0   50 Km

Eocene-Oligocene

■ Ophiolites (basic-ultrabasic)
▦ Andesite volcanoes
═ Mixed shale facies
∴ Flysch
⩗ Marine (conglomerates) breccias
▥ Deep-water sediments
▨ Continental sediments
▦ Shallow-water sediments

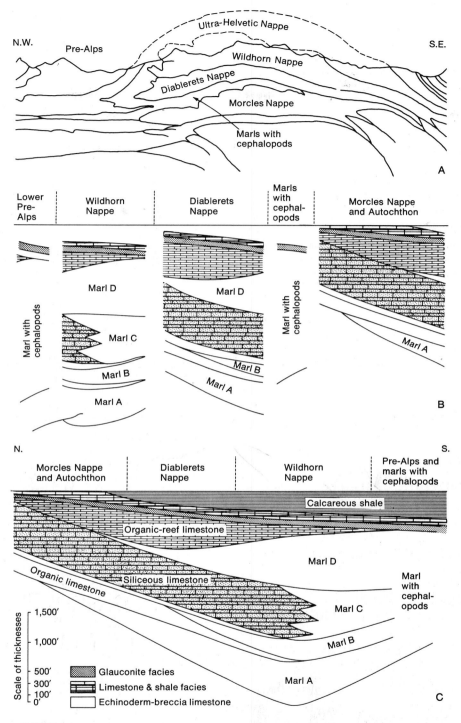

FIGURE 9-15

Cretaceous facies in the Helvetic nappes of western Switzerland. (A) Diagrammatic cross section of the Helvetic nappes. (B) Succession of facies in the principal structural units of the Helvetic nappes; notice the complete disorder of the facies arrangement. (C) Restoration of the segments in *B* to their position before folding. (Part A after Bearth and Lombard, 1964. Parts B and C after Gignoux, 1955.)

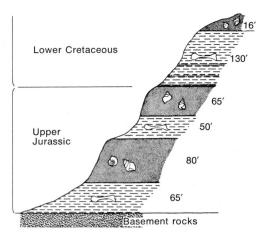

FIGURE 9-16

Upper Jurassic and Lower Cretaceous formations
in the vicinity of the Tendaguru Hills,
Tanzania. The alternating sequence of terrestrial
dinosaur-bearing formations and marine fossiliferous
formations probably represents a continuous
depositional history. (Data from E. Hennig, 1913.)

carbonates were deposited. Another transgression took place from the Gulf of Guinea, and at least intermittently during the Cretaceous the South Atlantic Ocean was connected with the Tethys across Nigeria, the French Sudan, Algeria, and Libya. A narrow coastal strip from Nigeria to Angola was also inundated by transgressive seas.

Much of the eastern coast of Africa, from Somaliland to South Africa, had marginal seas during at least part of Cretaceous time. The facies are a mixture of clastic and carbonate rocks, generally fossiliferous. One of the most famous Cretaceous localities is at Tendaguru, in Tanzania, where there exists a continuous succession of alternating sequence of marine and nonmarine strata, ranging in age from Middle Jurassic to Lower Cretaceous (Fig. 9-16). The nonmarine formations are famous for the fossils of gigantic dinosaurs that were collected by the German Tendaguru Expedition in 1909–1912. The marine strata contain invertebrate fossils that can be related closely with the standard European sequence. It is partly on the correlation with the vertebrate faunas in these strata that the Morrison Formation of western North America is considered to be uppermost Jurassic.

North of a line from the Gulf of Guinea to the headwaters of the Nile and south of the shorelines of the marginal transgressive seas a vast continental sandstone complex, called Nubian sandstone, was deposited. The area of this formation includes much of the Sahara Desert and the Anglo-Egyptian Sudan (Figs. 9-17 and 9-18). The formation consists of both wind- and water-laid deposits. On its outer margins (north and east) this sandstone complex intertongues with Late Cretaceous marine formations. Note, in Figure 9-17, that the underlying Paleozoic formations are of the same facies.

9-3 THE MIDDLE EAST— THE CENTRAL TETHYS

The great central belt of the Tethys, extending from Turkey to India, has had a very active Mesozoic history. The few data available make possible an outline of the main features of the era. The northern boundary of the Tethys graded onto the

FIGURE 9-17

Sections of Paleozoic and Mesozoic continental formations exposed in Wadie Karkur Murr in the border region of extreme southern Libya and Egypt. (After Sandford, 1935.)

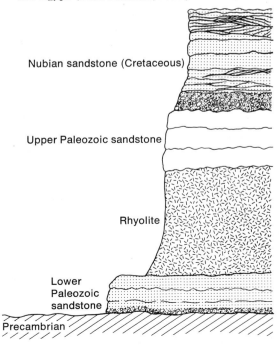

FIGURE 9-18

Second cataract of the Nile, south of Abu Sir, Sudan. Continental Upper Cretaceous sandstones (Nubian Series) rest unconformably on crystalline rocks (Precambrian). In the foreground, crystalline rocks forming the bank of the Nile; in the middle distance, eroded surface; in the background, buff and cross-bedded sandstones (Nubian Series) with fossil wood. In this area the middle and lower members of the Nubian Series and all of the Paleozoic are absent. (Photograph by G. A. R. Sandford; published with the permission of Geol. Soc. London.)

Russian platform and onto the shelf-sea areas just north of the Black Sea that extended eastward through the Caspian Sea to India. The southern shore of the Tethyan sea infringed on the Ethiopian shield.

The Triassic Period

In cental, northern, and eastern Iran Lower and Middle Triassic strata are shallow-water limestone and dolomite deposits, reflecting a continuation of the Paleozoic pattern. In the Zagros Mountains area shallow-water carbonate facies of much greater thickness developed. The trough-like or geosynclinal aspect of this region began in the Permian and continued into Cenozoic time. Herein accumulated nearly continuous sequences of marine limestone and shale (Fig. 9-19).

The area of Iran northeast of the Zagros Mountains, which had been of platform facies from the late Precambrian through Middle Triassic, was folded and faulted in the Late Triassic. On the eroded surface, Late Triassic and Early Jurassic coalbearing sandstone and shale accumulated, reflecting a continental regime with rare marine incursions. These rocks differ fundamentally from the correlative marine limestone and shale in the Zagros trough. The facies are uniform throughout central, north and east Iran, but change abruptly in thickness from a few tens to more than 6,000 feet.

The Jurassic Period

In the area of Iran northeast of the Zagros Mountains Late Triassic diastrophism was followed by a phase of continental deposition that continued well into Early Jurassic time. Continental facies of this age are also well developed in Afghanistan (Fig.

FIGURE 9-19

Generalized stratigraphic section of the Arabian peninsula and the Zagros Mountains and foothills. (After J. Law, *Bull. Am. Assoc. Petrol. Geol.*, 1957.)

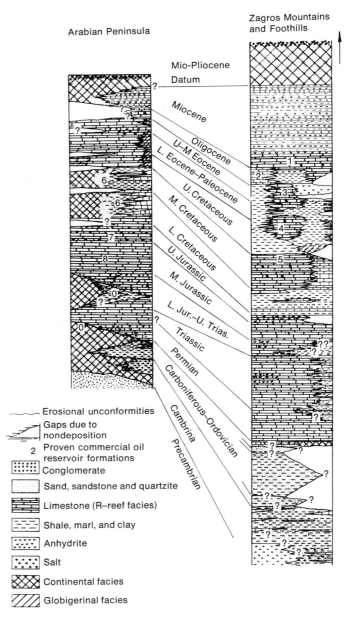

9-20). A transgression in the late Early Jurassic gave rise to extensive marine limestone and shale formations. Deposition of these facies continued in Middle and Late Jurassic time in northern Iran, but in east-central Iran extensive evaporites and reef formations accumulated. Some areas in central Iran continued to receive continental deposits; elsewhere no deposits accumulated. Evaporites were also extensively de-

veloped in the Zagros trough and over vast areas of the Arabian platform. Upper Jurassic strata south of Kerman in central Iran form thick sequences of radiolaritic limestones, shales, and abundant volcanic rocks, overlain by fossiliferous Late Jurassic limestone.

Along the southern shore of the Tethys geosyncline, shallow-water clastic and carbonate formations formed. In northern Is-

FIGURE 9-20

Nonmarine plant-bearing sandstone, shale, and conglomerate of Jurassic age overlain by Middle Cretaceous marine limestones, Hindu Kush Mountains, eastern Afghanistan. Note the thrust fault just above the lower escarpment. (Photograph by H. H. Hayden, Geol. Surv. India.)

rael about 5,000 feet of mainly carbonate facies, including some sandstone beds, thin rapidly to the south, where they wrap around the ancient Ethiopian shield. The shoreline probably extended eastward through the Sinai peninsula to north-central Saudi Arabia and thence swung south to Ethiopia and the coastal region of Tanzania. A large island covering eastern Somaliland and southeastern Aden lay off the Jurassic coast of Africa.

The thickness of the Jurassic deposits in the area northeast of the Zagros Mountains ranges from about 3,000 to more than 16,500 feet. The average thickness, about 10,000 feet, nearly equals the average total thickness of the Late Precambrian to Middle Triassic platform cover. This thickness also contrasts with the Zagros trough maximum of less than 5,000 feet and an average of

less than 3,000 feet. Although during the Jurassic the region northeast of the Zagros was more active and subsided more, it remained continental, whereas the Zagros trough, though continually marine, received far thinner sediments.

A second diastrophic phase affected Iran northeast of the Zagros in the Late Jurassic-Early Cretaceous, as discussed in the next section.

The Cretaceous Period

The diastrophic movements of the Late Jurassic-Early Cretaceous were more intense than those of the Late Triassic and were accompanied in central Iran by more widespread folding, some granitic intrusions, and low-grade metamorphism. Movement also commenced at this time in the salt domes of central Iran. A period of erosion followed, and later widespread but thin continental red beds were deposited. In the Zagros trough, however, deposition appears to have continued from the Jurassic into the Cretaceous. Not until the mid-Early Cretaceous did widespread marine transgression recur over much of central Iran. The Middle and Upper Cretaceous strata of most of the Middle East are mainly limestone and marl (Fig. 9-21).

In the Late Cretaceous, thrusting and folding began in the northeast part of the Zagros trough, accompanied by deposition of radiolarian oozes and *Globigerina* marls and by the intrusion of basic igneous rocks (ophiolites). Similar radiolarian-ophiolite complexes were formed in the Iran-Turkish border area, in the region of the Makran Ranges of southern Iran, in the East Iranian ranges, and in central Iran (Fig. 9-22). The detritus from the post-orogenic phase of this belt has been preserved in several frontal depressions immediately to the southwest, locally as much as 8,000 feet of clastic deposits, including thick massive chert conglomerates and reef limestones. In southeastern Turkey Late Cretaceous diastrophism in the Taurus fold belt—continuation of the Zagros fold belt—led to massive gravity slides southward into marine basins. The slide material is composed of eugeosynclinal sediments and basic igneous rocks. They are intercalated in a shaly-sandy sequence of Upper Cretaceous age. One gravity-slide body in southeast Turkey is 10,000 feet thick.

9-4   THE EASTERN TETHYS— INDIA, PAKISTAN, BURMA

The Triassic Period

In the Himalayas, the Triassic formations are best known in a belt that stretches from Kashmir southeast to Nepal (Fig. 9-23). One locality in particular—Spiti—is famous for the completeness of the Mesozoic section exposed in the vicinity (Fig. 6-28, p. 201).

FIGURE 9-21

Cretaceous limestone, Zagros Mountains, southwestern Iran. (Photograph by J. V. Harrison.)

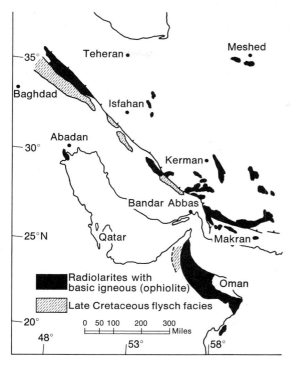

FIGURE 9-22

Major facies trends of the Late Cretaceous for the Middle East. (After N. L. Falcon, *Advancement of Sci.*, **24**, 1967.)

In the region of Spiti more than 4,000 feet of dark, richly fossiliferous limestones and shales comprise the Triassic record. This section, in which all stages of the Triassic are represented, is gradational with the overlying Jurassic formations. The Lower and Middle Triassic formations are thin, each being about 100 feet thick. The Upper Triassic formations are much thicker, up to nearly 4,000 feet. The upper half of the Upper Triassic sequence is nearly everywhere a light-colored, massive dolomite and limestone facies without cephalopods, in contrast to the underlying dark, cephalopod-rich shales and limestones. Aside from minor variations in thickness, the lithology and faunal zones are remarkably uniform.

The invertebrate faunas of these formations, especially the ammonoids, are different from the Alpine faunas. In certain Triassic stages most of the ammonoid species

in the Alpine region have no ornamentation; in strata of the same age in the Himalayas, however, the ornamented species far outnumber the smooth. The reason for this variation is not clear. Whereas the Alpine facies is pure carbonate with little or no clay, the Himalayan facies is also carbonate but with a high clay content. This suggests that environment may have been decisive. On the Tibetan plateau, only ten miles from some of the classic exposures of the typical "Himalayan facies," Triassic blocks of limestone, from a few cubic feet to a few cubic miles in size, are found lying in complete disorder on late Mesozoic and early Cenozoic lavas (Fig. 9-24). These *exotic blocks* include limestones, from Permian to Jurassic in age, in a facies completely unlike that of the normal Himalayan facies and very similar to the limestone facies in the Alps. The faunas are also different, particularly the Upper Triassic ammonoid faunas that occur in red limestones. The faunas and lithologies of these particular blocks are nearly identical with certain Upper Triassic limestones of the Alps. The ammonoids are predominantly of the smooth (Alpine) type. The same very strong similarity between these blocks and Alpine facies is seen in the Lower Jurassic limestones. But an anomalous situation is found in the Lower and Middle Triassic blocks: these are of the same red limestone as the Upper Triassic blocks, but the enclosed faunas are typically Himalayan (ornamented) and not Alpine in affinities.

A question immediately arises: How did such strikingly different facies with such different faunas come to be only ten miles apart? Two suggestions have been put forward: (1) the blocks, resulting from a gigantic volcanic explosion, were transported by lava flows to their present positions; (2) the blocks are remnants of a great thrust sheet from the north.

In northwestern Kashmir, volcanic activity and partially emergent conditions persisted from late Paleozoic time well into the Triassic. Here only the Upper Triassic is represented by marine strata; the Lower

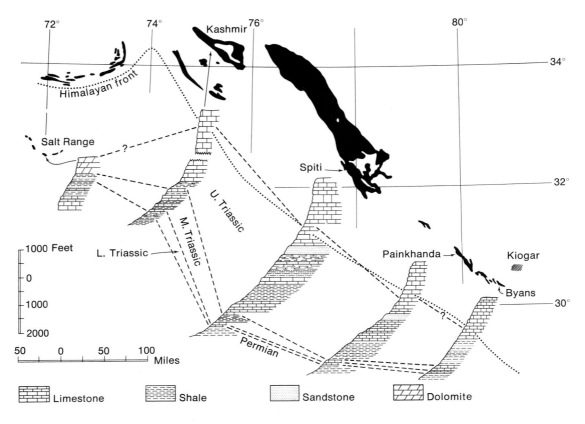

FIGURE 9-23

Selected stratigraphic sections of Triassic formations in the Salt Range
and the Himalayas. (Data from C. Diener, 1912.)

and Middle are volcanic. In the region
of Srinigar the Triassic consists entirely
of limestone and shale with no volcanic
rocks (Fig. 9-23). About 200 miles south
of the front of the western Himalayas lies
the Punjab Salt Range, another important
Triassic locality. Here the whole Triassic
sequence is less than 1,500 feet thick and
of a limestone and clastic near-shore facies.
The Lower Triassic consists of approximately 600 feet of richly fossiliferous sandstones and limestones, completely conformable with the underlying Upper Permian
shale. Because of the richness of the invertebrate faunas in these beds and the underlying Upper Permian strata, the Salt Range
is a principal area for study of the great

faunal changes that took place between the
two periods. Although there is complete
conformity between the Permian and Triassic formations there is a sharp break in
lithofacies. The uppermost beds of the
Permian are white calcareous sandstone,
overlain by a much recrystallized brown
dolomite of lowermost Triassic age. New
data on Permian fauna and strata in southern
U.S.S.R. and South China now suggest that
the Permian formations of the Salt Range
are not youngest Permian, as was generally
thought, but slightly older. The stratigraphic
gap thus entails several millions of years.
The Middle and Upper Triassic formations
of the Salt Range consist of sandy limestones
and dolomites and are not fossiliferous.

Himalayan facies

☐ Cretaceous flysch   ☰ Jurassic shale   ▦ Triassic sh. & ls.   ▨ Lower Trias. to Silurian

Tibetan facies

⊞ Basic igneous rock   ▦ Mesozoic limestone   ■ Exotic blocks

FIGURE 9-24

Geological sketch map of the Kiogar-Chirchun region on the frontier of India and Tibet. (After Heim and Gansser, 1939.)

The Jurassic Period

Upper Triassic deposits in the Himalayas consist of massive limestone and dolomite (similar to the Alpine section); deposition of this facies continued into Middle Jurassic time with no significant breaks. The carbonate strata are followed by 300–500 feet of black-to-gray micaceous fossiliferous shales of Late Jurassic age. These two Jurassic facies were widespread throughout the Himalayan area, including central and southern Tibet. The upper shale facies passes conformably upward into Lower Cretaceous sandstones.

In Baluchistan is a thick sequence of carbonates and fine-grained clastics from Early to early Late Jurassic in age. Through most of Late Jurassic time the area was undergoing erosion.

During the Jurassic peninsular India, which had long been a persistent continental block, receiving only terrestrial sedimentation, underwent subsidence in the Cutch and Kathiawar districts (western India). By Late Jurassic those districts were covered by a marginal sea, which extended northward through the Thar desert of Rajputana and the Salt Range to the Tethys proper. The highly fossiliferous sediments are mostly fine-to-coarse clastics with some limestone. Some sandstone beds contain fossil plants with Gondwana affinities.

The Cretaceous Period

The Cretaceous is widely distributed in India and Pakistan. Besides a continuation of the Himalayan-Tethyan geosynclinal development, marine encroachments took place on the coast of peninsular India. The present form of India was beginning to take shape. Toward the end of the period, enormous volcanic activity occurred throughout much of the Himalayan and peninsular regions.

At Spiti the Jurassic shales pass upward without recognized break into sandstone a few hundred feet thick, which contain Early Cretaceous ammonoids. Overlying the sandstone are 100 feet of white limestone of latest Cretaceous age. Between the sandstone and the limestone is a large time break. Such a large break probably indicates withdrawal of the Tethyan sea during most of Middle Cretaceous time. The limestones are overlain by unfossiliferous sand-and-shale facies of Cretaceous age. This whole sequence at Spiti clearly indicates a shallowing of the Tethyan sea and regressions and transgressions during Cretaceous time. What a contrast to the facies of the early Mesozoic!

In the Kiogar-Chirchun region, on the edge of the Tibetan plateau, the Lower Cretaceous clastics, correlative with those at Spiti, are followed by sandstone-and-shale facies, the limestone being missing but the large stratigraphic gap still recognizable. The upper sand-and-shale facies, including red and green siliceous deposits and radiolarian beds of Late Cretaceous age, is overlain by more than 1,000 feet of volcanic tuff and breccia. In this breccia are found the famous exotic blocks, composed mainly of limestone and ranging in age from Permian to Jurassic.

A diagrammatic cross section of the region of these exotic blocks is shown in Figure 9-25. During Late Cretaceous, while the clastic and siliceous facies were being developed, submarine volcanic activity commenced, and more than 1,000 feet of tuffs and breccias accumulated. Later in this volcanic episode a large thrust movement from the north began to override these submarine volcanoes, and isolated blocks of limestone became embedded in the tuff and breccia to form the exotic blocks (Figs. 9-24 and 9-25).

Little Cretaceous history is known, of the western Himalayas, in Kashmir, except that fossiliferous limestones and shales are intercalated with volcanic rocks, and that granites and other igneous rocks intrude Cretaceous formations. Somewhat similar rocks occur in Afghanistan.

In Baluchistan a decided unconformity separates Jurassic and Cretaceous. There is also an unconformity that includes all of Middle Cretaceous time. Early Cretaceous seas deposited fossiliferous limestone and shale, and after long erosion, marine conditions returned in Late Cretaceous with mostly clastic facies with intercalated tuff, agglomerate, basalt, etc.

Peninsular India has been remarkably emergent since Precambrian time. Marine Cretaceous strata in the interior and coastal regions mark one of the few times of transgression. Along the Narbada valley, just above Bombay, more than 200 feet of a lower clastic facies follow a richly fossiliferous carbonate facies of Middle Cretaceous age. Marginal to these marine beds fresh-water clastics contain plant and dinosaur remains.

In the Trichinopoly district on the eastern coast of India a rather thick coastal wedge is exposed that consists mainly of clastic facies of middle Early to Late Cretaceous age.

At the head of the Bay of Bengal, in Assam, nearly 1,000 feet of clastic formations contain Late Cretaceous fossils allied to those of the Trichinopoly district and not to those of the Tethys, showing that the Bay of Bengal was then not connected with the Tethys and that much of the former Tethyan area between the eastern Himalayas and northern Burma was emergent.

The Cretaceous of Burma is not well known, but fossiliferous limestone, shale, and sandstone are present. Cretaceous igneous activity produced serpentine, peridotite, gabbro, etc. in the western belt of Burma and on the Andaman and Nicobar islands (§ 13-5).

9-5 THE SOUTHWESTERN PACIFIC

At one time or another during the Mesozoic, the ocean inundated most continental areas bordering the southwestern Pacific.

Australasia

With the close of the Permian, geosynclinal conditions ended in Australia and probably (temporarily) in Indonesia. The Mesozoic depositional history thus differs from that of the Permian. In Australia the Triassic was a time of nearly complete emergence. In the earliest Triassic there was a small

FIGURE 9-25

Diagrammatic structural cross section of the Kiogars. The summits consist of Triassic and Jurassic limestone of Tibetan facies, floating on and in a sheet of basic igneous rocks, which is thrust over underlying autochthonous sediments. (Slightly modified from Heim and Gansser, 1939.)

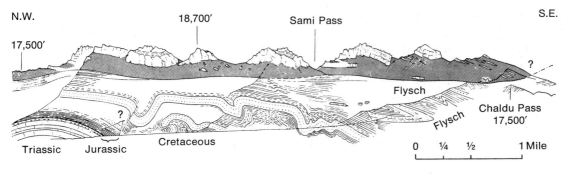

marine embayment on the east and three on the west coast. Inward from the eastern coastal region, terrestrial sandstone and shale, mostly red, accumulated in intermontane basins and in those inherited from the Permian. Some of these areas were volcanically active. No Middle or Late Triassic seas transgressed Australia. Terrestrial clastic sediments continued to accumulate in the eastern basins; in some, coal is abundant. At the close of the period, mild folding affected this eastern area, but throughout the rest of Australia, Triassic history is documented only in a few small isolated basins of terrestrial clastic sediments in central and western Australia.

The Triassic history of Australia contrasts markedly with that of Indonesia, where Triassic deposits are widespread. Timor is one of the most remarkable Triassic localities in the world; approximately 1,000 species of marine invertebrates have been described. These beautifully preserved fossils have Tethyan and circum-Pacific affinities and represent a nearly complete sequence of Triassic stages. They include a remarkable multiplicity of facies (for example, red limestone cephalopod facies; red limestone pelecypod facies, without cephalopods; clastic formations with plant remains and interbedded limestones and shales; oolitic and clastic limestone facies; and reef limestone facies). From a six-foot bed of limestone 462 species of ammonoids, 62 of nautiloids, 26 of gastropods, 2 each of brachiopods, hydrozoans, crinoid stems, and worm tubes have been described—a mixture of species characteristic of two ages of Late Triassic time. Such mixtures are *condensed* sections. They are common elsewhere; for instance, thin, extremely fossiliferous sections, often with faunas of more than one stage, are recognized in the Triassic of the Himalayas and in the Jurassic of Europe. Condensed sections imply extremely slow deposition and perhaps redeposition.

Unfortunately, all these wonderful Triassic fossils occur in isolated blocks within Cenozoic formations. In western Timor, where most of the faunas described come from, the geologic structure is extremely complex and has not been completely deciphered. The problem is to distinguish the Triassic rocks transported from far away during diastrophism from those in the approximate position of original deposition. Outcropping extensively over eastern Timor is a 3,000-foot-thick formation of Middle and Upper Triassic age, consisting primarily of fine-grained calcareous muds, calcareous shales with a little fragmental limestone, bituminous limestone, and quartz sandstone. These rock types and their limited fauna suggest they were deposited in shallow water probably far from any lands. The highly bituminous limestones at several horizons suggest stagnant bottom conditions at various times and places. In Sumatra, Borneo, Celebes, and elsewhere. Triassic strata are mainly clastic facies, some nonmarine. A paleogeographical map of the Upper Triassic of Australasia is shown in Figure 9-26, A. The central area around Timor was one of shallow-water carbonate deposition with very little detrital sediments. Marginal to this area are mainly clastic facies that in the more peripheral areas are both marine and nonmarine. It had long been thought that an arcuate geosyncline occupied the Timor-Ceram-Celebes area during the Triassic, but new data do not support this interpretation.

The main patterns of land and sea deposition of the Triassic continued into the Jurassic, during which time progressive changes began to take place. Australia remained emergent into the Jurassic. Seas encroached into marginal embayments in Western Australia, but elsewhere, especially in the eastern half, coal-bearing terrestrial deposits were laid down. East of the Precambrian shield a vast complex basin was formed by subsidence of the central part of eastern Australia and slight uplift of the margins. Within this lowland numerous lakes and swamps received centripetal drainage from the surrounding lands. The sediments in these lakes, swamps, and alluvial plains were mostly fine-to-coarse clastics, which attained a thickness of 5,000 feet or more. The only fossils in these formations are fresh-water clams, plants, and dinosaurs.

During the Triassic Indonesia underwent regression, which continued until the middle Jurassic, so that the area of nearly pure limestone deposition was smaller than during the Triassic. Likewise, more of the archipelago — Sumatra, Java, and Borneo — was emergent. The upper half of the Jurassic, however, is marked by widespread marine transgressions and local orogeny that produced islands from Sumatra to Timor. All the Upper Jurassic sedimentary rocks indicate shallow-water deposition just as in the earlier Mesozoic periods (Fig. 9-26, B).

The great fresh-water lakes occupying much of eastern Australia toward the close of the Jurassic persisted into the Cretaceous. Not until the middle of Early Cretaceous time did shallow seas invade a wide sector through the east-central part of the continent; and somewhat later the western coast was occupied by marginal seas, as in the Jurassic. The sequence in east-central Australia begins with non-marine clastic formations, transitional to the underlying Jurassic series, followed by a few thousand feet of coarse-to-fine clastic marine sediments, which grade upward into a fine nonmarine clastic series containing some coal. Terrestrial basin deposition, invasion by shallow seas, regression, and a return to continental deposition is perfectly recorded in the Early Cretaceous (Fig. 9-27).

More uniform conditions prevailed along the western coastal region. The encroaching sea first deposited glauconitic sandstones, generally followed by chalk. Shoreward facies of deltaic clastics are also recognized in places. Marine conditions prevailed along the coast well into the Late Cretaceous, after regression of the interior seas.

Cretaceous strata are known from two isolated regions in extreme eastern Queensland, where more than 5,000 feet of terrestrial and lacustrine clastic series, a thick volcanic series, and some coal beds are dated by thin, fossiliferous, marine members as Early Cretaceous in age.

The Late Jurassic diastrophic movements that affected Indonesia continued into the Early Cretaceous, developing uplifted islands and shifting transgressive seas (Fig. 9-26, C). During the Late Cretaceous significant changes developed in this region (Fig. 9-26, D). A deeply subsiding geosynclinal belt developed in the Timor-Celebes region (the Timor geosyncline), and another developed along northern New Guinea, northern Celebes, and northern Borneo (the Borneo-New Guinea geosyncline). The Upper Cretaceous of the Timor geosyncline consists of radiolarian shale, radiolarite, limestone, and chert with marl, clay, and sandstone, interpreted as of fairly deep-water origin. Away from the geosynclinal belts the sedimentary record consists mainly of shallow water clastics. Widespread orogeny began here in the Late Cretaceous and continued into the Eocene (Chap. 13).

New Zealand

The middle Phanerozoic history of New Zealand, from Late Paleozoic through Jurassic, is dominated by rapid deposition in the New Zealand geosyncline. A reconstruction for the Triassic is shown on Figure 9-28. Two major facies can be recognized in the Permian, Triassic and Jurassic rocks of the New Zealand geosyncline. One is a shelf facies along the margins of the geosyncline, and the other consists of poorly fossiliferous graywackes deposited along the axis. The source of the sediments was probably to the west and south. The deformed early Paleozoic and possibly Precambrian rocks of western South Island were part of a larger land area that extended

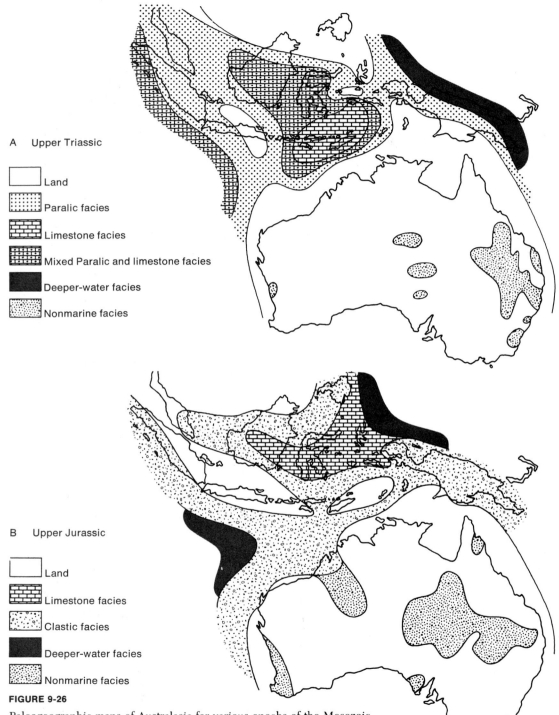

A   Upper Triassic

  Land
  Paralic facies
  Limestone facies
  Mixed Paralic and limestone facies
  Deeper-water facies
  Nonmarine facies

B   Upper Jurassic

  Land
  Limestone facies
  Clastic facies
  Deeper-water facies
  Nonmarine facies

FIGURE 9-26

Paleogeographic maps of Australasia for various epochs of the Mesozoic.
(Date on Australia adapted from Brown *et al.*, 1968; data on Indonesia
adapted from M. G. Audley-Charles, 1966.)

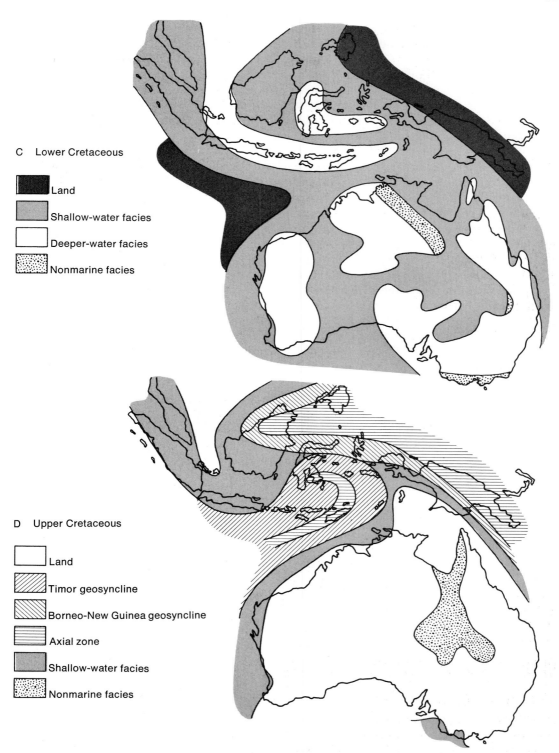

C   Lower Cretaceous

- Land
- Shallow-water facies
- Deeper-water facies
- Nonmarine facies

D   Upper Cretaceous

- Land
- Timor geosyncline
- Borneo-New Guinea geosyncline
- Axial zone
- Shallow-water facies
- Nonmarine facies

westward in the Tasman Sea. The presence of marginal facies in southernmost South Island and axial facies in Chatham Island to the east suggests a land area to the south. The Auckland Islands have a geological history that ties them to the foreland area of western South Island. A schist on Campbell Island and granites on Bounty Island are both considered part of this foreland region.

Early Jurassic deposits of New Zealand follow directly on the facies pattern of the preceeding Triassic and Permian formations. Much of the New Zealand geosyncline was highly deformed, metamorphosed and uplifted in Middle or Late Jurassic (Fig. 9-29) and early Cretaceous time. Geosynclinal conditions persisted in the eastern part of North Island but over much of the remainder of New Zealand, later Jurassic time is represented by nonmarine and marine sediments in isolated basins within the uplifted regions.

An interpretation of the Cretaceous geography of New Zealand is shown in Figure 9-30. Geosynclinal conditions prevailed for a time in eastern North Island, but erosion during most of the Cretaceous led to peneplanation of the lands uplifted in the Jurassic. The sea transgressed across lands of very low relief along the eastern margin of both South and North Island. Marine strata on the west coast of South Island offer the first evidence of Cretaceous on that side of New Zealand.

A major feature of New Zealand geology is the Alpine fault (Fig. 6-38). It is believed that since the Jurassic there has been about 300 miles of lateral movement along this fault. The primary evidence for this interpretation is the correspondence of regional sequences of Upper Paleozoic and Mesozoic strata on opposite sides of the Alpine fault in the north and south of South Island.

9-6 JAPAN

Throughout the Paleozoic the Japanese Islands were in a complex geosynclinal zone. However, the orogenies of the Late Permian and Early Triassic, accompanied by extensive plutonic intrusions in the Inner zone, completely altered this pattern. Geosynclinal conditions ceased and this area became a platform except on the Pacific margin of the Japanese Islands. The major geologic provinces for the Mesozoic of Japan are shown in Figure 9-31.

Lower and Middle Triassic strata laid

FIGURE 9-27

Flat-topped hills of horizontal marine Cretaceous beds, Neales River, South Australia. (Photograph by C. G. Stevens.)

down before the orogenic climax were almost entirely marine. They consist of sandstone and shale with minor limestone. The post-orogenic depositional pattern was quite different; the relief was high, and deposition was confined to many intermontane basins (Fig. 9-32, A). Sandstones with conglomerates are very common. Many basins were nonmarine, and in some coal deposits accumulated. In the Outer zone of Japan a Triassic eugeosynclinal belt probably persisted, but as yet no fossils of this age have been recorded there.

The paleogeography of Japan changed considerably during the Jurassic (Fig. 9-32, B). For the first time marine transgressions appeared on inner Japan, confined to large isolated basins in which various clastic sediments accumulated. In the lower half of the Jurassic these basin deposits were largely marine, but in the upper half mainly nonmarine, including coal deposits. Some basins contain as much as 30,000 feet of Jurassic strata. On the Pacific margin of Southwest Japan geosynclinal conditions prevailed. The facies are mainly clastic, but a remarkable series of carbonate formations is interpreted as a great reef complex something like the Great Barrier Reef off eastern Australia. The outermost part of this Pacific marginal zone was a typical eugeosyncline, in which shale, graywacke, and volcanics were deposited. These deposits are particularly thick in northern Honshu Island and southwestern Hokkaido Island. The axial region of Hokkaido Island was the site of thick accumulations of eugeosynclinal sediments and volcanics, but the relation of this zone to that of southwest Hokkaido Island is not known.

Cretaceous formations are more widely distributed and of more diverse facies than those of the Jurassic or Triassic. The Cretaceous was also a time of considerable igneous activity. In fact most granites now exposed in Japan are of Cretaceous age, concentrated mainly in the inner side of Southwest Japan.

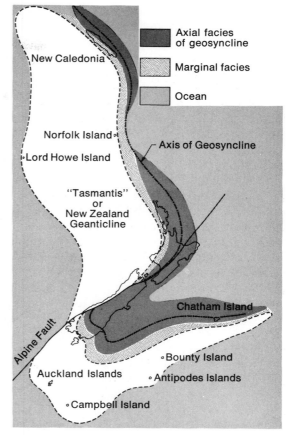

FIGURE 9-28

Major features of New Zealand geosyncline for the Triassic. (After C. A. Fleming, *Tuatara*, **10**, 1962.)

Cretaceous sedimentary formations are mainly clastic. On the inner side of Japan Cretaceous deposits are confined to structural basins wherein accumulated very thick sequences of continental sediments (Fig. 9-32, C). These basins were also centers of widespread volcanism. Deposition in these basins reflects active movement— folding and block faulting—and is confined to the Lower Cretaceous. In the mid-Cretaceous began an intrusion of granites throughout the area on a very large scale. On the Pacific margin eugeosynclinal conditions prevailed in the outermost zone and nonvolcanic clastic deposits in the inner

FIGURE 9-29

Isoclinally folded geosynclinal graywacke and argillite of the Torlesse Group (Triassic?), Malte Brun Range, Chudleigh Peak, South Island, New Zealand. (Photo courtesy of New Zealand Geol. Surv.)

zone. In northeastern Honshu Island a significant orogeny occurred in the mid-Lower Cretaceous. Younger Cretaceous strata here are thin, shallow-water deposits that are flat lying except where disturbed by block faulting. In the axial area of Hokkaido Island geosynclinal conditions persisted throughout the Cretaceous, and 18,000 feet of sediments accumulated (Fig. 9-32, D).

9-7 EASTERN ASIA

The southeastern border region of the Angaran continent underwent striking changes during the Triassic. Some physical events (depositional and tectonic) represent continuation of Paleozoic conditions that ceased during the Triassic; but depositional and tectonic patterns characteristic of the rest of the Mesozoic were also established.

The broad Paleozoic depositional basin extending from southwestern China to Manchuria persisted into the Triassic. A seaway occupied most of the southern half of this basin and connected with the Tethyan geosyncline to the southwest, probably through Burma. North of this marine embayment continental redbeds with fossil plants (coal beds) were deposited. The marine embayment received fine clastic and carbonate sediments. The Lower and Middle Triassic has a rather wide distribution, but the Upper Triassic is known only in the

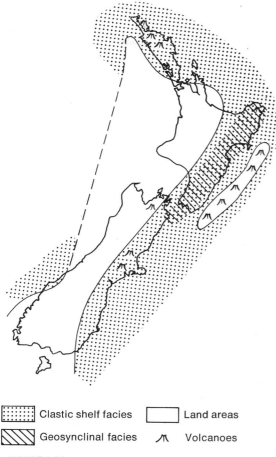

☐ Clastic shelf facies ☐ Land areas

⬚ Geosynclinal facies ⋀ Volcanoes

FIGURE 9-30

Paleogeographic map of New Zealand for the Cretaceous. (After C. A. Fleming, *Tuatara*, **10**, 1962.)

Across eastern Siberia, connecting the Pacific and Arctic Oceans, a broad Triassic sea extended from the Vladivostok region off the Sea of Japan to the area opposite the Sea of Okhotsk. The western edge of this sea lapped the Angaran shield. A fairly complete sequence of Triassic strata, mainly clastic sediments with some carbonates, records this ancient seaway. Triassic deposits are also present off the coast of Siberia in the New Siberian Islands.

The Jurassic history of China is written mainly in terrestrial coal-bearing sediments deposited in many isolated basins. Only two poorly known localities of marine Jurassic strata are found, one at Hong Kong and the other in western Yunnan province. An orogenic phase toward the close of the Jurassic affected most of China, and it was upon tilted Jurassic strata that the final deposition of the Mesozoic took place.

Most of northern Asia, including China, Mongolia, Korea, and Russia east of the Urals, was land throughout Cretaceous time except for a few marginal marine basins along the Arctic and Pacific oceans. The tectonic disturbances of the Jurassic left an undulating landscape consisting of many basins separated by higher areas. The Cretaceous here is chiefly continental sediment deposited by stream, lake, and wind—deposits almost exclusively of coarse-to-fine clastics, many red. Individual deposits show enormous variability in facies and sedimentary structures such as bedding. Some thin coal beds are present in places. The lakes and streams were populated by freshwater clams, snails, various arthropods and fish.

The depositional history of one of these Cretaceous basins—in the Gurbun Saikhan mountain region of Mongolia—is illustrated in Figure 9-33. Pre-Cretaceous rocks are highly disturbed and delimited by striking unconformities; the Cretaceous and Cenozoic sediments are nearly flat except near block faults. The history of this basin is typical of central Eurasia in Cretaceous and Cenozoic time. The dinosaurs from

southwestern part of the basin. A diastrophism that affected most of southeastern Asia interrupted deposition in Late Triassic time; however, even later Triassic beds were laid down unconformably over these folded strata. Since Jurassic deposits are generally unconformable on all Triassic formations, it appears that two orogenies took place in China during the Late Triassic. The regression of the seas in still later Triassic time finally closed the great cycle of marine transgressions and regressions in China. The post-Triassic history of China is read almost wholly in continental sediments.

these formations, especially those of Mongolia, have been made famous by the American Museum Mongolian Expedition of the 1920's, during which the famous dinosaur eggs were found.

The coasts of Asia were probably more humid than the interior, for Cretaceous deposits there are sandstone and shale, commonly including coal and plant beds. During Early Cretaceous time marine embayments widened at the mouths of most of the longer rivers flowing into the Arctic Ocean. One such embayment was the Ob estuary. By Late Cretaceous time intermittent connections had been established between it and the Tethys to the south along a narrow seaway between the Urals to the west and the Angaran land to the east. The Cretaceous deposits accumulated in this sea were mainly thin clastics.

Much of the Anadyr peninsula, of extreme eastern Siberia, was also inundated by Cretaceous seas. The record of these marine invasions consists of alternating marine and non-marine clastic formations, many with coal beds. Many stratigraphic gaps in the sequence and the alternating facies indicate highly fluctuating conditions. The eastern coast of Siberia south of the Anadyr peninsula was likewise marked by a few embayments. The Anadyr Sea, at least at times, joined the Pacific and Arctic Oceans, and much of the Kamchatka peninsula was submerged. A rather large embayment extended southwestward from the Sea of Okhotsk toward Manchuria, covering most of Sakhalin Island. The deposits of these basins are, like those in the Arctic embayments, alternating sequences of marine and nonmarine clastic facies with coal beds.

Throughout much of Cretaceous time volcanoes were intensely active in most of Asia. Lava flows, ash beds, and agglomerate deposits are widely distributed. The former South China platform (Fig. 6-22) was an especially active center of volcanic activity and underwent vast granitic intrusions.

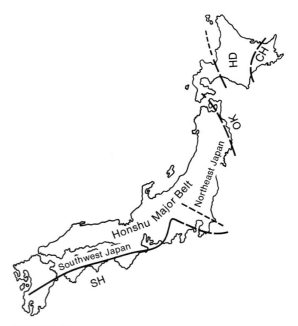

FIGURE 9-31

Major geologic provinces in the Mesozoic of Japan. *CH:* Chishima Major Belt; *HD:* Hidaka Major Belt; *OK:* Outer Kitakami Belt; *SH:* Shimanto Major Belt. (After K. Ichikawa, *J. Geosciences*, Osaka City Univ., 1964.)

9-8 THE CIRCUM-ARCTIC REGION

The existence of a Triassic Arctic Ocean can easily be surmised from the distribution of Triassic deposits of various ages on the continental margins of the present Arctic. Triassic faunas are known on the eastern Siberian coast and New Siberian Islands, referred to above. Faunas of Triassic age have also been collected on Bear Island, Spitsbergen, the eastern and northern coasts of Greenland, Ellesmere Island, Bathurst Island, Prince Patrick Island, and the northern coast of Alaska.

The Triassic of Spitsbergen is represented by most of its stages, nearly 3,000 feet thick in some places; mostly fine-grained clastic and carbonate facies, with plant-bearing continental beds in the Middle Triassic, even including some thin coal beds (Fig. 9-34). The Early Triassic formations rest

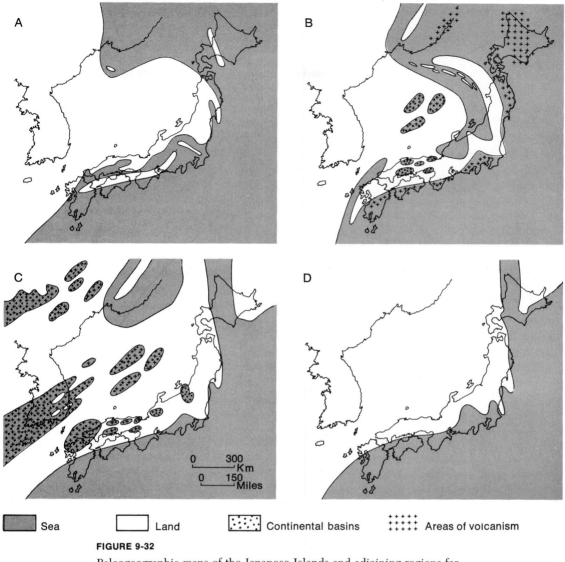

FIGURE 9-32

Paleogeographic maps of the Japanese Islands and adjoining regions for (A) the Late Triassic, (B) the Late Jurassic, (C) late Early Cretaceous, and (D) the Late Cretaceous. (After Minato *et al.*, 1965.)

Legend: Sea · Land · Continental basins · Areas of volcanism

unconformably on Permian formations. Because the Upper Triassic in some places on Spitsbergen and on Bear Island rests on pre-Triassic rocks, uplift and erosion probably took place before Late Triassic time over at least part of these areas.

The Jurassic and Cretaceous history of much of the circum-Arctic region was included in the discussion of North America and Eurasia. In Spitsbergen continuous deposition of shale continued from Jurassic into Cretaceous time. The clay shale and carbonate bottom sediments preserve a rich molluscan fauna and establish that an Early Cretaceous sea covered most of Spitsbergen and Novaya Zemlya but only the southernmost part of the Franz Josef Islands. This marine depositional phase, ended in middle Early Cretaceous time, and continental plant-bearing sandstones and shales were

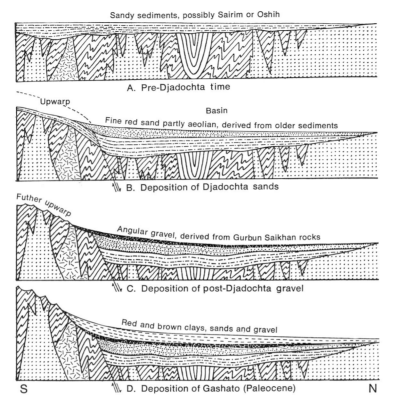

Sandy sediments, possibly Sairim or Oshih

A. Pre-Djadochta time

Upwarp

Basin

Fine red sand partly aeolian, derived from older sediments

B. Deposition of Djadochta sands

Futher upwarp

Angular gravel, derived from Gurbun Saikhan rocks

C. Deposition of post-Djadochta gravel

Red and brown clays, sands and gravel

S   D. Deposition of Gashato (Paleocene)   N

FIGURE 9-33

Four stages of Cretaceous and early Cenozoic sedimentation in the Gurbun Saikhan Range, Mongolia: (A) a cover of Lower Cretaceous sediments resting on a complex old-rock floor; (B) initial upwarp of the Gurbun Saikhan Range and corresponding deposition of Upper Cretaceous formations; (C) renewed upwarp and deposition of angular pebbles from the newly exposed Gurbun Saikhan rocks; (D) deposition of the Paleocene rocks. Since this last stage there have been renewed deformation and extensive erosion, and alluvial fans from the mountains are now encroaching upon the remnants of all three formations. (From C. P. Berkey and F. K. Morris, *Natural History of Mongolia*, 2, American Museum of Natural History, 1927.)

then deposited on Spitsbergen and the Franz Josef Islands, which, along with Norway and northern Russia, must have been land at this time. Another marine transgression touched much of Spitsbergen, producing mainly clastic deposits, but by the end of the Early Cretaceous this sea had also receded, and these Arctic islands had become land undergoing erosion for the remainder of Cretaceous time. This long spell of erosion (including nearly all of Late Cretaceous time) exposed early Paleozoic formations in many places, especially in the north.

9-9 ANTARCTICA

The Mesozoic history of East Antarctica is confined to continental sedimentation and extensive volcanism. These events are

an intimate part of the Gondwana problem discussed in Chapter 11. In West Antarctica, Mesozoic rocks crop out extensively in the Antarctic Peninsula. These consist of clastic and volcanic formations of Jurassic and Cretaceous age, some very fossiliferous. One nonmarine Jurassic formation, for instance, has yielded 61 species of plants. Although many isolated studies of these Mesozoic formations have been made, data are insufficient as yet to allow significant generalizations to be made. Major orogeny prior to Jurassic time intensely deformed and weakly metamorphosed the late Paleozoic geosynclinal clastic rocks — an orogeny perhaps related to the formation of the primitive Scotia Arc. Geosynclinal accumulation continued into Late Cretaceous time, when the region underwent further orogeny, uplift, and intrusion by various igneous bodies that brought geosynclinal conditions to a close. The Mesozoic rocks of the Antarctic Peninsula are very similar to those of the Andean region and are probably part of one continuous geosynclinal and orogenic belt connected by the Scotia Arc. The geologic history of southern Victoria Land (East Antarctica) and the Antarctica Peninsula (West Antarctica) are summarized in Table 9-1.

9-10 SOUTH AMERICA

The Triassic Period

The emergent state of the South American continent that had been established in the Late Permian prevailed well into the Triassic. Lower and Middle Triassic rocks have been identified only in small areas of western Argentina and central Chile. In western Argentina the Early Triassic is represented by thick sandy-conglomeratic continental deposits and in central Chile by volcanic

complexes with interbedded conglomerate beds. The Middle Triassic seas encroached on a narrow coastal strip of central Chile, and volcanic activity was particularly active through much of Chile and western Argentina. Not until the Late Triassic did seas again transgress over much of the Andean region (Fig. 9-35, A). Thick limestone formations of this age extend from central Columbia, south through Ecuador, Peru, and into Bolivia. Along the northern Peruvian coast thick volcanic sequences with some sediments contain Triassic fossils. This sea extended for a brief time through Columbia and western Venezuela, but most of this region was one of continental redbed deposition. Southward in Chile, seas encroached on the continent only in a few discontinuous isolated basins. In western Argentina a number of isolated basins accumulated continental deposits. In Brazil the whole Paraná basin may have been the site of eolian accumulations under desertic conditions in the Late Triassic (Chap. 11).

FIGURE 9-34

Lower and Middle Triassic shales and sandstones exposed along Nordfjord, western Spitsbergen (Svalbard). (Photograph by H. Frebold.)

TABLE 9-1

Generalized geologic histories in two contrasted regions of Antarctica

Era	Period (Ages in millions of years)	Southern Victoria Land events	Antarctic Peninsula events
Cenozoic	Quaternary (1)--------	Local extrusions of basalt and trachyte / Continental glaciation —?—?————————— ?—————	Glaciation, local sedimentation (pecten conglomerate), and volcanism ————— ?—————— ?—————
Cenozoic	Tertiary	(No record)	Local sedimentation and volcanism — – ?— —— – ?— ——— – ?— — Orogeny and widespread intrusion of gabbro, diorite, and quartz diorite
Mesozoic	(63)———————		——— – ?—————— – ?— —
Mesozoic	Cretaceous (135)-------	——— – ?————— – ?—— Deposition of tillite (?)	Deposition of thick marine beds — — — — — — — — Andesitic to rhyolitic volcanism
Mesozoic	Jurassic	Intrusions of diabase sills and basaltic volcanism ——— – ?————— – ?—	Local deposition in lakes
Mesozoic	(181)-------		Orogeny and intense folding with weak metamorphism of older rocks
Mesozoic	Triassic	Deposition of sands, silts and coal beds (Beacon Sandstone)	
Paleozoic	(230)———————		——— – ?————— – ?— —
Paleozoic	Permian (280)------- Carboniferous periods		Deposition of geosynclinal graywacke, shale, and conglomerate (Trinity Peninsula series)
Paleozoic	(345)-------		——— – ?————— – ?— —
Paleozoic	Devonian (405)-------	——— – ?————— – ?— —— Time of erosion following Mountain uplift	Granitic intrusion
Paleozoic	Silurian (425)-------	——— – ?————— – ?— ——— Folding and low-grade(?) metamorphism of older rocks, accompanied by widespread granitic intrusions	(Central part of peninsula) Volcanism
Paleozoic	Ordovician		
Paleozoic	(500)-------	Limestone deposition	——— – ?————— – ?— —
Paleozoic	Cambrian (600?)———————	Long history of sedimentation, local high-grade regional metamorphism, and plutonic intrusion	Sedimentation and metamorphism of basement rocks (central part of peninsula)
Paleozoic	Precambrian (+ 3000?)		

Source: After E. B. Ford, 1964.

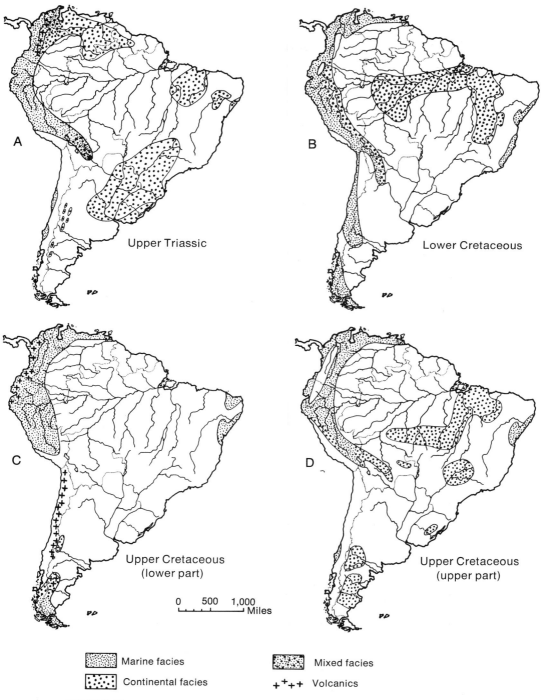

A — Upper Triassic

B — Lower Cretaceous

C — Upper Cretaceous (lower part)

D — Upper Cretaceous (upper part)

0   500   1,000
|————|————| Miles

Marine facies

Mixed facies

Continental facies

+ + + Volcanics

FIGURE 9-35

Paleogeographic maps of South America during various periods of the Mesozoic: (A) Late Triassic, (B) Early Cretaceous, (C) early Late Cretaceous, (D) late Late Cretaceous. (After Harrington, 1962.)

The Jurassic Period

The pattern of the South American west-coast marginal geosynclines was well established by the Jurassic. Thick sequences of fine clastic and carbonate facies were deposited in this geosyncline from Ecuador through Chile (Fig. 9-36). Numerous gaps in the sequence, reflected in the faunal zones, indicate that the sea margins probably migrated extensively. North of the geosyncline, in Colombia and Venezuela, continental redbeds were being laid down. Similar clastic facies are found in eastern Ecuador and Peru.

In Late Jurassic time orogenic movements led to a regression of the seas, isolating many marginal basins from the sea; in these basins anhydrite, gypsum, and other sediments were laid down. This orogenic phase was accompanied by much volcanic activity, especially on the western edge of Chile. Another transgression of the seas followed the orogeny, depositing shale and carbonate sequences.

The Late Jurassic orogenic movements roughly correspond to those of western North America at this time.

The Cretaceous Period

The Cretaceous Andean sea was the most widespread of the South American Mesozoic seas, following the pattern of transgression observed on other continents. Nearly the whole western margin of South America was inundated. In Peru and Chile the Lower Cretaceous strata are completely conformable on Upper Jurassic formations. In Venezuela Cretaceous seas, the first transgressions of the Mesozoic, spread across a variety of older formations. The most widespread and deepest invasion took place in Peru, where the initial transgressive facies are fine-to-coarse sandstone, in places containing plants and coal, followed by mainly carbonate facies and some shales. The east-

ern shore lay in western Brazil, against the Brazilian shield, where clastic facies persist throughout the Cretaceous System. Westward the sandstone facies tongues out into the carbonate and finegrained clastics. Thus along much of the western edge of the shield an intricate pattern of transgressive and regressive facies developed. On the shield itself vast deposits of fluvial and lacustrine sandstone accumulated.

In the region from northern Venezuela south to Peru the maximum extent of the transgression came in the mid-Cretaceous (Fig. 9-35). At this time there was very extensive volcanic activity in the geosynclinal region of western Ecuador and Columbia. However, all of Bolivia and most of Chile appear to have been emergent and were the site of numerous volcanoes, especially Chile. This volcanic activity was followed in the early Late Cretaceous by intense orogenic activity accompanied by vast batholithic intrusions (Fig. 8-10), marking the beginning of the formation of the Andean Cordillera. The first phase of this orogeny uplifted the Andes of western Argentina and Chile, the Western and Central Cordillera of Peru, the Cordillera Real of Ecuador, and the Central Cordillera of Columbia and formed two subparallel troughs in the region from Peru to Venezuela (Fig. 9-35, D). The uplifted medial areas continued to undergo diastrophism for the remainder of the Cretaceous.

Figure 9-37 shows the sequence and distribution of Mesozoic facies across northern Peru. Many of the major features of the Mesozoic history of the Andean geosyncline are summarized in this diagram.

In the Early Cretaceous, for the first time in the Mesozoic, the sea transgressed southward across northern Venezuela. The varied mixture of clastic and carbonate facies deposited in this shallow shelf sea clearly show the transgressive nature of the advancing sea. In the Dutch Leeward islands of Aruba, Curaçao, and Bonaire, more than 15,000 feet of tuff, chert, graywacke, and

FIGURE 9-36

Folded Mesozoic and Cenozoic formations in the central Andes of Peru. In the foreground is the town of Oroya, a famous smelting center. The contorted rocks across the upper central part of the picture are Jurassic limestones. The alternating hard and soft units in the left foreground are Lower Cretaceous sandstones; to the right of these are Cretaceous limestones; the dark beds in the right center are Cenozoic redbeds. (Photograph by G. C. Amstutz.)

limestone of Early Cretaceous age accumulated. The Venezuelan shelf seas were clearly miogeosynclinal. The Leeward islands, with their volcanics, were within a eugeosyncline that included Hispaniola, Cuba (except for the most northern part), Jamaica, and probably the island of Tobago. On all of these islands thick sequences of tuff, breccias, cherts, lavas, and sediments testify to eugeosynclinal conditions.

During the middle of the Late Cretaceous the coastal region of Venezuela, from Trinidad west to within 150 miles of Lake Maracaibo and curving north and west round the present west coastal region of Venezuela to Panama, became a zone of intense tectonic activity, accompanied by igneous extrusions and intrusions (Fig. 9-38). Sedimentation and volcanic eruptions proceeded at the same time as deformation. Most of this unstable belt was under water, but deformation undoubtedly raised various parts to form islands, and many of the volcanoes that rose from the sea floor were probably emergent. Within this tectonic zone the deposits consist of conglomerates, coarse-to-fine graywackes and quartzites, and silty shales, all generally dark-gray or black.

Interbedded with these are flows, from basaltic to andesitic in composition, tuff, and agglomerates. South, from this mobile belt to the Orinoco River, the Upper Cretaceous facies are shale, limestone, sandstone, and an abundance of chert at various horizons.

The culmination of the orogenic deformation in the mobile belt of northern Venezuela came toward the close of the Cretaceous, when the Cordillera de la Costa was formed. Volcanism and sedimentation continued into the Paleocene but on a much more restricted scale. In the shallow sea south of the deformed belt sedimentation was continuous from the Late Cretaceous into the Paleocene. Deformation and intrusion of quartz diorite or granodiorite took place in the Greater Antilles and the Virgin Islands toward the close of the Cretaceous. Intense volcanic activity appears to have prevailed in Panama during the Late Cretaceous. Most rocks of this region are Cenozoic, but some volcanic complexes in the northwestern part of the country near Costa Rica have yielded Cretaceous foraminifers from siliceous limestones interbedded with the volcanic rocks.

Two small marine embayments in eastern Brazil and one in Patagonia are the only marine transgressions of the Mesozoic coast of eastern South America. The interior of Brazil was the site of extensive fresh-water lakes, where clastic sediments and some carbonates accumulated. Similar fresh-water and terrestrial sediments accumulated in southwestern Uruguay and in Patagonia.

9-11   MESOZOIC RECAPITULATION

At the end of the Mesozoic Era, the present structural and morphological features of the earth's crust were roughly outlined. A glance at a physiographic map of the world shows a close coincidence between modern

FIGURE 9-37

Selected stratigraphic columns of Mesozoic formations across northern Peru. (Partly after A. G. Fischer, 1956.)

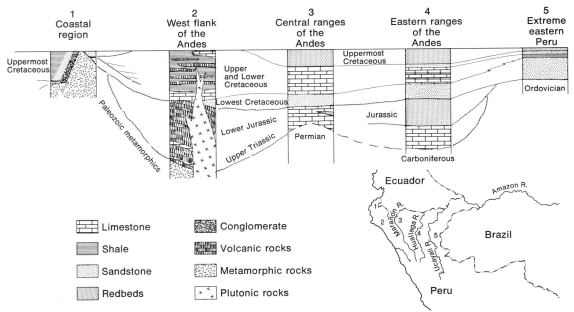

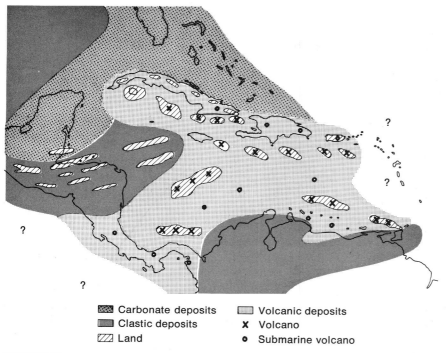

Carbonate deposits   Volcanic deposits
Clastic deposits   X Volcano
Land   ⊙ Submarine volcano

FIGURE 9-38

Caribbean region during Late Cretaceous time. (After Woodring, 1954.)

mountain ranges and the Mesozoic geosynclines just discussed. Much of the structure of these mountain ranges had already begun to form during the Mesozoic. Additional orogenic phases completed that formation in the Cenozoic, and, of course, constant erosion continues, along with diastrophism, to sculpture our modern landscape. Many of the earlier Mesozoic physical patterns are transitional from those of the late Paleozoic, and those of the late Mesozoic are transitional to those of the Cenozoic.

In North America early Mesozoic deposition was confined to the western geosyncline. Early Triassic geosynclines continued late Paleozoic patterns. By Middle Triassic time the miogeosyncline had become the site of terrestrial depositional basins, but marine conditions persisted in the eugeosyncline. Such remained the con-

ditions in the Early Jurassic. By Late Jurassic time, however, the eugeosyncline in the western United States had been broken by the Nevadan orogeny, accompanied by granitic intrusions. Eugeosynclinal conditions prevailed along the Pacific coast. Two new continental seas began to develop during the Jurassic: one came down through western Canada from the Arctic Ocean and formed the Sundance Sea; the other marked the beginning of the Gulf of Mexico. By Cretaceous time these three areas of Jurassic deposition had been further developed, and a continuous seaway had been formed between the Arctic Ocean and the Gulf of Mexico. Diastrophism in the land area between the central sea and the Pacific coastal embayments furnished an active source of clastic sediments for these basins. The Laramide orogenic phase, beginning in the Late Cretaceous, folded and faulted the sediments of this interior sea. Cretaceous seas also occupied the Atlantic seaboard, but the sediments there were not folded.

In this pattern of geosynclinal development we can see a complete change in the tectonic evolution of the continent. The symmetrical disposition of Paleozoic geosynclines and the concentration of orogenic phases in the eastern geosyncline had set the stage for Mesozoic history.

In Eurasia we have much the same kind of history. The dominating geosyncline was, of course, the Tethys. This great mobile belt lying between two stable zones was the center of a very active and diverse depositional history. From time to time there was volcanic activity of one sort or another through the central parts of the Tethyan region. To the north and south of this central zone nonvolcanic sedimentary sequences grade into foreland facies, often continental in origin.

Europe eastward to the Urals was mainly a site of continental deposits during the Triassic, a depositional phase started back in the Permian. During the Jurassic and Cretaceous continental seas spread across much of Europe. In the southern foreland of the Tethys, in Africa, transgressive phases were much more restricted and did not reach any appreciable extent until Late Cretaceous time.

The central and eastern zones of the Tethys had much the same kind of depositional history. Foreland facies next to the shields grade outward to normal geosynclinal deposits. The wide diversity of facies and volcanic activity in the Late Cretaceous indicate tectonic movements within the geosyncline and a beginning of the breakup of the Tethys.

In the southwestern Pacific the Indonesian area, as far as New Zealand, was an active center of deposition and tectonism. Australia remained mainly a site of continental deposition with only minor (and mainly) marginal transgressions. In New Zealand a great orogenic phase in the Middle or Late Jurassic brought to a close a tectonic and depositional pattern that had existed since the late Paleozoic. Cretaceous deposits were laid down in shallow marginal seas.

The vast land mass of Asia north of the Tethys was also the site of terrestrial depostion during Mesozoic time except for marginal embayments from the Arctic and Pacific Oceans. Several orogenic phases interrupted the depositional history. The Mesozoic history of Japan was marked by two great orogenies, the axis of the later one lying just east of the earlier (Triassic) orogeny. The Mesozoic sediments of Japan consist of complex interrelated facies, mainly clastics, with, at times, abundant volcanic materials. The Japanese area appears to have been an eugeosyncline marginal to the Asiatic continent.

The distribution of Mesozoic sedimentary rocks around the margins of the present Arctic Ocean is ample testimony to the existence of this water mass throughout Mesozoic time. The stratigraphic record of northern Alaska suggests that the deep subsidence of the Arctic to its present level began about Middle Cretaceous time.

South America had a single mobile belt, along the western side of the continent. With each succeeding period of the Mesozoic there were additional transgressions. A volcanic and a nonvolcanic portion of this geosyncline can be recognized. The Andes, accompanied by intrusions of great granitic batholiths, began to rise in the Middle Cretaceous. Cretaceous orogenic movements also affected the coastal ranges of Venezuela and the adjoining West Indian islands.

Figure 9-39, in which the Mesozoic folded belts and their general alignment are plotted, clearly shows that the circum-Pacific belt was the principal zone of diastrophism of this era. Comparing this map with

FIGURE 9-39

Location of the principal folded belts formed during Mesozoic orogenies. (After J. H. F. Umbgrove, *The Pulse of the Earth*, Nijhoff, The Hague, 1947.)

a similar one for late Paleozoic orogenies (Fig. 6-35), we can see that, in general, the Mesozoic belts lie outside the late Paleozoic zones. The Pacific coastal margins of South America, North America, and Asia are essentially the results of these Mesozoic orogenies. What were once mobile depositional belts are now, in part, rigidly incorporated into the continental frameworks. Further orogenic phases will continue in these belts in the Cenozoic, but these will be, in some respects, an anticlimax. In the remaining mobile belt, the Tethys (including the Indonesian area to New Zealand), the major Cenozoic orogenic phases were concentrated.

SUGGESTED READINGS

See the list at the end of Chapter 4.

10

THE MESOZOIC LIFE

The faunas and floras of the Mesozoic differ so strikingly from those of the Paleozoic that the two eras were differentiated more than a hundred years ago. The most drastic change is in the marine invertebrate faunas, reflecting the late Paleozoic wave of extinctions, which we commented on in § 7-13.

Throughout the world, wherever the lowest Triassic strata are present their fossil faunas consist of a fairly homogeneous group of cephalopods and pelecypods. Brachiopods, crinoids, and gastropods are sparsely represented. The marine faunas began to expand and diversify by the Middle Triassic, and the diversification continued throughout the remainder of the Mesozoic Era. One of the chief reasons for the acceleration of evolution was the great expansion of the continental seas, which gave marine faunas widespread new environmental niches to occupy.

During the Triassic the continental seas were largely confined to the Tethys and to the marginal areas of continents that faced the Pacific and Arctic oceans. But in the Jurassic extensive shelf seas again began to develop across the continents. With minor fluctuations the shelf seas continued to grow through the Jurassic and the Cretaceous, reaching their broadest spread in the Late Cretaceous. At the close of the Cretaceous a drastic reduction in their size resulted in a pattern not unlike that of today. At the same time some invertebrate faunas died out, but there was no vast extinction like that at the end of the Paleozoic. The most notable extinction at this time was that of the ammonoids.

Vertebrate animals evolved phenomenally during the Mesozoic. The dominant vertebrates of the era, the reptiles, evolved into a great many forms, each adapted to an environmental niche on land, in the sea, or even in the air. Two other important groups also made their first appearance in the Mesozoic, evolving from reptiles: the birds and the mammals.

The great climatic change toward the close of the Paleozoic and in the early Mesozoic led to the extinction of much of the lush flora that had characterized the Carboniferous swamps. Late Triassic and Jurassic floras were quite different because of the appearance and predominance of new groups. A most significant change in the evolution of plants during the Early Cretaceous was the appearance of the flowering plants—the angiosperms. Ever since the Late Cretaceous they have been the predominant plants, all other classes either becoming extinct or being relegated to a minor position in modern floras.

10-1 MARINE INVERTEBRATES

The dominant invertebrate phylum of the Mesozoic seas was the Mollusca, comprising the cephalopods, gastropods, and pelecypods. Within the Mollusca the most characteristic group, in the Mesozoic, was unquestionably the ammonoids, which underwent a phenomenal evolutionary radiation and became extremely abundant and widespread. In the Jurassic and Cretaceous another group of cephalopods, the belemnoids, with internal shells, became widespread and abundant.

A large and diverse fauna of pelecypods also evolved during the Mesozoic. Because of their slow evolution, the pelecypods cannot be generally used for refined correlations, but many special adaptive types are excellent as ecologic indicators. Gastropods were just as abundant and diverse but are less fully understood. The Mesozoic representatives of both groups closely resembled those living today.

The long history of the colonial rugose and tabulate corals ended during the late Paleozoic extinction. Our modern reef-building colonial corals belong to a new order, the Scleractinia, and made their first appearance in the Middle Triassic. The distribution of reef-building corals during the Mesozoic gives us a fairly good insight into the probable climatic zones and into the nature of geosynclinal and continental seas.

Ammonoids

No other invertebrate group was so abundant, so widely distributed, or so characteristic of the Mesozoic Era as the ammonoids. The sequence and distribution of these organisms form the basis of the refined chronologic divisions used for Mesozoic formations. One reason they are so valuable in correlation is that they suffered three great crises, the last of which led to their extinction at the close of the Cretaceous. The ammonoids were not exceptionally abundant at any time in the Paleozoic, but by Early Carboniferous time they had become greatly diversified and of world-wide distribution. They underwent their first near-extinction in the Late Permian, only a single stock managing to survive into the Triassic. From this single stock evolved a fauna that was larger, more diverse, and more abundant than that of the Paleozoic. In the Late Triassic, however, the group suffered its second crisis, and again all but a single stock became extinct. Again there was a great evolutionary radiation, which lasted through the Jurassic and most of the Cretaceous, on a scale far grander than that of any previous periods, but this was followed by the third crisis, which ended in the ammonoids extinction by the end of the Cretaceous.

These great retrocessions are the principal criteria for three boundaries of the geologic column. The first marks the boundary between the Paleozoic and the Mesozoic, the second the boundary between the Triassic and the Jurassic, and the third the boundary between the Mesozoic and the Cenozoic. When the boundaries between these geologic divisions were first proposed, however, they were recognized more on lithologic criteria and unconformities in the type areas than on the fossil content of their deposits. Later, however, recognition of the rapid evolution and wide geographic spread of the ammonoids led to the establishment of the best chronologic system in the whole geologic column.

The ammonoids (Fig. 10-1) were all strictly marine animals with a single, generally planimetrically coiled shell. Since only the shells have been preserved as fossils, the soft structures of the animal are unknown but are assumed to have been similar to a modern pearly nautilus. The ammonoid shell had two main divisions, a generally large forward chamber occupied by the animal and a coil of smaller chambers filled with gas and liquid. Such a shell must have been a rather effective hydrostatic apparatus. A fleshy tubular extension of the posterior portion of the body extended back through the chambered portion of the shell. The function of this tube is not known for certain, but it may have been instrumental in generating gas to equalize the pressures inside and outside the shell. To strengthen the shell, the septa enlarged their lines of attachment (sutures) to the inside of the shell. The sutures of most Paleozoic ammonoids are simple wavy lines with smooth forward and backward inflections (Fig. 7-36, p. 254). The forward inflections are called saddles, the backward inflections lobes. This simple type of suture, with smooth lobes and saddles, is called goniatitic. The next grade of sutural development has smooth saddles but lobes with denticulations. This type, found in many Triassic ammonoids, is called ceratitic. The most highly developed sutures, which have both lobes and saddles denticulated, and are found in Jurassic and Cretaceous ammonoids, are called ammonitic sutures. When the terms *goniatitic, ceratitic,* and *ammo-*

nitic were first introduced, more than a hundred years ago, they had significance for purposes of classification, and all Paleozoic ammonoids were placed in the genus *Goniatites;* it was thought that the sutural patterns were restricted stratigraphically—ceratitic patterns to the Triassic, and so on. This was an error: the terms have no taxonomic or stratigraphic significance, but they are excellent descriptions of the basic suture patterns.

The ammonoid shell shows an almost infinite diversity in size and shape. The largest ammonoids had a diameter of nine feet, the smallest only a few millimeters. Some shells were fat, globular, and tightly coiled, some were slim and loosely coiled, some forms were coiled like gastropods, others could uncoil, and still others seem to have had no symmetry of coiling at all. Ornamentation is uncommon in Paleozoic ammonoids, but in Mesozoic forms it is very common, in the form of ribs, nodes, spines, and other growths. This great diversity in form is understandable only against the background of the evolution of the group during the Mesozoic.

A very early Triassic ammonoid that is close to the surviving stock from the Permian is illustrated in Figure 10-2, C. The lowest Triassic strata have ammonoids similar to this, which has a compressed, smooth, loosely coiled shell with a simple ceratitic suture. Since in Early Triassic time a wide range of environmental niches was left vacant by the extinction of most of the Paleozoic ammonoid stocks, we find in the record numerous types that evolved directly or indirectly from the basic type illustrated. This evolution is expressed in the diverse shapes of the shell and in the suture. By the close of the Early Triassic most of the adaptive types had become extinct, but some lines persisted or gave rise to distinct evolutionary lines (superfamilies) that persisted to the Late Triassic. Thus we find in the Middle and Upper Triassic ammonoids that evolved along clearly marked trends;

FIGURE 10-1

Reconstruction of a Late Cretaceous sea bottom in the central United States (Coon Creek, Tennessee). The scene includes long straight ammonoids (*Baculites*) and a large coiled form (*Placenticeras*) and has also a wide variety of pelecypods and gastropods. (Chicago Museum of Natural History.)

FIGURE 10-2

Representative Mesozoic ammonoids.

A. Side and front view and suture of *Lytoceras*, one of the persisting root stocks of the Jurassic and Cretaceous. (From A. d'Orbigny, 1840–49.)

B. Side and front view and suture of *Phylloceras*, the other persisting root stock of the Jurassic and Cretaceous. (From A. d'Orbigny, 1840–49.)

C. Side and front view of *Ophiceras*, a common earliest Triassic form. (From C. Diener, 1895.)

D. *Douvilleiceras*, an ornamented Early Cretaceous ammonoid. (From A. d'Orbigny, 1840–49.)

E. Side and back view of *Protrachyceras*, an ornamented Late Triassic form. (From E. von Mojsisovics, 1893.)

at the same time, however, within each of these groups there was a great multiplication of genera. Most of the Early Triassic ammonoids were unornamented forms with rather simple ceratitic sutures; most Middle and Late Triassic ammonoids are highly ornamented with ribs, nodes, or both, and the suture is ceratitic or ammonitic. A common Late Triassic ammonoid is illustrated in Figure 10-2, E.

All but one of these diverse evolutionary lines became extinct in the Late Triassic, and with the Jurassic a new and varied pattern of evolution began. Two of the new groups showed clear-cut evolutionary lines of forms that changed very little throughout the rest of the Mesozoic. These are the lytoceratid line and the phylloceratid line, of which typical representatives are illustrated in Figure 10-2, A, B. The shells are generally loosely coiled, and ornamentation is either weak or absent. The lytoceratid suture has acute denticulation of the lobes and saddles; the phylloceratid suture has a phylloid (leaf-like) pattern in the minor denticulation of the lobes and saddles. Most Jurassic and Cretaceous ammonoids, however, were neither phylloceratids nor lytoceratids; most were highly ornamented and had variable types of sutures. A typical representative is illustrated by Figure 10-2, D. As a general rule, the phylloceratids and lytoceratids are most abundant in the Tethyan geosyncline, and the remaining ammonoids (the great majority of the genera) are found in deposits laid down in the shelf seas that adjoined the Tethys. There is, of course, some mixing; and phylloceratids have been found far to the north and south, but they are far fewer there than in the Tethys. The classification of Jurassic and Cretaceous ammonoids is based on the hypothesis that the phylloceratids and lytoceratids are conservative, persisting stocks common to the deeper waters of the Tethys, and that the stocks that inhabited the shelf seas bordering the Tethys evolved from them. These newer, adaptive stocks died

out as environmental conditions changed, leaving ecological niches to be filled by fresh derivatives from the persistent stocks in the deep waters. This complicates the evolutionary picture; there is a definite limit to the number of possible variations in the ammonoid shell and a definite limit to the number of possible combinations of environmental factors to which the ammonoids had to adapt themselves. Accordingly, similar adaptations must have recurred repeatedly throughout the Jurassic and Cretaceous, and this is just what the fossil record seems to indicate.

Ammonoids lend themselves beautifully to detailed analysis of evolutionary changes through time. One of the best and most detailed examples is the phylogeny of the Liparoceratidae, an Early Jurassic family. The specimens used in this study were collected along the Dorset coast of southern England in the strata illustrated by Figure 9-6 (p. 30). The ancestral liparoceratids had evolute (openly coiled) or involute (tightly coiled) shells with fine bituberculate ribs (Fig. 10-3, A). The more evolute forms resemble closely the members of the family from which the liparoceratids evolved but the more involute forms are like their own descendants. Slightly later, types that were even more involute appeared; in them the ribs had become coarsened, the inner tubercle diminished, and the outer tubercle accentuated (Fig. 10-3, B). From this type diverged two evolutionary lines, the first having still more involute forms with numerous ribs, the second having coarse and simple ribs in the early whorls (Fig. 10-3, C). In this second line the inner whorls had blunt ribs, and in successive species these blunt, widely separated ribs gradually appeared on the middle and eventually on the outer whorls of the shell (Fig. 10-3, D, E). The final step in this evolutionary series was the slight projection of the blunt ribs on the ventral region of the shell (Fig. 10-3, F). All these complex changes can be found in the short span of only three

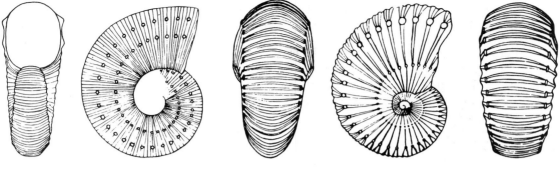

A-1   A-2   B-1   B-2   C-1

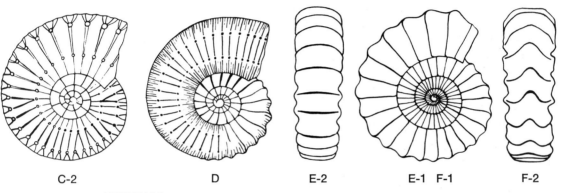

C-2   D   E-2   E-1   F-1   F-2

FIGURE 10-3

Evolution of the Early Jurassic ammonoid family Liparoceratidae.

A. *Tetraspidoceras reynesi*, ancestral form.
B. *Liparoceras (Parinodiceras) parinodus.*
C. *Androgynoceras sparsicosta.*
D. *Androgynoceras henleyi.*
E. *Androgynoceras lataecosta.*
F. *Oistoceras figulinum.*

(From *Ammonoidea*, Vol. L of *Treatise on Invertebrate Paleontology*,
Geol. Soc. Am. and Univ. of Kansas Press, 1957.)

Lower Jurassic stratigraphic zones. There are seventeen ammonoid fossil zones recognized in the Lower Jurassic of northwestern Europe.

Nautiloids

The nautiloids were a rather unimportant group during the Mesozoic; but because their evolutionary history was so different from that of the ammonoids, they warrant a few words. The greatest development of this group, both in the number of species and in the extent of their distribution, was in the Ordovician and the Silurian; after that it declined in size and importance. By the Carboniferous the many forms of nautiloid shells had decreased to only two—either straight conical shells or coiled shells that in outward appearance, at least, are not unlike those of the modern *Nautilus,* the only surviving genus. Yet when most of the other invertebrate groups underwent complete or partial extinction at the close of the Paleozoic, the nautiloids continued to evolve, elaborating on lines that had begun in the Early Carboniferous. During the Late Triassic some groups developed highly specialized ornamentation and sutural complexity, but at the close of the Triassic they, like the ammonoids, nearly became extinct; only a single genus survived into the Jurassic. This group underwent, during the Early Jurassic, an extensive evolutionary radiation, which produced a great number of species all belonging to the same genus. Although a number of new lines evolved in the Middle and Late Jurassic, none was particularly successful, and the fossil record shows a gradual decrease in the number of evolving lines. Only a single new family appeared during the Cenozoic. Just why creatures as similar as the nautiloids and the ammonoids had such different histories we do not know.

Belemnoids

The two-gilled cephalopods called belemnoids were particularly common and widespread during the Jurassic and Cretaceous. Their shell, unlike that of the ammonoids or nautiloids, was internal and consisted of a cigar-shaped mass of calcium carbonate with only a small chambered portion. As shown by impressions preserved in the

FIGURE 10-4

Many Jurassic seas were populated by great numbers of belemnoids, as illustrated in this reconstruction. The bottom has a dense bank of oyster-like pelecypods. In the background is the marine reptile ichthyosaur. (Chicago Museum of Natural History.)

rock, the belemnoids resembled the modern squid in outward appearance and in at least one internal structure, the ink sac; presumably they had the same mode of life (Fig. 10-4). Gregarious animals, they generally traveled in large groups. Many Jurassic formations of Europe are made up almost entirely of the shells of belemnoids.

Pelecypods

Next to the ammonoids, the pelecypods were the most widely distributed and diverse of Mesozoic invertebrates. Although in general they are not as useful as ammonoids for long-range correlation there are some rather startling exceptions. For example, fossils of *Monotis subcircularis* occur at a definite horizon in the Upper

Triassic and are found in nearly all countries bordering the Pacific Ocean in which rocks of this age crop out (Fig. 10-5). It is a guide fossil *par excellence*. Fossils of another species, a large, oyster-like pelecypod, *Exogyra cancellata,* occur in North America in the Upper Cretaceous strata that crop out in the coastal plains of the Atlantic and Gulf of Mexico. With some minor interruptions, this species has been traced from New Jersey to Mexico, a distance of 2,500 miles. In one place the fossiliferous zone is 115 miles wide. The fossils of both species are nearly always extremely abundant, indicating large populations.

Pelecypods are particularly good ecologic indicators and hence are extremely useful to the historical geologist. One unusual group of Mesozoic pelecypods, the rudistids, was made up of species that adopted a sedentary

FIGURE 10-5

Circum-Pacific region, showing distribution of the Late Triassic pelecypod *Monotis subcircularis.*

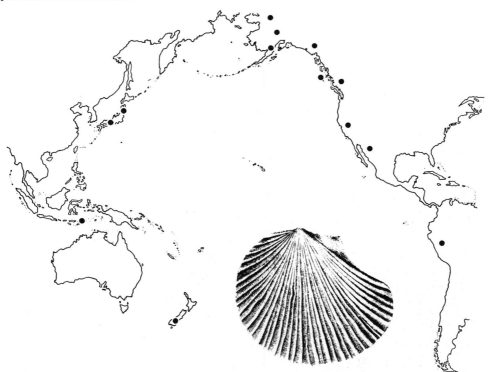

mode of life and became attached to the sea bottom. One valve of the rudistid's shell became long and cone-shaped and was cemented to any hard object on the bottom of the seas; the other valve was highly modified to form an articulated lid. The shell wall in some types became extremely thick. In outward appearance they resembled solitary corals (Fig. 10-6), and they were extremely numerous in a circumtropical belt during the Cretaceous. They lived in densely populated banks in shallow warm waters, very much like corals. The fossil records of the rudistids, solitary corals, and richtofenid brachiopods (Fig. 2-9, p. 29) constitute one of the classic examples of convergent evolution (§ 2-3).

In recent years extensive studies of the Pacific Ocean bottom have revealed the presence of numerous flat-topped **sea mounts**, submerged to a depth of 4,000–7,000 feet. Dredgings from the tops of two of these mounts brought up a Cretaceous fauna of reef corals, stromatoporoids, gastropods, pelecypods (including some rudistids), and echinoids. This faunal assemblage was one that was adapted to warm shallow waters and was characteristic of the Cretaceous Tethyan region. The evidence available indicates that the sea mounts were a chain of basaltic islands during the Cretaceous. They were then wave-eroded to flat banks on which a coral-rudistid fauna became established and grew into reefs but never developed into atolls. During the later part of the Cretaceous, the sea mounts were submerged to a depth that was below the zone where reef coral could grow, and finally sank to their present depth.

Corals

The wave of extinctions of invertebrate phyla toward the close of the Paleozoic included the dominant coral groups (Rugosa and Tabulata) that were characteristic of that era. A strange enigma of the fossil record is that no corals have ever been found in strata of Early Triassic age.[*] The first reappearance of corals is in Middle Triassic formations of Germany, the southern Alps, Corsica, and Sicily, but these belong to a completely new order (Scleractinia). The first Mesozoic corals are represented by only a few genera. There is still considerable debate about their ancestry: one school has them evolving from some branch of the rugose corals of the Paleozoic, another school claims their ancestors were Paleozoic corals that left no record because they had no calcareous skeleton. Both hypotheses have merit, but the data now available are insufficient to indicate clearly which is correct; possibly neither is. The main feature of the scleractinian corals is the insertion of new septa in cycles of six. The group was and is a highly successful one, and it constituted the bulk of the Mesozoic and Cenozoic coral faunas (Fig. 10-7).

By the Late Triassic the corals had become almost world-wide in distribution, and at that time there was greater similarity among them than there has ever been since. Late Triassic coral reefs are known from 60° north latitude to 10° south. In the passage from the Triassic to the Jurassic the corals showed little change: Early Jurassic coral faunas were very similar to those of the Late Triassic. A big change in evolutionary tempo, however, occurred in the Middle Jurassic. A new radiation was marked by the beginning of a greater diversification and multiplication of genera and species, a phase that continued through the remainder of the Mesozoic and the Cenozoic. The principal centers of this diversification were the western region of the Tethys and the shelf seas that covered much of Europe. The Late Jurassic was marked by extensive reef-building activity from about 55° north latitude (England, Germany) to 5° south (eastern Africa). This

---

[*] Rugose corals have been reported from Lower Triassic strata in Armenia, U.S.S.R. There is some question, however, whether these strata are really of Mesozoic age.

FIGURE 10-6

Representative Jurassic and Cretaceous pelecypods.

A. *Inoceramus*, Late Cretaceous.
B. *Trigonia*, Late Jurassic.
C. *Gryphaea*, Early Jurassic.
D. *Ostrea* (*Lopha*), Late Jurassic.
E. *Mytilus* (*Falcimytilus*), Late Jurassic.

F. *Hippurites*, Late Cretaceous.
G. *Caprinula*, Cretaceous.
H. *Barbatia*, Early Cretaceous.
I. *Pecten* (*Chlamys*), Late Jurassic.

(A, D, E, H, and I from memoir of Geol. Surv. Great Britain; B, C, and F from booklet by British Museum, Natural History; G from R. T. Jackson, 1890.)

latitudinal range is nearly the same as that of modern reef-building corals, but it is 20° farther north.

The extensive regressions of the shelf seas that took place in the earliest Cretaceous on most continents affected the corals so much that there was then little or no reef-building, but by the late Early Cretaceous reef-building had again become widespread. Reefs of some consequence were built at this time in France, Switzerland, Capri, Tunisia, Algeria, Catalonia, the western Balkans, Bulgaria, Romania, the Crimea, Kenya, Tanzania, Japan, Mexico, and Texas. Scattered coral banks existed as far south as the province of Neuquén, Argentina, in Colombia, Peru, and Venezuela, and in Europe north to the Isle of Wight. The reefs lie in a belt between 47° north latitude and 10° south. This period of wide-spread reef-building was followed in the Middle Cretaceous by a period of adversity for reef-builders, but again in the Late Cretaceous suitable conditions for reef-builders

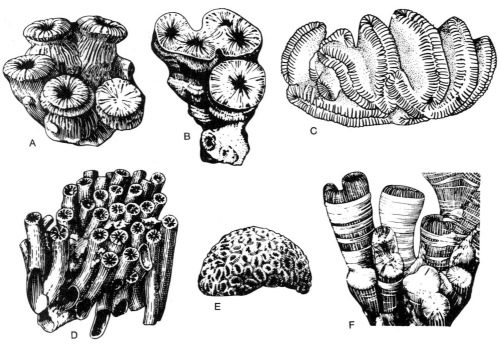

FIGURE 10-7

Representative Mesozoic corals.

A. *Margarosmilia* (Middle and Late Triassic).
B. *Latomeandra* (Middle Jurassic and Early Cretaceous).
C. *Rhipidogyra* (Late Jurassic).
D. *Stylosmilia* (Middle Jurassic and Early Cretaceous).
E. *Favia* (Cretaceous–Recent).
F. *Thecosmilia* (Middle Triassic–Cretaceous).

(From *Coelenterata*, Vol. F of *Treatise on Invertebrate Paleontology*, Geol. Soc. Am. and Univ. of Kansas Press, 1956.)

were again widespread, though to a lesser extent than in the Late Jurassic and the Early Cretaceous. These Late Cretaceous reefs are found between 50° north and the equator, but the principal reefs are between 47° north and 30° north.

With the close of the Mesozoic no great changes took place in the coral faunas, and most of the families continued into the Cenozoic. It was not until the mid-Cenozoic that many of the modern elements of the coral faunas first appeared and the older elements became extinct.

10-2 PLANTS

Upper Permian and Lower Triassic terrestrial deposits are mainly redbeds, many containing evaporites, which seem to indicate that semi-arid or arid climatic conditions prevailed in many regions. The great floras of the Carboniferous swamps had disappeared, and many groups had become extinct before the close of the Paleozoic. The fossil record of early Mesozoic floras is very sparse and indicates that the flora was greatly impoverished, probably because of unfavorable climates, and that it consisted mainly of survivors from the Paleozoic. A reconstruction of an Early Triassic landscape somewhere in Germany is shown in Figure 10-8. The scene is a vast, semi-arid wasteland containing a few pools and lakes, around which a limited flora grew. Fair-sized conifers (*Voltzia*) were fairly common, and their remains are among the most widespread fossil plants in the Bunter Series of Germany. The slender, stilt-like *Pleuromeia*, related to the Carboniferous *Sigillaria*, grew to a height of seven feet, was crowned by short, tough leaves, and terminated in a cone. The trunk

is marked by transverse scars that were left when leaves fell off. The low, bush-like plant in the left foreground is a fern of the genus *Neuropteridium*. The plants living in the shallow waters are representatives of the horsetails (Equisetales); this particular genus (*Schizoneura*) was very common in Gondwana strata in the southern hemisphere during the late Paleozoic. (We discuss the Gondwana formations in the next chapter.)

Middle Triassic floras were much the same as those of the Late Permian and Early Triassic. It was not until the Late Triassic that the land plants first showed a distinctly

FIGURE 10-8

Reconstruction of an Early Triassic landscape in Germany during the time of deposition of the Bunter formation. Reptile footprints are very common in the Lower Triassic strata of Europe, but no fossil bones have been found. The reptile on the horizon is a reconstruction made only from the footprints. The sparse vegetation living in this semi-arid wasteland consisted of species that had survived from the late Paleozoic. (Drawing by Z. Burian under the supervision of Prof. J. Augusta.)

Mesozoic stamp. The floras became more varied and abundant and a large number of new groups appeared. From the Late Triassic through the Early Cretaceous the land flora was surprisingly uniform throughout the world. The principal components of this flora were the cycads, ginkgos, conifers, and ferns. This was the time of the great evolutionary radiation of the gymnosperms.

The most characteristic of all plants of the Mesozoic were the cycads. The living cycads still to be found in certain tropical regions are almost identical with the Mesozoic forms in all features except reproductive organs. Although the evolutionary relationship of the Recent and fossil forms is still a mystery, it is convenient to go on using the term "cycad" for both. A reconstruction of a typical Mesozoic land scene is shown in Figure 10-9. The low plant in the foreground with a barrel-shaped trunk and a crown of palm-like leaves is a cycad (Cycadeoidea). The trunk bore large flowers. Another cycad (shown at right center) was *Williamsonia*, which had a slender trunk six or seven feet tall and marked by spiral rows of rhombohedral leaf scars; the crown bore long, slender, palm-like leaves and thin, slender flowers. Still another is the short bush-like plant, next to the barrel-trunked cycad, called *Williamsoniella;* this plant had a forked stem, and the flowers were borne in an upright position in the angle of the fork. These three types of cycads are just a sample of a group that included hundreds of different forms reflecting a rapid and diverse radiation.

The larger trees in Figure 10-9 are conifers, ginkgos, and the first representatives of the huge sequoias. Ferns no longer occupied a dominant position in the flora, but were still represented by numerous groups. The waters of the pool supported various horsetails, but these were far smaller than their Paleozoic ancestors.

The characteristic forms of the mid-Mesozoic floras were strikingly different from the Paleozoic floras discussed in Chapter 7. The dominance of the cycad-ginkgo assemblage ended, toward the close of the Early Cretaceous, with the almost explosive radiation of the flowering plants, the angiosperms, which have been, ever since, the predominant plant group covering the land. Angiosperms were fairly widely distributed and abundant in the Early Cretaceous, but the gymnosperms were still the predominant element of nearly all floras. Beginning in the Late Cretaceous and continuing through the Cenozoic, the fossil record of flowering plants becomes increasingly more abundant, and the modern aspect of the flora is unmistakable.

10-3 REPTILES

Unlike many of the invertebrates, the land-living animals—which at this time consisted of only amphibians and reptiles—showed no significant change in evolutionary tempo or mode in the passage from Permian to Triassic time. With the appearance of the reptiles the amphibians had declined greatly in numbers and variety. The cotylosaurs are looked upon as the stem reptiles—that is, the ancestral stock from which most later groups evolved (Fig. 10-10). Even in the Permian, adaptive lines of marine forms, turtles, mammal-like reptiles, and thecodonts (ancestors of the dominant reptiles of the Mesozoic) were already well defined. The thecodonts were probably the most important of these radiations, for they gave rise to the dinosaurs, the flying reptiles, the crocodilians, and finally the birds.

The reptiles were the first vertebrates to take full advantage of the increasing quantities of oxygen in air and ocean. They invaded a wide range of adaptive niches on the land, in the water, and in the air. In the rest of this chapter we shall investigate this evolutionary radiation, with emphasis on the position of each group within the environmental community and the adaptive significance of various structural features. Since the thecodonts are the root of a wide evolutionary tree, let us begin with them.

The Thecodonts

A typical early thecodont was *Euparkeria*, illustrated in Figure 10-11. It was a small animal that may have progressed on all fours but seems to have been adapted for speedy bipedal locomotion. Its forelimbs were about half as long as its hind limbs. Its skull shows many primitive features but the overall architecture of the skeleton is of a pattern from which all other ruling reptiles could have evolved. Some thecodonts were quadrupeds with fore- and hind limbs of approximately equal length. The more advanced forms, however, evolved toward a bipedal pattern of locomotion, reflected in the shortening of the forelimbs and lengthening of the hind limbs.

An Upper Triassic form, *Ornithosuchus*, (Fig. 10-12) was long classified as an advanced thecodont but is now considered to have been a primitive dinosaur. It was rather small, having a total length, including the tail, of only three or four feet. It was clearly adapted for walking on two legs, in contrast to the four-legged walk of many of its ancestors. The hind legs were strong and muscular, capable of rapid running, but the forelegs were greatly reduced and the front feet were somewhat hand-like, adapted for

FIGURE 10-9

Reconstruction of a Mesozoic landscape illustrating the flora of the Jurassic and Early Cretaceous. (Drawing by Z. Burian under the supervision of Prof. J. Augusta.)

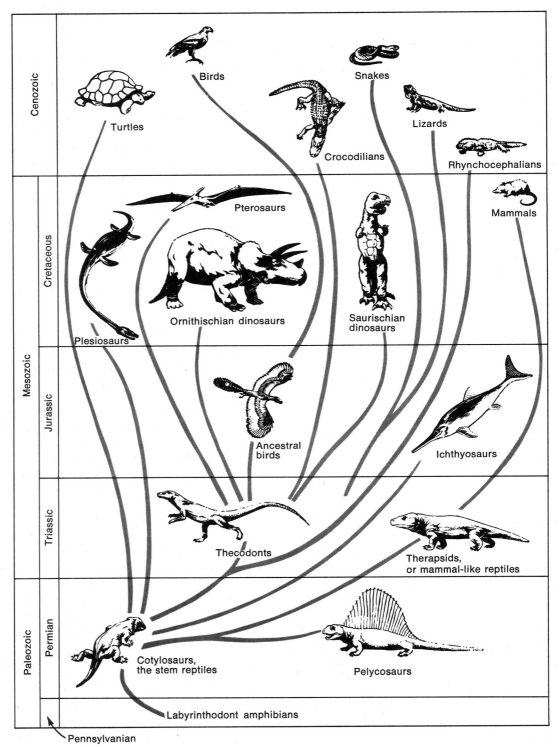

FIGURE 10-10

Family tree of the reptiles. (From Colbert, 1945, American Museum of Natural History.)

grasping. The long tail served as a balancing mechanism. The striking change in posture and mode of locomotion necessitated changes in the hip-bone structure both to support the body and to permit firmer articulation of the legs. The general structure and anatomy of *Ornithosuchus* provide us with a clear blueprint of the basic body plan of the dinosaurs.

The evolution of the thecodonts, however, is exceedingly complex; for, besides giving rise to the dinosaurs, flying reptiles, and ichthyosaurs, they were also the ancestors of several short-lived groups. One of these comprised strange animals that returned to four-legged locomotion and a water-dwelling life. These animals, called phytosaurs, were remarkably similar in appearance to modern crocodiles. They were, of course, not crocodiles at all, but a group of animals that occupied an ecologic niche that was later reoccupied by the crocodiles. Most of the thecodonts were small, but the phytosaurs grew to a large size, some fossil specimens twenty feet long having been collected.

The thecodonts never included more than a few genera and species, and their life span was confined to the Triassic.

The Dinosaurs

The ruling reptiles during the Mesozoic were the dinosaurs, whose evolutionary history is one of the most interesting in the annals of the vertebrates. The name "dinosaur," meaning "terrible lizard," was proposed more than a hundred years ago for certain large reptiles whose fossil remains had recently been discovered. Since then it has been well demonstrated that there are not one but two distinct groups of these reptiles, and that not all dinosaurs were gigantic; they ranged in size from the eighty-foot giants down to creatures of the size of chickens. Even so, the word "dinosaur" is so well implanted in the literature

that it still serves as a useful designation although it no longer has any formal systematic significance.

The dinosaurs comprised two great orders of reptiles, the Ornithischia and the Saurischia, whose primary basis of distinction lies in the construction of the hip bones. In the Saurischia the hip bones (illium, ischium, and pubis) were arranged triradially according to the typical reptilian plan (Fig. 10-13); the name "Saurischia," in fact, means "reptile hips." The Ornithischia (meaning bird hips) had the bones of the hip arranged like those in birds, with the pubis brought back parallel to the ischium and its forward part greatly extended. Both groups evolved from thecodont ancestors. A diagrammatic sketch of dinosaur evolution is shown in Figure 10-13. The Saurischia include large herbivorous sauropods and ferocious two-legged carnivores as well as some other forms. The Ornithischia include a variety of bizarre forms, all of which were plant-eaters.

Among the various evolutionary trends exhibited by these groups two are worth a brief mention. The ancestral thecodonts were adapted for bipedal locomotion, with long, well-developed hind legs and short fore legs. This pattern was retained in the large carnivorous saurischians and in a few minor groups. However, many plant-eating saurischians and ornithischians returned to using all four legs in locomotion. In nearly every case, though, the forelegs were shorter than the hind legs, reflecting the ancestry of the bipedal thecodonts. Another trend shown by most groups of dinosaurs was a general increase in size.

Saurischian Dinosaurs. The oldest dinosaurs, of Late Triassic age, had reptile-like hip bones and belonged to the Saurischia. One of the best-known of these early forms is *Coelophysis* (Fig. 10-14), of which several complete skeletons have been collected from Upper Triassic formations in northern New Mexico. The earliest forms

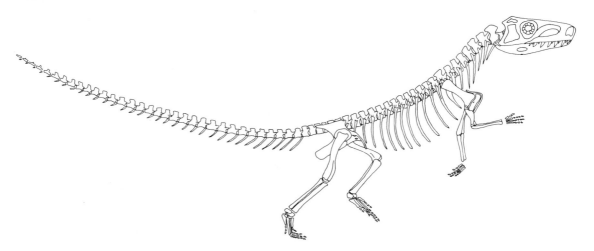

FIGURE 10-11

Euparkeria, a Lower Triassic thecodont reptile representing the stock from which the dinosaurs, birds, and flying reptiles arose. (After Ewer, 1965, with modifications suggested by A. S. Romer.)

are exceedingly close in structure to the ancestral thecodonts, like *Ornithosuchus* (Fig. 10-12).

This creature was about eight feet long and was bipedal, with long hind legs and short fore legs. Since the bones of the skeleton were hollow, the animal probably did not weigh more than fifty pounds. The jaws were equipped with numerous sharp teeth well adapted for eating meat. The major skeletal features of *Coelophysis* clearly show that it was from this adaptive type— a meat-eating animal equipped for rapid running—that the later saurischians developed. The major feature of the evolutionary line proceeding from *Coelophysis* was the tendency toward giant size, with parallel changes in anatomical structure. The culmination of this trend was the mightiest carnivore of all times, *Tyrannosaurus*, which lived during the Cretaceous (Fig. 10-15). This awesome creature stood nearly twenty feet tall and was fifty feet long from the tip of the nose to the tip of the tail. Its body was extremely massive, and its tail was long and heavy; mature individuals must have weighed eight or ten tons. The hind legs were extremely stout and muscular, with feet capped by large

claws; the forelimbs, in contrast, were extremely small, so small that it is hard to imagine that *Tyrannosaurus* had much use for them. There was also a reduction in the number of digits, each hand having only two functional clawed fingers. The head, however, attained an enormous size and, equipped as it was with a formidable battery of large teeth, was the main offensive weapon. Although *Tyrannosaurus* represents the highest development of the

FIGURE 10-12

Ornithosuchus, a primitive Triassic saurischian reptile. (From G. Heilman, *The Origin of Birds*, Witherby, London, 1926.)

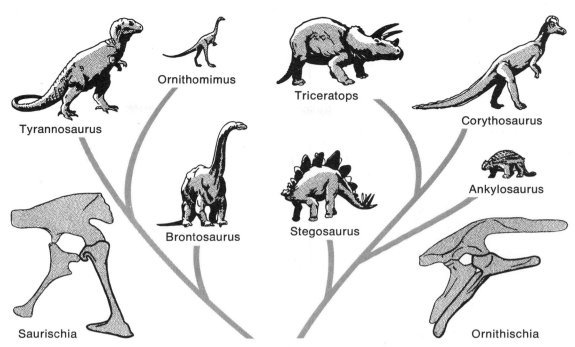

Ornithomimus

Tyrannosaurus

Triceratops

Corythosaurus

Brontosaurus

Stegosaurus

Ankylosaurus

Saurischia

Ornithischia

FIGURE 10-13

Simplified evolutionary tree of the dinosaurs, showing just a few of many divergent
adaptive lines. Also shown are the hip-bone constructions of the two principal orders,
the Saurischia and the Ornithischia.

FIGURE 10-14

Coelophysis, an early saurischian dinosaur of Late Triassic age.
(American Museum of Natural History.)

FIGURE 10-15

Tyrannosaurus, the largest carnivorous land animal that ever lived, here shown attacking the horned dinosaur *Triceratops.* (C. R. Knight, American Museum of Natural History.)

carnivorous trend, there were many other carnivores living during the Jurassic and the Cretaceous. Most of these were built on the same basic plan as *Tyrannosaurus,* varying in details of limb proportions, size, shape of the skull, or other features. An early and unusual carnivorous dinosaur is *Ceratosaurus,* illustrated in Figure 10-16 in company with two stegosaurians, which we shall discuss later. *Ceratosaurus* lived in western North America during the Late Jurassic. It never attained the size of *Tyrannosaurus,* a mature adult measuring about twenty-four feet long. In conformation, if not size, this creature was very similar to the later *Tyrannosaurus,* but its most striking feature was the bony horn on its nose.

The trend toward large size in carnivorous dinosaurs was by no means the only evolutionary trend among the bipedal saurischians: the group had an extremely wide range of adaptive radiation. One evolutionary line remained rather modest in size. A common Late Jurassic member of this line was *Coelurus,* illustrated in Figure 10-17. This animal illustrates well the main features of the group: a light, graceful body; a long, slender tail; slender, bird-like legs; and short, grasping arms. Although they were not large enough to prey on the larger plant-eating dinosaurs, types like *Coelurus* must surely have caused the smaller dinosaurs plenty of trouble. Their small size (five or six feet long) and light build suggest that they were very agile creatures, well able to avoid predators and to run down their prey. There were even smaller members of this group, such as *Compsognathus* (seen in company with the earliest known birds in Fig. 10-35). This small creature was only as big as a Mallard duck and lived along lagoonal shores in Bavaria during the Late Jurassic. Its food consisted mostly of insects and other small invertebrates. The size range of these bipedal saurischians—between *Compsognathus* and *Tyrannosaurus*—is impressive and reflects the extensive adaptive radiation of the group.

One other group of bipedal dinosaurs is of unusual interest because certain of its structural features varied greatly from those of the forms we have mentioned. This group is well exemplified by *Ornithomimus* (Fig. 10-18). The basic body plan at first appears to be quite like that of such other bipedal forms as *Coelurus*—a gracefully built body, bipedal mode of locomotion, and long, slender tail. In this form, however, the neck is long and sinuous and the head comparatively small. The most unusual feature of the head is the beak-like jaws that have no teeth. A common name for this animal is "ostrich dinosaur." *Ornithomimus*, obviously, was not a flesh-eater; it subsisted on insects and fruits and probably on the

FIGURE 10-16

Ceratosaurus, another large meat-eating saurischian dinosaur, had a high comb on the nasal bone but did not grow up to the size of *Tyrannosaurus*. It is here contemplating an attack on *Stegosaurus*, the famous plated dinosaur. (Drawing by Z. Burian under the supervision of Prof. J. Augusta.)

FIGURE 10-17

Not all the carnivorous saurischian dinosaurs were large, like *Tyrannosaurus;* one trend, illustrated here by *Coelurus,* remained close to the ancestral ornithosuchians. (American Museum of Natural History.)

eggs of other dinosaurs. Its general physical structure suggests that it was capable of running fast—a good thing since this was its only mode of defense.

The other main adaptive trend of the saurischians evolved large, four-legged, plant-eating dinosaurs known as sauropods. This group included the largest vertebrates that ever walked this earth.

The carnivorous *Coelophysis* was by no means the only dinosaur extant in the Late Triassic. Some forms were of fairly large size (twenty feet long) with a basic body design like that of *Coelophysis* but with most features more massive. The bones were solid rather than hollow, the hind legs heavy and broad, and the front legs enlarged for walking on all fours. The skull was fairly small; the blunt and flattened teeth found as fossils clearly show that they belonged to a plant-eater and not to a flesh-eater. It

FIGURE 10-18

Cretaceous "ostrich dinosaur," *Ornithominus.* (From G. Heilman, *The Origin of Birds,* Witherby, London, 1926.)

was from such early types that the famous giant sauropods developed during the Jurassic and Cretaceous. One of the best-known of these giant sauropods is *Brontosaurus* (Fig. 10-19), which attained a length of eighty feet from the tip of its nose to the tip of its tail. Much of this length was taken up by the long neck and the even longer tail. The body weighed many tons, and the legs, especially the hind legs, were very thick. Even more remarkable was the smallness of the head. *Brontosaurus* was a plant-eating animal, and it is hard to believe that it could take in enough plant food per day to keep alive. But it did. It seems most likely that the giant sauropods spent much of their time in swamps, lakes, and rivers, where there would be an ample supply of soft plants to feed on. This habitat also provided excellent protection against the large carnivores of the time. Retreating far out from shore in a lake or swamp, *Brontosaurus* was completely safe from types like *Tyrannosaurus*. The remains of the largest sauropod known were collected from beds at Tendaguru, eastern Africa (§ 9-1) that are of Late Jurassic and Early Cretaceous age; other specimens have been found in the Morrison formation of western North America (Fig. 10-20). This sauropod, *Brachiosaurus*, had body measurements that are quite impressive. A specimen mounted and exhibited in the Museum of the University of Berlin has a length of 81 feet 8 inches; its neck is 18 feet 7 inches long; its height is 42 feet 4 inches; and its body is 10 feet

FIGURE 10-19

Brontosaurus in its normal marshy habitat. (C. R. Knight, American Museum of Natural History.)

FIGURE 10-20

Brachiosaurus, a giant plant-eating dinosaur that lived in the Late Jurassic swamps of Colorado and eastern Africa. (Drawing by Z. Burian under the supervision of Prof. J. Augusta.)

in circumference. Even though it has been estimated that this great animal may have weighed as much as 50 tons, this skeleton is not the largest known; scattered bones have been found that are longer and thicker by far than the corresponding bones of the mounted skeleton.

Brachiosaurus was somewhat different from *Brontosaurus* in its body plan. For one thing, since the forelegs were longer than the hind legs, the body sloped strongly downward to the rear. The tail was rather short, but the neck was long and heavy, and the head was small. The nostrils were on a raised dome on top of the head. This last feature, plus the long neck and massive body and legs, suggests that *Brachiosaurus* spent much of its time in water. *Brachiosaurus* was able to live in much deeper waters than the brontosaurians; it was so tall, in fact, that with lifted neck it could, without any effort, have looked in at the attic windows of a three story house.

The sauropod dinosaurs were a very successful adaptive group, and flourished till the end of the Mesozoic; they then became extinct, as did all the other dinosaurian groups.

Ornithischian Dinosaurs. This order of dinosaurs includes all forms with bird-like hip bones, all of whom were herbivorous.

In many respects these dinosaurs were more highly evolved than the Saurischia. Their advanced evolution was especially marked in the development of their hip bones. The saurischian hip bone was almost exactly like that of the ancestral thecodonts, but that of the Ornithischia had become greatly modified. The evolutionary history of this order shows a great adaptive radiation, manifested in the appearance of a large number of distinct types. The highest evolutionary development of the Ornithischia came in the Cretaceous, unlike the Saurischia, whose highest development occurred in the Jurassic, and whose Cretaceous forms were merely continuations of structural types evolved in the earlier period.

The Ornithischia comprised a heterogeneous assortment of odd-looking dinosaurs, including the well-known "duck-billed" forms, heavily armored types, horned specimens, and several others. All these odd and rather aberrant forms evolved from a basic, primitive, unspecialized stock, of which *Camptosaurus* (Fig. 10-21) is a good example. Members of this stock were still bipedal, with large, heavy hind legs, but with forelegs sufficiently elongated so that they probably often moved around on all fours. They were heavier than saurischian dinosaurs of equal size.

The skull was rather long and flattened and had features characteristic of nearly all the Ornithischia. There were no front teeth, instead, the front of the jaw was a bird-like beak; the lateral teeth were flattened, blade-like, and adapted for eating plants; and the

FIGURE 10-21

Camptosaurus, one of the first ornithischian dinosaurs. (American Museum of Natural History.)

jaws were so hinged that when the mouth was closed all the teeth came together at nearly the same time, thus increasing the grinding surface working on the plant food at one time (Fig. 10-22). In contrast, the jaws of the carnivorous Saurischia closed with a scissors action, the upper and lower teeth sliding past each other.

Another ornithischian dinosaur, whose unspecialized body resembled that of *Camptosaurus*, was *Iguanodon* (Fig. 10-23). This dinosaur is well known from skeletons found in England and Belgium and is of special interest because it was the first dinosaur to be scientifically studied. Although *Iguanodon's* structure was like that of *Camptosaurus*, it tended toward large size, some skeletons more than thirty feet long having been collected. Of interest are the large spike-like thumbs on the fore limbs, which must have been rather effective defensive weapons.

Specialization of the basic camptosaurid type of body resulted in some very strange patterns. One of the most unusual and diffi-

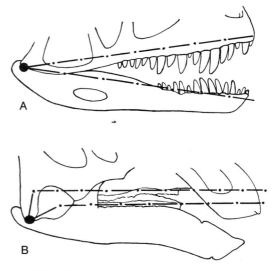

FIGURE 10-22

Comparison of the mechanics of biting in the two orders of dinosaurs. In the Saurischia (A) the hinge of the jaw is approximately in line with the tooth sockets, but in the Ornithischia (B) the hinge is below the line of the tooth sockets so that, when the jaws close, the teeth clamp together almost at the same time. (After E. H. Colbert, 1945, American Museum of Natural History.)

FIGURE 10-23

Idealized landscape of southern England during the Early Cretaceous. The large dinosaur in the center is *Iguanodon*. (Courtesy of Her Majesty's Geol. Surv., British Crown copyright.)

cult to explain took place in a group that developed extremely thick skull roofs. The culmination of this evolutionary trend was *Pachycephalosaurus,* which had a skull roof of solid bone some ten inches thick (Fig. 10-24). To enhance the beauty of the beast, the nose and back of the head had an array of spikes and nodes of various sizes and shapes. Just what adaptive advantage a head like this could have had is a mystery.

The most successful and widespread evolutionary line that evolved from a camptosaurid type of ancestor was the famous duckbilled dinosaur of Cretaceous age. The commonest and central type of this group was *Anatosaurus* (Fig. 10-25). Its body plan was quite like that of *Camptosaurus,* but on a larger scale, and with head and feet that had become highly specialized. The

FIGURE 10-24

Reconstruction of the head of *Pachycephalosaurus.* (American Museum of Natural History.)

jaws were greatly flattened and duck-like in appearance. The teeth were simple leaf-like structures, arranged in several parallel rows closely pressed against one another in each jaw. Some forms had as many as 2,000 teeth. The duckbilled dinosaurs were clearly adapted for living much of the time in water, and many of them undoubtedly could swim. Some remarkable skin impressions of these dinosaurs that have been found in western Canada show, among other things, that the feet were webbed. The anatomical features and the nature of the deposits in which these fossils are found strongly suggest that the duckbills fed by grubbing along the bottom of shallow waters and occasionally feeding on land. In case of danger they could move out into deeper waters and swim away.

The head developed a most surprising range of adaptive variations. Figure 10-25 shows two duckbills besides *Anatosaurus:* one, *Parasaurolophus,* had a greatly enlarged, tubular, bony crest extending back of the head, which probably supported a flap-like piece of skin that extended down the back of the neck; the other, *Corythosaurus,* had a long, narrow crest on the top of the head. Each of these bony crests was formed by the nasal bones and actually consisted of a greatly extended nasal passage. Such an adaptation would be ideal for an aquatic animal. The enlarged nasal passages would make excellent air-storage chambers, enabling the animal to stay submerged for a considerable time.

An early offshoot of the ancestral ornithischian camptosaurid stock was the plated dinosaur, *Stegosaurus* (Fig. 10-16), which lived only in the Jurassic. The stegosaurs reverted completely to four-legged locomotion, but the front legs were so short that the head was carried close to the ground. The most striking feature was a double row of large, bony, triangular plates down the middle of the back, extending from the head to the tail, which was equipped with four rather formidable

FIGURE 10-25

Assorted Cretaceous dinosaurs. On the right are duckbilled dinosaurs, *Anatosaurus* (trachodon). In the left background are crested duckbills, *Parasaurolophus,* and in the water is *Corythosaurus.* In the center background is the "ostrich dinosaur," *Ornithomimus,* and in the center foreground is the armored dinosaur, *Ankylosaurus.* (C. R. Knight, Chicago Museum of Natural History.)

spikes. Presumably these bony plates and the tail spikes were protective mechanisms, but the whole side of the body was unprotected. The stegosaurs are famous for their extremely small brain, which was about the size of that in a small kitten— not much of a brain for an animal that attained a length of twenty feet. As a possible compensation for the tiny brain there were enlargements of the spinal cord in the hips and shoulders (Fig. 10-26); the enlargement in the hips was twenty times the size of the brain. This has given rise to the idea that the dinosaurs had two brains, an idea developed in a most scientific manner some years ago by Bert Leston Taylor in the *Chicago Tribune.*

The Dinosaur

Behold the mighty dinosaur,
  Famous in prehistoric lore,
Not only for his power and strength
  But for his intellectual length.
You will observe by these remains

The creature had two sets of brains—
One in his head (the usual place),
  The other at his spinal base.
Thus he could reason "A priori"
  As well as "A posteriori."
No problem bothered him a bit;
  He made both head and tail of it.
So wise was he, so wise and solemn,
  Each thought filled just a spinal column.
If one brain found the pressure strong,
  It passed a few ideas along.
If something slipped his forward mind,
  'Twas rescued by the one behind,
And if in error he was caught,
  He had a saving afterthought.
As he thought twice before he spoke,
  He had no judgment to revoke.
Thus he could think without congestion
  Upon both sides of every question.
Oh, gaze upon this model beast,
  Defunct ten million years at least.

The armored dinosaurs, known as the ankylosaurs, appeared in the Cretaceous, and they are the easiest of all adaptive forms for us to understand. They were completely covered by an overlapping pavement of bony plates (Fig. 10-25). They were very squat and broad; the larger species grew to a length of twenty feet. Aside from the heavy armor, most species had numerous large, pointed spikes on the sides of the body, and in many the tail had a club-like bony process at the end. All these features were merely additional deterrents for would-be predators.

The last of the major dinosaur groups to appear were the horned dinosaurs, or ceratopsians. The first dinosaur eggs ever discovered (Fig. 10-27) were laid by a member of this group and preserved in the Upper Cretaceous Djadochta beds of Mongolia (see § 9-7). In some of the eggs were found the bones of unhatched embryos.

The richly fossiliferous Upper Cretaceous formations of Asia and North America have yielded a complete series of fossil ceratopsians, the earliest of which were small bipedal animals, rather primitive in their structure and quite close to the ancestral camptosaurid stock. The skulls of these early forms had already developed the features that were to become characteristic of the group; they were rather narrow and deep, so that the jaws resembled a large, pointed beak. From this ancestral type evolved the first "frilled" ceratopsian, *Protoceratops* (Fig. 10-28). The fore part of the head was characteristically narrow and deep, terminating in a pointed beak, but the back was a wide flange of bone forming a shield, or frill, over the neck. *Protoceratops* walked on all fours and attained a maximum length of about six feet. These

were the ceratopsians that laid the eggs found in Mongolia, and it was from this general type that the remaining known forms evolved. Evolution brought an increase in size and the development of large, formidable horns. The best known of the more advanced ceratopsians is *Triceratops* (Fig. 10-15), whose most striking feature, as the name implies, was the three large horns. These animals grew to a length between twenty and thirty feet and stood some eight feet high at the hips, the head measured fully one-third of the length, and the frill that extended back of the head was exceptionally large. When we remember that *Triceratops* was roaming the plains at the same time as *Tyrannosaurus*, the adaptive significance of the head frill and horns is easily understood. In addition to *Triceratops* there were other Late Cretaceous forms, similar in proportions but differing in the shape of the frill and the pattern of the horns. One of these had a single large horn on the nose and a smaller horn just above the eyes; the frill was modest in size. Another type had a frill of approximately the same pattern and size as *Triceratops'*, but the borders of the frill had large horns; there were also a pair of small horns just above the eyes and a large horn on the nose.

The ceratopsians had a rather short history, since they lived only during the Late Cretaceous, but in the course of that history they underwent tremendous evolutionary changes. Toward the close of the Cretaceous they were among the most numerous of dinosaurs, but by the end of the period they had become extinct.

FIGURE 10-26

Brain and spinal cord in *Stegosaurus*. Note that the enlargements of the spinal cord are much larger than the brain. (After E. H. Colbert, 1945, American Museum of Natural History.)

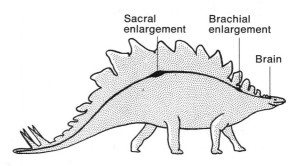

FIGURE 10-27

A neat dozen of eggs of *Protoceratops*, the ancestral horned dinosaur, as they were discovered in the Upper Cretaceous Djadochta beds of Mongolia by the Central Asiatic Expedition of the American Museum of Natural History. (American Museum of Natural History.)

Marine Reptiles

One strange aspect of the evolutionary story is that no sooner had reptiles evolved from amphibians and developed a group of landliving animals, than other evolutionary lines, evolved from the same ancestors, became water-dwellers, completely adapted to an aquatic environment. There were four distinct radiations, three of which probably evolved from the stem reptiles (cotylosaurs) in the late Paleozoic, the other from lizards during the Mesozoic.

The oldest aquatic reptile is a form known as *Mesosaurus,* which figures prominently in the Gondwana problem discussed in the next chapter. It was a small reptile with a long slender body, a long neck, an elon-gated jaw bearing a large number of teeth (Fig. 10-29), short legs, and feet like broad paddles. *Mesosaurus* is known from Early Permian strata only in Brazil and South Africa. The character of the rocks enclosing the fossils suggests that the environment was not open marine, but lagoonal, bay, or possibly even fresh water.

Probably the best-adapted to their environment of all the aquatic reptiles were the marine ichthyosaurs, which looked greatly like modern sharks or porpoises. When the first ichthyosaurs appeared in the Middle Triassic they were already highly adapted forms not unlike their descendants, which were abundant during Jurassic and Cretaceous. The body was definitely fish-like and very streamlined. The limbs were reduced and had become modified into paired fins; there was even a large central fin on the back. The head had large eyes, and jaws that were greatly elongated and armored with a formidable battery of teeth, which, interestingly, show the same labyrinthine structure as the teeth in the stem reptiles and the labyrinthodont amphibians (Fig. 7-23, p. 243).

For many years there was some question about how these reptiles reproduced. After all, fertilizing eggs laid in the open sea would be quite a trick. The question was finally resolved by the discovery of a fossil specimen, now in the American Museum of Natural History in New York, that shows the skeletons of seven very small ichthyosaurs partly within and partly outside of the body cavity (Fig. 10-30) of a much larger one. Since the small specimens do not show any marks made by teeth or any dissolution by gastric juice they cannot be interpreted as the remains of a meal, and detailed study has shown that they were unborn. Apparently the females retained the fertilized eggs inside their bodies and the young were born alive. The ichthyosaurs were a highly successful group through the Mesozoic, but like many reptiles, became extinct at the close of that era. Fossil ichthyosaurs are

FIGURE 10-28

A primitive horned dinosaur, *Protoceratops*. The eggs laid by this small dinosaur are illustrated in Fig. 10-26. (C. R. Knight, Chicago Museum of Natural History.)

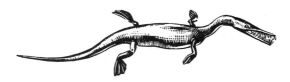

FIGURE 10-29

Mesosaurus, an early Permian aquatic reptile. (From E. H. Colbert, 1945, American Museum of Natural History.)

often beautifully preserved, a carbonized impression of the body outline being clearly indicated.

Another radiation from the stem reptiles in the late Paleozoic comprised the plesiosaurs and their kin (sauropterygians). The best-known and most highly evolved of this group were the plesiosaurs, but two other groups entirely confined to the Triassic are also of particular interest—the nothosaurs and the placodonts (Fig. 10-31).

The nothosaur was widespread during the Triassic and was not fully adapted to a marine life; rather, it was capable of some movement on land, like our modern sea lions. It was an elongated reptile with long tail, body, neck, and jaws. The feet were short and modified to short paddles.

Living at the same time as the nothosaurs was a highly specialized group, the placodonts, adapted for eating mollusks (Fig. 10-31). The body was rather heavy-set, with a short neck, and the jaws contained greatly modified teeth. The front teeth extended almost straight forward; the back teeth were

huge blunt grinding mills capable of crushing the toughest mollusk shells. The limbs were modified to paddles, but apparently the placodonts were not fast swimmers. Their food habits did not make speed necessary; they probably moved slowly along the bottom of shallow waters, eating mollusks as they found them. For this type of life the ability to swim fast was no advantage; the oysters were not going to run away.

Both the nothosaurs and the placodonts disappeared by the end of the Triassic. In the Late Triassic the first plesiosaurs appeared, and they became dominant members of the marine environment during the Jurassic and Cretaceous. In many respects the plesiosaurs resembled enlarged versions of the Triassic nothosaurs. There was a clear tendency toward giantism among the plesiosaurs, some forms growing to fifty or more feet in length. Their bodies were generally large and rather bulky, not fish-like as were those of the ichthyosaurs. The limbs were modified into huge paddles rather than into the smaller balancing planes

FIGURE 10-30

Female ichthyosaur skeleton with skeletons of seven young (unborn) ichthyosaurs, partly within and partly drifted out of the body cavity. The seven young skeletons are in heavy shading, the mother skeleton in outline. (American Museum of Natural History.)

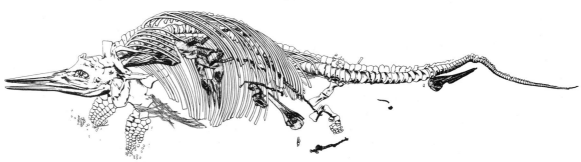

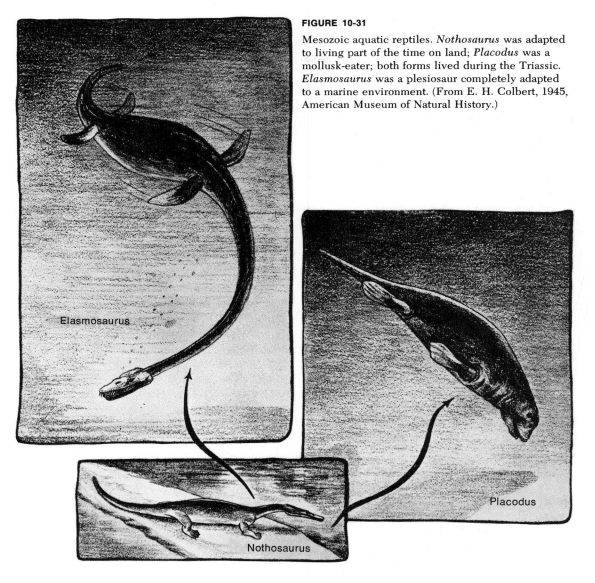

FIGURE 10-31

Mesozoic aquatic reptiles. *Nothosaurus* was adapted to living part of the time on land; *Placodus* was a mollusk-eater; both forms lived during the Triassic. *Elasmosaurus* was a plesiosaur completely adapted to a marine environment. (From E. H. Colbert, 1945, American Museum of Natural History.)

of the ichthyosaurs. The plesiosaurs came in two styles, one with a long, sinuous neck and a small head, the other with a short neck but a high, elongated head (Fig. 10-32). The long-necked forms apparently could flex their necks rapidly from one side to the other for catching fish; in the short-necked forms the elongated head and jaws compensated for the loss of flexibility of the neck. One early worker on these strange creatures described them as snakes strung through the body of a turtle.

The last marine reptiles we shall discuss had a completely different ancestry from any of the other groups. These are the mosasaurs, which were actually lizards that became adapted to a marine life. The mosasaurs lived only during the Cretaceous, but in their short history they became numerous in the sea. Like practically all the marine reptiles, the mosasaurs grew to a large size. Their land-living lizard ancestors were generally small creatures; but, once they had adapted themselves to marine life, they grew to lengths of twenty or more feet. The body of the mosasaur was greatly elongated and slender, and the tail was narrow and deep. The limbs were modified into strong swimming paddles. The neck was short; the head and jaws were elongated and equipped with numerous sharp teeth. The meeting of a long-necked plesiosaur and a mosasaur must have been a spectacle, even to a flying reptile, which was probably the only creature around to see it. Figure 10-32 is an artist's impression of such an encounter.

Flying Reptiles

The last uninhabited environmental domain was the air, which was not invaded until the Jurassic. Among the earliest flying reptiles was *Rhamphorhynchus* (Fig. 10-33), a strange-looking creature with a body about two feet long, a short neck, a small but elongated skull, and jaws fitted with an array of sharp, forward-directed teeth. The long, slender tail was tipped with a diamond-shaped membrane. The hind limbs were rather short and weak, but the fore limbs, were highly modified into a framework for the membranous wing. The fourth finger was tremendously elongated to form the principal support of the wing membrane. The remaining fingers were reduced to small hooks. The back part of the wing membrane was probably attached to the hind legs.

A characteristic feature of all pterosaurs (the ordinal name of all flying reptiles) was the light weight skeleton. Most of the bones were hollow and rather delicate, but so modified that they were strong at the same time: the girdle and the backbone accommodated strong muscles for moving the wings.

Living during the Late Jurassic with *Rhamphorhynchus* were small bird-saurians of the genus *Pterodactylus*. The size of these small pterosaurs varied from that of a sparrow to that of a large hawk; they had a very small tail, wide wings, and an elongated skull with teeth only in the anterior portion of the jaws.

Types like *Rhamphorhynchus* represent the primitive pterosaurs and are confined to the Jurassic. The advanced pterosaurs can be exemplified by the well-known *Pteranodon*, which was widespread during the Cretaceous (Fig. 10-32). It grew to a large size, attaining a maximum wingspread of about thirty feet, and showed considerable modification from *Rhamphorhynchus*, although the basic structure of the most striking feature was, of course, the enormous wingspread; the body was only as big as a turkey's. There was no tail to speak of. The head was large for the body, but the jaws had no teeth, and the back part of the skull was extended to form a crest. The hind legs were so weak that there is some question whether *Pteranodon* could move around at all on level ground. The life pattern of *Pteranodon* has been compared to that of the present-day albatross, which inhabits the open sea of the southern hemisphere.

One of the most famous collecting localities for pteranodons is the Upper Cretaceous deposits of western Kansas. The pterosaurs thrived during the Jurassic and Cretaceous, but became extinct before the Cenozoic.

Miscellaneous Reptile Groups

So far our discussion of reptiles has concentrated on the dominant groups of the Mesozoic, all of which became extinct at the close of that era. Living along with their spectacular relatives there were a few orders of small reptiles that had one thing in common—they survived at the close of the Mesozoic and constitute our modern reptile fauna. These orders include the turtles, lizards, snakes, crocodiles, and rhynchocephalians.

FIGURE 10-32

Elasmosaurus, one of the long-necked plesiosaurs, in a duel with *Tylosaurus*, a marine lizard. Witnesses of this fight are the large flying reptiles *Pteranodon*. (Drawing by Z. Burian under the supervision of Prof. J. Augusta.)

FIGURE 10-33

Rhamphorhynchus. (Drawing by Z. Burian under the supervision of Prof. J. Augusta.)

The turtles first appear in the geologic record in the Upper Triassic, and their fossils indicate that they had already become a distinct adaptive type that was not unlike modern turtles in structure and form. The ancestry of the turtles is somewhat obscure, but we think they evolved from the stem reptiles (cotylosaurs) at some time in the late Paleozoic. In the long span of time from the Late Triassic to the Recent the turtles have been extremely conservative: their evolutionary changes have been merely refinements of the basic adaptive pattern. They have been and are a successful group, and they display a great range in size and food habits.

Most groups of surviving reptiles are represented today by only a few species — some groups by a single one — but not the lizards and snakes, which are today the most numerous and diverse of all reptiles: they belong to the same order (Squamata). Neither group has much of a fossil record, and neither is of great paleontological importance. Lizards first made their appearance in the Triassic. The marine mosasaurs and the snakes evolved from the lizards in the mid-Mesozoic. Snakes are merely modified lizards that have lost their legs. They are a most remarkable group, though, because so many species are poisonous—an adaptation acquired in the mid-Cenozoic.

The crocodiles, alligators, and gavials are the largest modern reptiles and apparently have always been tropical or subtropical creatures living in or near swamps and rivers. The earliest crocodiles known were of Middle Triassic age and, though primitive, show the basic features found in all later forms. The anatomy of these early forms clearly reflects a thecodont ancestry. All later crocodiles are pretty much alike, differing only in details of their skeleton, in size, and in mode of life. One Jurassic group, the geosaurs, became adapted to marine life. The ancestry of the modern forms goes back to the Cretaceous. Increase in size was a marked and common evolutionary tendency. The largest fossil crocodile discovered so far has a skull six feet long; the total length of this monster must have been forty or fifty feet. This specimen came from Upper Cretaceous strata along the Rio Grande in Texas.

The last of the surviving reptile groups are the rhynchocephalians, now known only on a few islands off the coast of New Zealand. The only genus extant is a small lizard-like animal named *Sphenodon* (Fig. 10-34). The rhynchocephalians first appear in the geologic record of the Triassic, in which period they underwent a fair degree of evolutionary radiation and became nearly worldwide in distribution. Since the Triassic the group has become more and more restricted, and it survives today only in an isolated habitat in which it is not in competition with other groups.

10-4 BIRDS

The most startling fossil discovery in the nineteenth century was that of fossil birds in the Solenhofen limestone beds of Bavaria, Germany. In 1860 a feather was found, and in the next year an incomplete skeleton. Then, in 1877, a beautifully preserved complete skeleton, with impressions of feathers in the surrounding rock, was uncovered. This specimen is in the Museum of Natural History, Berlin, and the incomplete specimen is in the British Museum in London. These tell us just about all we know about the first bird, which has the name *Archaeopteryx* (Fig. 10-35). The anatomical features were clearly reptilian—so much so that, without the feathers, it would have been classified as a reptile without any hesitation.

Archaeopteryx was about as big as a crow; its head was elongated, and the front part was narrowed into a beak with teeth. The body was lightly but strongly constructed, as it must be for a flying animal. The long tail was bony—a reptilian characteristic. The pelvic girdle was constructed on a plan similar to that of the ornithischian dinosaurs. In contrast to the pterosaurs, this Jurassic bird had strong hind legs. The bones of the forelimbs were elongated, and the hands, composed of three digits, were also elongated and free.

It does not seem likely that *Archaeopteryx* was a very good flyer; but the big step in bird

FIGURE 10-34

The one surviving rhynchocephalian, *Sphenodon*, now living in New Zealand. (From E. H. Colbert, 1945, American Museum of Natural History.)

evolution had taken place—the evolution of feathers. The airways at that time were pretty well dominated by the pterosaurs, but the birds' structural plan was much more efficient than that of the flying reptiles, and birds eventually became dominant.

The fossil record of birds is much greater for the Cretaceous than for the Jurassic, but still is only fragmentary. The fossil materials available clearly indicate that the emergence of birds of modern type took place largely during the Cretaceous.

10-5 MAMMAL-LIKE REPTILES

The dinosaurs, flying reptiles, and swimming reptiles are all extremely interesting but, from a strictly anthropocentric viewpoint, the most important of all the reptile groups is probably the one that gave rise to the mammals. This group, which may be called the mammal-like reptiles, was one of the many radiated from the stem reptiles (cotylosaurs) in the late Carboniferous. At that end of their evolution they were structurally very close to the ancestral cotylosaurs; at the other end they were just a step away from the mammals in their anatomical structures. An early member of this evolutionary trend was the genus *Ophiacodon* (Fig. 10-36), of Permian age. This creature grew to a length of eight feet; it had a long, slender tail and a narrow, deep, rather pointed head. The jaws were equipped with numerous sharp teeth. The body was rather lizard-like in appearance, especially because of its short legs. *Ophiacodon* was probably a fish-eating reptile that lived near streams and ponds, as depicted in Figure 10-36.

From an ancestral type like *Ophiacodon* evolved two evolutionary lines of the most bizarre-looking late Paleozoic reptiles—reptiles with odd sail-like structures on their backs (Fig. 10-37). One had a large head and

FIGURE 10-35

Archaeopteryx, the earliest known bird, looking down on *Compsognathus*, one of the smaller dinosaurs. (Drawing by Z. Burian under the supervision of Prof. J. Augusta.)

FIGURE 10-36

Ophiacodon, a primitive ancestor of the mammals. (Drawing by Llewellyn Price, courtesy of A. S. Romer.)

formidable dagger-like teeth (*Dimetrodon*) and was obviously an aggressive carnivore. The other, with a large body and very small head (*Edaphosaurus*), had small uniform teeth and was a plant-eating animal. The unique feature of their fossils is the greatly extended vertebral spines, which in life must have supported a membranous sail. In *Edaphosaurus* these spines had enlarged knobs or cross-bars.

Reptiles of the types represented by *Ophi-*

acodon, Edaphosaurus, and *Dimetrodon* belong to a group called the pelycosaurs. The evolutionary relationships of the pelycosaurs are diagrammatically shown in Figure 10-10. From an early pelycosaur stock the true mammal-like reptiles, the therapsids, evolved, and it is within this group that we find the ancestors of the mammals. Among these reptiles we can recognize many anatomical features, in the skull and postcranial skeleton, that are characteristic of

FIGURE 10-37

Among the most bizarre of Permian reptiles are the "ship lizards." Those with the large head and formidable row of teeth are *Dimetrodon*, a carnivore. The one with the small head is the plant-eating *Edaphosaurus*. (C. R. Knight, Chicago Museum of Natural History.)

mammals. The teeth, for one thing, were beginning to become differentiated, and in advanced forms there were sharply contrasted incisors, canines, and cheek teeth, in contrast to the uniformity of reptilian teeth. The legs also changed; rather than extending from the side of the body, as in the typical reptiles, they were under the body, raising it farther off the ground and increasing the efficiency of the legs for locomotion.

The therapsids fell into two groups, plant-eating and animal-eating, the first serving as the food supply of the second. The plant-eaters tended to be large and bulky and the meat-eaters tended to become even more mammal-like in their basic anatomy. An early plant-eater was *Moschops*, a common reptilian fossil in Permian formations of South Africa (Fig. 10-38). This was a large, powerful reptile with strong limbs. The front legs were longer than the hind ones, and the back sloped like that of a giraffe. The teeth were but slightly differentiated and were rather peg-like in form.

Another genus of therapsid plant-eaters was *Kannemeyeria* (Fig. 10-39). The group to which this genus belonged first appeared in the Middle Permian and survived until the close of the Triassic. During this time they became nearly world-wide in their distribution and extremely numerous. There was a wide range in size among this group, and *Kannemeyeria* is one of the larger forms. Like *Moschops*, it had a massive body with very stout legs. A peculiar feature was the presence of two large "tusks" in the upper jaw. The fact that in large collections of this type of reptiles the tusks appear in only about half of the specimens suggests that the tusks were a sexual character, presumably male. The remainder of the teeth in the jaw were reduced, peg-like forms, suggesting that *Kannemeyeria* was a herbivore.

The evolutionary line that includes the carnivorous therapsids is by all odds the most interesting, for this was the group that eventually gave rise to the mammals. A typical and well-known member of this group, all of which were carnivores, is *Cynognathus*, illustrated in Figure 10-39. The skull

FIGURE 10-38

Moschops, a large, plant-eating, mammal-like reptile that lived in South Africa during the Permian. (Drawing by Z. Burian under the supervision of Prof. J. Augusta.)

FIGURE 10-39

Mammal-like reptiles from the Early Triassic of South Africa. The large animal to the right is a plant-eater, *Kannemeyeria*, being attacked by three flesh-eaters of the genus *Cynognathus*. (C. R. Knight, Chicago Museum of Natural History.)

and post-cranial skeleton show many features that are directly antecedent to those of typical early mammals. The dog-like skull had highly differentiated teeth, anticipating the basic tooth plan of mammals. The lower jaw was made up mostly of the bone that holds the teeth (dentary), and the other bones were much smaller than those of normal reptiles, whose lower jaw is made up of several coequal bones. The legs were beneath the body, and the feet were well formed, features that greatly increase the efficiency of locomotion of land-living animals. In the Triassic another group of these meat-eating reptiles (ictidosaurs) appeared, one that was even more mammal-like than *Cynognathus*. It is, in fact, as perfect an intermediate form between reptiles and mammals as we could ever hope to find.

The mammal-like reptiles were common

and widespread during the Permian and Triassic. By the close of the Triassic, however, the group had become extinct except for a few stragglers that persisted into the Jurassic. Before their extinction the meat-eating therapsids gave rise to an evolutionary offshoot that was eventually to become the dominant group of vertebrate animals; these were the mammals, which first appear in the paleontologic record in the Upper Triassic.

10-6 THE FIRST MAMMALS

Mammals are today the dominant land-living animals, and so they have been since the early Cenozoic. With the close of the Mesozoic and extinction of the ruling reptiles of the Cretaceous, the mammals underwent an evolutionary radiation unparalleled by any other vertebrate group except perhaps the fishes. Even before the Cenozoic, however, the mammals had had a long history, extending back at least to the Late Triassic, but the Mesozoic evolution of mammals was in striking contrast to that of the Cenozoic. Next to Jurassic birds, probably the rarest of fossils are the remains of Triassic, Jurassic, or Early Cretaceous mammals. The fossil material available consists mostly of isolated teeth, a few jaw fragments, and practically no other parts of the skeleton. That teeth are the commonest fossils is indeed fortunate, for they, more than any other anatomical part, reflect many fine degrees of adaptation and evolution. We shall postpone discussion of the anatomical and physiological features of mammals, as contrasted with those of reptiles, until Chapter 14, when we discuss

FIGURE 10-40

Geologic distribution and evolutionary relationships of Mesozoic mammals, with a crown and side view of the lower molars of each group. (Data from Patterson, 1956, Simpson, 1929, and Butler, 1939.)

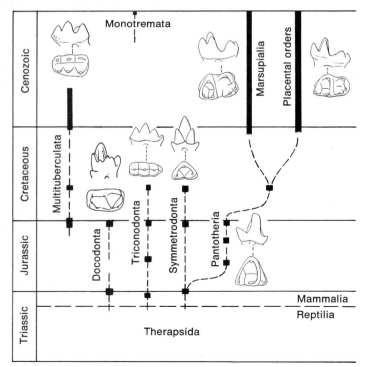

Cenozoic life. For the moment let us concentrate on the paleontological record and relationships of Mesozoic fossil mammals.

The ancestry of the mammals is to be found within the group of mammal-like reptiles (therapsids) that we discussed in § 10-5. The paleontological data, however, are still much too incomplete for us to be able to analyze in detail the great evolutionary transformation from reptile to mammal. One of the strange aspects of the problem is that, though we know a great deal about the skull and post-cranial skeleton of the therapsids, we know little about the teeth; for the Mesozoic mammals just the opposite is true—practically all the available data come from teeth. Whether several groups of mammals were evolving from various therapsid ancestors at the same time or there was a single origin is not yet known.

Fossil mammals from Upper Triassic deposits make up only three distinct orders, but by the Late Jurassic five orders were in existence. Each order is characterized by a distinctive pattern of the molar teeth. The names, geologic distribution, relationships, and diagrammatic figures of the molar teeth of these early mammalian orders are shown by Figure 10-40. All the early mammals were small, ranging in size from a mouse to a cat.

The triconodonts had molar teeth with three cusps in a longitudinal row. The group appears to have been unsuccessful and died out during the Middle Cretaceous. The multituberculates were a highly specialized group that survived into the early Cenozoic. Their molar teeth had two parallel rows of cusps. One very interesting specialization of the multituberculates was that the last lower premolar was greatly enlarged and had vertical ribs on the sides. This tooth formed an excellent shearing blade. This group was clearly adapted for plant-eating and probably lived a life similar to that of our modern rodents. The docodonts had molar teeth of squarish outline with three main cusps on the outer side of the lower molar, flanked by a series of smaller cusps on the inner side. This group first appears in the

fossil record in the highest Triassic strata but had become extinct by the close of the Jurassic. In the symmetrodonts the three cusps of the molar teeth were arranged in a rather symmetrical pattern on the crown. This was another group that did not survive beyond the Middle Cretaceous.

Of all the orders of early Mesozoic mammals, the pantotheres were without question the most important and interesting; this was the group that gave rise to the marsupial and placental mammals. The molar teeth were of a general triangular outline with cusps arranged also in a triangular pattern. The upper molars were arranged with the apex of the triangle directed inward, and the lower molars were set so that the apex of the triangle pointed outward. Thus, when the jaws were closed, these molar teeth were a complex shearing and grinding mechanism (Fig. 10-41). This functional arrangement of the

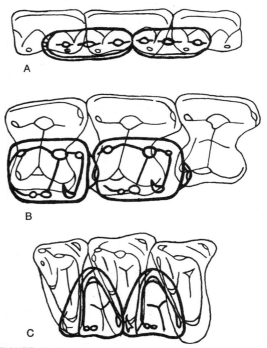

FIGURE 10-41

Diagram of occlusive relations in (A) triconodonts, (B) docodonts, and (C) pantotheres. The lower molars are represented by heavier lines. Not drawn to scale. (After Simpson, 1929.)

molars is found in some later primitive mammals, a likeness that strengthens the conclusion of some authorities that the pantotheres were the ancestors of the marsupials and placentals.

In the first phase of mammalian evolution, covering nearly 100 million years, the mammals remained small, almost insignificant animals. Of the five orders that are recognized early in their history, mainly by the characters of the molar teeth, each represents an early experimental adaptation. Only one of these groups, the multituberculates, survived into the Cenozoic, and for only a brief period. The remaining orders had become extinct by the Middle Cretaceous; but before extinction one of these orders, the pantotheres, gave rise to the two mammalian orders that dominated the Cenozoic land faunas. In the Late Cretaceous marsupials and placentals become commoner and more widely distributed, though they were still very small animals. Among the placental mammals a fairly well-known Late Creta-

ceous type is *Zalambdalestes,* from Mongolia. This had a skull less than two inches in length (Fig. 10-42). The shape of the molar teeth and the pattern of the cusps appear to be intermediate between those of the pantotheres and more advanced mammals. *Zalambdalestes,* which belongs to the order Insectivora, the most primitive of all placental mammalian orders, may be considered a fair approximation of the early placental ancestors.

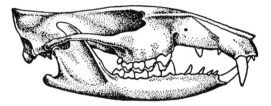

FIGURE 10-42

Late Cretaceous insectivore from Mongolia, *Zalambdalestes.* (From A. S. Romer, *Vertebrate Paleontology* (3rd ed.), Univ. Chicago Press, 1966.)

SUGGESTED READINGS

See list at end of Chapter 2.

11

GONDWANA FORMATIONS

The mathematician and geographer Ptolemy wrote of a great southern land, which he called Terra Australis Incognita and which was sought by many explorers of the eighteenth century. In 1775, when Captain James Cook returned to England from his second voyage round the world, he was able to report that no such vast southern continent existed. A hundred years later this southern continent was recreated in the minds of some geologists as a vast area that has girdled more than half the circumference of the earth through the greater part of geologic time; it was renamed Gondwanaland.

One of the most startling impressions we get from a study of the geological history of the southern hemisphere is the apparent stratigraphical and biological uniformity of that area, during a certain part of geologic time, and the great contrast it offers to what we have already seen in the northern hemi-

sphere. As early as 1875, the distribution of *Glossopteris*, a genus of seed fern that grew in South Africa, India, and Australia raised the question: "How did this plant get to all these areas that are now widely separated by oceans?" Since the question was first posed, many additional paleontological data seem to suggest that there were once geographical connections between the now widely scattered southern continents. Several aspects of physical-geological history support the conclusions drawn from the fossil flora and fauna. Most striking is the presence of similar tillite formations of Permo-Carboniferous age in South America, the Falkland Islands, Antarctica, South Africa, Madagascar, India, and Australia. Overlying the glacial tillite beds are thick series of predominantly terrestrial sedimentary rocks, including thick basaltic lava over enormous areas. The faunas and floras of these formations differ from those of the northern hemisphere. The fact that the geologic time scale was based on the marine faunas and strata in Europe has made precise correlation of these southern strata difficult and uncertain. These formations are the Gondwana "System" in peninsular India, the Karroo "System" in South Africa, and the Santa Catharina "System" in South America.

Of the many fascinating and perplexing problems in the earth's geologic history, few have stimulated the imagination and caused more heated debate than the Gondwana strata. The search for a clearer understanding of the faunal and floral distribution patterns that show an apparent similarity in the geological history of these regions has led to the formulation of several hypotheses, each of which has strong supporters. It is difficult to be entirely objective on this subject, for any ultimate conclusion rests on how the available data are evaluated—a process that depends on education, experience, and scientific philosophy. Let us first explore the physical stratigraphy of the southern continents; then we can proceed to the biological data, and from there to some of the theories that have been advanced to explain what we find. In this chapter we shall discuss peninsular India, southern Africa, Madagascar, central and eastern South America, the Falkland Islands, and Antarctica. Australia is also an area where Gondwana formations are known, but, as we have discussed this continent in Chapters 6 and 9, we shall not do so here.

11-1 INDIA

Few large areas of the world remained so uniformly underformed and emergent during Phanerozoic time as peninsular India. The great complex of horizontal strata belonging to the peninsula's Gondwana System cover an age span from the Late Carboniferous to the Early Cretaceous. Throughout this long range of time minor marine inva-

sions occurred briefly in the Early Permian and in the Cretaceous; aside from these two exceptions the history of peninsular India is deciphered from continental rather than marine deposits. Gondwana formations rest on Precambrian formations with a striking unconformity; all of the early and middle Paleozoic Era was a time of erosion, not a single sedimentary formation being deposited in, or at least preserved from, this interval. The Gondwana formations are limited to linear, rather narrow fault troughs in the Precambrian basement (Fig. 4-14, p. 103). It seems certain that Gondwana deposits were once more widespread. The faulting may have begun during late Gondwana time. The present tectonic relations of the Gondwana rocks are illustrated in Figure 11-1. The maximum thickness of the sequence in India is about 20,000 feet. Because of the large coal deposits, the system is of great economic importance.

The Gondwana formations are almost entirely clastic. Two main divisions—the Lower and the Upper Gondwana—are separated by an unconformity and distinguished by paleontological criteria.

The basal formation of the Lower Gondwana is a tillite, the Talchir Formation, the basal member of which is typically a boulder bed up to 100 feet thick. Above is a greenish shale, which is overlain by beds of silty shale and sandstone; boulder-bearing strata are found in many of these beds. The formation usually fills in the hollows of an uneven topography developed on the underlying rocks. The glaciogene character of part of this formation has been recognized since the nineteenth century (Fig. 11-2); in a few places the basal boulder bed rests on a striated pavement. This formation contains a *Glossopteris* flora, and marine invertebrate fossils are found at a few localities, generally at horizons above the basal bed, but actually in the basal bed at one place. The marine fauna of the Talchir Formation has been shown to be Lower Permian by correlation with well-studied

Gneiss — Vindhyan sandstones

Gondwana rocks with coal seams

FIGURE 11-1

Tectonic relations of Gondwana rocks in peninsular India. (After
D. N. Wadia, *Geology of India*, Macmillan, 1953.)

sequences of late Paleozoic faunas of western Australia. The fauna, which is considered to be that of a cold-water environment, consists of brachiopods, pelecypods,
gastropods, crinoids, and scyphozoa, the
pelecypod *Eurydesma* being one of its most
characteristic members.

The first Phanerozoic deposits on peninsular India are thus glacial, with marginal
marine deposits. Figure 11-3, A shows the
main features of Talchir time. The craton
shore areas are regions where marine invertebrates are present in clearly glaciogene facies. In the areas of marginal

FIGURE 11-2

Basal Talchir tillite bed, Adjae River, Raniganj coal fields, peninsular India.
(Photograph by E. R. Gee, Geol. Surv. India.)

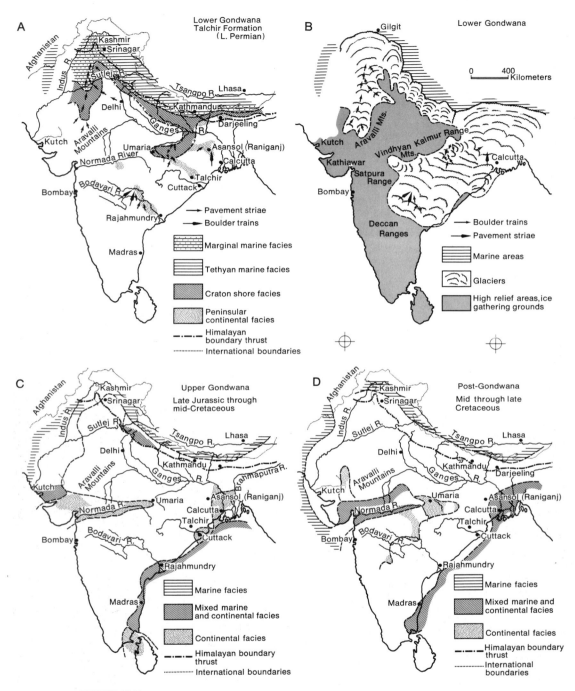

FIGURE 11-3

Paleogeographic maps of peninsular India. (A) Major geologic features for lower Gondwana (Talchir Formation) time, (B) Diagrammatic interpretation for same period of time, (C) Upper Gondwana (Late Jurassic through Early Cretaceous) time, (D) Post-Gondwana (Upper Cretaceous) time. (After unpublished maps kindly furnished by P. L. Robinson.)

marine facies, the glaciogene facies are directly overlain by marine formations. Figure 11-3, B represents an interpretation of this Indian region during the period of early Permian glaciation.

The rest of the Lower Gondwana deposits overlying the Talchir Formation consist of sandstone, shale, and clay ironstones, together with the chief coal horizons of India, some of which are 90 feet thick (Fig. 11-4).

The Upper Gondwana contains similar clastic formations with fewer coal-bearing strata and locally fairly extensive basalt flows. Reptiles, amphibians, and plants are the main fossils. The lower formations of the Upper Gondwana are of Triassic age, as shown by a rather extensive fauna of reptiles and amphibia, and are the first extensive continental redbeds in the Gondwana sequence. Lower Jurassic strata are known only in a small area of the central peninsula, where they consist of arkosic sandstone, shale, and lenses of fresh-water limestone

FIGURE 11-4

Outcrop of coal bed in a Gondwana formation, peninsular India. (Photograph by L. L. Fermor, Geol. Surv. India.)

containing a veritable cemetery of dinosaur bones and logs of fossil wood as much as 10 feet long. Most bones are of a sauropod dinosaur whose thigh bones are as much as five feet long.

The Upper Gondwana formations of Late Jurassic through Lower Cretaceous age are much more restricted. They are confined to the eastern coastal region and to an east-west tract along the Narbada River in central-west India (Fig. 11-3, C). In the southeastern coastal region these formations consist of clastic wedges of mixed marine and nonmarine facies, whose fossil plants are typical of the Gondwana. The youngest marine fossils are Lower Cretaceous ammonites. Further north along the coast only continental facies are present in basins near Cuttack and Calcutta (Fig. 11-3, C). The Narbada Valley area also accumulated continental clastics. The Upper Cretaceous was a time of local marine transgressions (Fig. 11-3, D), confined to small embayments along the east coast and up the Narbada Valley. These formations are a mixture of marine and continental facies. The marine facies are rich in invertebrate fossils, and the continental facies in some places contain abundant plant remains. Other fossils worth mentioning are a tree trunk 86 feet long and 4.5 feet wide, broken dinosaur bones, and a reptilian egg.

11-2 SOUTH AFRICA

The equivalent of the Gondwana "System" in Africa is the Karroo "System." The formations are almost horizontal, lithologically distinct, and occupy vast areas from the equator to the Cape (Fig. 11-5). The area exposed today is but a small erosional remnant of what once existed. The formations range in age from Permian through Early Jurassic. The thickest and most complete development is in the Cape region of the Union of South Africa, where more than

35,000 feet of Karroo rocks have been measured. North of the Cape, in Transvaal, the Karroo is much thinner. It can, in fact, be conveniently divided into a southern area with thick, conformable formations, beginning with a typical tillite, and a northern area extending to the equator, wherein the strata are much thinner, with many disconformities; the basal tillite beds are thin and scattered. The Karroo formations are almost entirely continental lake, stream, and wind deposits, a single marine unit in the lower Karroo of Southwest Africa being the only exception.

The Karroo "System" in the Cape region rests unconformably on very thick clastic formations, the Cape "System," which in turn rests nonconformably on a granite whose radiometric age suggests Middle Cambrian. The Table Mountain Series, the basal unit of the Cape "System," is a quartzite some 4,000 feet thick that covers an area roughly 1,000 miles E-W and 250 miles N-S. In the upper third of the unit tillite beds indicate a glacial episode. The sediment of the Table Mountain Series came from the north. A small fauna of brachiopods from the upper part of the series indicates a Late Ordovician age. The middle unit of the Cape System (the Bokkeweld Series) is a sequence of quartzite and shale that has yielded a marine invertebrate fauna of Lower Devonian age. The uppermost unit of the Cape System (the Witteberg Series) is a plant-bearing sequence of quartzite and shale that is late Devonian, possibly early Carboniferous in age.

The Karroo "System" is separated into four groups. In the Cape region the basal unit is a shale-and-tillite, the Dwyka Group. Above this is a complex shale-sandstone-coal facies, the Ecca Group, which in turn is overlain by a thick sandstone complex, the Beaufort Group. The uppermost group, the Stormberg, begins with mainly sandstone facies but ends with a very thick sequence of basaltic lavas. The correlation

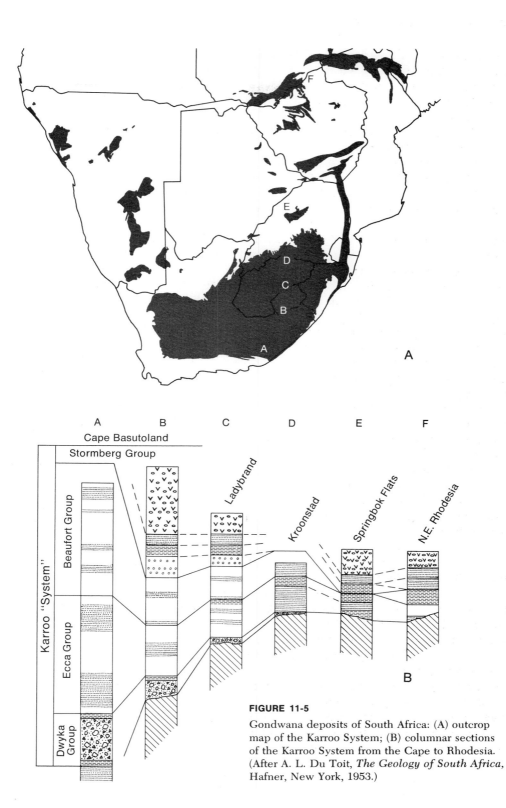

FIGURE 11-5

Gondwana deposits of South Africa: (A) outcrop
map of the Karroo System; (B) columnar sections
of the Karroo System from the Cape to Rhodesia.
(After A. L. Du Toit, *The Geology of South Africa*,
Hafner, New York, 1953.)

FIGURE 11-6

Glacial striae on a polished rock basement, formed during the glacial epoch represented by the Dwyka tillites, near Kimberley, South Africa. (Photograph by J. H. Wellington.)

of the Karroo formations from the Cape region north to Rhodesia is illustrated in Figure 11-5.

The Dwyka Group is largely a tillite resting unconformably on the underlying formations. Most tillite strata are morainal deposits except in the southern region of the Cape, where more than 2,000 feet of strata are probably glacial sediments that were "dumped" into a body of fairly deep water. In Southwest Africa the tillites spread into a shallow sea. North of the Cape region the tillites are much thinner ground moraines resting on a rock floor with a relief of more than 1,000 feet that in places has well-developed roche moutonnées and striated surfaces (Fig. 11-6). The upper part of the Dwyka Group is a shale unit that includes a bed of black carbonaceous shale that weathers white, forming the so-called

"White Band." This unusual bed has yielded an aquatic reptile, *Mesosaurus*, discussed later. In Southwest Africa *Mesosaurus* has also been found near the top of the Dwyka Shale, and within the glaciogene Dwyka two shale horizons have yielded marine fossils, including fish, *Eurydesma*, gastropods, and crinoids. *Eurydesma* is the Permian pelecypod typically associated with the glaciogene Talchir Formation of peninsular India; it is also present in western Australia.

The Dwyka Group is overlain by the Ecca Group, a widespread fine-to-coarse clastic formation developed in four distinct but gradational facies. In the Cape region are some 10,000 feet of sandstone and shale, probably alluvial fan deposits, whose source was to the south. In the central part of the Union of South Africa this group consists

FIGURE 11-7

Cliff of Cave sandstone in the Orange Free State of South Africa. The lower part of this cliff is water-laid, the upper part wind-laid. (Photograph by L. C. King.)

FIGURE 11-8

The Great Escarpment of South Africa in the Eastern Buttress at the Royal Natal National Park. More than 4,000 feet of the Drakensberg lavas are exposed in the giant precipices. (Photograph by L. C. King.)

entirely of bluish and greenish shales averaging 2,500 feet thick. East of this belt, in the Orange Free State, the Transvaal, and Natal, sandstone and shale with prominent coal beds represent the Ecca Group. In Southwest Africa red sandstones and shales accumulated. In the northern areas, where the Dwyka is discontinuous, the Ecca Group rests directly on the older rock floor with a striking unconformity. The succeeding Beaufort Group is noteworthy because of its rich fauna of reptiles. Here again the group is thickest in the Cape region, where it comprises 9,000 feet of arkose and shale of fluvial origin, coarsest to the south and finest to the north. The Beaufort Group is not represented in Southwest Africa or in most of the Transvaal. During the deposition of the Cape sediments, the region underwent intense compression. The sediments derived from this folded and raised mass spread as a huge, wedge-shaped alluvial fan northward as far as the Transvaal—the initial deposit of the Stormberg Group. These formations contain some coal beds, which suggest that humid conditions prevailed at times. The overlying formations are first red sandstone and shale and then a massive sandstone that bears all the characteristics of desert origin. Much of southern Africa was then a vast arid plain accumulating dune-deposited sandstone (Fig. 11-7). The final phase of Karroo history was the outpouring of a vast plateau of basalt, some from volcanoes, but most are apparently from fissures. This basalt is as much as 4,000 feet thick (Fig. 11-8).

11-3 MADAGASCAR

Gondwana faunas and floras on Madagascar have very complex facies, correlated with the Karroo "System." The basal unit is a tillite known only in a small area in southernmost Madagascar. The overlying formations consist of typical continental sandstone and shale and several fossiliferous marine shale units with Permian, Triassic, and Jurassic faunas of Tethyan affinities. The continental formations contain the *Glossopteris* flora and fossil reptiles typical of the Karroo. The whole Karroo in Madagascar comprises 25,000 feet of strata. The interrelations of the multitude of facies are known only in the broadest reconnaissance fashion.

11-4 SOUTH AMERICA

Gondwana deposits in South America are found in Brazil, Paraguay, Uruguay, and Argentina, the most extensive being in Brazil. A generalized stratigraphic column for southern Brazil is shown in Figure 11-9. Here, as elsewhere, the sequence begins with a glacial deposit, which contains some marine interbeds (Fig. 11-10). In São Paulo five separate tillites are recognized. Most Brazilian coal deposits are interglacial. The glacial series is about 3,500 feet thick in São Paulo but is generally much thinner. Glacial beds are also present in southern Brazil and Argentina. Ammonites found in strata between the tillite beds closely resemble species from the middle Pennsylvanian of North America. If the Carboniferous age of the main glaciation here is therefore accepted and it is recognized that in Australia the main glaciation was no older than early Permian, it follows that the Late Paleozoic glaciation occurred at different times in different parts of the Southern Hemisphere. In the Sierra de la Ventana, 375 miles southwest of Buenos Aires, several thousand feet of glaciogene Permian strata contain *Glossopteris*, and one bed contains *Eurydesma*, the only occurrence of this marine Permian Gondwana fossil known in South America.

Overlying the glacial strata are bituminous limestones and dark shales that contain marine invertebrates and aquatic reptiles (*Mesosaurus*). The remaining formations of the Santa Catharina "System" are sandstone and

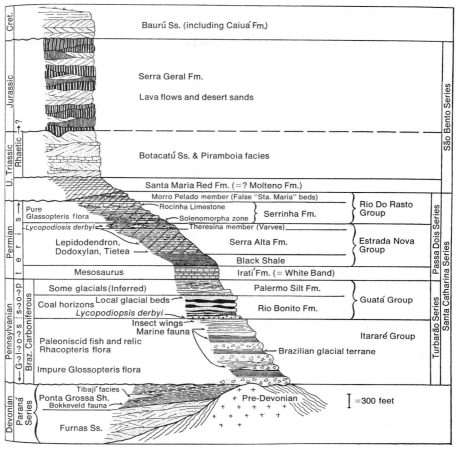

FIGURE 11-9

Schematic column illustrating Gondwana formations in southern Brazil. (From K. E. Caster, *Bull. Am. Museum, Nat. Hist.*, 1952.)

shale, mainly continental but with some marine interbeds containing Permian invertebrates. The uppermost formations are sandstone containing rich vertebrate faunas of Middle and Late Triassic age. A profound erosional unconformity separates the Triassic from the underlying *Glossopteris*-bearing formations. The last deposit was a vast effusion of lava flows, possibly the most voluminous in the earth's history. The main part spread over a maximum width of nearly 600 miles and a length of fully 1,200 miles in southern Brazil, eastern and southern Paraguay, northern Uruguay, and northeast-

ern Argentina. The thickness ranges from 300 to more than 3,000 feet. The age of these lavas has long been considered earliest Jurassic, the same as that of the South African lava flows in late Karroo time.

Recently radiometric dates have become available, pointing to a mid-Cretaceous age for the greatest basalt effusions, though they possibly began in latest Jurassic time. As the lavas contain intercalated sandstone resembling the underlying eolian Botucatu Sandstone, this formation may likewise be of early Cretaceous age rather than uppermost Triassic, as previously believed.

11-5 FALKLAND ISLANDS

The rock succession and facies of Gondwana and pre-Gondwana age on the Falkland Islands strikingly resemble those of South America and South Africa. The pre-Gondwana formations, which rest on Precambrian basement rocks, consist of nearly 10,000 feet of sandstone and shale containing Devonian and Early Carboniferous fossils. The lowest Devonian (?) formation is unfossiliferous, but the overlying formation contains a rich fauna of Early Devonian age, closely comparable to South African and South American faunas. The highest formations of this pre-Gondwana series are apparently terrestrial, for they contain remains of land plants.

The Gondwana strata on the Falkland Islands begin with a tillite, in some places underlain by sandstone and shale of marine or fluvio-glacial origin and conformable on the Lower Carboniferous. The tillite, similar to that of South Africa, probably is partly marine and elsewhere ground moraine. The boulders include not only older rocks of the Falkland Islands but many igneous and metamorphic rocks entirely foreign to the island. The Gondwana formations overlying the tillite include approximately 10,000 feet of terrestrial sandstone and shale. No vertebrate remains have been found, but a rich

FIGURE 11-10

Fluvio-glacial sandstone of the Carboniferous (Itararé group, Fig. 11-9), Vila Velha State Park, Paraná, Brazil. The people on the top of the formation give the scale. (Photograph by Lange da Morretes, courtesy of K. E. Caster.)

Glossopteris flora has been collected from several formations. Correlation of these formations with South African stratigraphy is based mainly on paleobotanical evidence.

11-6 ANTARCTICA

Throughout much of the Transantarctic Mountains thick clastic sedimentary rock formations, ranging in age from middle Paleozoic to early Mesozoic, are commonly referred to as the Beacon Sandstone. In the middle of the Beacon Sandstone is a conspicuous disconformity. The rocks immediately below contain Devonian fish remains; those above contain *Glossopteris* and coal beds. In several places in the Transantarctic Mountains the upper sequence includes thick tillite deposits considered to be of Permian age. Above the tillite the formations are generally sandstones containing varied *Glossopteris* floras. On the opposite side of Antarctica in the region of the Amery Ice Shelf there are sandstones with minor interbeds of coal, whose microflora indicate a Permian age. Thus the glossopterid flora was widespread over most of East Antarctica during Permian time. The glossopterid flora of Antarctica is much like the floras of South America, South Africa, and India, but does not contain any non-Gondwana elements, which the other floras do.

The first record of tetrapod life from Antarctica is a fragment of an amphibian jaw discovered in 1967. This fossil came from part of the Beacon Sandstone, which has been dated on the basis of plants as Triassic in age (near the Beardmore Glacier, Fig. 4-19). In 1969 remains of mammal-like reptiles were discovered.

The Gondwana episode in Antarctica ended with an extensive outpouring of basaltic lavas onto the Beacon Sandstone, which was subsequently intruded by dykes and sills. Radiometric dates on these igneous rocks yield a mid-Jurassic age.

11-7 FOSSIL FLORAS OF THE GONDWANA FORMATIONS

The earliest land plants had a world-wide distribution. The oldest were of Silurian age, and from that time through the Early Carboniferous the world flora was amazingly homogeneous. In the Late Carboniferous distinct floras can first be recognized — a distinction continuing to the beginning of the Triassic. Of the several floral provinces recognized, the most striking is the *Glossopteris* flora of mainly the southern hemisphere but also present in India, Siberia, and France. Some of its principal members are illustrated in Figure 11-11. The distribution of this flora in South America, the Falkland Islands, Antarctica, South Africa, Madagascar, India, and Australia furnished one of the arguments for continental connections between or juxtaposition of, the southern continents. The *Glossopteris* flora is mainly confined to the areas south of the Tethys. North of the Tethys three main floral provinces, the European, the Angaran, and the Cathaysian, are generally recognized (Fig. 11-12). The European flora is recognized in Permo-Carboniferous rocks of the eastern United States, throughout Europe to the Urals, and into Iran and Turkestan. The Angaran flora is found in the vast area of Asia east of the Urals and north of the Tethys except for the eastern portion of China. The Cathaysian flora is found in Korea, eastern China to Sumatra, New Guinea, and western North America as far as Colorado, Oklahoma, and Texas.

The striking individuality of the Gondwana flora can possibly be related to the cooling of the southern hemisphere during the Late Carboniferous and to the widespread glaciation. The cosmopolitan flora that had occupied the area became extinct with the change to a cold-temperate climate, to which the Gondwana flora was adapted. Northern floral elements mixed

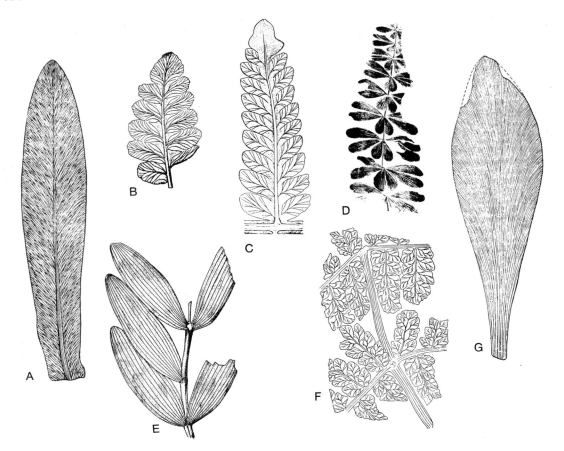

FIGURE 11-11

Representatives of the Gondwana flora.

A. *Glossopteris decipiens.*   E. *Schizoneura gondwanensis.*
B. *Merianopteris major.*   F. *Sphenopteris polymorpha.*
C. *Sphenopteris polymorpha.*   G. *Gangamopteris cyclopteroides.*
D. *Sphenophyllum speciosum.*

(From E. A. N. Arber, 1905, British Museum Catalogue of the Fossil Plants of
the *Glossopteris* Flora in the Department of Geology, British Museum, London.)

with the Gondwana flora are found in a few localities in South America, South Africa, and Madagascar.

The Gondwana flora, wherever found, is remarkably uniform but quite different from the northern floras. It first appears immediately above the tillites, and spores of *Glossopteris* have been recovered from some tillite beds. Proponents of land connections between, or juxtaposition of, the southern continents contend that the vast tracts of water would prevent a migration of land plants from one area to another. The spores of *Glossopteris*, however, are a winged type adapted to dispersal by the wind, and many other plants and animals, at different times and in different regions, have achieved wide dispersion over vast bodies of water. The restriction of *Glossopteris* in distribution seems, therefore, to have resulted chiefly from climatic barriers.

11-8 VERTEBRATE FAUNA OF GONDWANA FORMATIONS

The distribution and relations of the vertebrate fauna of Gondwana strata have strongly influenced the various interpretations of the geology of the southern hemisphere. The incompleteness of the geologic record, especially in continental formations, and the rarity of vertebrate fossils handicap the study. As has been noted before, more factors limit the distribution of land-living reptiles than limit that of aquatic reptiles or plants.

The most abundant and diversified reptilian fauna is in South Africa, where the Upper Dwyka shales contain the swimming reptile *Mesosaurus*. The Ecca Group bears a *Glossopteris* flora but has yielded no reptilian remains. The overlying Beaufort Group, however, is a veritable graveyard of ancient reptiles in which no less than

600 species have so far been discovered. Within the Beaufort, in fact, six distinct reptilian zones have been recognized, and they greatly aid in correlation. The overlying Stormberg Group also contains a rich reptilian fauna, sharply differentiated from that of the Beaufort.

Reptiles are apparently not common in the Gondwana strata of Madagascar. The middle part of the lower group (corresponding to Ecca) has some genera also present in Tanzania. The fauna of the upper part of the series has affinities with that of the Upper Gondwana of India.

We know much less of the land reptiles of South America. No land reptiles of Permian age are so far known there. The Irati formation (Permian), which immediately overlies the tillite formations, has yielded *Mesosaurus*. A rather large fauna of seventeen genera has been collected from the Upper Triassic Santa María Formation.

FIGURE 11-12

Distribution of late Paleozoic floras. The continents are shown in their present position and size. The three arrows point to the localities where mixed floras of southern and northern elements have been found. Mixed floras have also been reported from Patagonia, Mozambique, and Madagascar. (After T. K. Just, *Bull. Am. Museum, Nat. Hist.*, 1952.)

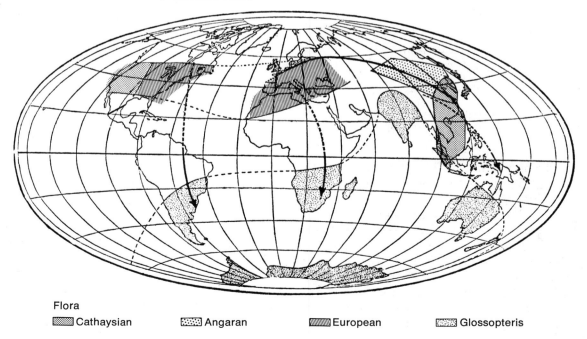

Flora

▨ Cathaysian   ▨ Angaran   ▨ European   ▨ Glossopteris

The Gondwana strata of the Falkland Islands have yielded no reptilian remains. In peninsular India fragmentary reptilian remains from the Lower Gondwana include some characteristic of South Africa and northwestern China. The Upper Gondwana reptiles are similar to Upper Triassic forms of the northern hemisphere, and one Lower Jurassic bed has been found to contain an abundance of dinosaur bones. This find is particularly important, as no other well-preserved terrestrial vertebrates are known from the Lower Jurassic, though excellent dinosaur faunas are known from both Upper Triassic and Upper Jurassic. This discovery should help fill in "missing links" in dinosaur evolution. Amphibians of Gondwana age have been known from Australia for some time, and recently some reptiles, as yet unstudied, have been discovered.

In the northern hemisphere Permian and Triassic land-living reptiles are known in North America, western and central Europe, European Russia, China, and Indo-China.

Now let us briefly examine the time-space relations of the Permian and Triassic land-living reptiles. The reptile groups living at this time comprise the stem reptiles (the cotylosaurs), the fin-backed forms (the pelycosaurs), the mammal-like reptiles (the thecodonts), the ancestors of the dinosaurs, and the rhynchocephalians. In the Late Carboniferous only the cotylosaurs and the pelycosaurs were extant, and they are known only from localities in North America and Europe. Both groups are found in North America but only the pelycosaurs in Europe. The same restricted distribution persists into the Early Permian, but at that time cotylosaurs were also present in Europe. Middle Permian fossil reptiles are known only in Russia and South Africa. The Russian fauna comprises pelycosaurs and cotylosaurs; these same orders are present in the South African fauna, plus various groups of mammal-like reptiles. Late Permian reptiles are known in Russia, Scotland, and South Africa, and the faunas of these widely separated regions are fairly homogeneous, being composed of the same orders of reptiles (Fig. 11-13, top). The orders include cotylosaurs, plant-eating and flesh-eating mammal-like reptiles, and an early group (eosuchians) that gave rise to the thecodonts. There is an appreciable increase in the fossil record of reptiles for the Early Triassic, both in the number of localities where they are known and in the number of orders they represent (Fig. 11-13, middle). Reptile faunas of this age are known in Europe, China, southeastern Asia, peninsular India, eastern Africa, South Africa, and Madagascar. There is an extraordinarily close relationship between certain specialized plant-eating mammal-like reptiles and certain species of thecodonts in the Lower Triassic deposits of South Africa, India, northwestern China, and Russia—that is, in areas on both sides of the Asian end of the Tethys. The distribution of reptile orders during the Middle and Late Triassic is shown in Figure 11-13, bottom. The impressive feature of this map is the homogeneity of the reptile faunas on all the continents.

There is no agreement among vertebrate paleontologists as to the meaning of the above data. To account for the distribution pattern, one group insists that the data strongly suggest that the continents were joined. Others do not feel that the data support the existence of a Gondwanaland; instead they favor an arrangement of the continents similar to that of today. These specialists are convinced by the distribution of the reptile faunas that there was ample opportunity for migration between the southern and northern hemispheres and that direct east-west connections in the southern hemisphere were therefore not necessary.

Probably the strongest evidence from vertebrate fossils that suggests a connection between South America and South Africa is *Mesosaurus* (Fig. 10-29, p. 370), a small reptile measuring about two feet in length, tail and all, is peculiar in structure and its phylogenetic position doubtful. It has no known close relatives, and probably evolved independently as a short side-branch from the basic reptilian stock of the Carboniferous. *Mesosaurus* was amphibious, had well developed limbs, and was perfectly able to walk on land, but it obviously spent most of its time in the water.

Although *Mesosaurus* was a water-dweller, no one considers it very possible that it could have crossed the South Atlantic. An additional strange fact is that it is known only in South America and South Africa, where it is found in marine shale immediately above the tillite, and has never been found elsewhere.

The amphibians do not help solve the

FIGURE 11-13

Distribution of reptiles in the upper Permian (top), Lower Triassic (middle), and Middle and Upper Triassic (bottom). (A) Stem reptiles, cotylosaurs. (B) Plant-eating mammal-like reptiles. (C) Carnivorous mammal-like reptiles. (D) Eosuchia (ancestral to the thecodonts). (E) Rhynchocephalians. (F) Thecodonts. (G) Dinosaurs. (After S. H. Haughton, 1953.)

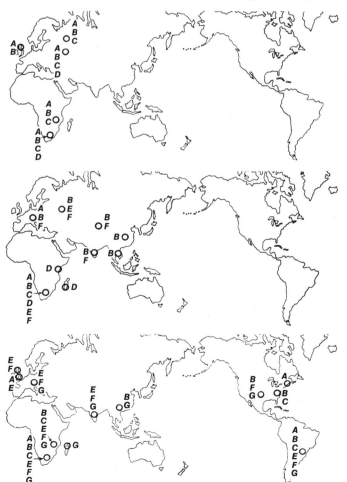

problem; they do not require intercontinental connections to explain their distribution. The Triassic amphibians of South America and South Africa, for instance, belong to groups widely distributed in the southwestern United States, India, Australia, and central Europe.

Data on the fresh-water Gondwana fish are very incomplete. The fresh-water fish genera of South America and South Africa are present in strata of the same age in North America and Europe. The lungfish of this age, for instance, has an almost world-wide distribution.

11-9 PALEOGEOGRAPHICAL CONCLUSIONS

The physical and biological data of the Gondwana strata have strongly stimulanted the imagination of many geologists. The hypothesis of Gondwanaland was early put forward to explain the history of the southern hemisphere. Over the years many different hypotheses have been proposed.

Gondwanaland

The early students of the Gondwana strata, especially members of the Geological Survey of India, were greatly impressed by the similarities of the faunas and floras in India and the various southern continents and by the similarities of their geologic histories. This homogeneity of biologic and stratigraphic development, they felt, could be explained only by a vast continental area, including much of South America east of the Andes, most of Africa, Madagascar peninsular India, and Australia. This great continent was given the name Gondwanaland. Figure 11-14 shows the area included in it. The intercontinental portions of Gondwanaland in the Indian Ocean were believed to have broken up and submerged during the Jurassic and those in the Atlantic during the Cretaceous. Gondwanaland was bordered on the north by the Tethys and was the southern counterpart to the Eurasian continental mass lying north of the Tethys.

The abundant geophysical evidence now available for the South Atlantic and Indian

FIGURE 11-14

Gondwanaland. The authors of this hypothetical continent regarded all the present southern continents as connected by broad continuous lands, including what is now the South Atlantic and Indian oceans.

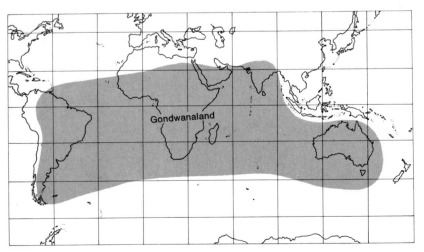

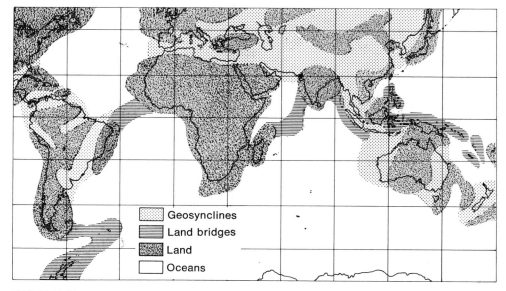

Geosynclines
Land bridges
Land
Oceans

FIGURE 11-15

Map of the southern hemisphere during Permian time, illustrating the land-bridge hypothesis as a solution of the Gondwana problem. (After C. Schuchert, *Geol. Soc. Am. Bull.*, 1932.)

oceans shows them to have typical oceanic crust. No evidence supports the suggestion that these areas were once high, mountainous, continental regions.

Gondwana Land Bridges

Some geologists objected to the idea of a vast Gondwana continent, mainly because of the difficulty of submerging continental connections of such great size, but they still held to the need of land connections between the various continental areas where the Gondwana System is developed. They solved this problem with narrower land connections, real land bridges. Proponents of this hypothesis felt that it was easier to sink small masses than much larger ones. Figure 11-15 illustrates the location and general configuration of the land bridges suggested for Permian time. The abundant new data on the oceans has pretty well destroyed this proposal.

Continental Drift

One of the most imaginative attempts to explain the biological and physical data of the Gondwana "System" suggests that the continents of the eastern and western hemispheres were joined early in the earth's history and have subsequently drifted apart. The supporters of this hypothesis point to the fact that much of the continental border of the Pacific basin consists of late Mesozoic and early Cenozoic folded belts. The westerly drift of South America and North America would readily account for these folded belts on only the western sides of the continents, and the supposed structural continuity of Paleozoic folded belts bordering each side of the Atlantic could be easily explained by assuming that when they were formed the present widely separated belts were part of a continuous geosynclinal and orogenic zone.

Continental drift was first suggested in the nineteenth century, again in 1910 by

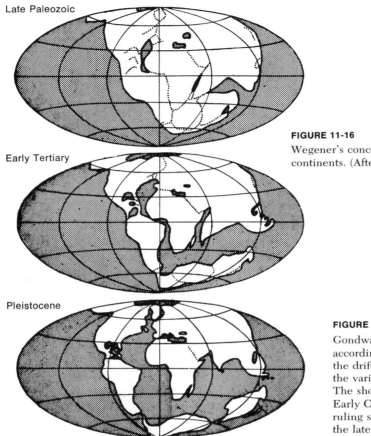

Late Paleozoic

Early Tertiary

Pleistocene

FIGURE 11-16

Wegener's conception of the drifting of
continents. (After A. Wegener, 1912.)

FIGURE 11-17

Gondwanaland during the Paleozoic Era
according to another strong advocate of
the drift hypothesis. The space between
the various portions was then mostly land.
The short lines indicate the Precambrian or
Early Cambrian structural trends. Diagonal
ruling shows the Samfrau geosyncline of
the late Paleozoic. Stippling marks regions
of late Cretaceous and Cenozoic compression.
(After Du Toit, 1937.)

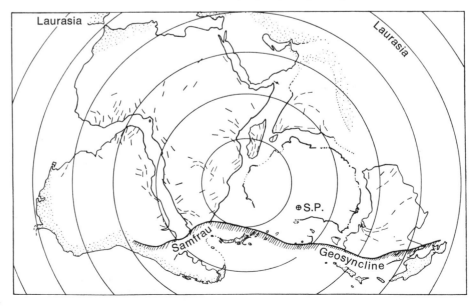

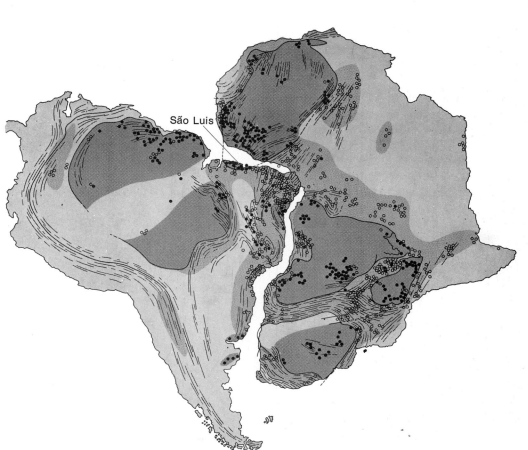

São Luis

FIGURE 11-18

Tentative matching of South America and Africa with sites of radiometric age dates; the solid dots represent ages of at least 2,000 m.y., the open dots ages of 450–650 m.y., but some represent ages as old as 1,100 m.y. The stippled areas are the ancient cratons we discussed in Chapter 4. The inference is that some portions of South America, for instance, the region of São Luis on the coast of Brazil, was adjacent to part of Africa. (After P. M. Hurley, "The Confirmation of Continental Drift." Copyright © 1968 by Scientific American, Inc. All rights reserved.)

the American geologist F. B. Taylor, and in a more elaborate form in 1911, 1915, and later years by Alfred Wegener, Professor of Meteorology and Geophysics at the University of Graz, who developed and presented in a very inclusive manner the first comprehensive picture of the earth's development within the framework of the hypothesis. Wegener's writings became the focal point of a long, heated debate over the meaning and interpretation of the Gondwana formations and their flora and fauna.

The essence of Wegener's ideas is illustrated in Figure 11-16. At the beginning of Gondwana time all the continents were parts of a single unit called Pangaea. The breaking-up of Pangaea did not begin until the early Cenozoic and did not cease until the Pleistocene. Wegener felt that the zonal distributions of climates of the past and of life could be explained only by reassembly of the lands as indicated in his map.

Figure 11-17 presents another interpretation of the relations of the continents under the drift hypothesis, bringing out an alignment of the geosynclines of eastern Australia, Antarctica, the Cape region of Africa, and the ranges south of Buenos Aires, to form a "Samfrau geosyncline." These geosynclinal belts, however, had quite different histories. The Tasman geosyncline had a complex Paleozoic history. The West Antarctica geosyncline was Mesozoic and Cenozoic, closely related to the Andean geosynclinal belt. The Cape Region of South Africa is a thick, narrow miogeosynclinal complex of Paleozoic age overlain by the Karroo formations, and was deformed in the late Permian or Early Triassic. The so-called fold belt south of Buenos Aires is hardly geosynclinal. It is, however, a region of fairly thick late Paleozoic glaciogene deposits, also containing *Eurydesma*. The joining of these belts into a single geosynclinal complex is highly dubious.

The apparent fit between South America and Africa has always been a strong argument that these two continental masses were once joined. Recent computer analysis of the middle of the continental slopes does indeed show a remarkable fit can be postulated. The average error is no greater than one degree over most of the boundary. Many attempts have been made to correlate geological features between the two continents, but the results have generally been inconclusive. The most recent approach has been from Precambrian geochronology. Africa is largely a continent of Precambrian rocks. Intensive radiometric studies of these Precambrian terranes has shown that (1) some yield dates of at least 2 billion years and (2) some dates of 450 million to as much as 1 billion years. The older terranes are considered ancient continental blocks, the younger troughs of sediments and volcanics whose major diastrophism were within the time span indicated. A plot of many radiometric age dates suggests that the patterns in Africa and South America are parts of a unit complex (Fig. 11-18),

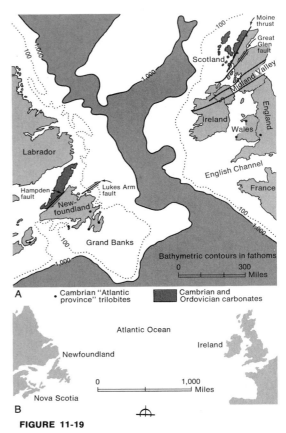

FIGURE 11-19

(A) Relationships of Newfoundland and Ireland following the hypothesis of continental drift. (B) Map of present North Atlantic Ocean. (After Marshall Kay and E. H. Colbert, *Stratigraphy and Life History*, Wiley, New York, 1965.)

and that parts of coastal South America are in fact "broken" off from Africa.

For the North Atlantic, similarities in the early Paleozoic histories of Newfoundland and Great Britain have suggested to some geologists that the two areas were parts of a continuous geosynclinal belt and once were much closer together (Fig. 11-19). Northwestern Newfoundland has Cambrian and Ordovician limestone formations overlying Precambrian gneiss. Central Newfoundland contains thick complex eugeosynclinal suites of rocks intruded by Paleozoic plutons. Similar suites of rocks are present over much of Great Britain. Finally, southeastern Newfoundland has Cambrian shale of platform facies, whose unique trilobite faunas differ radically from those of the rest of Newfoundland. These Atlantic Province trilobite faunas are found in Wales! Thus, there are apparent similarities in the threefold division of early Paleozoic formations and faunas between Newfoundland and Great Britain. There are, however, differences as well as similarities in the geology of the two regions; not all students are willing to regard the evidence as conclusive.

The widespread continental glacial deposits on all the continents with Gondwana-type deposits is another focal point in support of continental drift. The movements of the ice sheets as shown by glacial striae raise several problems. The movements in South Australia are northward, suggesting a land area in the seas south of Australia;

in southeastern Africa, near the coast, striae indicate a source in the Indian Ocean. The dominant direction of glacial movement in Argentina and Brazil is from east to west. One authority believes that 70 percent of the clasts in the tillites of South America have no possible known source on that continent. To account for the widespread distribution of these glacial deposits and for the direction of movement of the ice, many authors have postulated a grouping of the continental areas concerned around the South Pole during the Late Carboniferous and early Permian (see Figure 11-20). Reassembly of the southern continents as shown in this figure does offer a plausible explanation for the glacial deposits, but is this arrangement compatible with all the other geological data for that time and the immediately following periods? Another factor is that we really do not know what causes widespread continental glaciation,

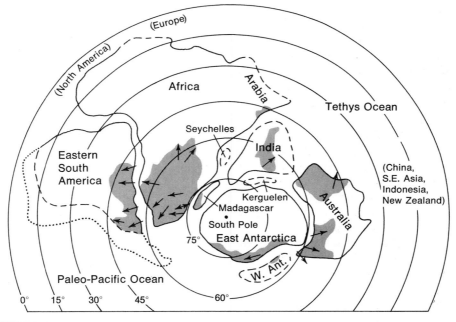

FIGURE 11-20

Reconstruction of Gondwanaland at beginning of Permian time, showing paleolatitudes. Stippling regions of known tillite, and arrows show directions of ice flow determined from glacial pavements. (From W. Hamilton and D. Krinsley, *Geol. Soc. Am. Bull.*, 1967.)

as for instance in the Pleistocene. Many explanations have been put forth, but none as yet has been widely accepted.

In the past two decades the study of rock magnetism has been developed. Most rocks owe their magnetism to small amounts of various iron-bearing minerals. Igneous rocks, on cooling, acquire a direction of magnetism that may be experimentally determined. Such a rock therefore reveals the direction of the magnetic field that existed when the rock solidified. The direction of magnetization of sedimentary rock, produced by the orientation of small grains of magnetic material, can also be measured, but it is susceptible to change produced by chemical and physical alteration of the rock after deposition. It is difficult to determine the extent and significance of the changes that may have occurred since the rock was originally deposited.

The magnetic field of the earth today is assumed to act as though a dipole (a linear magnet) at the center of the earth were roughly parallel to the axis of rotation of the earth. If this assumption is true, it should be possible to determine the positions of the magnetic poles at past times by measuring the magnetism in ancient rocks. Several thousand such measurements have been made, and they yield startling results — results that are still to be explained.

The earth's magnetic field, from Oligocene to Holocene time, has very closely approximated that of today. Before the Oligocene there appears to have been considerable wandering of the magnetic poles, if the assumption is correct that the field averages to a dipole coinciding with the earth's axis. The paths of this wandering for Europe, North America, Australia, India, and Japan are shown in Figure 11-21. The positions of the poles should be the same from these different continents, but it can readily be seen that they are not and that very large relative drifts are required to bring the paths into coincidence. Figure 11-22 is a plot of the magnetic poles during the Permian and Carboniferous from Eu-

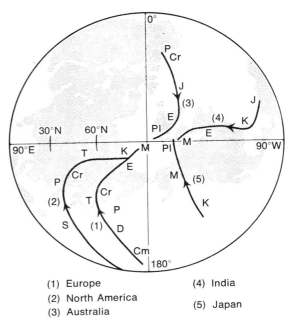

(1) Europe (4) India
(2) North America (5) Japan
(3) Australia

FIGURE 11-21

Postulated paleomagnetic polar-wandering curves for Europe, North America, Australia, India, and Japan: Cm, Cambrian; S, Silurian; D, Devonian; Cr, Carboniferous; P, Permian; T, Triassic; J, Jurassic; K, Cretaceous; E, Eocene; M, Miocene; Pl, Pliocene. (After Cox and Doell, 1960.)

FIGURE 11-22

Permian and Carboniferous magnetic poles from Europe, North America, and Australia: North America, squares; Europe, circles; Australia, triangles. (After Cox and Doell, 1960.)

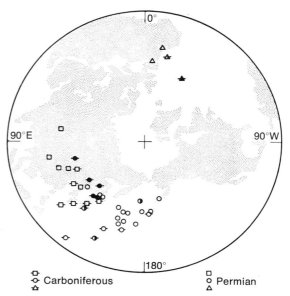

Carboniferous   Permian

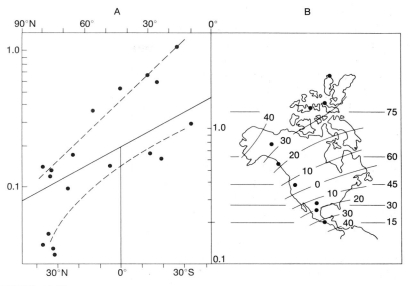

FIGURE 11-23

Diversity data for nine Permian faunas of North America. (A) Upper graph diversity plotted against present latitudes, lower graph diversity plotted against best assessment of proposed Permian latitudes from paleomagnetic data. (B) Location of nine Permian faunas used for Part A, and present and Permian latitudes. (Data from F. G. Stehli.)

rope, North America, and Australia. Notice how the North American poles are displaced far to the west of the European poles, whereas they should be coincident if the axial dipole theory is correct.

The whole thesis in using paleomagnetic data to arrive at the conclusions that the continents have wandered rests on the assumption that the earth's magnetic field is the product of a simple dipole that is parallel with the earth's rotational axis. Furthermore, it is assumed that this condition has always prevailed. The earth's field, as viewed today, is dipolar but distinctly non-axial. The north magnetic pole lies about 70°N, 100°W, and the south magnetic pole lies at about 68°S, 143°E. Is it possible that the above assumption may not be valid? One test of that assumption is currently being investigated using paleontologic data of marine invertebrate faunas. Modern marine invertebrate faunas display clear-cut diversity gradients, with largest numbers of species in the equatorial belt and decreasing numbers in the higher latitudes. These

modern gradients are controlled by temperature. Particular attention has been paid to the Permian fossil record to see if diversity gradients exist and if so, what pattern they display. Good data on Permian faunas are so far available only from the northern hemisphere, but analysis of faunas from North America is very instructive. A summary of faunal data based primarily on brachiopods is shown in Figure 11-23, A. The ordinate of this graph, labeled Permian ratio, is a proportion developed to express faunal diversity. The details of arriving at these figures need not bother us here, but the higher the number, the greater the diversity. By analogy to distribution patterns of modern marine organisms, the low Permian ratios are believed to be cold-water faunas. In the upper graph of Figure 11-23, A the faunal data for nine stations are plotted according to their present latitudes, which are shown in Figure 11-23, B. The graph shows a distinct rise in faunal diversity from north to south. Unfortunately there are no data for the southern hemisphere. In the

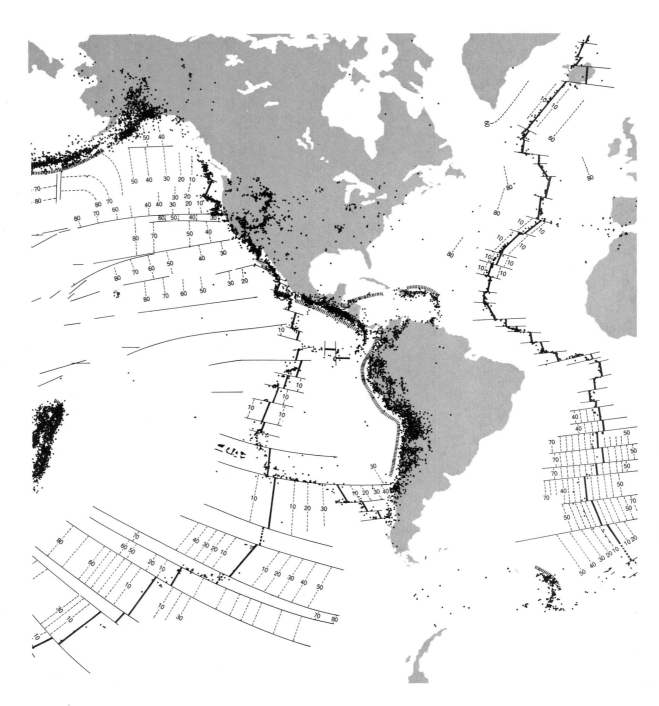

FIGURE 11-24

Summary of major features of oceanic geology. New ocean bottom is continuously being extruded along the crest of a worldwide system of ridges (*thick black lines*). The present position of material extruded at intervals of 10 million years, as determined by magnetic studies, appears as broken lines parallel to the ridge system, which is offset by fracture

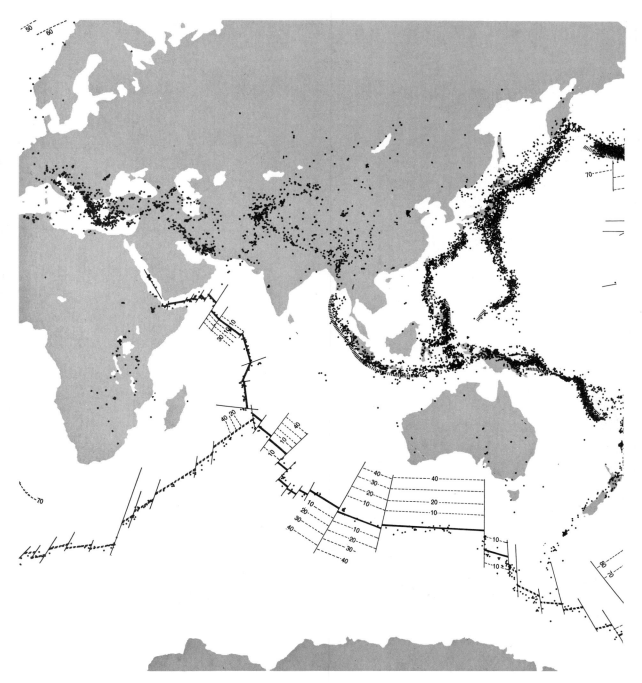

zones (*thin black lines*). Earthquakes (*black dots*) occur along the crests of ridges, on parts of the fracture zone, and along deep trenches. These trenches, where the ocean floor dips steeply, are represented by hatched bands. (From Sir Edward Bullard, "The Origin of Oceans," Copyright © 1969 by Scientific American, Inc. All rights reserved.)

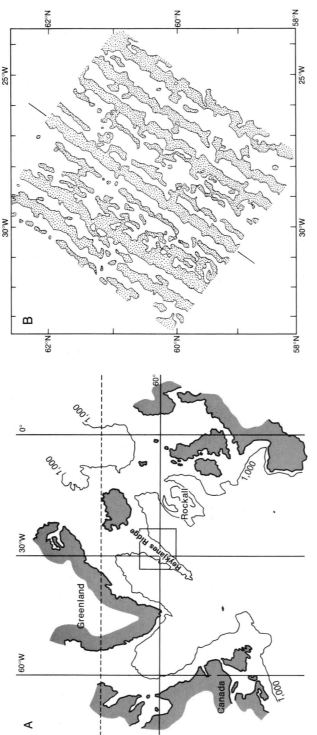

FIGURE 11-25

(A) The location of Reykjanes Ridge, southwest of Iceland. The 1,000-fathom submarine contour is shown together with the 500-fathom contour for Rockalt Bank. (B) Summary diagram of the magnetic anomalies observed over Reykjanes Ridge. The straight lines indicate the axis of the ridge and the central positive anomaly. (From F. J. Vine, *Science*, **154**, Dec. 15, 1966. Copyright © 1966 by the American Association for the Advancement of Science.)

lower graph of Figure 11-23, A these same stations are located on the Permian latitudes as determined from paleomagnetic data. The ratios obviously continue to rise across the supposed Permian equator. If one accepts the modern faunal gradients as being applicable to Permian faunas, the above example clearly shows that the faunal data agree more with present latitudes than with those suggested for the Permian. Much more data of this sort are needed, of course, but this example does raise the valid question whether the basic assumption behind the paleomagnetic approach is correct. Perhaps it is in the earth's magnetic field that has wandered.

The most recent, and perhaps the most significant, contribution to the problem of continental and oceanic stability was the discovery in the late 1960's that the ocean floors are apparently spreading! This discovery came with the recognition of the apparent relationship between the topography of the ocean floor, the localization of seismicity of the earth, and the periodic reversal of the earth's magnetic field.

The deep ocean basins have great mid-oceanic ridges. The loci of the topographic crests of these ridges are indicated in Figures 11-24. Note that the northwest Pacific is devoid of such a ridge or rise, there is, however, evidence that such a ridge existed in the past but has subsided. These mid-oceanic ridges are disrupted by prominent linear fracture zones, typically expressed by narrow ridges and troughs transverse to the mid-oceanic ridges and offsetting the ridge crests. Finally, in the circum-Pacific region we find marginal trenches and island arcs.

Earthquake activity is most intense in the former Tethyan geosynclinal region, in the circum-Pacific region, and along the mid-oceanic ridges. Recent advances in seismology have made possible much more accurate location of earthquake epicenters. New plots define these belts of seismicity in remarkable detail. The current seismicity of the earth is largely confined to very restricted linear zones (Fig. 11-24). Consideration of the depth to which these earthquakes extend reveals that the activity associated with ridge crests is essentially confined to very shallow depths, probably not exceeding 10 or 20 kilometers. In the young fold mountains and trench systems, however, shallow-focus earthquakes also occur, but in addition, intermediate and deep-focus earthquakes, occurring to a maximum depth of 700 kilometers are recorded (Fig. 11-24). The hypothesis of sea-floor spreading proposes that these two seismic provinces—those in which only shallow earthquakes occur and those that are characterized by deep-focus earthquakes—reflect quite different but complementary processes at work in the upper mantle and accommodated in the earth's crust. The mid-oceanic ridges are extensional features along which new oceanic crust is created, and the trench systems are regions in which oceanic crust is partially resorbed. The oceanic crust is thus considered to be a surface expression of the mantle, derived from it by partial fusion and chemical modification beneath ridge crests and in part returned to the mantle beneath the trench systems.

Further evidence for spreading of the sea floor lies in magnetic data obtained from the rocks at and near the ridge crests. The towing of a magnetometer by a ship or plane reveals that there are linear belts of alternating normal and reversely magnetized materials symmetrically paired on each side of the ridge axis. The data on one such study on Reykjanes Ridge southwest of Iceland is illustrated in Figure 11-25. Of utmost importance is that these magnetic episodes can be dated. We have fairly detailed studies of young lava flows from many parts of the world that have yielded a reversal time scale for the past 3.5 million years. Additional data from deep-sea sediment cores have extended the reversal time scale back to about 10 million years. There

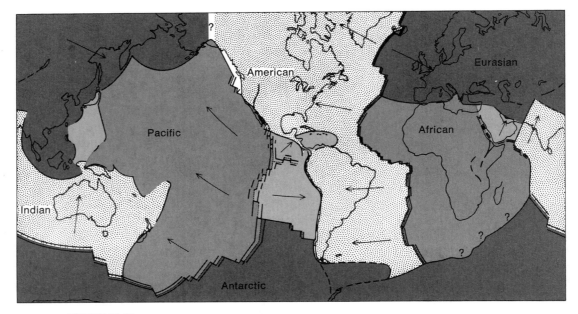

FIGURE 11-26

The present pattern of continental drift is inferred to be taking place by six major plates. In this model the African plate is assumed to be stationary. Arrows show the direction of motion of the five other large plates which are generally bounded by ridges or trenches. (From Sir Edward Bullard, "The Origin of Oceans." Copyright © 1969 by Scientific American, Inc. All rights reserved.)

are now 171 reversals known of the earth's magnetic field. Extrapolating back from the data we have on the more recent reversals, it is possible to assemble a reversal time scale going back 77 million years. Although geological data support this extrapolation, it should be noted that the extrapolation is based on the assumption of a constant rate of spreading. The rates of spreading determined from this reversal time scale range from 1 cm per year per flank near Iceland to 6 cm per year per flank in the equatorial Pacific.

Figure 11-24 summarizes the available data obtained so far on spreading along the ocean ridge systems; the lines drawn parallel to the ridge crests are ten-million-year "growth lines." The present pattern of continental drift is inferred to be taking place in six major plates (Fig. 11-26). It is likewise thought unlikely that drifting began only during the past 100 million years but rather

was an integral part of the earth's history throughout geologic time.

As mentioned earlier in this chapter the subject of continental drift has captured the imagination and attention of a great many earth scientists. New data are appearing monthly in hundreds of different scientific journals. The enthusiastic proponents of continental drift accept it as a fact, and with only details yet to be worked out. There is no question that an impressive list of data supports this interpretation, but there are also a disturbing number of inconsistencies which indicate that caution is warranted. We are a long way from fully understanding and integrating the geological record of the continents and ocean basins. It seems best to recognize that some form of continental drift may indeed be part of the earth's history, but it can only be considered a hypothesis worthy of serious attention, not an established fact.

SUGGESTED READINGS

Bigarella, J. J., R. D. Becker, I. D. Pinto, eds., *Problems in Brazilian Gondwana Geology* (1st Intern. Symp. on the Gondwana Stratigraphy and Paleontology, Argentina, 1967).

Blackett, P. M. S., Sir Edward C. Bullard, S. K. Runcorn, eds., "A Symposium on Continental Drift," *Phil. Trans. Roy. Soc. London*, **A258** (1965).

Cox, Allan, and R. R. Doell, "Review of Paleomagnetism," *Geol. Soc. Am. Bull.*, **71**, No. 6 (1960).

Darlington, Jr., P. J., *Biogeography of the Southern End of the World* (Harvard Univ. Press, Cambridge, 1965).

Du Toit, A. L., *A Geological Comparison of South America with South Africa* (Carnegie Inst. Washington, Publ. 381, 1927).

————, *Our Wandering Continents: An Hypothesis of Continental Drifting* (Oliver & Byrd, Edinburgh, 1937).

Kay, Marshall, ed., *North Atlantic—Geology and Continental Drift—a Symposium* (Am. Assoc. Petrol. Geol., Mem. 12, 1969).

Irving, E. M., *Paleomagnetism and its Application to Geological and Geophysical Problems* (Wiley, New York, 1964).

Mayr, Ernst, ed., "The Problem of Land Connections Across the South Atlantic, with Special Reference to the Mesozoic," *Bull. Am. Museum Nat. Hist.*, **99**, Art. 3 (1952).

Meyerhoff, A. A., "Continental Drift: Implications of Paleomagnetic Studies, Meteorology, Physical Oceanography and Climatology," *J. Geol.*, **78** (1), (1970).

Munyan, A. C., ed., *Polar Wandering and Continental Drift* (Soc. Econ. Paleont.-Mineral., Special Publ. 10, 1964).

Nairn, A. E. M., ed., *Descriptive Palaeoclimatology* (Wiley, New York, 1961).

Piel, Gerard, ed., "Gondwanaland Revisited: New Evidence for Continental Drift," *Proc. Am. Phil. Soc.*, **112** (1968).

Runcorn, S. K., ed., *Continental Drift* (Academic Press, New York, 1962).

Takeuchi, E., S. Uyeda, and H. Kanamori, *Debate About the Earth*, rev. ed. (Freeman, Cooper, San Francisco, 1970).

Wegener, Alfred, *The Origin of Continents and Oceans*, trans. from 4th German ed. by J. Biram (Methuen, London, 1966).

Wilson, J. Tuzo, ed., *Symposium on Continental Drift in the South Atlantic Region* (UNESCO, Paris, in press).

Scientific American **Offprints**

863 George A. Doumani and William E. Long, *The Ancient Life of the Antarctic* (September, 1962).

868 J. Tuzo Wilson, *Continental Drift* (April, 1963).

874 Patrick M. Hurley, *The Confirmation of Continental Drift* (April, 1968).

875 J. R. Heirtzler, *Sea-Floor Spreading* (December, 1968).

877 Björn Kurten, Continental Drift and Evolution (March, 1969).

880 Sir Edward C. Bullard, *The Origin of the Oceans* (September, 1969).

883 H. W. Menard, *The Deep-Ocean Floor* (September, 1969).

12

THE CENOZOIC ERA

I. North America

The modern world as we know it—its great mountain systems, its plains, its coast lines— unfolded during the Cenozoic Era, during the past 70 million years. As later features, the formations of this era are more widespread and less disturbed than those of any earlier era in the earth's long history (Fig. 12-1). The geological and paleontological data are therefore vastly more abundant than for any other era. Moreover, the Cenozoic formations can be more closely related to past and present geographic features of the continents than can those of the earlier eras. The similarities of Tertiary to Recent faunas and floras also allows much greater precision in paleoecologic and paleogeographic interpretation. Partly for these reasons there are a great many students of the Cenozoic, but probably the most important reason is that more than 50 percent of the world's pro-

duction of petroleum comes from Cenozoic strata.

Obviously, if the various segments of the geologic column were to be treated in proportion to the data available, the Cenozoic chapters would have to be many times as long as those for any other era. For the sake of balance, however, we give the Cenozoic only about as much space as was allotted to the other eras.

With the retreat of the Cretaceous seas from the continent the last great marine invasion of North America ended. Cenozoic seas were restricted to mere marginal invasions of the present coasts. Diastrophism was extremely active throughout this era, but orogeny—resulting in folds and thrust faults—was not as extensive as epeirogeny —resulting in upwarps and downwarps.

The Laramide orogeny, which formed the

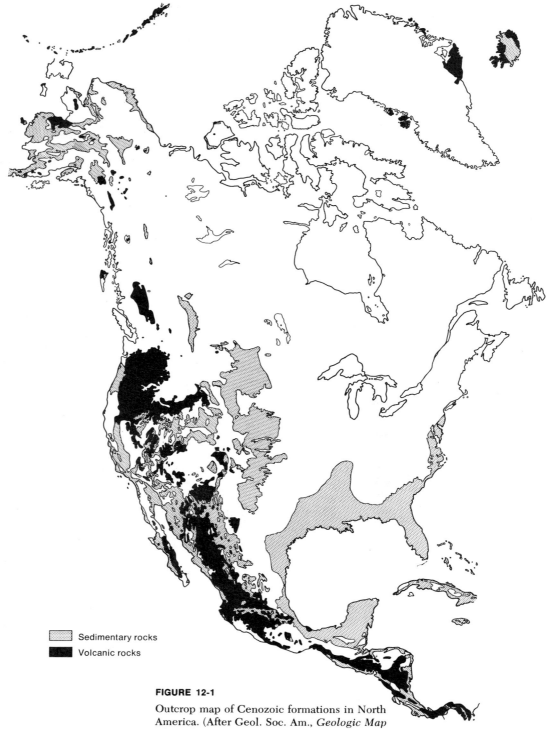

Sedimentary rocks

Volcanic rocks

FIGURE 12-1

Outcrop map of Cenozoic formations in North America. (After Geol. Soc. Am., *Geologic Map of North America*, 1946.)

Rocky Mountain region, began in the Late Cretaceous and continued into the Eocene. The only other active area of orogenic deformation during this era was the Pacific coastal belt from California to Alaska, where folding, which is still going on, was strong in the middle Pleistocene. During the Cenozoic, most of North America underwent one or more epeirogenic movements, which, along with erosion, were principally responsible for the present shape and elevation of the mountains and the courses of the major streams. The Atlantic and Gulf coastal plains were subsiding following the pattern established in the Cretaceous. The Pacific coastal belt was the site of deposition in marginal troughs and basins on a pattern like that established in the Cretaceous.

The next to the last phase of Cenozoic history was the Pleistocene ice age, discussed in Chapter 15. To treat the Tertiary, we discuss the physiographic provinces of North America in sequence.

12-1 THE APPALACHIAN MOUNTAINS

The earlier geosynclinal and orogenic history of the Appalachian region have been discussed in § 5-6. The present topography, elevation, and scenic features of the Appalachians are due to erosion and further upwarping in the Cenozoic. A fairly satisfactory sequence of events for this phase of Appalachian history has been worked out, but the timing of events are not well enough known to relate specific geomorphic events in the Appalachians with dated Cenozoic deposits of the Atlantic coastal plain.

Figure 12-2 illustrates in a very simplified manner the classic interpretation of the nature and sequence of events in the Cenozoic history of the Appalachians. Rough topography characterized most of the Appalachian region at the close of the Triassic through the cumulative effect of the Alleghanian and Palisade disturbances. By the close of the

Mesozoic, however, erosion had reduced this area to the low, undulating plain that we call the Schooley peneplain (Fig. 12-2). Broad arching of a few hundred feet, caused by another upwarp of the region, reactivated the streams, and flat, plains-like areas were carved in the belts of nonresistant rocks, producing a second erosional surface, which is called the Harrisburg erosion surface. This sequence of events was repeated once more: uplift, erosion by streams, and production of another erosional surface, called the Somerville erosion surface.

The present surface of the Schooley peneplain is, in places, at an altitude of about 4,000 feet. Only remnants of this surface are still preserved; if it were complete, it would form a broad arch a few hundred miles wide. In Figure 12-2 the main features of the Cenozoic history are clearly illustrated. The main feature of the Cenozoic history of the Appalachians consists of erosional sculpturing of a gently uplifted area of already highly folded and consolidated geosynclinal rocks.

12-2 THE ATLANTIC COASTAL PLAIN

On the eastern seaboard of the United States the pattern established during the Cretaceous — transgression and intermittent regression of the seas — continued through the Cenozoic. The deposition of great wedges of sediments on the coastal plain and continental shelf was going on at the same time that uplift and erosion were sculpturing the Appalachian Mountains. There is, however, no agreement on correlation of specific phases of uplift and erosion in the mountains with specific depositional phases on the coastal plain. The profiles of Figure 8-14 (p. 299) show three cross sections of the coastal plain and continental shelf. In the cross section constructed from well-drilling records (Fig. 8-16, C) the wedge-like profile of Cretaceous and Cenozoic sediments is clearly demonstrated. The deep well at Cape Hatteras

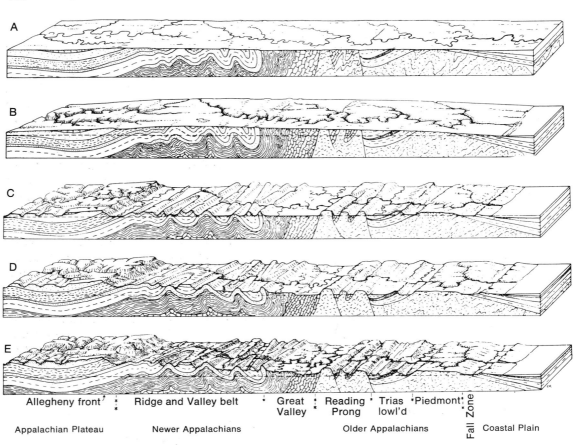

Allegheny front | Ridge and Valley belt | Great Valley | Reading Prong | Trias lowl'd | Piedmont | Fall Zone

Appalachian Plateau   Newer Appalachians   Older Appalachians   Coastal Plain

FIGURE 12-2

Physiographic evolution of the Appalachian Mountains. (A) Schooley peneplain, truncating older folded rocks of the Appalachians and Cretaceous formations of the coastal plain. (B) Arching of the Schooley peneplain. (C) Dissection of the Schooley peneplain and development of the Harrisburg erosion surface. (D) Uplift and dissection of the Harrisburg erosion surface and development of the Somerville erosion surface on the belts of weakest rocks. (E) Uplift and dissection of the Somerville erosion surface, producing the present conditions. (From D. W. Johnson, *Stream Sculpture on the Atlantic Slope*, Columbia Univ. Press, New York, 1931.)

penetrated 10,000 feet of sedimentary rocks, more than half of which was Cretaceous. In the other profiles, constructed from seismic data (Fig. 8-16, A, B), the boundary between the semi-consolidated deposits and the unconsolidated deposits is taken to be a horizon in the lower part of the Upper Cretaceous. The Cenozoic formations are unconsolidated. The most interesting feature of the northernmost profile (Fig. 8-16, A) is the fact that the basement rises toward the edge of the continental shelf.

Pleistocene and Recent sediments form the surface cover of most of the continental shelf and slope. To date, boring, drilling, and dredging have shown that the pre-Pleistocene geology is highly varied. A highly simplified summary map is shown in Figure 12-3. Most of the continental shelf has Pliocene or Miocene strata immediately beneath the Pleistocene and Recent sediments, but in the Gulf of Maine this level is of Eocene or greater age. Rock samples recovered from the continental slope and from submarine

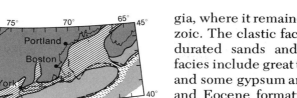

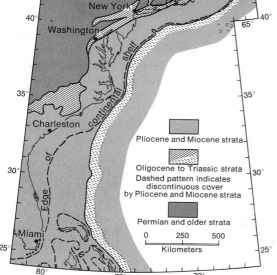

FIGURE 12-3

Map of continental shelf and slope
showing distribution of pre-Pleistocene strata.
(From K. O. Emery, 1966.)

canyons cut into it include strata of Pliocene,
Oligocene, Eocene, Paleocene, and Cretaceous ages. The Pliocene rocks are mainly
from near the top of the continental slope,
and the older rocks are from greater depths.

12-3 FLORIDA AND THE BAHAMA BANKS

The platform carbonate facies that covered
Florida during the Cretaceous continued
throughout the Cenozoic. In § 8-3 we pointed
out that the cross section of Cretaceous and
Cenozoic formations from Georgia to southern Florida (Fig. 8-17, p. 282) shows a twofold facies, with clastic sediments to the
north and carbonate facies to the south. In
the Early Cretaceous this facies boundary
was in northcentral Florida, but by Late
Cretaceous time it lay across southern Geor-

gia, where it remained throughout the Cenozoic. The clastic facies are soft, partially in-
durated sands and shales; the carbonate
facies include great thicknesses of limestone
and some gypsum and anhydrite. Paleocene
and Eocene formations covered the whole
state; but, beginning in Oligocene time, portions of the northern part of the peninsula
were emergent for varying times, forming
islands off the mainland. One of these islands supported a rich mammalian fauna
during the Miocene, the only known land
fauna of this age east of the Mississippi
River. Figure 12-4 illustrates the thickness
pattern and distribution of these Cenozoic
deposits. All of the carbonate facies indicate shallow-water environments; in effect,
Florida was a long, narrow platform with a
rich marine fauna much like that of the
Bahama Islands today. The volume of Cenozoic sediments in southern Georgia and
Florida has been calculated at about 50,000
cubic miles.

FIGURE 12-4

Isopach map of Cenozoic sediments in Florida
and adjacent states. (After L. D. Toulmin,
Geol. Soc. Am. Bull., 1952.)

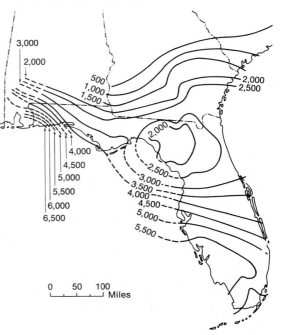

The Bahama Islands, separated from Florida by a channel about 50 miles wide, are among the largest sites of modern carbonate deposition. The Bahamas consist of many islands, cays, and rocks, on a platform covered by water only 15–25 feet deep. Great submarine valleys cut into the platform (Fig. 12-5). The eastern margin drops off abruptly to the Atlantic Ocean basin, and the southern margin is separated by a deep channel from Cuba and Hispaniola. The surface rocks of the Bahamas are Pleistocene or Holocene reef limestones, oolitic dune deposits, calcareous ooze, and shelly limestones.

A deep test well drilled on Andros Island to a depth of 14,585 feet encountered the Cenozoic-Upper Cretaceous contact at a depth of 8,760 feet. The well bottomed in Lower Cretaceous strata. The whole sequence that was penetrated consisted of limestone, dolomite, and anhydrite and was remarkably similar to the Florida section.

FIGURE 12-5

Great Bahama Banks and the surrounding area, showing the general form of the channels and platforms and the location of the principal deposits of lime ooze and oolitic sand. (After N. D. Newell, *Geol. Soc. Am.,* Special Paper 62, 1955.)

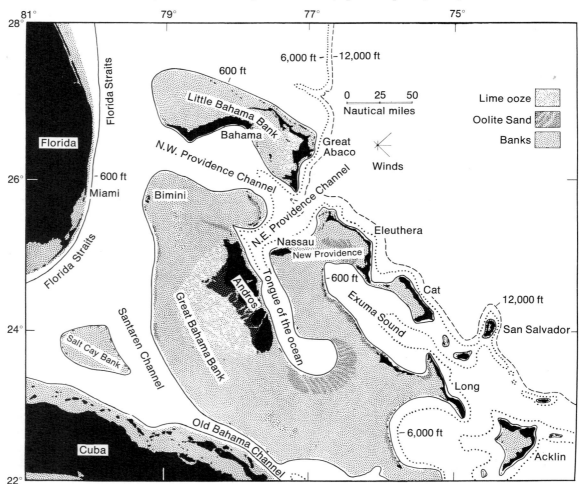

Clearly, the platform character of this region, with its abundant organic reefs and with physical conditions ideal for precipitation of calcium carbonate, has persisted throughout Cretaceous and Cenozoic time to the Holocene.

The present topography, along with the character and distribution of the carbonate sediments, offers abundant evidence as to the history of the Bahama Banks. The platforms are surrounded by elevated rims of rocky shoals, reefs, sand ridges, cays, and islands composed mainly of pure oolitic limestone of Pleistocene age and calcareous sand and mud of Holocene age. The production of both organic and inorganic calcium carbonate is greater in these peripheral areas of the Banks than over the central platform. The living coral reefs are merely thin accretions over submerged shelves and terraces. Since the volume and extent of the living reefs are only fractions of what the existing favorable conditions would allow, the present reefs are apparently young. The sediments now forming in the peripheral areas are largely sand- and silt-size particles of plants and animals, derived from erosion of the coral reefs and especially from the abundant lime-secreting organisms that populate the protected waters behind the reefs. Over the platform, nonskeletal sediments—chemically precipitated ooze, oolitic sand, aggregate grains, and fecal pellets—are abundant. On much of the shelf west of Andros Island the deposits consist of carbonate ooze, mainly aragonite precipitated as microscopic acicular crystals. Cross sections of two typical banks are illustrated in Figure 12-6. Note the steep marginal escarpment that clearly defines the outer limits of the platform and the distribution of the various carbonate facies. The Bahama Banks are regarded as Cenozoic coral atolls that have gradually spread laterally until they have joined the continental shelf. The steep marginal escarpments and the deep channels are thought to be constructional features formed by coral

reefs, as in typical modern atolls. The falling temperature and the lowering of the sea level during the Pleistocene locally exterminated the coral reefs, which are just now getting reestablished. Also during the Pleistocene the area was mantled by a thin veneer of oolite, which still masks the fundamental character of the Banks.

12-4   THE GULF COASTAL PLAIN

Few places in the world have been so intensively explored for petroleum as the Gulf coastal plain. The tremendously thick Cenozoic deposits of this region have been more thoroughly studied than any other geologic province in North America. Most of the studies are of the subsurface. The pattern of outcrops of the Cenozoic systems, illustrated in Figure 12-7, clearly shows that the maximum inundation of this region came during the Eocene, when the seaways briefly reached southern Illinois. The remainder of the Cenozoic has seen a gradual recession Gulfward. Most surface exposures of the Cenozoic rocks in the Gulf coastal region are continental or brackish-water clastic facies of sand and clay. Each of these units grades Gulfward into marine shale and sandstone, which thicken greatly; these in turn grade farther southward into deeper-water shales and thicken still more.

The earliest Cenozoic formations on the outcrop are separated from the underlying Cretaceous by an unconformity but are similar clays and marls. In the overlying Eocene we find the first coarse clastic continental facies, which reflect the rising of the Rocky Mountains and the Appalachians. The whole sequence of Cenozoic deposits indicates eight major transgressions and regressions, which closely controlled the various facies deposited. At the south, marine facies are continuous (Fig. 12-8) and consist largely of shale. Each major unit thickens toward the Gulf at two distinct rates: on land at a fairly constant rate in feet per mile, in the

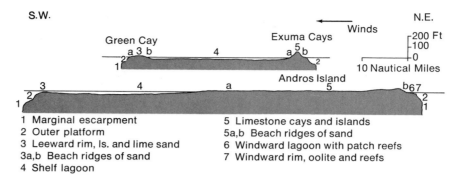

1 Marginal escarpment
2 Outer platform
3 Leeward rim, ls. and lime sand
3a,b Beach ridges of sand
4 Shelf lagoon
5 Limestone cays and islands
5a,b Beach ridges of sand
6 Windward lagoon with patch reefs
7 Windward rim, oolite and reefs

FIGURE 12-6

Profiles across typical Bahamian platforms, showing the barrier rim
and the shelf lagoon. (From N. D. Newell, *Geol. Soc. Am.*, Special Paper 62, 1955.)

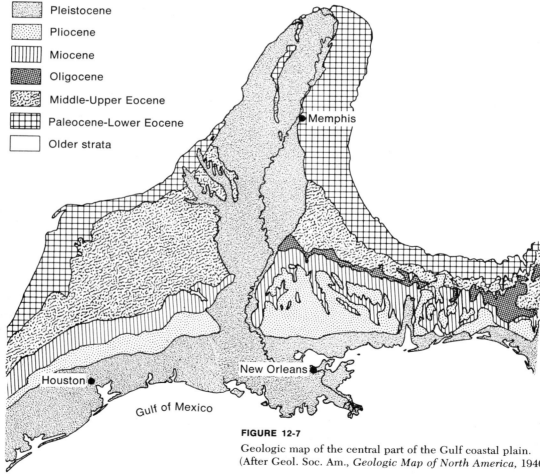

FIGURE 12-7

Geologic map of the central part of the Gulf coastal plain.
(After Geol. Soc. Am., *Geologic Map of North America*, 1946.)

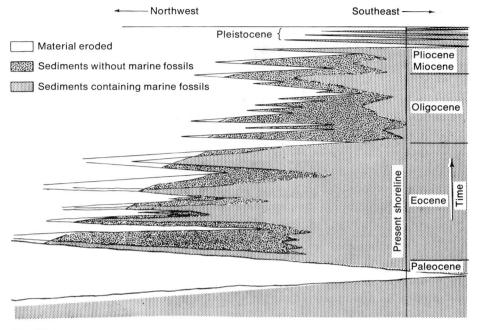

FIGURE 12-8

Sedimentation-time diagram of Cenozoic formations in the Gulf coastal plain. Beyond the position of the present shoreline, deposition went on continuously, but inland from this position the seas advanced several times, bringing with them tongues of sediments. Since retreats of the sea exposed these deposits to erosion, the tip of each wedge or tongue was generally removed by erosion. During the Pleistocene, and possibly during the late Pliocene, sedimentation was sometimes interrupted beyond the present shore line. (After L. W. Storm, *Bull. Am. Assoc. Petrol. Geol.*, 1945.)

Gulf at a rate that increases geometrically. The narrow zone between the areas characterized by the differing rates of thickening is called the flexure. The flexure migrated Gulfward with each successive major sedimentary cycle (Fig. 12-9). The sediments in the area of the flexure are either continental shelf facies or continental shelf interbedded with brackish facies.

The thickening of the Cenozoic strata toward the Gulf is quite impressive. One of the deepest wells in the world, near the Louisiana coast, was still in the middle or lower Miocene at a depth of 22,570 feet. Wells drilled at the tip of the Mississippi delta produce oil from probably Pliocene sediments at a depth of 9,000–10,000 feet. In these wells the Pleistocene and Holocene

sediments are about 4,500 feet thick. Some other wells near the present coast of Louisiana have penetrated as much as 15,000 feet of Pleistocene sediments, and farther out on the shelf these marine Pleistocene sediments contain oil.

All stratigraphical and geophysical data indicate that the regional dip and thickening toward the Gulf of Mexico extend to the coast line without any indications of a synclinal axis at or immediately beyond the coast. The thickness of sediments at the coast line is at least 40,000 feet (Fig. 12-10).

We saw in § 8-2 that the general pattern of deposition in the Gulf coastal region, so far as we can tell from the available record, was first established in the Jurassic. Cretaceous transgressions on the margins of the Gulf of

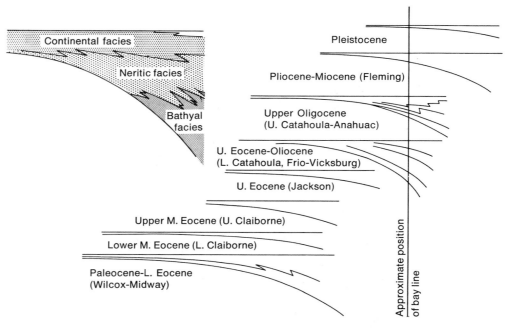

FIGURE 12-9

Diagrammatic representation of the major cyclical sedimentary units in the Gulf coast Cenozoic, showing the position of their rapid thickening (flexure) relative to the bay line. The upper left shows the distribution of facies within a cyclical sedimentary unit. (After S. W. Lowman, *Bull. Am. Assoc. Petrol. Geol.*, 1949.)

Mexico were very extensive, especially in the southcentral United States. The Cenozoic seas were not as extensive as those of the Cretaceous, but the rate of sedimentation was almost as great. It is estimated that there are 300,000 cubic miles of Cenozoic sediments and approximately 400,000 cubic miles of Mesozoic sediments in the emerged portion of the Gulf coastal plain between Georgia and Mexico, and more than 500,000 cubic miles of Mesozoic and Cenozoic sediments in the submerged (offshore) portion of the coastal plain. The total volume of Mesozoic and Cenozoic sediments in the emerged and submerged portions of the coastal plain is therefore about 1,200,000 cubic miles.

The Cenozoic formations of the Gulf coastal plain are the richest petroliferous zone in North America. The principal structure with which oil is associated here is the

salt dome. Such a dome is a vast plug of salt that has protruded up into or through Cenozoic formations. Drilling in this region has encountered more than 200 salt domes, and many more exist on the continental shelf. The upward movement of the salt is due to the difference in density between the salt and the overlying sediments. This difference is attained only after the salt is buried at a depth of 2,000–3,000 feet, as the density of the overburden increases with depth. Being lighter, the salt tends to rise along any outlet or fracture that may be present. The rate and time of the salt movement vary greatly. In the dome illustrated in Figure 12-11 the salt has reached and elevated the surface 190 feet. This is the highest dome on the Gulf coast; salt has been mined on it since the Civil War, and today the mines are 540 feet deep. This deposit will not soon be

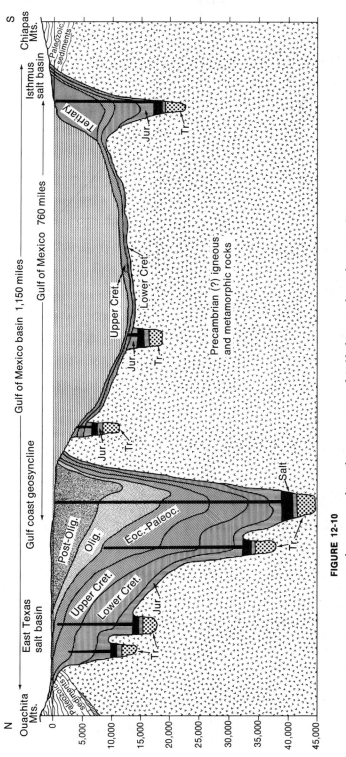

FIGURE 12-10

A schematic north-south cross section of Gulf of Mexico basin from Ouachita Mountains of southern Oklahoma to the Chiapas Mountains of southern Mexico. (After E. H. Rainwater, *Trans. Gulf Coast Assoc. Geol. Societies,* 1967.)

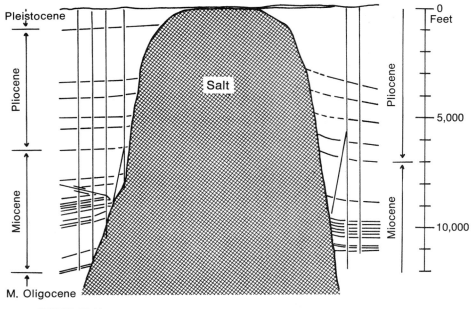

FIGURE 12-11

East-west cross section of the Avery Island salt dome, Iberia Parish, Louisiana. (After J. B. Carsey, *Bull. Am. Assoc. Petrol. Geol.*, 1950.)

mined out, for it is about three miles wide and from five to seven miles deep.

Many geologists think the salt domes originate from Upper Jurassic beds. As we pointed out in § 8-20, the subsurface of the Gulf coastal region contains evaporite deposits of Late Jurassic age, and advocates of this theory see these as the source of the salt domes.

Another theory holds that the salt is from Permian deposits, for fossiliferous cores from strata that include the salt sequence in northern Louisiana and southern Arkansas have been identified as late Paleozoic, probably Permian, in age. A large area in western Texas during the Late Permian was a great evaporite basin, and thousands of feet of chemical deposits accumulated there (§ 5-6). The advocates of this theory point out also that the Gulf coastal area itself, during the Late Permian, was probably a large evaporite basin with a narrow connection to the basin in western Texas. Data are still too few to substantiate either theory.

12-5 THE ROCKY MOUNTAINS

The structures of the present Rocky Mountains were largely formed during the Laramide orogeny (Fig. 12-12). The present topography of the Rocky Mountains, however, is the result of early Cenozoic deposition, followed by a long period of erosion. The Laramide movements along the western margin of the Late Cretaceous interior sea were largely great thrust faults. Farther east many broadly folded, anticlinal mountains, which today make up the Uinta, Wind River, and Big Horn mountains arose (Fig. 12-13). Between the mountains were broad basins that trapped detritus from the surrounding highlands. The sediments brought into these basins were largely fine-to-coarse clastics deposited in alluvial fans on flood plains and in lakes (Fig. 12-14). Active deposition continued through Paleocene and Eocene time. Most basin deposits contain tuff and ash beds. Lava flows were abundant in the Absaroka Range and Yellowstone National

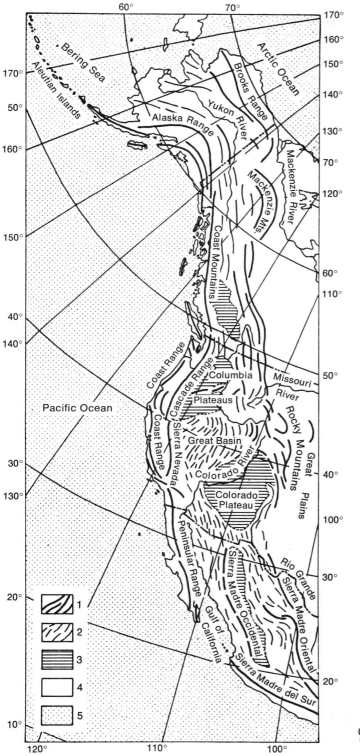

60° 70°

170°
160°
150°
140°
130°
70°
120°
60°
110°

Arctic Ocean

Bering Sea

Aleutian Islands

Brooks Range

Yukon River

Alaska Range

Mackenzie Mts

Mackenzie River

Coast Mountains

Pacific Ocean

Coast Range

Cascade Range

Columbia

Plateaus

Sierra Nevada

Great Basin

Colorado River

Coast Range

Colorado Plateau

Missouri River

Rocky Mountains

Great Plains

Peninsular Range

Sierra Madre Occidental

Gulf of California

Rio Grande

Sierra Madre Oriental

Sierra Madre del Sur

50°

40°

30°

20°

10°

170°
160°
150°
140°
130°

50°
40°
30°
20°

120° 110° 100°

50°

40°

30°

20°

1
2
3
4
5

Generalized map of western
North America, showing the
main topographic features: 1,
principal ranges; 2, minor
ranges; 3, plateaus; 4, lowlands,
including plains of the
continental interior; 5,
submerged areas, mainly ocean
basins, but including
continental shelves. (From
P. B. King, *in* C. L. Hubbs,
ed., *Zoogeography*, Am.
Assoc. Adv. Sci., Publ. 51,
1958.)

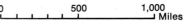

0   500   1,000
⌐ ¦ ¦ ¦ ¦ ¦ ¦ ¦ ¦ ¦ ¦ ⌐ Miles

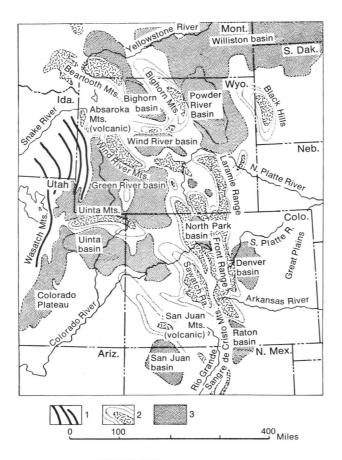

FIGURE 12-13

Map of the central and southern Rocky Mountains, showing early Cenozoic uplifts and basins: 1, folded and thrusted mountain ranges; 2, uplifts with Precambrian rocks exposed in higher parts; 3, basins that received early Cenozoic sediments. (After P. B. King, *in* C. L. Hubbs, ed., *Zoogeography*, Am. Assoc. Adv. Sci., Publ. 51, 1958.)

FIGURE 12-14

Bridger Formation (Eocene), Grizzly Buttes, eastern Wyoming. (American Museum of Natural History.)

Park through Eocene and Oligocene time. The Absaroka Range, in fact, consists largely of early Cenozoic lava flows (Fig. 12-15).

Composite sections of the formations laid down in these various basins during Paleocene and Eocene time show thicknesses of more than three miles. By Oligocene time the level of the basins had been raised by deposition nearly to that of the surrounding eroded mountains; and a low, monotonous, peneplaned terrane was established with an elevation of only a few thousand feet above sea level (Fig. 12-16). Epeirogenic uplift began in Miocene time, and active erosion followed. With uplift of the region the erosive action of streams intensified and removed thousands of cubic miles of the sediments that had been deposited earlier in the early Cenozoic basins. Uplift continued intermittently through Pleistocene time. Thus the present topography of the Rocky Mountains is due primarily to stream erosion since the Miocene arching and epeirogenic uplift of the region. The basic structure of the mountains is due to Laramide folding. This sequence of events in the Cenozoic history of the Rocky Mountains can be interpreted from (1) the succession and facies of Cenozoic formations in the basins, (2) the superposition of many of the rivers in the area, (3) the presence of peneplaned remnants in many of the ranges, (4) the preservation of Cenozoic deposits around the high margins of the basins, and (5) the sequence and facies of Cenozoic formations on the Great Plains.

12-6 THE HIGH PLAINS

Active deposition in the basins of the Rocky Mountains took place during the Paleocene, Eocene, and Oligocene and was followed by erosion for the remainder of the Cenozoic. The High Plains have a generally different history. Paleocene deposits are known only in the northern High Plains, and Eocene deposits are only local. Oligocene and later deposits are widespread throughout the High Plains, from Canada to Mexico. For convenience the northern, central, and southern High Plains are discussed separately.

The northern High Plains, eastern Montana, North Dakota, and the adjacent parts of South Dakota and Canada have Paleocene and Eocene nonmarine sandstone, shale, and lignite. In North Dakota a marine or brackish-water formation consisting of 300 feet of buff-to-green sandstones and dark-gray shales contains a fauna of mollusks and foraminifers. The age of this formation has been debated for many years; some hold that it represents a last phase of the Cretaceous interior sea, but the foraminifers suggest that it is more likely Paleocene, for they closely resemble Paleocene foraminifera of the Gulf coast.

The Eocene formations of the northern High Plains consists of fluvial and lacustrine fine-to-coarse clastic sediments. Unconformably overlying the Eocene formations are isolated remnants of Oligocene shale, sandstone, and a bentonitic clay that is one of the most conspicuous formations of the central High Plains of South Dakota, Nebraska, Kansas, Oklahoma, Colorado, and Wyoming. The Cenozoic deposits consist of a broad sheet of sediments derived from the Rocky Mountains. Over most of the area the oldest Cenozoic formations are Oligocene clay and sand formations containing abundant mammals (Fig. 12-17). These sediments make up the Badlands of South Dakota. An idealized profile of the Badlands, showing the sequence of deposits, is presented in Figure 12-18. The *Titanotherium* sandstone, *Metamynodon* sandstone, and *Protoceras* sandstone are river-channel deposits. The rest of the Oligocene deposits are made up of floodplain silt and clay. The overlying Miocene formations are generally coarser and include volcanic ash layers.

The Panhandle of Texas and adjoining regions make up the southern High Plains. The oldest Cenozoic formations are Pliocene and rest with marked unconformity on

FIGURE 12-15
Horizontal beds of volcanic breccia and flows, Table Mountain, Absaroka Range, northwestern Wyoming. (U.S. Geol. Surv.)

Cretaceous and older formations. The Pliocene and Pleistocene fluvial sediments, as much as 500 feet thick, are chiefly of early and middle Pliocene age. In late Pliocene time deposition ceased or greatly declined, and caliche (a whitish accumulation of calcium carbonate in a soil profile) was widely formed. Pleistocene deposits consist of windblown sands, bentonitic clays, and lake deposits.

12-7 THE COLORADO PLATEAU

In the vast plateau drained by the Colorado River the Cenozoic history differs greatly from that of the High Plains. The Colorado Plateau (Fig. 12-19), consists of virtually horizontal strata. The rich deposits of uranium have greatly intensified geologic study of the area. The present physiographic and structural character of the region began to develop in mid-Cenozoic time; earlier the region had been a basin, probably not far above sea level and surrounded by newly formed mountains. Chas. B. Hunt, formerly of the U.S. Geological Survey, constructed a series of diagrams of the Colorado Plateau for various times during the Cenozoic. Slightly simplified adaptions of his diagrams are reproduced here as Figures 12-20, 12-21, and 12-22.

By Late Cretaceous time the Colorado Plateau was a vast floodplain fronting the

FIGURE 12-16

Rocky Mountain peneplain northwest of Colorado Springs. Pikes Peak is a great monadnock rising above this uplifted erosional plain. (Photograph by T. S. Lovering, U.S. Geol. Surv.)

FIGURE 12-17

Oligocene strata (Brule Clay) exposed at Scotts Bluff, western Nebraska. (Photograph by Fairchild Aerial Surveys, Inc.)

Cretaceous sea, which at that time lay in Utah and Colorado. Highlands stood in Nevada and southern Arizona. By latest Cretaceous time the sea had almost completely retreated, and the first local phase of the Laramide orogeny had begun. Folding continued well through Paleocene time, developing folded highlands and volcanic mountains on the plateau (Fig. 12-20, A). The new topography diverted the earlier eastward-flowing drainage system into an interior system draining into basins between the newly formed highlands. The oldest Cenozoic rocks in the plateau are thick sequences of variegated shale, sandstone, conglomerate, volcanics, and freshwater limestone. Lower Paleocene formations are locally more than 2,500 feet thick. Through Paleocene time the western portion of the plateau was downwarped and became the main drainage center of the region, eventually forming a large lake in which carbonates accumulated. By Eocene time this lake had become enormous, covering most of eastern Utah and parts of western Colorado and southwestern Wyoming (Fig. 12-20, B). The lake sediments—the Green River Formation—consist of laminated organically

FIGURE 12-18

Diagrammatic representation of
Oligocene and Miocene formations
exposed in the Bad Lands of South
Dakota. (U.S. Geol. Surv.)

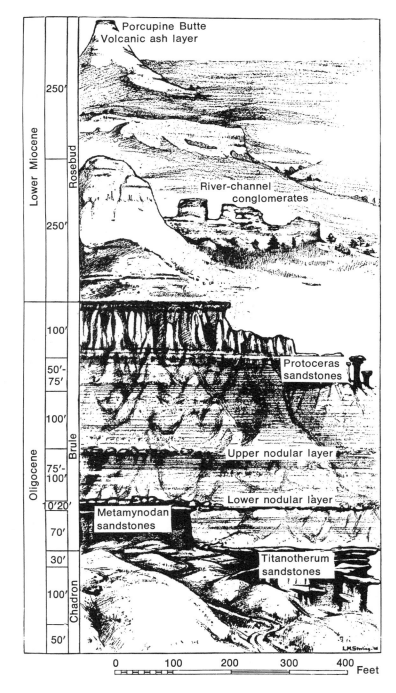

FIGURE 12-19

Wasatch Formation (early Eocene), Bryce Canyon, south-central Utah. (Photograph by Union Pacific Railroad.)

rich shale, fine-grained sandstone, siltstone, algal reef, marl, and coarse clastics, in places about 5,000 feet thick. The organically rich shales may some day become an important source of petroleum. By middle Eocene time the Green River Lake came to an end, probably by simple filling; extensive deposition of fluvial sediments followed through the remainder of Eocene time. Renewed uplift of some of the highlands also took place during the late Eocene. By the end of Eocene time the topography was more subdued, many highlands being so eroded that they approached the level of the basins. These conditions persisted through the Oligocene (Fig. 12-21, A).

During the early Miocene the plateau was epeirogenically uplifted higher than the surrounding basins, so that deposition on the plateau stopped, and erosion, through exterior drainage, began (Fig. 12-21, B). The Little Colorado River presumably drained westward approximately along the present course of the Colorado River. These conditions continued into middle Miocene time, when volcanoes became active around the margins of the plateau while intrusive stocks and laccoliths were emplaced on the interior. By this time the Grand Canyon had become a considerable gorge, but the drainage system was upset in late Miocene time when the plateau was again epeirogenically uplifted and tilted northeastward, again establishing large basins with interior drainage (Fig. 12-22, A). Into these basins poured a wide assortment of conglomerate, sandstone, shale, tuff, and lava. The basins were eventually filled and the streams spilled out into the old canyons. Since late Pliocene time there have been no great changes in drainage (Fig. 12-22, B), though warping has continued, and the canyons have been deepened. Extensive lava eruptions continued along the southern and western rims of the plateau until about 1,100 years ago.

12-8 THE COLUMBIA PLATEAU

Great sheets of basaltic lava cover many thousands of square miles on the Columbia Plateau. In North America flood basalts began in Miocene time in eastern Washington and Oregon, laying down 5,000 feet of lava, which forms the Columbia Plateau. These lava beds cover most of southeastern Washington, eastern Oregon, and southern Idaho over to the Rocky Mountains. The lava issued from many fissures and accumulated on a land surface that had as much as 2,000 feet of topographic relief. Interruptions in the drainage, ponding of streams, and local depressions caused the accumulation of many sedimentary wedges within the basalt sheets. The fossils from these interbedded sedimentary units provide the chronology.

In general, the Columbia River lavas lie in a low, shallow downwarp of the crust, which may have formed as accumulation progressed. In southern Idaho areas like the Craters-of-the-Moon National Park show spatter and cinder cones and lava flows, which clearly indicate that volcanic conditions continued into Holocene time. After the main lava outpouring the region was

FIGURE 12-20

The Colorado Plateau (A) in early Paleocene time and (B) in early and middle Eocene time. (After C. B. Hunt, *U.S. Geol. Surv.*, Prof. Paper 279, 1956.)

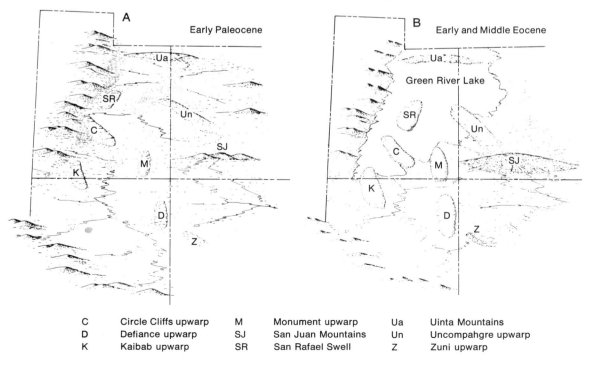

C	Circle Cliffs upwarp	M	Monument upwarp	Ua	Uinta Mountains
D	Defiance upwarp	SJ	San Juan Mountains	Un	Uncompahgre upwarp
K	Kaibab upwarp	SR	San Rafael Swell	Z	Zuni upwarp

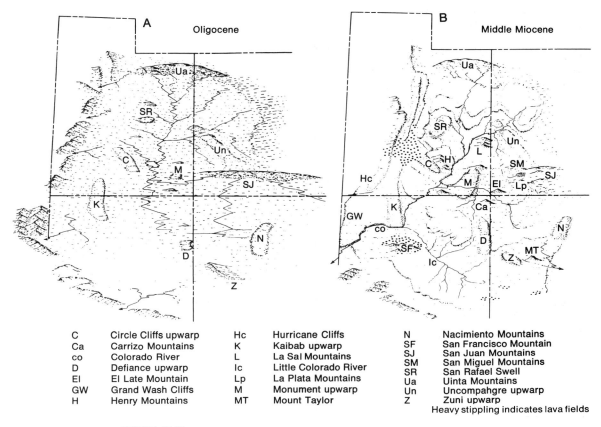

C	Circle Cliffs upwarp	Hc	Hurricane Cliffs	N	Nacimiento Mountains
Ca	Carrizo Mountains	K	Kaibab upwarp	SF	San Francisco Mountain
co	Colorado River	L	La Sal Mountains	SJ	San Juan Mountains
D	Defiance upwarp	Ic	Little Colorado River	SM	San Miguel Mountains
El	El Late Mountain	Lp	La Plata Mountains	SR	San Rafael Swell
GW	Grand Wash Cliffs	M	Monument upwarp	Ua	Uinta Mountains
H	Henry Mountains	MT	Mount Taylor	Un	Uncompahgre upwarp
				Z	Zuni upwarp

Heavy stippling indicates lava fields

FIGURE 12-21

The Colorado Plateau (A) in Oligocene time and (B) in middle Miocene time. (After C. B. Hunt, *U.S. Geol. Surv.*, Prof. Paper 279, 1956.)

epeirogenically uplifted, and today the elevation of the plateau varies from 3,000 feet to more than 8,000 feet.

12-9 THE BASIN AND RANGE PROVINCE

The present physiographic character of the Basin and Range province is shown by Figure 12-23. As the name implies, the province consists of many isolated, linear, faulted mountain ranges surrounded by flat basins of late Cenozoic and Holocene sediments. The present aspect of the province closely reflects its late geologic history.

Thick lacustrine deposits of Late Cretaceous age are known around Eureka in east-central Nevada. Several horizons of similar gravel accumulations were laid down during the early Cenozoic in eastern Nevada. In the eastern part of the province, during the early Cenozoic, lake, swamp, and flood-plain deposits were formed. The central and western parts of the province, in contrast, stood slightly higher and were the site of local volcanic deposits of tuff, agglomerate, welded tuff, and lava. By the mid-Cenozoic volcanism became widespread throughout much of the province. During one or more episodes of deformation after Eocene and before late Miocene time, the older Cenozoic sedimentary and volcanic sequences were faulted and tilted in many places. By late Miocene time much of the province was a low plateau with scattered mountains and many lakes and swamps. Erosional products of the mountainous areas accumulated in the adjoining basins. The sedimentary sequences were interrupted by occasional beds of volcanic ash, and in some areas beds of lava poured out across the basins. With topographic conditions like these during Pliocene time a wide variety of sedimentary deposits were formed, ranging from coarse alluvial fan deposits to fine muds and diatomaceous oozes in lakes and swamps. About middle Pliocene time large-scale block faulting and gradual regional uplift caused extensive deformation throughout much of the province. This structural break-up established the present structural and physiographic pattern of most of the province. Figure 12-24 illustrates the structural evolution of one of the ranges in northeastern Nevada.

12-10 THE PACIFIC COASTAL BELT

The most active site of orogeny during the Cenozoic was the Pacific coastal belt. All the main physiographic provinces of this region, each consisting of two parallel mountain ranges and, separating them, a broad linear basin, were formed during this era. The ranges are the Coast Ranges of California, Oregon, and Washington on the west and the Sierra Nevada and Cascade Mountains on the east; the intervening basin comprises the Great Valley of California and the Willamette-Puget trough in Oregon and Washington. The physiographic elements of California are not continuous with those of Oregon, but both merge into the Klamath Mountains of northern California and southwestern Oregon, which are an old terrane formed during the Nevadan orogeny.

California

The marginal embayments of the Nevadan land mass that marked California during the Cretaceous continued with only slight change into the Cenozoic. The shores of Paleocene and Eocene seas varied greatly, but the maximum spread of the seas was approximately that of the Cretaceous seas. Eastern California, including most of the present Sierra Nevada, and northern California were highlands (Fig. 12-25). In the region of the present Coast Ranges islands formed through growth of anticlines of Cretaceous and older rocks. The major sedimentary facies of the early Cenozoic are fine-to-coarse clastics derived mainly from the land mass to the east. In some places around the margins of the seas, swampy conditions prevailed, and some coal beds were formed. The great variability in thickness of the Paleocene and Eocene formations, from a few feet to more than 10,000 feet, reflects very uneven subsidence and the presence of many small basins.

The Oligocene was marked by regression of the seas, owing to regional uplift of much of California near the end of Eocene time. Large areas of the Coast Ranges were islands or peninsulas. The southern portion of the Oligocene embayment was the site of thick accumulations of red shales, sandstones, and conglomerates of continental origin. The sediments of the marine basins are clastic facies similar to those of the earlier Cenozoic periods.

In the Miocene a great change occurred in the tectonic framework of California. The earliest Miocene seas were small, with variable transitional shore lines (Fig. 12-25, B). The sediments brought into these seas were mostly fine-to-coarse clastics from the highlands of eastern California. Late Miocene seas were more widespread and significantly different in that they accumulated much nonclastic sediment (Fig. 12-26, A). Siliceous shale and diatomite are the main facies, though many clastic units of sandstone and even conglomerate were de-

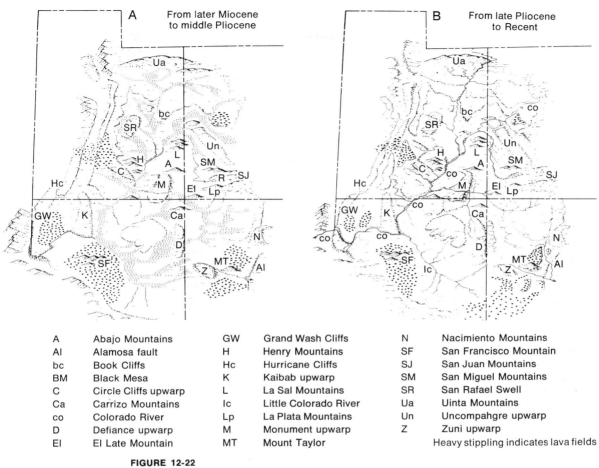

A	Abajo Mountains	GW	Grand Wash Cliffs	N	Nacimiento Mountains
Al	Alamosa fault	H	Henry Mountains	SF	San Francisco Mountain
bc	Book Cliffs	Hc	Hurricane Cliffs	SJ	San Juan Mountains
BM	Black Mesa	K	Kaibab upwarp	SM	San Miguel Mountains
C	Circle Cliffs upwarp	Ic	Little Colorado River	SR	San Rafael Swell
Ca	Carrizo Mountains	L	La Sal Mountains	Ua	Uinta Mountains
co	Colorado River	Lp	La Plata Mountains	Un	Uncompahgre upwarp
D	Defiance upwarp	M	Monument upwarp	Z	Zuni upwarp
El	El Late Mountain	MT	Mount Taylor		Heavy stippling indicates lava fields

FIGURE 12-22

The Colorado Plateau (A) from late Miocene to middle Pliocene time
and (B) from late Pliocene to Holocene time. (After C. B. Hunt,
U.S. Geol. Surv., Prof. Paper 279, 1956.)

posited round the basin margins. Volcanism was widespread throughout much of California at this time, as it was also in Nevada. The Miocene was a time of intense folding and faulting in the Coast Ranges and in eastern California. These diastrophic movements disrupted much of eastern California, establishing numerous local basins wherein terrestrial sediments of various types accumulated in late Miocene time, while volcanism continued to be widespread throughout this region. The adjoining sea basin accumulated great thicknesses of organic siliceous shales. Interbedded with these shales are beds of limestone, sand-

stone, and volcanic ash. In the Miocene the land area of eastern California ceased to be an important source of sediments deposited in the coastal seas, as it had been since the Jurassic, and broke into local basins with interior drainage systems.

Pliocene seas were much smaller than those of the late Miocene: many small basins and widespread areas of terrestrial deposition established the main pattern (Fig. 12-26, B). In the Los Angeles basin, for instance, more than 15,000 feet of Pliocene sediments accumulated. The facies throughout California are mainly fine-to-coarse clastic sediments.

FIGURE 12-23

Typical basin and range topography near Roach, Nevada.
(Photograph by Fairchild Aerial Surveys, Inc.)

In the San Joaquin valley the earliest Pliocene seas occupied only the central and deeper parts of the basin surrounded by lakes and flood plains. As the seas gradually spread, they transgressed highly tilted and truncated Miocene formations, forming a profound unconformity, which is one of the most widespread stratigraphic traps for oil accumulation in the state. From middle Pliocene time until the end of this period the seas gradually receded to the west, followed by accumulations of clastic terrestrial facies. Pleistocene deposits are largely nonmarine also, except for marginal basins around Los Angeles. The dominant facies are fine-to-coarse clastics, but some Pleisto-

cene chalks were deposited near San Pedro. The basins farther inland received much sediment from surrounding uplifted highlands.

The most intense orogenic phase affecting California after the Nevadan orogeny occurred in the middle Pleistocene, producing strong folding and both reverse and thrust faulting. Strongly tilted early Pleistocene and older formations are overlain by flat late Pleistocene deposits. The extensive raised marine terraces and the earthquakes along known faults (for example, the San Andreas fault) testify to continued movement.

The Sierra Nevada is one of the most imposing mountain ranges in western North America. After the Nevadan orogeny eastern California had become an important source of sediments to the embayments of central and western California during the Cretaceous and Cenozoic. The present structure and topographic relief of the Sierra Nevada were produced very late in the Cenozoic. The present mountain range is a huge tilted fault block about 100 miles wide and 400 miles long, with elevations of about 13,000 feet along the eastern side of the range which is marked by a great normal fault (Fig. 12-27). Miocene-Pliocene fossil floras from west-central California, the Sierra Nevada, and western Nevada are related to living communities whose topographic and climatic relations provide a basis for estimating the altitude of the Sierra Nevada and west-central Nevada during the Miocene and Pliocene. Miocene floras indicate only moderate differences in vegetation, climate, and topography across this area, where today we find great environmental diversity. A suggested topographic profile for this area during the Miocene is shown in Figure 12-28. The present elevation of the Sierra Nevada is due to uplift and tilting away from the eastern boundary fault. During the late Miocene and the Pliocene, volcanism was particularly active in the Sierra Nevada, and the deposits in some places are 3,000 feet thick. This thickness does not necessarily mean, however, that the mountains were raised by this amount, for isostatic adjustments of the crust probably depressed the basement to accommodate such a great blanket of volcanic rocks. To maintain isostatic balance, the basement would have had to sink and displace a mass of sima equal to the added volcanics. Such a blanket of volcanics, 3,000 feet thick, would have had to

sink approximately four-fifths of its thickness to come into equilibrium, and the surface would thus be only about 600 feet higher than before the volcanism.

Oregon and Washington

The Coast Ranges and Cascade Mountains of Oregon and Washington are products of the same late Pliocene and Pleistocene orogenic phase that was important in California. These northern states, however, have somewhat different Cenozoic histories; the most notable feature is the great quantity

FIGURE 12-24

Sections showing fault-block evolution of the Ruby-East Humboldt Mountains, northeastern Nevada. (From R. P. Sharp, *Geol. Soc. Am. Bull.*, 1939.)

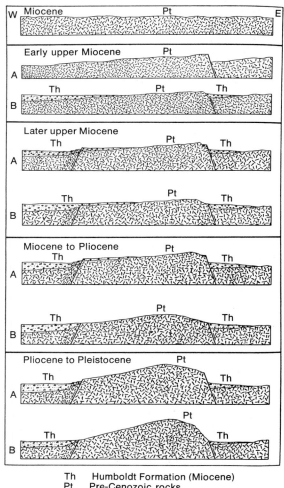

Th Humboldt Formation (Miocene)
Pt Pre-Cenozoic rocks

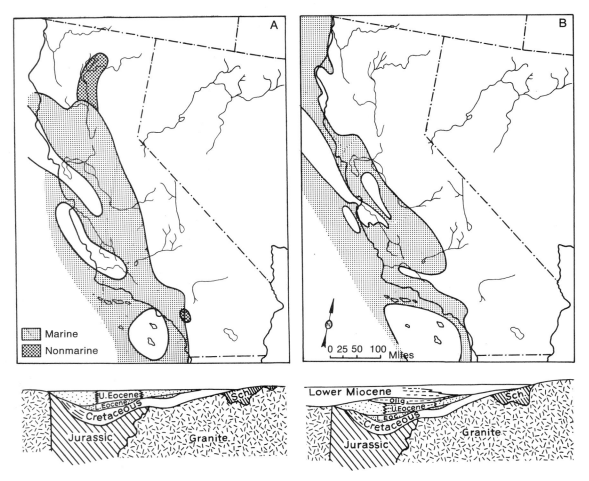

FIGURE 12-25

Paleogeography of California, with cross section of San Joaquin Valley, (A) for Late Eocene time and (B) for Early Miocene time. (Maps after R. D. Reed, Am. Assoc. Petrol. Geol., 1933; cross sections after Hoots, Bear, and Kleinpell, 1954.)

of volcanic rocks that accumulated.

Cenozoic rocks in Oregon and Washington rest on a basement of mainly igneous and metamorphic rocks that had been extensively eroded. In the earliest Cenozoic time the area of these two states was probably a vast coastal plain of gently undulating topography with westward flowing streams. Subsidence commenced in the early Cenozoic, and the region of the Coast Ranges was marked by violent submarine volcanism

that produced andesitic and basaltic flows, together with tuff, agglomerates, and many intrusive plugs. This lava field extends from Vancouver Island to the Klamath Mountains, where more than 5,000 feet of lacustrine and fluvial clastic sediments accumulated in broad valleys opening to the west.

Near the close of this volcanic phase, in early Eocene time, coastal Oregon and Washington were upwarped while a parallel north-south downwarp formed just to

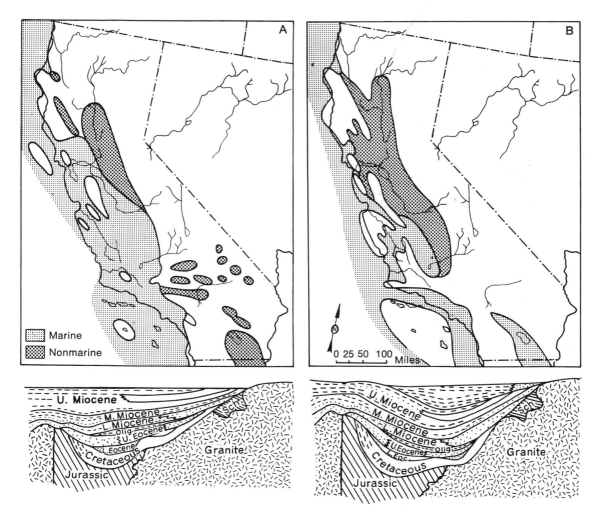

FIGURE 12-26

Paleogeography of California, with cross section of San Joaquin Valley, (A) for late Miocene time and (B) for Pliocene time. (Maps after R. D. Reed, Am. Assoc. Petrol. Geol., 1933; cross sections after Hoots, Bear, and Kleinpell, 1954.)

the east. The upwarped coastal arch of volcanic rocks formed a peninsula extending from Vancouver Island to southern Oregon. The adjoining downwarp was a narrow, linear basin, closed in the area of Puget Sound but opened to the Pacific through a wide portal in southern Oregon, forming a gulf in which more than 15,000 feet of clastic marine, brackish-water, lacustrine, and fluvial sediments accumulated. The freshwater sediments contain commercial coal seams. By late Eocene time the peninsula subsided below sea level and became the site of thick accumulations of sediments. Farther east, in the Cascade Mountains, fluvial, lacustrine, and thick volcanic deposits were laid down. These conditions continued through Oligocene time.

FIGURE 12-27

Eastern front of the Sierra Nevada. (Photograph by W. C. Mendenhall, U.S. Geol. Surv.)

During early and middle Miocene time several large marginal embayments formed, and there was much oscillation of the shore lines. The extensive volcanism of the Columbia Plateau is recorded by great quantities of lavas interbedded with the marginal coastal sediments. A major orogenic phase strongly folded and uplifted the Coast Ranges in northwestern Washington in northwest-southeast folds; farther south the axial trend was nearly north-south, like that of the present Coast Ranges. During the Pliocene the new mountains underwent vigorous erosion, whose products accumulated on coastal plains with numerous embayments to depths of several thousand feet.

FIGURE 12-28

Topographic profile across the central Sierra Nevada today and inferred profile of the time just before Mio-Pliocene volcanism. (After D. I. Axelrod, *Geol. Soc. Am. Bull.*, 1957.)

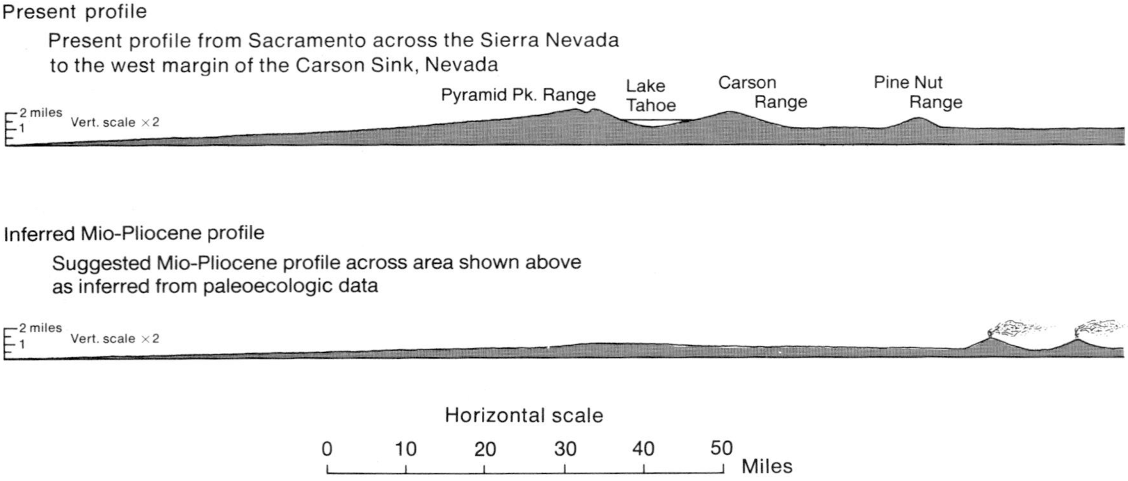

Present profile

Present profile from Sacramento across the Sierra Nevada to the west margin of the Carson Sink, Nevada

Pyramid Pk. Range   Lake Tahoe   Carson Range   Pine Nut Range

2 miles
1
Vert. scale ×2

Inferred Mio-Pliocene profile

Suggested Mio-Pliocene profile across area shown above as inferred from paleoecologic data

2 miles
1
Vert. scale ×2

Horizontal scale

0   10   20   30   40   50   Miles

The Coast Ranges and the Cascade Mountains resulted from an orogeny at the close of the Pliocene that affected the entire western coast of America. During this episode the famous volcanic cones of Baker, Rainier, St. Helens, Adams, and Hood were formed. The present topographic features of western Oregon and Washington were largely produced at this time, but they were somewhat modified by glacial action and erosion during the Pleistocene.

12-11 ALASKA

Since the major structural features of Alaska had been developed during the late Mesozoic orogenic phases, the Cenozoic formations are products of erosion and volcanism, which were mostly confined to low intermountain valleys and coastal regions. The commonest and most widespread facies of this era are non-marine coal-bearing clastic formations and volcanics.

In the region of the Gulf of Alaska there are about 25,000 feet of Cenozoic strata; the lower 7,000 feet consists of interbedded marine and coal-bearing nonmarine formations of Eocene age; the upper 18,000 feet are marine. In the Panhandle of southeastern Alaska several areas show lower Cenozoic sandstone and conglomerate resting unconformably on Lower Cretaceous; these deposits grade upward into volcanic rocks. Along the western Gulf of Alaska, around Cook Inlet, another extensive region of Cenozoic formations is composed of nonmarine sandstone, shale, and conglomerate, with numerous coal beds.

The Alaskan peninsula and the Aleutian Islands were the scene of very intense volcanism during the Cenozoic, and even today there are scores of active or recently active volcanoes along this belt (Fig. 12-29), which is remnant of the eugeosyncline that has characterized western North America throughout Phanerozoic time.

In interior Alaska local basins have ac-cumulated Cenozoic coarse-to-fine clastic sediments of nonmarine facies. Throughout the region enormous volcanic deposits of various types accumulated. Most of the lowland areas adjoining the major stream systems of central Alaska are underlain by Pleistocene and Recent alluvial deposits.

The Arctic coastal plain of Alaska is largely covered by Pleistocene and later deposits, but the few data available show that at least in some places there are 7,000 feet of middle and upper Cenozoic marine formations that thicken northward.

Throughout most of Alaska uplift and deformation occurred in the Pliocene. The most severe folding and faulting were in the region bordering the Gulf of Alaska.

12-12 THE ARCTIC ISLANDS

Cenozoic formations in the Arctic islands are largely confined to the Sverdrup Basin, the Arctic coastal plain, and the Eureka Sound fold belt (Fig. 5-11, p. 133). They consist mainly of nonmarine and poorly consolidated sandstone and shale, commonly interbedded with lignite. In the Eureka Sound belt there are perhaps 7,000 feet of openly folded sandstone and shale with apparently Paleocene and Eocene plants and lignite. Part of the sequence of these strata indicates cyclic deposition. Carbonized logs and stumps stand upright in some strata. In the Central Ellesmere fold belt (Fig. 5-11) exposures up to 200 feet thick include shale, mudstone, sand, grit, conglomerate, sandstone, limestone, nodular limestone, seams of impure peaty lignitic coal, sandstone with carbonized plant fragments, and lignified trees that indicate an age not greater than Late Cretaceous. These strata are folded and appear to have been laid in catch basins on the eroded surface of lower Paleozoic beds.

The time-stratigraphic range of the Cenozoic formations is still poorly known, as is the time of folding. The last orogeny in this

FIGURE 12-29

Mount Shishaldin (elevation 9,387 feet), one of the great chain of active volcanoes of Cenozoic and Holocene age that extends through southwestern Alaska and the Aleutian Islands. (U.S. Geol. Surv.)

area apparently commenced not earlier than Late Cretaceous time, possibly as late as Miocene, and may have ended before the close of the Miocene or more recently. We do not know whether it was restricted to one epoch or whether it extended to many, possibly occurring in various phases in different areas.

12-13 GREENLAND

Throughout the Paleozoic and most of the Mesozoic, Greenland was an integral part of the Canadian shield. The presence of marine Cretaceous rocks along the central west coast of Greenland, with faunas closely related to those of Montana and not to those of the Atlantic coastal plain Cretaceous, indicates open connection with the Arctic Ocean and not with the Atlantic, and at least a partial separation from the Canadian shield. The separation of Greenland from the shield was probably the result of faulting that began in Late Cretaceous and continued into the Cenozoic. Late Cretaceous and early Cenozoic deposits form a completely gradational and continuous series of alternating marine and non-marine strata. This depositional phase ended with the extensive development of basaltic flows and plutonic intrusions.

Along the eastern coast of Greenland volcanism occurred on an even grander scale. Overlying the basaltic flows, in some places, are mainly continental plant-bearing facies with some marine horizons. Following this depositional history, peneplanation, faulting, and uplift marked the remainder of Greenland's history. The Pleistocene ice age is discussed in Chapter 15.

SUGGESTED READINGS

See the list at the end of Chapter 4.

13

THE CENOZOIC ERA

II. The World Outside North America

Toward the close of the Mesozoic (§ 9-10), the seas retreated from the continents on a grand scale leaving the extended Tethys as the only inland sea. Tectonic instability and orogeny — which originated late in the Mesozoic on the margins of the Tethys, along the eastern margin of the Pacific, and in the Andean geosyncline, including its extension into the Antarctic peninsula — continued during the Cenozoic and produced the great Alpine, Himalayan, and Andean mountain systems. As in the early Mesozoic, marine transgression took place, but on a small scale, the maximum inundation coming in the Eocene. After that the continental and geosynclinal seas gradually withdrew, until in the Pliocene the continents assumed their present form. (Figs. 13-1, 13-2).

During most of the era the Tethys lay to the south of Europe, as it had during the Mesozoic. South of the Tethys, and continuous with it, epicontinental seas spread across North Africa, where thick sequences of nummulitic* limestones accumulated, especially in Libya and Egypt. Northern Europe was only intermittently subject to marine invasion; at different times the North Sea transgressed upon southeastern England, northwestern France, Belgium, Holland, northern Germany, and Denmark, and the Atlantic Ocean encroached upon western France. In eastern Europe a shallow continental sea extended along the eastern side of the Ural Mountains from

*Nummulites were large foraminifers.

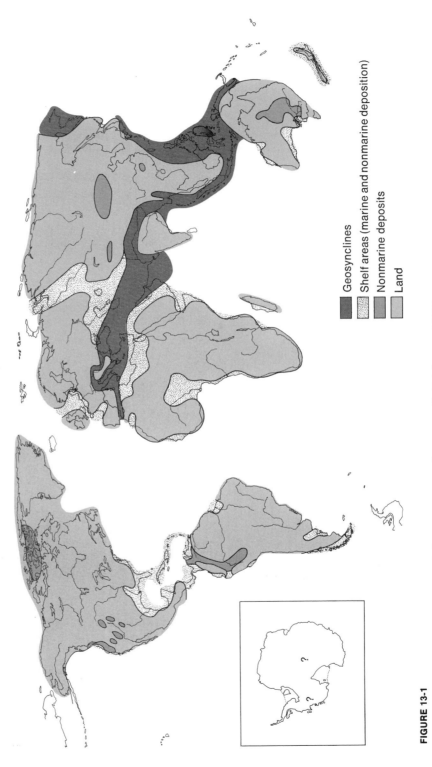

FIGURE 13-1

Paleogeographic map of the Eocene. (After many sources but especially Brinkmann, 1954.)

Geosynclines

Shelf areas (marine and nonmarine deposition)

Nonmarine deposits

Land

the Arctic Ocean to the Tethys, much as in the Late Cretaceous (§ 9-7). By the mid-Cenozoic this sea connection was broken, and the Eurasian land mass was again continuous. The Black, Caspian, and Aral seas were formed during the breakup of the Tethyan seaway.

During the Cenozoic, a series of orogenies profoundly affected the Tethyan realm. In western North Africa, along the southern margin of the Tethyan geosyncline, orogeny produced the Atlas mountain system. At the northern margin a series of orogenies produced the Alps, Carpathians, Balkans, Apennines, and Pyrenees. Patterns of deposition and the orogenies

were closely related and extremely complex in detail. At the same time, volcanism was particularly active over much of the Middle and Far East, as it was also in the region from Indonesia to New Zealand and along the Pacific border in the Philippines, Japan, and Kamchatka. All these areas were tectonically unstable throughout much of the Cenozoic (Fig. 13-3).

In Australia the Cenozoic was an era of comparative stability, although there was strong volcanic activity and extensive lava flows in the eastern part during the Eocene and Oligocene, and the lacustrine deposition that had begun in the late Mesozoic continued until the lakes disappeared in

FIGURE 13-2

Paleogeographic map of Europe for the Eocene. (After Wills, 1951, von Bubnoff, 1935, and Brinkmann, 1954.)

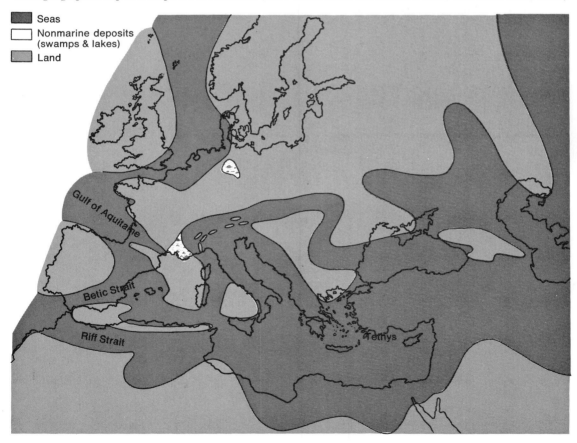

the late Miocene. Marine incursions took place along the southern and western coasts in the Miocene and early Pliocene, but they were minor and the sea had withdrawn by the end of the Pliocene. The only diastrophic movements were fairly gentle but broad upwarps and downwarps.

Africa, except for the Tethys and its contiguous continental sea, had only minor marine transgressions, which took place along the western and eastern coasts, following the Cretaceous pattern. The most spectacular event on this continent was the continuing development of the great rift (fault) valleys, which extended north from the Zambezi River through eastern Africa, the Red Sea, the Dead Sea, and the Jordan Valley. This phase of faulting, which was accompanied by much volcanism, had begun in the Jurassic in Malawi and Tanzania.

In South America the main Cenozoic event was the formation and uplift of the Andes. This event had begun in the Cretaceous, and a complicated series of orogenies and epeirogenies followed during the Cenozoic. The sedimentational patterns of most Cenozoic formations of South America were controlled by, and intimately related to, the uplift of the Andes. Along the eastern coast of the continent minor marine incursions followed the Cretaceous pattern.

In the West Indies a depositional pattern inherited from the Late Cretaceous was completely unlike that in adjoining parts of North and South America.

13-1 EUROPE

The Mesozoic ended with a great regression of the seas from Europe, without any significant orogenies. European geography then resembled that of the late Paleozoic

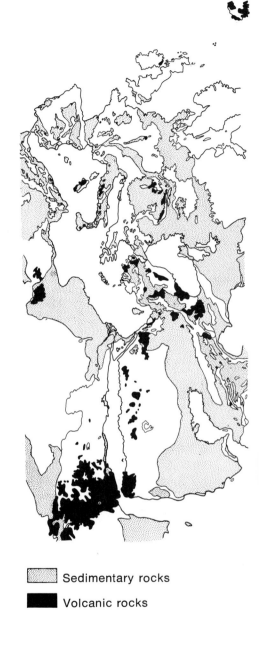

☐ Sedimentary rocks

■ Volcanic rocks

FIGURE 13-3

Outcrop map of Cenozoic formations of Eurasia. (Redrawn from *Geologic Map of Eurasia*, Russian Geol. Surv.)

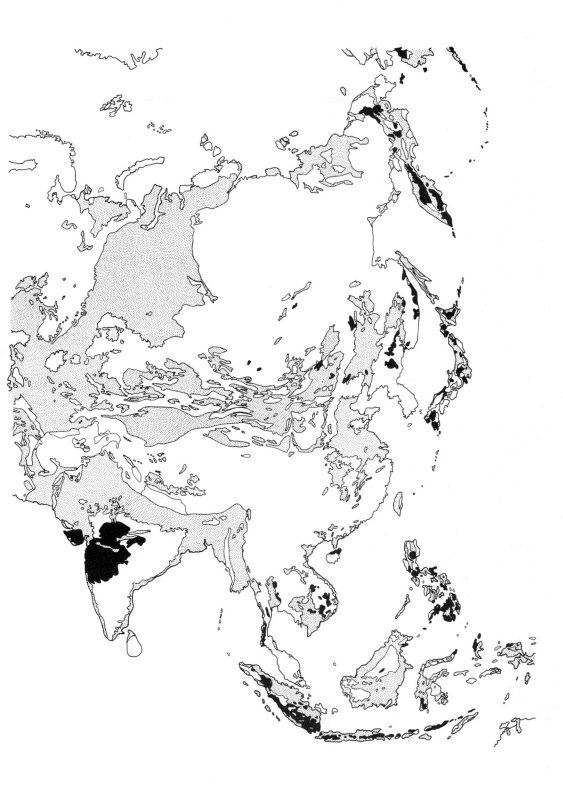

and early Mesozoic. In the Cenozoic a new cycle of greatly reduced transgressions took place. During the first half of the era (Paleogene), the seas were largely confined to the Tethyan geosyncline. A northern sea transgressed across Denmark, Holland, Belgium, northwestern France, and southeastern England. In the Oligocene a wide seaway extended across Russia and through Poland and Germany, joining the northern sea with the Tethys. Toward the close of the Oligocene came widespread regression and lake and stream sediments were deposited in ephemeral and isolated basins. There were thus three main depositional provinces: the Tethys, the North Sea, and the zone of middle Europe between these seas, here discussed separately.

The West European Tethys

After the Cretaceous regression the seas were largely confined to the Tethys, and as yet no severe orogeny had affected the Alpine belt except for continued movement along the tectonic ridges within the geosyncline. This instability in the geosyncline is reflected in the facies in its central regions, which consist of very thick shale and sandstone (*flysch*). There were also conglomerate and breccia as in the Cretaceous and Jurassic of this zone (Fig. 9-14). The northern margin of the Tethys extended from southeastern France round the northern side of the present Alps, then east to Austria, and continued in a sinuous course to the Middle East (Fig. 13-2). A large area in the central Balkans was emergent during the early Cenozoic. In the shelf regions of this sea, north of the southern Alpine geosynclinal tract with tectonic ridges, were shallow-water carbonate facies with nummulites and calcareous shale facies. Late in the Eocene sands from the emerged geosyncline spread across these marginal shelf areas.

The end of the Tethys in the Alpine region began during the Oligocene when some of the principal folds of the southern zones of the Alps arose and expelled the seas from the Alpine belt, into a narrow sea with adjoining lagoons to the north. This narrow foredeep to the rising Alps is the **peri-Alpine depression**. During the Oligocene and Miocene it was the site of very thick accumulations of deltaic, lagoonal, lacustrine, and fluvial sediments. Marine conditions persisted only in southeastern France and to the east of the peri-Alpine depression (Fig. 13-4).

The general recession of the seas from this area in the Oligocene was followed by a new transgression, which marked the beginning of the Miocene (Fig. 13-5). However, the early Miocene seas encroached only upon the western and eastern ends of the peri-Alpine depression; that is, marine strata are found in a narrow belt in southeastern France and in Austria around Vienna, and the intervening area of the depression, covering much of northern Switzerland, was still the site of nonmarine deposition. Not until the early middle Miocene did the seas once again extend along the front of the Alps from southeastern France to Austria. The seas occupied the peri-Alpine depression only during the middle Miocene, for in the Swiss part of the depression nonmarine conditions returned in the late Miocene, while marine conditions persisted in southeastern France and in Austria. The Miocene sediments of the peri-Alpine depression consist of sandstone and shale, which toward the Alps grade into conglomerates; *molasse* of European geologists. The spatial relations of the Miocene facies in the peri-Alpine depression are diagrammatically shown in Figure 13-6.

The late Miocene and Pliocene history of the area east of the peri-Alpine depression is a unique chapter in the history of the Tethys, and a separate section (§ 13-1) is

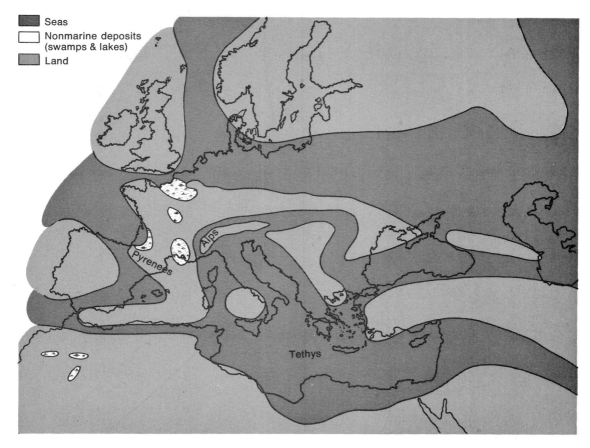

FIGURE 13-4

Paleogeographic map of Europe for the Oligocene.
(After Wills, 1951, von Bubnoff, 1935, and Brinkmann, 1954.)

devoted to it. The western end of the peri-Alpine depression, in southeastern France, became an area of large fresh- and brackish-water lakes in the late Miocene and Pliocene. The Mediterranean transgressed upon southern France to a limited extent only in narrow river valleys during the Pliocene (Fig. 13-7).

Toward the close of the Miocene further orogenic movements affected the Alps and the Carpathians. These movements were followed, during the Pliocene, by epeirogenic movements, which changed the face of this part of Europe and molded it closer and closer to the present. A very generalized map of the Alps is shown in Figure 13-8.

The origin of the Appennine Mountains has been a geological mystery for many decades (Fig. 13-9). They are enormously complex. In recent years a group of Italian geologists have developed a hypothesis, supported by most who have worked on the problem.

From Late Triassic to Oligocene this region underwent continuous marine sedimentation. The Upper Triassic strata rest

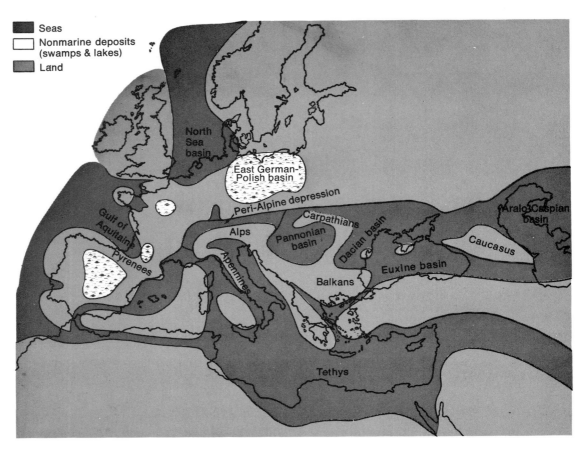

FIGURE 13-5

Paleogeographic map of Europe for the Miocene.
(After Wills, 1951, von Bubnoff, 1935, and Brinkmann, 1954.)

unconformably on crystalline schists, phyllites, quartzites, fine conglomerates, and volcanic rocks of Permo-Carboniferous age. The Mesozoic and early Cenozoic facies are mostly carbonates about 6,000 feet thick. The facies indicate gradual deepening of the region, and a maximum depth of perhaps several thousand feet was reached in the Late Jurassic or Early Cretaceous. The first influx of clastic sediments came in the Oligocene: more than 10,000 feet of sandstone and shale that show well-developed graded bedding. These clastic sediments came from erosion of an orogenic land emerging in the Tyrrhenian Sea. Deformation proceeded eastward and northeastward, spilling forward a series of successively younger sandstone and shale formations derived from it.

In the early Miocene, ridges bounded by faults began to rise under the sea near the western part of the Apennines. As these ridges formed, land slips on an enormous scale developed, producing a chaotic sedimentary mass of sheets and slabs of more or less consolidated rocks mixed up and

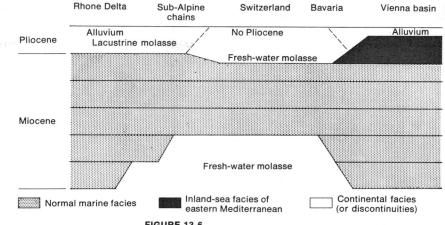

FIGURE 13-6

Diagrammatic cross section of Miocene and Pliocene facies in the peri-Alpine depression. (After M. Gignoux, 1955.)

FIGURE 13-7

Paleogeographic map of Europe for the Pliocene.
(After Wills, 1951, von Bubnoff, 1935, and Brinkmann, 1954.)

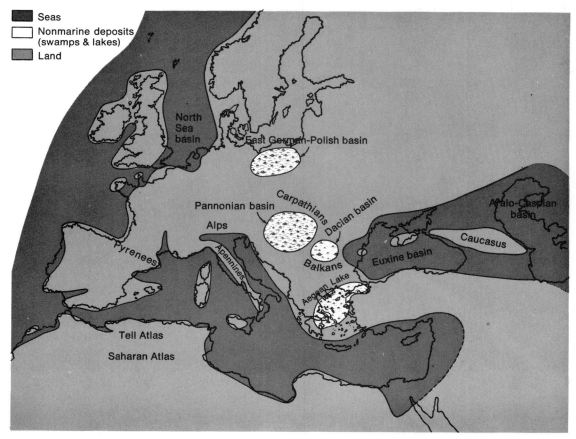

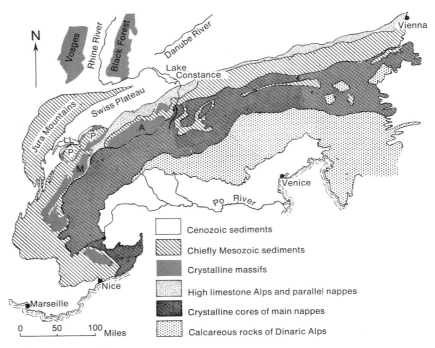

FIGURE 13-8

Generalized map of the Alps: *A,* Aar massif; *LL,* Lake Geneva (Léman); *M,* Mont Blanc and Aiguilles Rouges massifs; *P,* Prealps. (After R. Staub, redrawn from L. W. Collet, *Structure of the Alps,* Arnold, London, 1927.)

FIGURE 13-9

Cenozoic mountains of the western Mediterranean. (After L. U. DeSitter, *Structural Geology,* McGraw-Hill, New York, 1956.)

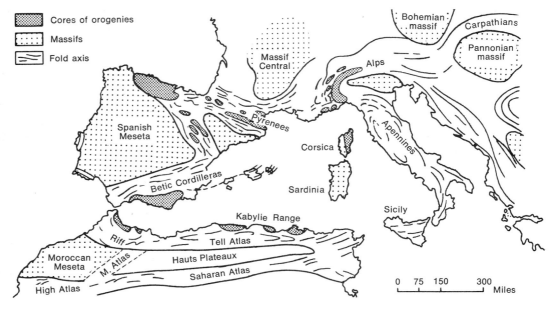

bound together by an argillaceous matrix, the **argille scagliose**. The size of the included masses varies from less than a six-inch cube to whole mountains (for example, Monte Morella near Florence). The sliding of the argille scagliose started in the Tyrrhenian area in the Late Cretaceous as the first ridges were formed and spread eastward and northeastward with the successive ridges of the Oligocene and subsequent orogenic phases. Thus the argille scagliose has neither a definite thickness nor a definite time of formation; the thickness varies from nothing to thousands of feet; the time of formation coincides, in any area, with the local time of the upheaval that caused the slides. The exotic blocks of the argille scagliose consist of pre-Triassic granite, Upper Jurassic or Lower Cretaceous intrusive rocks, cherts, and cherty limestones, Upper Triassic dolomites, Upper Cretaceous sandstones, and Eocene, Oligocene, and Miocene sandstones, marls, and breccias. The exotic igneous rocks have no roots in the Apennines. Similar tectonic and depositional phases are present all along the Apennines, extending into Sicily. The Miocene rocks of Sicily—6,000 feet of marine shale and sandy shale with two distinct horizons several hundred feet thick—are a chaotic assemblage of exotic blocks, from Paleozoic to Miocene in age, within a fine-grained clay matrix. One of these exotic blocks of the Upper Triassic dolomite is shown in Figure 13-10.

Pliocene seas covered most of the Italian peninsula and Sicily except for the Apennines and an area in northern Sicily (Fig. 13-7). In these coastal seas, sand, shale, and limestone were deposited in wedges that thicken seaward. Basins within the Apennines accumulated thick terrestrial sediments. With the close of the Pliocene the seas generally regressed from the peninsula and from Sicily. The Pleistocene history of Italy is marked by much volcanism in the southern part of the country; Mount Vesuvius and Mount Etna are the most famous of these late Cenozoic volcanoes.

The East European Tethys

The termination of marine conditions in the peri-Alpine depression during the Late Miocene, coupled with orogeny that caused uplift and emergence in the Balkans and Turkey, isolated the eastern part of the European Tethys as a large inland sea. The salinity of this sea decreased, and a very special fauna, quite distinct from the normal marine Tethyan fauna, appeared. Study of the late Cenozoic deposits of this basin and their fossil faunas indicates that this great inland sea was made up of four distinct basins that at times were separated and at other times were connected (Figs. 13-5 and 13-7): the western Panonian Basin, which lay between the Alps and the arc of the Carpathian and Balkan Mountains (Lake Balaton in Hungary is a remnant of this basin); Dacian Basin, along the lower Danube, east of the Carpathian Mountains and north of the Balkan; Euxine Basin; the present Black Sea; and the most eastern basin, the Aralo-Caspian, which extended northward far beyond the boundaries of the present Caspian and Aral Seas.

During the earliest phases after this inland sea had become isolated, normal marine conditions prevailed, and there were connections between all the basins, as we know from the uniform fauna. As is typical of inland-sea faunas, the variety of species was small, but the number of individuals was at times enormous. Soon after this eastern inland sea was defined, the Pannonian Basin became isolated; the salinity of its water began to diminish, and at the same time the fauna began to change. In the late Miocene the salinity of the waters in the three other basins diminished greatly, they also became brackish, and their faunas changed considerably. There was a short episode in the late Miocene when all four basins were again connected and a similar fauna was distributed throughout the area, but soon the Pannonian Basin again became isolated and was occupied by fresh-water lakes. This phase was followed in the latest

FIGURE 13-10

Exotic block of Triassic dolomite in central Sicily, lying in the midst of crumpled
shales. The exotic blocks vary from giants like this to small fragments and range in
age from Paleozoic to Miocene. Mussomeli Castle, dating from the Middle Ages,
is built on this dolomite block. (Photograph by M. P. Marchetti.)

Miocene by the complete separation of the
three eastern basins as a result of a general
regression of the seas.

Early in the Pliocene connections were
partly restored with the rejoining of the Euxine and Aralo-Caspian Basins, and brackish water again prevailed. In the western
basins, however, the Pliocene brought considerable changes. The Pannonian Basin
had been a region of fresh-water lakes in
the late Miocene and continued so into
the Pliocene. In the early Pliocene the
Dacian Basin became an area of fresh-water
lakes, which expanded to become one vast
fresh-water lake, and Aegean lake, that
covered the Sea of Marmora and a good part
of the Aegean Sea of today (Fig. 13-7). All
these fresh-water lakes and brackish seas
persisted with some modifications through
the Pliocene. During the Pleistocene the

Aegean lakes disappeared, and the Mediterranean invaded the Euxine Basin, but
the Caspian and Aral Seas had already been
separated and were not affected. During
this time the lakes in the Dacian and Pannonian Basins gradually disappeared. Thus
the fauna of the present Black Sea is largely
Mediterranean, but those of the Caspian
and Aral Seas are relics from the Pliocene
brackish-water sea.

The Atlantic Coast of France

Early in the Cenozoic the Atlantic over-
flowed upon southwestern France and
northeastern Spain in two gulfs separated
by the Pyrenees Mountains. The Aquitaine
Gulf in France apparently never connected
with the Tethys; but a Spanish gulf or strait

connected the Atlantic and the Tethys (Fig. 13-2). The pattern of facies of the early Cenozoic sediments deposited in the Aquitaine Gulf is much like that in the Paris Basin, discussed in § 13-1; that is, in the eastern portions of the gulf nonmarine facies are prevalent, and westward an alternating sequence of marine and nonmarine facies is encountered. These variations in facies reflect a continuously shifting shore line in the gulf.

The Pyrenees are an old Hercynian mountain range, which underwent orogeny from the Late Devonian to the end of the Carboniferous. During the Jurassic and the Early Cretaceous it was covered by the sea, but it remained high enough so that the deposits laid down were thin. Between the Early and the Late Cretaceous it underwent another orogenic phase, which established, parallel to the northern and southern sides of the range, troughs in which great thicknesses of Upper Cretaceous and lower Cenozoic sediments accumulated. At the close of the Eocene another orogenic phase affected the Pyrenees, but mostly in the area of the marginal troughs, which were strongly folded and faulted. During this phase there was a great influx of conglomerates into the parts of the Aquitaine Basin adjacent to the rising Pyrenees. These are the last folded strata in the region, as the overlying Oligocene formations are horizontal (Fig. 13-9).

The depositional pattern that characterized the Aquitaine Gulf during the early Cenozoic continued on a more restricted scale through the Miocene, after which the seas completely regressed from this part of France. The marine rocks are almost entirely sandstones. The non-marine formations near the Pyrenees consist of thick alluvial conglomerates derived from the erosion of this mountain system. The northern part of the basin, close to the Massif Central, has, not coarse clastic rocks, but fine-grained sediments and lacustrine marls and limestones.

The northern French Atlantic coast was inundated in Brittany, where a series of narrow gulfs formed across the lowlands east of the Armorican massif during the Miocene and Pliocene (Figs. 13-5 and 13-7). The sediments deposited in these gulfs are mainly sandstones containing an abundant fossil fauna.

The North Sea Basin

The classic area for the study of the lower Cenozoic of the North Sea Basin is the vicinity of Paris, where a remarkably complete and thoroughly studied series of strata crop out. This region was a large bay of the North Sea, which also spread across southeastern England, Belgium, Holland, Denmark, and extreme northern Germany (Fig. 13-2). The alternating sequence of marine and non-marine facies indicates a constantly changing shore line. The marine facies consist of fossiliferous shallow-water sandstones and organic limestone, the non-marine facies of lacustrine limestones and clays and alluvial and swamp deposits. Figure 13-11 shows the relations of these marine and non-marine formations. Toward the close of the early Cenozoic, in the late Oligocene, a final extensive transgression of the sea spread beyond any of the earlier shore lines of the Cenozoic. The facies developed in this sea are uniformly marls and sands. At the close of the Oligocene the sea retreated for the last time from the Paris Basin, and terrestrial deposition began.

A western embayment of the North Sea Basin extended eastward into Demark and northern Germany, lying between the Baltic shield to the north and the Hercynian massifs of central Europe (Fig. 13-2). The sediments laid down in this gulf were mainly clays and sands, but there were some marls. In the central part of the gulf deposition seems to have been continuous, but on the margins there are numerous hiatuses and thin formations of more variable facies.

With the advent of the Oligocene a great change took place in the paleogeography of this northern region (Fig. 13-4), for seas spread from southern Russia across Poland and northern Germany and joined the North Sea. During the early Cenozoic a seaway on the eastern side of the Ural Mountains had joined the Arctic and the Tethys. The Oligocene transgression across northern Europe resulted in the introduction of many new faunas to the North Sea Basin. Similar geographic changes took place in the Jurassic and Cretaceous (§ 9-1).

The southern shores of this Oligocene sea encroached upon the northern margins of the Rhenish and Bohemian massifs and extended farther south in a broad gulf between these massifs as far as the Jura Mountains in eastern France. The sediments laid down in this great sea were mainly clay, sand, and marls.

The western margin of the North Sea spread across southeastern England during several transgressive phases that alternated with phases of regression. The rest of England, Wales, and Scotland was emergent at this time and probably was of low relief. River systems that developed on this upland region emptied into the embayed regions of southeastern England. The sedimentary facies deposited range from fluvial sandstones to estuarine, deltaic, and normal shelf marine formations. At the close of the Eocene the seas retreated completely from the southeastern coastal region of England, but brackish-water and lacustrine deposits accumulated during the Oligocene in the general region of the Isle of Wight during two brief intervals.

Another region of very active deposition during the Eocene was western Scotland and northern Ireland, where volcanic activity was particularly intense. Lava flows and ash beds more than 5,000 feet thick occur in this region. The lava flows are well known for the exceptionally fine columnar structures developed in some places; those called the Giant's Causeway, in northern Ireland, are particularly famous. Detailed study of

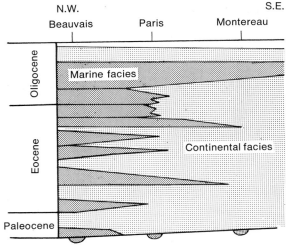

FIGURE 13-11

Diagrammatic cross section showing the relations of the marine and nonmarine formations of Paleocene-Eocene-Oligocene age in a northwest-southeast line through Paris. The length of the section is about eighty miles. (After M. Gignoux, 1955.)

the volcanic deposits shows that they probably came from several large volcanoes rather than from fissures. Occasionally there are thin interbedded units of sandstones, shale, lignite, and lacustrine limestones. The fossil plants preserved in these sedimentary layers indicate an Eocene age.

Upper Cenozoic marine formations of the countries adjoining the North Sea are a monotonous series of sand and clay deposits. The transgressions of these epochs were not great. During the Miocene the seas encroached only upon the lower portion of Denmark, the adjoining parts of Germany, and the coastal regions of Holland and Belgium, but did not reach the coast of England or Scotland. In the Pliocene the seas encroached upon southeastern England and parts of coastal Belgium and Holland, and appear to have barely touched Denmark. The fossil fauna of the English Pliocene formations contains a considerable number of warm-water types. This fact indicates that there must have been communication with the Atlantic at that time.

Central Europe

Nonmarine formations representing stream and lake deposits are known in several places on the continental region between the Tethys, the North Sea, and the Atlantic. Another rather common early Cenozoic deposit was red clay, which filled fissures in Mesozoic limestones. These clays are particularly important for their fossil mammalian faunas.

During the Oligocene sedimentation was especially active in faulted basins within and between some of the Hercynian massifs. Some of these deposits are more than 6,000 feet thick, reflecting a tremendous local subsidence during this period.

Isolated basins of lacustrine and fluvial deposition, such as existed during the early Cenozoic, persisted through the late Cenozoic. In eastern Germany and Poland, during the Miocene and Pliocene, there was a large basin where the deposits consisted of thick beds of lignite interbedded with sand and clay. The area appears to have been somewhat like the coastal plains of the lower Mississippi River today, with lush forest and abundant swamps.

There was some volcanic action on the continent during the Miocene and Pliocene. It was especially heavy in the Massif Central of France and in central Germany. The volcanics of the Massif Central are mainly Pliocene, but the activity began in the Oligocene, and some of it continued into the Pleistocene. Before these great volcanoes and their associated deposits had been attacked by erosion, volcanoes as grand as Mount Etna commanded the landscape of central France.

13-2   NORTH AFRICA

The southern margin of the Tethys had a Phanerozoic geological history very different from that of the northern margin. The only folded mountain ranges are in western North Africa — in Morocco, Algeria, and Tu-

nisia. The eastern half of North Africa, comprising Libya and Egypt, underwent no intense orogenic phases in the Mesozoic or Cenozoic, and the formations, which are mostly of carbonate facies, have been only moderately folded. Because of this great difference in the structural and depositional history of these two parts of North Africa, we discuss them separately.

Morocco-Algeria-Tunisia

The Cenozoic history of this part of North Africa is highly variable owing to the juxtaposition of different east-west structural provinces (Fig. 13-9). On the coastal belt a series of massifs composed of metamorphosed Paleozoic strata are overlain by a folded cover of Mesozoic and early Cenozoic formations. During the Mesozoic these massifs were anticlinal highs that clearly influenced the sedimentation both of that time and of the Cenozoic. This region makes up the present Kabylie Range. Immediately south of this structural unit are the Tell Atlas Mountains, which are a folded belt of Mesozoic and Cenozoic strata. These mountains are bordered on the south by the Hauts Plateaux of Algeria, which continue into the Moroccan Meseta to the west. This is a belt that was involved in Hercynian orogenies and that during the Mesozoic and Cenozoic acted as a positive block. To the south, the Saharan Atlas was an active geosyncline in the Mesozoic, but in the Cenozoic only its eastern part was inundated. The great Ethiopian shield has, since the Precambrian, been a stable bulwark over which the Tethyan geosyncline has rarely transgressed. Figure 13-12 summarizes the relations of the various structural units.

The early Cenozoic seas inundated a wide gulf, covering nearly all of Tunisia and extending westward halfway across Algeria along the trend now occupied by the Saharan Atlas (Fig. 13-2). The sediments deposited in this sea are mainly shallow-water facies consisting of marls and

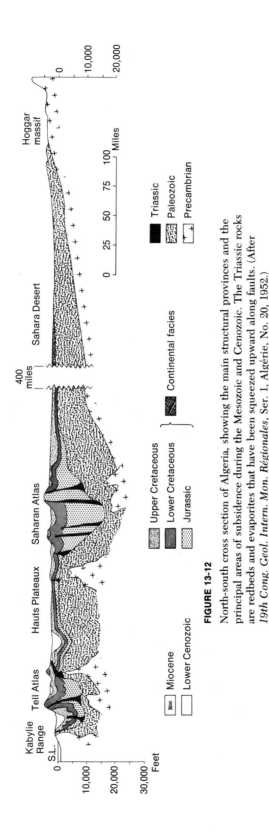

FIGURE 13-12

North-south cross section of Algeria, showing the main structural provinces and the principal areas of subsidence during the Mesozoic and Cenozoic. The Triassic rocks are redbeds and evaporites that have been squeezed upward along faults. (After *19th Cong. Geol. Intern. Mon. Régionales*, Ser. 1, Algérie, No. 20, 1952.)

lime-stones, which appear to be conformable to the underlying Cretaceous. Some of the limestone beds contain important commercial deposits of phosphate. Farther north seas covered the Tellian trough, where deeper-water carbonate facies were deposited. In the middle Eocene a slight orogenic phase raised a tectonic ridge that partitioned the Tellian trough into north and south components. The sedimentary deposits laid down in these two troughs were still mainly carbonates with an abundance of nummulites, a deposition that suggests that the tectonic ridge must have been of low relief. At the same time conditions were changing in the large gulf along the Saharan trend. The marginal areas of this gulf became lagoonal, and conglomerates and sandstones accumulated there, while varicolored gypsiferous shales became widely distributed in the main part of the gulf.

At the close of the middle Eocene, orogeny affected most of this region of North Africa. The Kabylie massifs (and their Moroccan counterparts) were uplifted in giant anticlinal flexures, and the sediments of the Tell and Riff troughs were compressed into structurally simple folds. The High Atlas was folded and overthrust, but its easterly extension, the Saharan Atlas, was only mildly deformed.

The late Eocene was ushered in by a transgression that covered only a restricted area north of the Hauts Plateaux. The Kabylie massifs were partially submerged and appeared as an island chain. The late Eocene and Oligocene sediments are mainly shale, marl, and sandstone, which reflect widely different source areas (Fig. 13-4). The most widespread facies unit, consisting of sandstones, shales, and marls, was derived from a metamorphic and sandstone terrain to the north (in the Mediterranean). The detritus directly derived from the Kabylie massifs also gave rise to a distinct facies unit. Another facies unit, confined largely to Tunisia, consists of detritus derived from Lower and mid-Cretaceous continental sandstones of the Saharan Atlas. Finally, in the western part of Algeria, just north of the Hauts Plateaux, are soft calcareous sandstones, marls, and limestones, all of which came from Jurassic sandstone and limestone formations of the Hauts Plateaux.

At the close of the Oligocene another important orogenic phase caused the complete emergence of Algeria and Tunisia, and Miocene strata rest on folded and eroded Oligocene and older formations except in eastern Tunisia, where the Miocene and Oligocene are conformable. During a short episode in the early Miocene a narrow seaway spread from Tunisia across Algeria along the old path of subsidence, the Tellian trough. The wide variety of facies, marine and nonmarine, consisting of marls, sandstones, and conglomerates, reflects continued tectonic instability and changing environmental conditions in this region. During this episode volcanoes, accompanied by the intrusion of granitic batholiths, appeared in the Kabylie massifs. Toward the close of the Miocene the region was deformed by a final orogenic phase, which was followed by epeirogenic uplift. A comparable level of tectonism persisted throughout the Pliocene and continued with diminished force into the Pleistocene.

Pliocene seas occupied only a few small embayments in Algeria and the coastal region of Tunisia, where clays and sands were deposited. In both countries the Pliocene strata have been folded in later orogenic phases.

The Gates of the Mediterranean

During the first half of the Cenozoic the western end of the Tethys connected with the Atlantic Ocean through two straits, one in southern Spain and the other in Morocco (Fig. 13-9). The Riff Strait, in Morocco, is a western continuation of the Tellian trough of Algeria and is bounded on the north by a Paleozoic massif and on the south by the Moroccan Meseta. The Betic Strait of southern Spain lies between the Spanish Meseta and another massif on the southern border of Spain. During the early Cenozoic the Spanish Betic massif and the Moroccan Riff massif were parts of a single positive block. Marine sediments ranging well up into the Miocene are found in both of these straits, clearly indicating sea connections between the Tethys and the Atlantic Ocean. In the late Miocene an orogeny folded and thrusted both of these sedimentary regimes. In the Riff zone intense deformation established a structural pattern similar to that in the Alps, where nappes developed and then moved south. The latest Miocene and Pliocene sedimentation in both regions consisted of nonmarine deposits. At the beginning of the Pliocene these two straits were definitely closed, but a new thoroughfare developed through the dislocation and collapse of the Betic-Riff massif to form the Strait of Gibraltar. Marine Pliocene formations are found at several places along the Strait of Gibraltar.

Egypt-Libya

Because Egypt and Libya lie on the southern border of Tethys, their geological history is marked by epeirogenic movements but no orogenic movements. This largely desert region contains a few high uplifted areas, but most of the country is low, undulating terrain, much of it covered by drifting sand. Almost all outcropping strata are horizontal or have very small dips. In general, outcrops in the northern parts of these countries expose Cenozoic strata, and those in the southern half expose Mesozoic and some Paleozoic strata. Geological exploration, by traditional means of surface mapping, is particularly difficult. Over the past two decades there has been an intense program of petroleum exploration by many private concerns with, it should be noted, spectacular success. The exploration has yielded large amounts of geological data, especially from

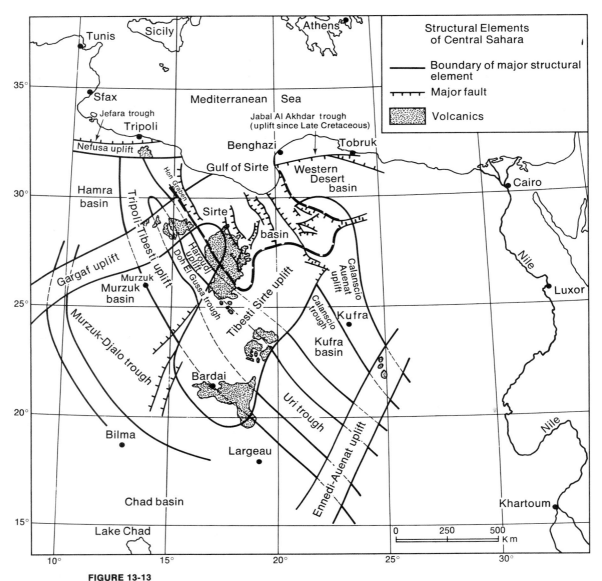

FIGURE 13-13

Map showing major structural elements in Libya and adjacent areas.
(After E. Klitzsch, Petrol. Exploration Soc. Libya, *10th Ann. Field Conf.,* 1968.)

geophysical surveys (both gravity and seismic) and drill holes. The geological history turns out to be extremely complex (Fig. 13-13).

In early Paleozoic time differential subsidence formed a system of troughs and uplifts striking northwest-southeast. The sediments deposited in these troughs are entirely sandstones and shales of both marine and continental facies. This early phase was followed in the period from the late Paleozoic to the Early Cretaceous by the development of a system of uplifts and troughs of northeast-southwest trend, approximately perpendicular to the earlier trend. The deposits of this phase are primarily clastics. In the northern part of the country they tend to be marine or marginal marine facies, but in the central and southern part of the country they are dominantly nonmarine facies. Finally, in late Cretaceous and Cenozoic time, northwest-southeast striking block-faulting accompanied the formation of a system of grabens and horsts in north-central Libya (Sirte basin), where the interaction of older uplifts had earlier formed a large structurally high area. During the early subsidence in the Sirte basin area, the Late Cretaceous seas advanced into north-central Libya. On most of the shallow shelf areas, limestone, dolomite, and marl were deposited, but the graben-type troughs of the Sirte basin were filled with bituminous shale. Carbonates, including reef developments, characterize the horst blocks in the Sirte basin. In early Cenozoic time the transgression that began in the early Upper Cretaceous reached its climax. The sediments consist mainly of limestone, dolomite, and marl. Many of the limestone formations are composed primarily of the remains of nummulites. The Sphinx and pyramids of Egypt are built of blocks of Eocene nummulitic limestones (Fig. 13-14). Tectonic movements were of less amplitude than during the late Cretaceous. By the close of the Miocene the seas had regressed from the coast of these north African countries.

13-3 THE MIDDLE EAST

Diastrophism is the main theme of the Cenozoic history of the Middle East. The Late Cretaceous orogenic phase continued until the early Eocene. Most of the structural features that characterize this region were developed in their present form during the Cenozoic. The main structural provinces are as follows (Fig. 13-15). On the southwest,

FIGURE 13-14

The Sphinx and pyramids of Egypt are built of blocks of Eocene nummulitic limestone. (Photograph by the author.)

the Arabian platform, which is part of the Arabian basement shield, is covered by sediments. No orogenic movements have occurred in this area since the Precambrian, only epeirogenic warpings, warpings due to salt flowage at depth, and structures attributed to basement faulting. Northeast of the Arabian Platform is the folded belt of the Zagros Mountains. In this region, sedimentation was nearly continuous from the late Precambrian to the Recent. From the late Precambrian through the Permian, platform facies were deposited; from the Triassic through the Miocene, geosynclinal facies were deposited; and from the Pliocene to the Pleistocene, synorogenic and postorogenic conglomerate beds were deposited. The folding in this region took place only in the latest phases of the Alpine orogeny in Pliocene-Pleistocene time. The rock formations of the Zagros folded belt show marked changes in thickness and facies. This indicates epeirogenic movements in preorogenic times, resulting in very gentle, large-scale undulations of the sea floor. These undulations were aligned partly parallel with the north-south trend of the Arabian Platform rather than parallel to the northeast-southwest trend of the Zagros thrust zone. A close relation of the folded belt with the Arabian Platform is indicated; the folded belt may be considered as a marginal, mobile, sedimentary trough superimposed on this platform.

The folded belt passes northeastward without a sharp boundary into a narrow zone of thrusting bounded on the northeast by the main Zagros thrust line. In this zone older Mesozoic and Cenozoic rocks and the Paleozoic platform cover were thrust southwestward on younger Mesozoic and Cenozoic rocks of the folded belt. It is therefore mainly in this thrust zone that the remarkable conformity of the sedimentary fill of the Zagros trough is displayed. The thrust zone represents the deepest part of the Zagros geosyncline, wherein accumulated 12,500 feet of Jurassic through Eocene ma-

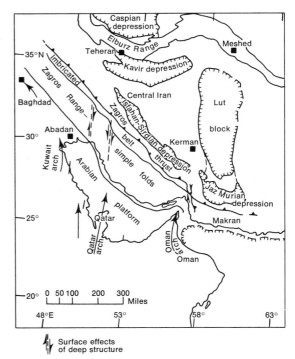

FIGURE 13-15

Main structural features of Iran. (After N. L. Falcon, *Advancement of Sci.*, **24**, 1967.)

rine deposits. Upper Cretaceous strata are *Globigerina* marl, radiolarite, and ophiolites; the Paleocene deposits are thick clastics (Fig. 9-22). These Late Cretaceous-Paleocene strata reflect important diastrophic movements in the thrust zone but did not affect the folded belt.

The structural zone of central Iran comprises a roughly triangular area limited by the Lut block on the east, the Elburz Mountains on the north, and the Zagros thrust zone on the southwest. The southeastern margin of this structural zone is a long, linear depression (the Isfahan-Sirdjan depression) that is continued in the Jaz Murian depression of Iranian Baluchistan. A similar area of depression (the Kavir depression) lies just south of the Elburz Mountains. Central Iran, like the rest of Iran, had a platform character during Paleozoic time, but became a very mobile orogenic zone in Mesozoic and Cenozoic time. In this zone the Mesozoic movements are expressed most clearly in pronounced unconformities, granite intrusions, and, in some localities, low-grade metamorphism. The later Alpine orogenic movements also were at least as intense in this zone as in the neighboring Elburz and Zagros ranges.

The Lut block is an irregularly outlined, essentially north-south-trending, rigid mass surrounded by the ranges of central and east Iran. The remarkable north-south trend of the bordering ranges apparently was enforced by the rigid block. Paleozoic and Mesozoic formations in the Lut block are thinner than in any other region of Iran. There are, however, upwards of 10,000 feet of Cenozoic volcanic rocks. The Lut block, which has been emergent since the end of the Cretaceous, is clearly a true median mass, though of much smaller extent than that assumed by earlier geologists for all the Iranian highlands between the Zagros and Elburz ranges.

The east-west Makran Range of southeastern Iran merge with the north-south-trending East Iranian ranges that lie east of the Lut block. The Makran Range is also apparently a continuation of the Zagros Mountains, but in the area of the Oman arch great changes in facies pattern take place, obscuring the relationship. The Makran and East Iranian ranges contain very thick Mesozoic and Cenozoic deposits and underwent several phases of diastrophism. The deep sea basin south of the Makran coast appears to be a large downfaulted feature of Cenozoic age.

The Elburz Range of northern Iran is closely related stratigraphically and structurally to central Iran. There was actually less mobility in the area of the Elburz during the Mesozoic than in central Iran. Granites, metamorphism, ophiolites are not present. The first physical expression of the Elburz Mountains came in the Eocene. Continued movements in the late Cenozoic

produced the range as we know it today.

In the Zagros geosyncline the Mesozoic pattern of nearly continuous limestone and shale sequences continues into the Cenozoic. The principal oil-producing formation of southwestern Iran is the Asmari Limestone, which is mainly Oligocene and Miocene in age. Similar oil-producing limestones in Iran are mainly Eocene and Oligocene in age. Overlying these limestone formations is a complex series of gray marls, anhydrite, and salt, with subordinate thin foraminiferal and shelly limestone formations. In the middle Miocene more and more clastic sediments entered the geosyncline and caused a southeastward withdrawal of the sea. This change in facies pattern reflects further orogenic movements in the Zagros Mountains. Continued uplift in the Pliocene contributed exceedingly thick fine-to-coarse terrestrial deposits to the basins to the southwest of the uplifted belt. The orogenic movements in the Pliocene also folded the Cenozoic formations that lay to the southwest of the Zagros Mountains.

The area of Iran to the northeast of the Zagros Mountains had a very different history marked by several orogenic phases during the Cenozoic. The most conspicuous feature of the early Cenozoic of this region is the widespread development of volcanic deposits, in places of considerable thickness (Fig. 13-16). These volcanic deposits were laid down in shallow marine seas. The Elburz Mountains make their appearance at this time as principally a volcanic range separating central Iran from the Caspian basin to the north. In central Iran the early Cenozoic was also a time of extensive accumulation of salt and other evaporite deposits. The last regional marine transgression onto central Iran was in the late Oligocene-early Miocene. The Kavir basin (Fig. 13-15) of central Iran began to take shape in Oligocene-Miocene time. In the earlier Cenozoic 7,500 feet of marine marl and limestone were deposited in it. During

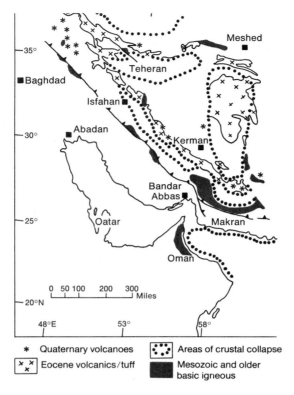

* Quaternary volcanoes   :•: Areas of crustal collapse

x x Eocene volcanics/tuff   ■ Mesozoic and older
x   basic igneous

FIGURE 13-16

Areas of major subsidence and volcanic activity during the Cenozoic in Iran. (After N. L. Falcon, *Advancement of Sci.*, **24**, 1967).

Miocene movements this area became land again and, after a second evaporitic phase another 15,000 feet of continental Miocene-Pliocene redbeds were deposited in it. Both the evaporite sequences contain salt and have formed salt plugs.

The youngest folding movements occurred in Pliocene-Pleistocene time and involved mountains and basins alike. Pliocene-Pleistocene conglomerate, spreading from the mountain ranges and gradually grading into finer clastics toward the basins, shows that the mountain-and-basin pattern created in central Iran by earlier movements was accentuated markedly by this latest tectonic phase. Tilted Quaternary terraces, the steplike raised beaches of the Makran coast, the well-known fluctuations of the Caspian shoreline, several sub-Holocene volcanoes, numerous active faults dissecting terraces and sub-Holocene alluvium, and the numerous earthquakes recorded in the recent history of the country are clear proof of continuing structural instability.

13-4 INDIA

The close of the Cretaceous was marked by a general regression of the seas from the Tethys and from the isolated embayments peripheral to the peninsula. During the early Cenozoic the seas again invaded the Tethyan realm in a wide belt covering West Pakistan and Kashmir and in the main Himalayan region to the latitude of Lhasa in Tibet. This pattern was very much like that of the Late Cretaceous seas in this area. The largest amount of subsidence and deposition in this early Cenozoic geosyncline was in Baluchistan. Here many thousands of feet of marine clastic and carbonate formations, with some coal-bearing horizons, document the Paleocene and Eocene. Many of the limestones contain abundant nummulites. Eastward, toward the shield of peninsular India, these early Cenozoic strata thin greatly and contain some horizons of evaporites. In the central Himalayas, some 100 miles east of Mount Everest, at Kampa Dzong, Tibet, about 1,000 feet of lower Cenozoic strata are present. Here the facies are a mixture of sand and shale with some nummulitic limestone.

Toward the close of the Eocene the Tethyan geosyncline underwent orogeny, accompanied by the intrusion of plutonic granite bodies. This orogenic phase marked the beginning of the formation of the Himalayan mountain chain. The compression, folding, and uplift of the Tethyan belt brought about profound changes in the geography of the region. Oligocene and Lower Miocene marine seas persisted only in southern West Pakistan; north of here and along the front of the newly folded mountain chain large fresh-to-brackish bodies of water persisted. In the western part of a marine gulf that occupied most of Baluchistan, and southern West Pakistan, great thicknesses of shallow-water sandstones and green arenaceous shales, with only minor limestone strata, accumulated (Fig. 13-17). In the eastern half of the gulf, however, the facies is much more calcareous and fossiliferous. In northern West Pakistan and adjoining parts of India Oligocene strata seem to be absent, but unconformably overlying the Eocene formations we find a very thick series (8,000 feet) of sandstones and shales of fresh-to-brackish water facies (Murree Series). The clastic sediments of the Murree Series were derived mainly from Peninsular India.

The most intense orogenic phase in the Cenozoic history of the Himalayas came in the middle Miocene after the deposi-tion of the Murree Series. This phase was marked by many great thrust faults and severe deformation of the Tethyan formations and by intrusion of great masses of granite into the axial region of the main Himalayas. Between the rising Himalayas and the Indian shield there developed a subsiding trough that extended all across northern India; in this trough accumulated 15,000–20,000 feet of fluvial and lacustrine sediments derived from the Himalayan Mountains (Fig. 13-18). This tremendous complex of terrestrial sediments, known as the Siwalik Series, gives a continuous record of sedimentation from the upper Miocene to the lower Pleistocene and has yielded many fossil mammals. The sediments and the fossils show that a wide range of environments existed within the

FIGURE 13-17

Miocene sandstones and shales (flysch) in southwestern Baluchistan. (Photograph by J. V. Harrison.)

depositional area of the Siwalik sediments, ranging from humid forest environments to open desert.

A simple hydrographic pattern has been suggested to explain the Siwalik sedimentation. The continuity of the deposits, together with their partial valley-flat character, suggests lateral distribution by a single large stream, named the Siwalik River. As originally conceived, this river flowed out of the Himalayas in Assam and proceeded to the northwest through the sedimented basin of the Himalayan foredeep to West Pakistan, where it turned south and followed the course of the Indus to the Arabian Sea. The present twofold drainage by the Indus system of rivers and the Ganges-Brahmaputra-Tsengpo system points to a partial reversal of drainage by stream piracy. There is, however, some lack of agreement as to when this separation happened and what stream was responsible for the capture of the upper part of the original single river. Today the two basins are separated only by an imperceptible divide near Delhi.

Continued movement was probably going on in the Himalayas during the Miocene and Pliocene, but toward the close of the Pliocene a severe orogenic phase folded the Siwalik Series, and it was overthrusted by older rocks of the Himalayas. The uppermost strata of the Siwalik Series were laid down in the early Pleistocene and consist of coarse sandstones and conglomerates. The foredeep in which the Siwalik Series accumulated is, for the most part, now covered by Holocene and sub-Holocene alluvial deposits up to 10,000 feet thick, deposited by the Indus and Ganges river systems (Fig. 13-18). Within the Himalayas there are extensive glacial deposits of Pleistocene age. During the Pleistocene there was a final uplift of the Himalayas, which in some places is thought to have been as much as 6,000 feet (Fig. 13-19). Uplift is still going on throughout the Himalayas. The main elevation of the Himalayas was an event witnessed by the earliest men. A famous Indian geologist, B. Sahni, has suggested that some of the earliest human migrations were facilitated by passages through a barrier of less forbidding height and steepness than the Himalayas of today.

Through the Miocene and Pliocene, while the Siwalik deposits were being laid down, marine conditions persisted in southern Baluchistan. Here a thick series of marine sandstones and shales, with some shelly limestones, accumulated. With the late Pliocene withdrawal of the last remnant

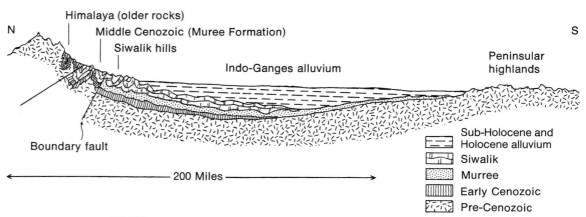

FIGURE 13-18

Diagrammatic cross section of the Cenozoic foredeep lying between the Himalayan Mountains and the stable shield of peninsular India. (After D. N. Wadia, 1953.)

FIGURE 13-19

K-2 in the Karakorum Range of Kashmir (20,449 feet). The present elevation of this
and the other ranges of the Himalayas resulted from epeirogenic uplift in the late
Cenozoic. (Photograph by Vittorio Sella, Gardner Collection, Harvard Univ.)

of the former great Tethys, marine episodes
came to an end in this region.

Deposition of sedimentary formations in
peninsular India during the Cenozoic was
confined to small isolated basins. During
the latest Cretaceous and the early Cenozoic, however, basaltic lavas, pouring out
through numerous fissures, accumulated in
some places to more than 5,000 feet in
thickness. These are the **Deccan traps**,
which, next to the Gondwana formations,
make up the most significant Phanerozoic
deposits, of peninsular India. Deccan is
a region of western India. "Trap," a vague
term for basaltic rock, is used in India to
mean steps or stairs—an allusion to the

usually step-like aspect of the weathered,
flat-topped hills of basalt in Deccan. The
Deccan traps now cover an area, on the west-
central half of the peninsula, of more than
200,000 square miles (Fig. 4-14, p. 103).
Isolated outlying remnants of the traps suggest that at one time they extended over
twice this area. The basalts rest on the undulating eroded surface of older rocks and
are thickest along the western coast, where
they are more than 7,000 feet thick; farther
east they thin to a thousand feet or less. The
individual flows that make up the total
thickness of the Deccan traps vary greatly in
thickness, from a few feet to 120 feet. The
traps originally extended for an unknown

distance to the west of India, but that part was faulted down, probably in or before the Miocene. The Bombay coast, for example, is a fault scarp on which the movement was 6,000–7,000 feet. Interbedded with the basaltic flows are occasional sedimentary bodies, representing lacustrine and alluvial deposits, which contain some fossils. The only movements that have affected the Deccan traps since their deposition have been some slight folding and faulting.

13-5 ASSAM AND BURMA

Throughout the Mesozoic Assam and Burma were part of the Tethys, but in the latest Cretaceous or possibly in the earliest Cenozoic orogenic activity greatly changed the geographic and tectonic relations of the region and established the pattern of Cenozoic sedimentation. This orogeny, which was accompanied by extensive igneous activity, uplifted the highlands of eastern Burma—the Shan plateau—and at the same time a long, narrow belt of western Burma— the beginning of the Arakan Yoma. A southward continuation of this tectonic ridge can be seen on the Andaman and Nicobar islands, which lie between Burma and Sumatra. To the north the Arakan Yoma joins the Naga Hills of eastern Assam. In the region between eastern Burma (Shan plateau) and the uplifted portion of the eastern Himalayas there were two large gulfs separated by the long, linear, tectonic highland of the Arakan Yoma and the Naga Hills (Fig. 13-20). During the Cenozoic this region underwent the sedimentary filling of the gulfs and continued tectonic movement of the adjoining highlands.

During the early Cenozoic (from Paleocene to Oligocene) the upper reaches of both the Assam Gulf and the Burmese Gulf were occupied by extensive deltas, which fanned out into seas occupying the lower parts of the gulfs. Most of the sediments laid down in these gulfs were clays and sands, but in western Assam during the Eocene nummulitic limestone facies were extensively developed. The greater part of the Assam Gulf was emergent during the late Eocene and the Oligocene, and thick, coalbearing, clastic formations were deposited. The thickest and most extensive of the coal beds are found in northern Assam. The lower Cenozoic formations in the Assam Gulf have a thickness of about 18,000 feet.

The sedimentary pattern for the early Cenozoic was similar in the Burmese Gulf. At the earliest stages the seas reached well into northern Burma, but the terrestrial delta deposits progressively encroached southward. The marine sediments are largely clay shales. A north-south cross-section of the Cenozoic strata of the Burmese Gulf is shown in Figure 13-21.

At the close of the Oligocene diastrophic movements in the Assam Gulf led to limited uplift and erosion so that the overlying Miocene strata were deposited on a well-marked unconformity. In the Burmese Gulf this horizon is marked only by a break in the paleontological record. Through the remainder of the Cenozoic, sandstones, conglomerates, and shales are the main sedimentary facies. The marine facies are confined to the southern regions of the gulfs. Many of these later sedimentary formations are correlative with the Siwalik strata of India and contain an abundant mammal fauna.

Orogenic activity was very pronounced in the Arakan Yoma during the middle Miocene and along the narrow belt of land in the middle of the Burmese Gulf. A final orogenic phase in the early Pleistocene raised the Arakan Yoma and the central ridge of the Burmese Gulf (Pegu Yoma) almost to their present elevation and drove the Burmese Gulf farther southward. During these later orogenic movements abundant volcanic activity was concentrated along the narrow central ridge of the Burmese Gulf.

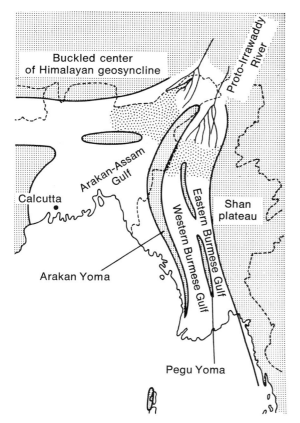

FIGURE 13-20

Paleogeographic map of Assam and Burma for the early Cenozoic. (After L. D. Stamp, redrawn from Chhibber, 1934.)

13-6 INDONESIA

With the advent of the Cenozoic Era, Malaya, Sumatra, Java, most of Borneo, the Timor-Celebes geosyncline, and the region between New Guinea and Australia were emergent. The early Cenozoic seas transgressed over much of this area, but the sediments are of shallow shelf or estuarine facies (Fig. 13-22). Malaya, Sumatra, western Borneo, and the area between them continued to be a land area undergoing erosion.

By the mid-Cenozoic marine transgressions had increased the sea areas considerably (Fig. 13-23). Most of Sumatra was in- undated, as was Borneo except for an axial land region. The remaining islands of the archipelago were likewise under water. Like the early Cenozoic deposits, all the mid-Cenozoic deposits clearly indicate deposition in shallow shelf seas. During the late Miocene came the greatest Cenozoic folding and thrusting.

The most intensely affected area in this Miocene folding was the curved belt comprising the Timor-Celebes geosyncline; other areas affected include the small islands off the western coast of Sumatra, a few areas in southwestern Sumatra, southern Java, and the Lesser Sunda Islands. New Guinea appears to have been folded during this same phase of compression.

During and after this late Miocene folding erosion was very active, and it eventually reduced the land to low relief. There then developed a remarkable series of subsiding basins that accumulated many thousands of feet of marine, fluvial, and lacustrine sediments. One of these basins was along the northeastern coast of Sumatra, and a very large one occupied east-central Sumatra. All of central and northern Java was an active area of deposition in another of these basins, as were the eastern part of Borneo and the southern tip of the Celebes Islands. The deposition within these various basins was variable: in some the lowest strata are terrestrial deposits that grade upward into marine beds; in others the lowest strata mark a marine transgression. The proportions of marine and terrestrial facies are as variable as their sequence, but generally the younger deposits (Pliocene) are more and more of terrestrial facies. Marine conditions persisted in northern Java and northern Sumatra through most of the Pliocene, but at the same time central and eastern Sumatra and the southern Celebes were low marshy areas (Fig. 13-24). During this period the islands off the western coast of Sumatra, parts of western Sumatra, and the Timor-Celebes area were again invaded by the seas. But in most of these areas sedimentation was confined to faulted troughs.

Toward the close of the Pliocene broad folding affected the several Miocene-Pliocene basins discussed above on eastern Sumatra, northern Java, eastern Borneo, and the southern tip of the Celebes. The oil fields of Sumatra are within the folded belts of these Miocene-Pliocene basins. At some time during the late Cenozoic there was much volcanic activity, which was accompanied by various plutonic intrusions that invaded a narrow belt of western Sumatra and Java. Whether this igneous activity began during the Miocene orogeny, during the Pliocene period of folding, or later remains unknown. The volcanoes active today are in this narrow zone of western Sumatra, Java, the Lesser Sunda Islands, and parts of the Celebes. Whatever the age of the volcanic activity, after the

late Pliocene folding western Sumatra and southern Java were greatly uplifted, and their relief was accentuated by the active volcanoes then present along this zone. The remaining parts of Sumatra and Java and eastern Borneo were eroded to low relief. Bathymetric (depth-contour) charts of the South China Sea and Java Sea clearly show the existence of drowned rivers. During a part of the Pleistocene the sea level was much lower than it is at present, and the areas of part of the South China Sea and the Java Sea were emergent. Streams from the highlands of western Sumatra and western Borneo flowed across this raised shelf and emptied into the South China Sea. Similarly, streams from Java and southern Borneo emptied into the eastern part of the present Java Sea. Late in the Pleistocene the rise of the sea level again inundated this area, and further epeirogenic

FIGURE 13-21

Diagrammatic cross section of Cenozoic formations of the Burmese Gulf, illustrating the intertonguing of the marine and non-marine facies. (After L. D. Stamp, *Geol. Mag.*, 1922.)

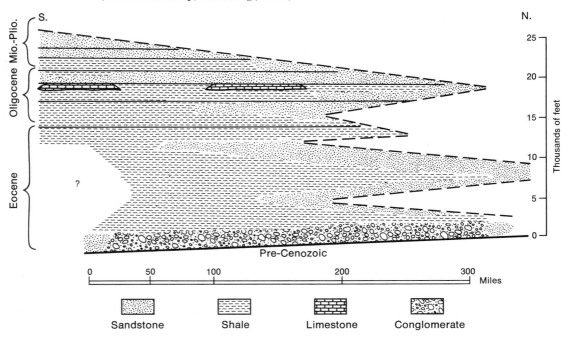

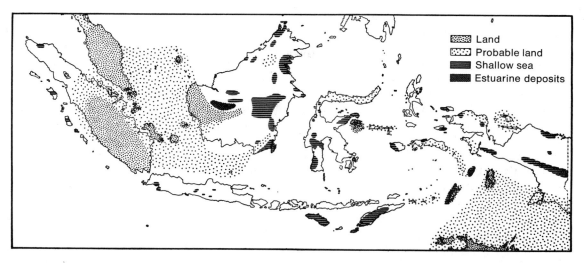

FIGURE 13-22

Paleogeographic map of Indonesia for the Eocene. (From Umbgrove, 1938.)

FIGURE 13-23

Paleogeographic map of Indonesia for the Miocene. (From Umbgrove, 1938.)

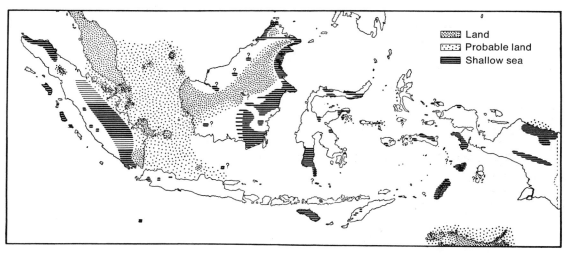

movements developed the modern Indonesian archipelago.

In summary, it may be well to point out the main geologic zones in the western part of the Indonesian archipelago (Fig. 13-25). The southwesternmost zone includes the islands off the western coast of Sumatra, which were folded during the Miocene but have no volcanoes. A continuation of this zone is the Timor-Celebes belt, the islands of which are characterized by block faulting, with deposition in faulted troughs during the Pliocene and subsequent elevation of their contents.

A second zone is represented by the highlands of western Sumatra, southern Java, and the Lesser Sunda Islands, which bear a conspicuous girdle of volcanoes. Northeast of this zone lie the Miocene-Pliocene depositional basins that were moderately folded toward the end of the Pliocene and then elevated.

Finally, in the third zone are Malaya, parts of Borneo, and the islands between Sumatra and Borneo. This region was involved in Paleozoic (Hercynian) and Mesozoic orogenies but throughout the Cenozoic was emergent and undergoing erosion.

Lying outside the first zone is a deep linear trough; another such trough lies between the first and second zones. Marginal deeps like these are characteristic of the circum-Pacific region. When these various structural zones are followed north into Burma, they are found to coincide with the structural zones already recognized in that country. Of special interest are the structural units in Burma that are comparable to the deep troughs of the Indonesian region—the marginal deep and the deep trough between the islands west of Sumatra and Sumatra itself (the Mentawei trough). In Burma these units are filled with Cenozoic sediments, but in Indonesia they are deep furrows on the sea floor. In Burma these structural units began to subside in the early Cenozoic, but it does not follow that the comparable Indonesian troughs began to subside at the same time. The structural and sedimentary evidence of the whole Indonesian region, in fact, suggests that these troughs did not begin to subside greatly until the late Cenozoic. The fact that they are deep indicates that subsidence has been much more rapid than sedimentation. In the Burmese troughs sedimentation was always extremely active and kept pace with the subsidence.

13-7 THE PHILIPPINES

The Philippine archipelago consists of eleven major islands and nearly 7,000 small islands between the lone island of Formosa on the north and the vast Indonesian archipelago on the south. The geologic histories of the Philippines and Indonesia are much alike in major orogenies and transgressions, and the surface and submarine morphology of the area between the archipelagoes clearly shows their relationship (Fig. 13-26).

The Philippine archipelago is set off from the Asiatic continent by the broad deep basin

FIGURE 13-24

Paleogeographic map of Indonesia for the Pliocene. (From Umbgrove, 1938.)

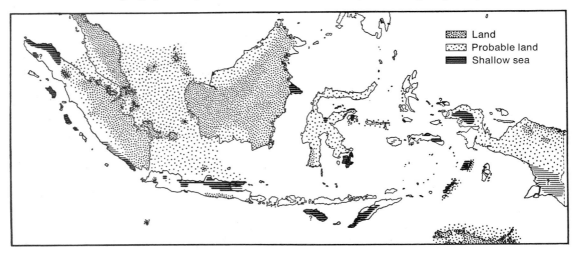

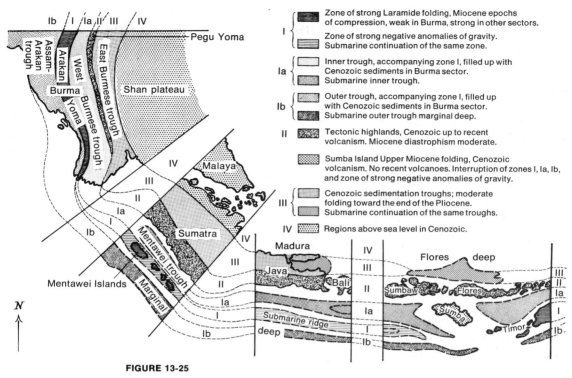

FIGURE 13-25

Summary map of the structural zones of western Indonesia and Burma. (Redrawn from J. H. F. Umbgrove, *Symphony of the Earth*, Nijhoff, The Hague, 1950.)

of the South China Sea on the west. On the east, the platform on which the archipelago rests is bounded by the great north-south Mindanao or Philippine deep, which reaches a depth of 34,218 feet. This great trench is clearly the tectonic boundary of the Philippine archipelago with the Pacific Ocean basin. To the south and southwest the Philippines connect with Borneo and the Celebes by three distinct ridges separated by oceanic deeps. The most northern of these ridges, the Palawan ridge, is composed largely of metamorphic and igneous basement rock. The surface expression of the central ridge, the Sulu archipelago, is largely volcanic islands, but some of the large islands and the southwestern peninsula of Mindanao are composed of basement rock, and for this reason the ridge also is thought to consist mainly of basement rocks.

The pre-Cenozoic formations cropping out in the Philippines consist mainly of igneous and metamorphic rocks, radiolarian cherts, and various types of volcanic deposits of unknown age. A single locality on Mindoro Island has yielded some ammonoids of Middle or Late Jurassic age. On Cebu Island a few localities have yielded some Cretaceous foraminifers. But a boulder from a late Cenozoic conglomerate in Mindoro has yielded fossil corals of middle or late Carboniferous age. Such a paucity of age data on the pre-Cenozoic formations makes it nearly impossible to reconstruct the earlier history and paleogeography of the archipelago.

Important orogenies must have affected the region in the late Mesozoic, for it was not until the late Oligocene or early Miocene that important marine transgressions

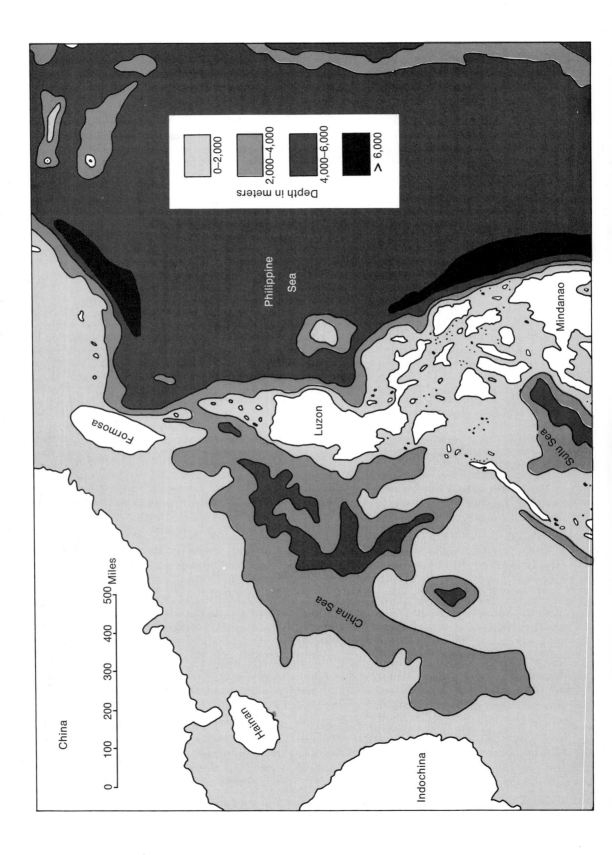

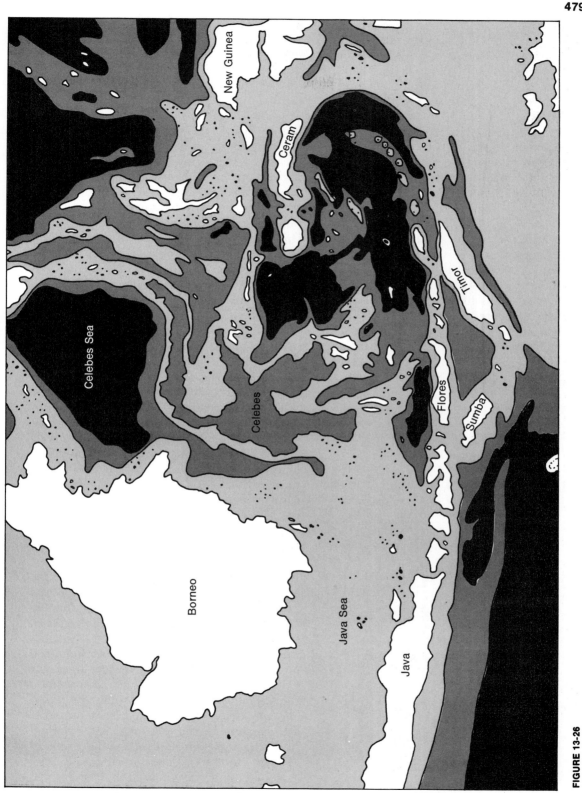

FIGURE 13-26

Bathymetric map of the southwestern Pacific, showing the continuity of the major relief elements extending from Indonesia into the Philippines. (From E. M. Irving, *Bull. Am. Assoc. Petrol. Geol.*, 1952.)

took place. The Cenozoic rocks of the Philippines are characterized by abrupt vertical and lateral lithologic changes, abrupt thickening and thinning, and, in general, poor sorting. Volcanic deposits are prominent in many regions. The rock record reflects, for the most part, a region of high relief, high tectonic activity, and archipelagic geography (Fig. 13-27).

Early Cenozoic strata (from Paleocene to middle Oligocene in age) are greatly restricted in distribution, mostly to the eastern margin of the archipelago. Most of the rocks of this age are carbonate facies deposited in the shallow waters of bays and shelves. A few localities on the western side of the archipelago have strata of this age, but it seems likely that they were not directly connected by a seaway across the archipelago to the eastern coast but were connected to the south with the Sulu Sea.

FIGURE 13-27

Areas of Cenozoic sedimentary rocks in the Philippine Islands. (After E. M. Irving, *Bull. Am. Assoc. Petrol. Geol.*, 1952.)

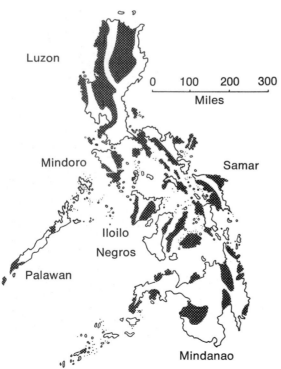

Widespread transgression occurred in the late Oligocene, and the distribution of the deposits suggests profound geographic changes to something resembling the archipelagic conditions of today. The facies of the late Oligocene and early Miocene are highly variable, ranging from limestones through coarse clastic formations and coal measures to volcanic rocks. The poor sorting, abrupt facies changes, and great thickness of the coarser clastic formations suggest deposition in narrow seaways that penetrated regions of high relief. Orogeny strongly uplifted the land during the late Miocene, and this orogeny was followed by the deposition of extensive, thick conglomerate formations. By the Pliocene, however, the land had been reduced to low relief, and the seas were more restricted in their distribution. The facies also reflect this change, in that sandstones are not widespread, but marls, limestones, and siltstones predominate. There was widespread reef development in the central Philippines.

The final molding of the Philippines came in the Pleistocene, when emergence and elevation of the high ranges came about. The deformation of this period was mainly epeirogenic and was accompanied by gentle folding, faulting, and much volcanism.

13-8 JAPAN

In latest Cretaceous time the Japanese Islands were highly emergent except for the persistent geosynclinal belt along the Pacific coast. The earliest Cenozoic rocks of the islands are in the same general areas as the latest Cretaceous formations but on a more limited scale. In most areas the early Cenozoic formations are unconformable on the Cretaceous. Paleocene strata are generally not recognized in Japan but probably are present in the marginal geosynclinal belt where 20,000 feet of Eocene to lower Miocene formations are present. Aside from this marginal geosynclinal belt the Japanese islands were mainly emergent and the site of nonmarine basinal deposition in the early Cenozoic (Fig. 13-28, A). A prominent embayment extended northward through western Kyushu Island, where a thick complex series of marine, estuary, lacustrine, and other deposits accumulated, some of which include coal seams. The facies are mainly nonmarine in the Eocene but became more marine in the Oligocene. The axial area of Hokkaido Island was also a center of thick accumulations of similar facies. In this area, however, an orogeny folded the basinal sediments, an event followed by extensive marine transgression in the Oligocene.

The long depositional history of the coastal geosynclinal belt (Mesozoic through Oligocene) was ended by an intense orogenic

FIGURE 13-28

Paleogeographic map of the Japanese Islands and adjoining regions for (A) the late Eocene, (B) the late middle Miocene, (C) the early Pliocene, and (D) the Plio-Pleistocene. (After Minato *et al.*, 1965.)

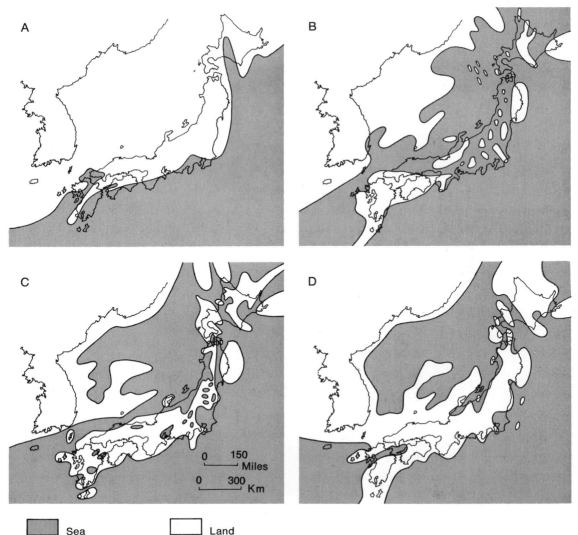

◼ Sea   ▢ Land

phase in the late Oligocene and early Miocene. Thereafter this belt was an integral part of the larger Japanese Island province. Miocene orogeny also affected the axial area of Hokkaido Island, and during the same epoch a transgression inundated most of Japan (Fig. 13-28, B), except for numerous islands, and parts of southeast Japan. The Sea of Japan also began to form as a result of faulting and general subsidence.

Volcanism in Japan occurred on a grand scale from the Miocene to the Holocene, generally along fault lines. During this episode Japanese Islands of today were developed. In general, Miocene and Pliocene deposits of Japan are developed in two distinct facies; in the Inner zones of Japan marine sediments include vast quantities of volcanic deposits, whereas in the Outer zone of Japan volcanic deposits tend to be absent or minor. Deposits of this age are especially well developed in the Inner zone of Northeast Japan. In this region thousands of feet of clastic sediments and volcanics occur in very complicated facies relations. Earth movements changed the geographic picture constantly. The great transgressive phase of the Miocene was followed by regression, which is still continuing. Pliocene seas were much less extensive than those of the Miocene (Fig. 13-28, C), and those of the Pleistocene still less so (Fig. 13-28, D).

Pleistocene deposits are volcanics or coarse clastics of mainly terrestrial origin. There are more than 200 principal Pleistocene volcanoes in the Japanese Islands, including thirty that have been active in historic time. Mount Fuji is a classic example of these Pleistocene volcanoes (Fig. 13-29).

FIGURE 13-29
Mount Fuji, one of the classic conical volcanoes of the world.
(Photograph by the National Park Association of Japan.)

13-9 NORTHERN ASIA

The vast area between the Ural Mountains to the west, the Pacific Ocean to the east, the folded Himalayas to the south, and the Arctic Ocean to the north was a remarkably stable block through the Mesozoic, and it continued so through the Cenozoic. Over most of this area Cenozoic sedimentation is confined to faulted basins lying between highlands. The only marine invasions were in a few small areas along the Pacific coast of Siberia, along the Arctic Ocean, and in a broad sea extending, in the early Cenozoic, along the eastern side of the Ural Mountains, connecting the Arctic Ocean with the Tethys.

The pattern of the early Cenozoic sea is similar to that of the Late Cretaceous. The sediments of this sea were mainly sands and clays. Toward the close of the Eocene there was slight epeirogenic uplift, and the sea was separated into a northern and a southern gulf, one opening to the Arctic Ocean and the other to the Tethys. In the Oligocene, widespread transgression again extended the sea into a continuous north-south body of water. This transgressive phase was short, however; during the late Oligocene regression again set in, and through the remainder of the Cenozoic there was a gradual retreat to the north and the south, bringing marine sedimentation in these regions to a close.

Terrestrial deposition occurred in numerous isolated basins. Such basins are particularly well known in China and Mongolia, where the pattern of deposition and deformation was much like that of the Jurassic and Cretaceous of this region (§ 9-7). The earliest Cenozoic formations in most Chinese basins are conglomerates that reflect Late Cretaceous uplift and generally high relief. Through the remainder of the early Cenozoic the clastic sediments became finer grained, and lacustrine deposits were numerous. Widespread epeirogenic uplift and faulting in the Miocene disturbed the lower Cenozoic formations so that the upper Cenozoic formations rest on them with angular unconformity. This disturbance is also reflected in the extensive upper Cenozoic conglomerates. These coarse clastic strata, however, were the result of a short phase, for they are overlain in a great many places by lacustrine facies. Some volcanic activity and the outpouring of basaltic lavas accompanied this depositional phase.

The Pleistocene was a period of erosion throughout most of China, but the deposits that remain are of great interest. In many areas cave and fissure deposits record Pleistocene events of particular significance, for these are the deposits that yielded the fossils and artifacts of prehistoric man discussed in Chapter 15. The most famous of these cave deposits is the Choukoutien site of Shansi, where *"Sinanthropus" pekingensis* was found. In northern China, north of the Yangtze River and extending to Korea, an area of thousands of square miles is covered by Pleistocene loess, in some places 1,500 feet thick.

The Cenozoic formations of Mongolia (Fig. 13-30) are especially well known for their fossil mammals, which have been widely publicized by the American Museum of Natural History.

Several regions in China have acted as rigid positive blocks, having a very thin sedimentary cover, throughout much of Phanerozoic history; and during the various late Paleozoic orogenic phases the Hercynian Mountains were wrapped round these rigid blocks. One such block is represented by the Tarim basin in western China, just north of Tibet (Fig. 6-22). To the north stand the Tien Shan Mountains and to the south the Kun Lun Mountains, the two ranges converging toward the ends of the basin. These mountain ranges are Hercynian and were formed when the Tarim basin was a rigid block. Since the late Paleozoic sedimentation has been intermittent in this basin. Seas occupied at least the southern half of the basin during the Eocene and Oligocene.

Its sediments consist of yellow clayey marine limestone interbedded with thick nonmarine marls, sandstones, and conglomerates. In the northern part of the basin all sediments of this age appear to be nonmarine. Disconformably overlying them are at least 3,000 feet of continental Pliocene siltstones and sandstones that grade upward into conglomerates. Overlying the fine-grained Pliocene sediments are more than 20,000 feet of conglomerates; even thicker conglomerates of this age are known along the foot of the Kun Lun Mountains.

The facies of the rocks and the flora of the Cenozoic formations indicate that the Tarim Basin was a large area of internal drainage throughout much of Cenozoic time. Rivers from the surrounding highlands deposited sands and silts in large deltas and flood plains, giving rise to a broad, flat surface on the floor of the basin, upon which the post-Pliocene and Holocene sediments were deposited. The younger, coarse, conglomeratic deposits clearly indicate that since the Pliocene the surrounding mountains have undergone extensive uplift, an event that accounts for the present topographic relief of the region.

13-10 NEW ZEALAND

The Cenozoic history of New Zealand contrasts strikingly with that of the Mesozoic. In the Mesozoic there was a single geosynclinal trough some 200 miles wide and 1,000 miles in length. In the Cenozoic a complex series of folds, welts, and troughs evolved on a much finer scale. The welts, which tended to be submarine ridges or land, were small, making New Zealand an archipelago in Cenozoic time. Changes in geography were frequent. Troughs sank rapidly but filled with sediment as they subsided.

FIGURE 13-30

Red clastic sediments of Eocene age overlain unconformably by bright-yellow Oligocene sandstones (the upper strata on the right), Uetyn Obo, Mongolia. (Photograph by the American Museum of Natural History.)

Welts rose in complementary fashion, but owing to constant erosion were not mountainous.

The paleogeographic maps of Figure 13-31 are interpretations of the geography of New Zealand during four episodes of the Cenozoic. By the close of the Cretaceous much of New Zealand was peneplaned, and marginal transgressive seas occupied the eastern part of the islands. The sea continued to lap further on to the peneplaned land in the Paleocene and Eocene (Fig. 13-31, A). The sediments deposited in these seas were muds, silts, and foraminiferal limestones. Some volcanic islands were present on the east side of South Island. This transgressive phase reached a climax in the Oligocene when seas flooded much of South and North islands (Fig. 13-31, B). The widespread development of organic limestone deposits suggest that the land areas were very low-lying. Many volcanic islands and submarine volcanoes were present on South Island. This phase of maximum transgression was followed in the Miocene by increased tectonic activity and great restriction of the overlapping seas on both South and North islands (Fig. 13-31, C). In general, the sediments were of clastic facies and coarser than those deposited in previous episodes. Volcanic activity was extensive, especially on North Island. The increased tectonic activity is also expressed in lateral movements of the Alpine fault. Unfortunately, the dating of movements along this fault are not settled to everyone's satisfaction. One school of New Zealand geologists holds that the movments on the fault occurred in the middle to late Cenozoic; another school holds that the major movements came earlier, during the orogeny that folded and metamorphosed the New Zealand geosyncline. The problem, too complicated to be discussed here, is a major unsolved problem of New Zealand geology.

Earth movements became even more intense during the Pliocene, approaching a climax in an orogeny to which New Zealand owes its present geography (Fig. 13-31, D).

Pliocene seas occupied narrow coastal areas on South Island and the southern part of North Island. Volcanic activity was very intense on the northern part of North Island. The major feature of Pliocene history, however, was uplift, which developed the major topographic features of New Zealand. The position of these areas of uplift are indicated in Figure 13-31, D.

13-11 AUSTRALIA

Throughout Cenozoic time the continent of Australia was emergent except for small incursions of the sea along the southern and western coasts. Extensive lake deposits and lava flows characterized Cenozoic deposition. During the early Cenozoic marine embayments existed in only a few isolated regions; not until the Miocene did they become significantly larger. In the Eocene and Oligocene extensive lake basins occupied eastern Australia and Tasmania (Fig. 13-32), and a wide variety of clastic rocks were deposited within these basins. The duration of individual lakes was highly variable, but most of them had ceased to exist by the end of the Miocene. During this episode of widespread interior basins, volcanic activity began and was especially severe in Queensland, where thousands of feet of lava flows accumulated, filling valleys and obliterating surface irregularities. Toward the middle of the Miocene there was a long period of erosion, during which a most remarkable lateritic soil, called duricrust by Australian geologists, developed over much of the continent, especially in western Australia.

During the Miocene a few large marine embayments developed along the southern coast, and smaller ones along the western coast. In these embayments sequences of alternating marine and brackish-to-fresh water sediments were deposited, testifying to minor variations in sea level.

Toward the close of the Miocene, epeirogenic uplift, accompanied by trough faulting, led to a temporary withdrawal of the seas

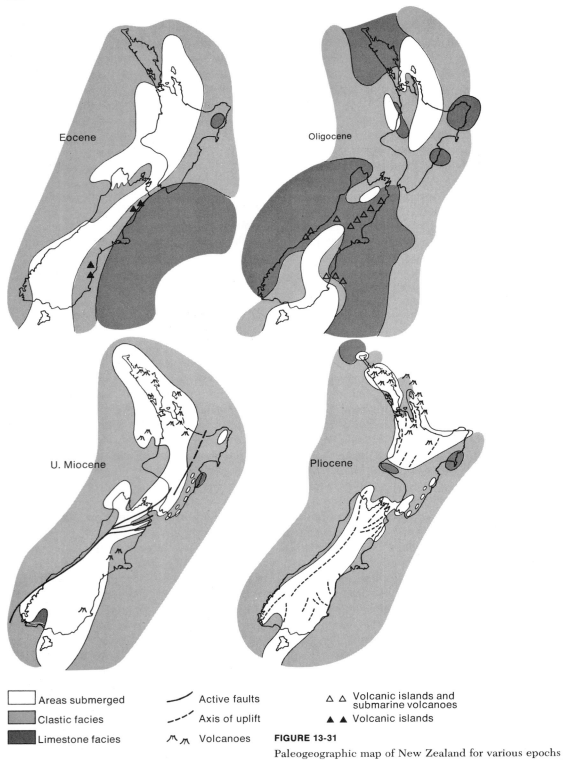

Eocene

Oligocene

U. Miocene

Pliocene

Areas submerged

Clastic facies

Limestone facies

Active faults

Axis of uplift

Volcanoes

△ △ Volcanic islands and submarine volcanoes

▲ ▲ Volcanic islands

FIGURE 13-31

Paleogeographic map of New Zealand for various epochs of the Cenozoic. (After C. A. Fleming, *Tuatara*, **10**, 1962.)

from the southern and western coastal embayments, to increased erosion of the land, and to more volcanic activity. In the Pliocene the seas partly returned along the previous embayments, but by the end of that epoch they had largely retreated from the continent.

13-12 AFRICA

The great Ethiopian shield continued as a stable land mass through the Cenozoic, just as it had done during most of the Phanerozoic. Cenozoic seas invaded the continent as small marginal transgressions, like those of the Late Cretaceous (Fig. 13-1), on the eastern and western coasts. Most transgressions were short-lived and confined to the early Cenozoic. Erosion prevailed in most of the continental interior. The highlands of eastern Africa, from the Zambezi region north to the Red Sea, were produced by fracturing, subsidence, and extensive volcanism, which created some of the most spectacular structural features known. This region of mountains, in contrast to most of the others discussed, was produced not by

folding but by faulting, as long linear systems of uplifted fault blocks and faulted troughs—the famous *rift valleys* of eastern Africa (Fig. 13-33). In addition to the block mountains, many volcanic mountains, of which Kenya and Kilimanjaro are the best-known, give added color to the landscape.

The southernmost rift valleys are in Zambia and Malawi, where faulting began during Karroo time. They spread north in the Cretaceous and Cenozoic, reaching Kenya and Ethiopia in the Miocene where volcanic activity and faulting have continued to the present. At the end of Lake Nyasa there is a bifurcation of the rift system; one branch extends north-northeast to form the Gregory rift valley in Tanzania, Kenya, and southern Ethiopia; the other branch curves west and then north to form the western rift valley between the upper Congo River and Lake Victoria. Another rift valley trends through Ethiopia.

Two features of these rift valleys are the active or recently active volcanoes on their floors and flanks, and the series of great lakes along their courses. The floor of any particular rift system did not subside equally, but in places transverse faults have caused very

FIGURE 13-32

Paleogeographic maps of Australia (A) during early Oligocene time and (B) during Miocene time. (After David, 1950.)

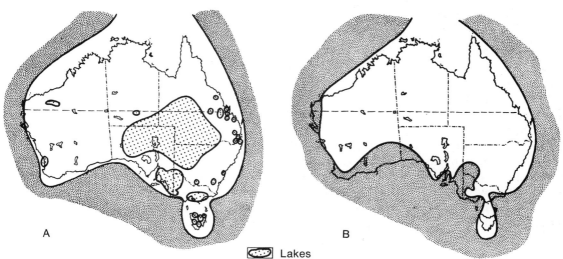

A   B

Lakes

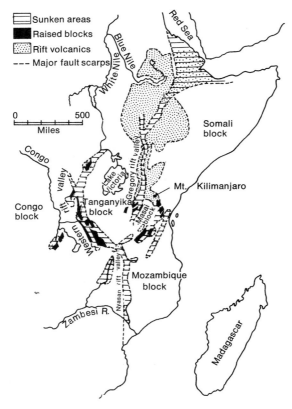

- Sunken areas
- Raised blocks
- Rift volcanics
- --- Major fault scarps

0 — 500
Miles

Blue Nile
White Nile
Red Sea
Congo valley
Somali block
Lake Victoria
Gregory rift valley
Mt. Kilimanjaro
Congo block
Tanganyika block
Masai block
Western rift valley
Mozambique block
Nyasan rift valley
Zambesi R.
Madagascar

FIGURE 13-33
Structural map of the African rift-valley system.
(After R. M. McConnell, 1951.)

unequal levels in the floor. Many blocks are tilted, and great complexes of volcanoes break the continuity of the valley floor.

13-13 ANTARCTICA

The orogenic activity that marked the Mesozoic history of West Antarctica and culminated in batholithic intrusions in the Antarctic Peninsula region continued into the Cenozoic. In contrast to the tight folding and deformation of the Mesozoic, Cenozoic movements were largely epeirogenic, accompanied by block faulting. The Cenozoic was also a time of extensive volcanism. Sedimentary formations of Cenozoic age on the Antarctic Peninsula have yielded fossil faunas and floras. The fossil faunas include abundant mollusks, brachiopods, nautiloids, fishes, and early penguins. The fossil floras, which include an ancestral stock of southern hemisphere conifer and the southern beech *Nothofagus* in association with other woody, temperate or subtropical species, reflects continuation of more genial climatic conditions.

In the Transantarctic Mountains the Cenozoic was a time of renewed mountain uplift and probable faulting, of local sedimentation and volcanism, and of widespread glaciation, continuing to the present. The only post-Jurassic rocks in this long mountain chain are scattered late Cenozoic volcanics. Younger strata, however, may be hidden by the polar ice cap, as suggested by morainal debris containing early Cenozoic microfossils near McMurdo Sound.

13-14 SOUTH AMERICA

The end of the Andean geosyncline in South America came in the Late Cretaceous, when orogeny, accompanied by batholithic intrusions, uplifted much of this belt. This marked the first step in the shaping of modern South America. A series of orogenic phases during the Cenozoic led to further deformation of the geosyncline and profoundly modified the pattern of deposition. Marine strata are mostly confined to small basins along the west edge of the growing Andes and to a few areas on the eastern coast, especially in Argentina. A great north-south belt along the eastern side of the Andes was the site of a tremendous accumulation of lacustrine and fluvial sediments, mostly from the Andes (Fig. 13-34). The Amazon valley was also the site of active deposition during the Cenozoic. In addition to this sedimentation there was also volcanic activity within the Andes.

During the early Cenozoic, within the newly uplifted Andean geosyncline, deposition was in structural depressions or

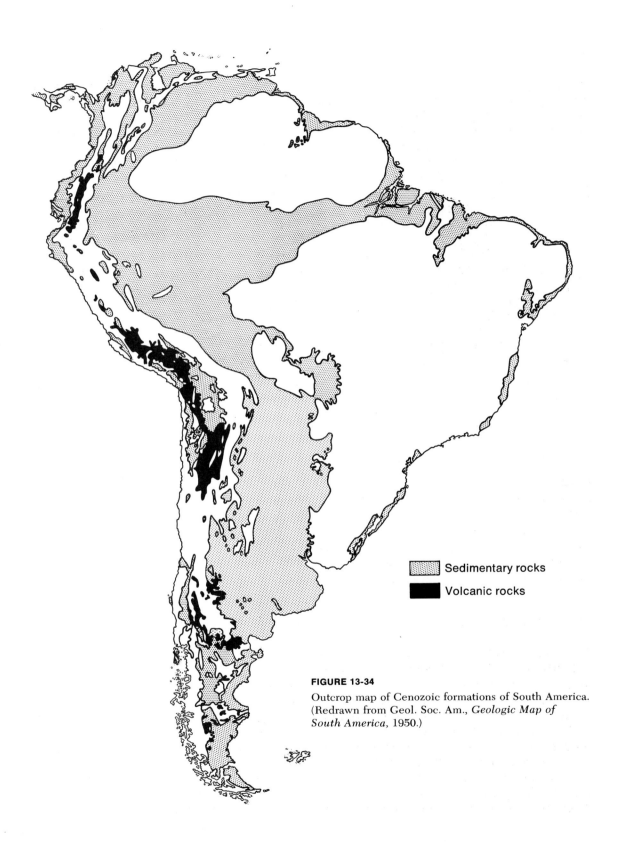

FIGURE 13-34

Outcrop map of Cenozoic formations of South America.
(Redrawn from Geol. Soc. Am., *Geologic Map of
South America*, 1950.)

Sedimentary rocks

Volcanic rocks

intermountain basins. The deposits consist largely of fine-to-coarse clastics, commonly red and, in southern Peru and Bolivia, enormously thick (Fig. 13-35, A). Around Lake Titicaca 20,000 feet of these early Cenozoic clastic formations are recognized (Puno Group). These consist of red-to-chocolate-colored sandstone and conglomerate with much gypsiferous chocolate shale and white volcanic tuff. Formations of equivalent age and facies are found throughout the Altiplano region of Bolivia (Corocoro Group). In northern Peru and probably in southern Ecuador the incipient Andes were broken intermittently by marine embayments that extended eastward from the Pacific. Marine-to-brackish conditions existed east of the early Andes in the Paleocene in a narrow zone from Ecuador south to Bolivia; the outlet of this sea was to the Pacific across northern Peru. Marine-to-brackish facies also developed east of the Andes in the Eocene and the Oligocene, in a restricted part of northern Peru and Ecuador (Fig. 13-35, B).

In the middle Cenozoic another orogenic phase folded the whole Andean region from the Pacific to Brazil. This phase was followed by the first important Cenozoic volcanics (Fig. 13-35, C). In Peru the volcanics are largely confined to a broad belt along the Cordillera Occidental in the central and southern parts of the country. Equally extensive volcanic deposits of the same age are found in Chile, Bolivia, and western Argentina. Orogeny continued during and after this phase of extensive volcanism. Then followed a period of erosion in which much of central and southern Peru was reduced to a late-mature or old-age surface not far above sea level. This erosional surface is named the Puna surface and was probably developed in the Pliocene. Pliocene upwarping of the Puna surface brought on a new surge of eruptive activity, especially in southern Peru and adjacent Chile and Bolivia. The uplift took place in several steps, and in the Andes of Peru, at

least, two later erosion surfaces are recognized. Within the Andes the Puna surface was broken by block faulting, which produced Lake Titicaca and other basins (Fig. 13-35, D).

The Andes reached approximately their present height at some time in the late Pleistocene. The greatest elevation of the Puna surface appears to have been attained in the western Andes of central Peru, where a general summit level of 15,000–16,000 feet is normal, and monadnocks rise to 19,000–21,500 feet. Peaks in southern Peru reach comparable elevations, but they are volcanoes that rest on the extensive lavas and pyroclastics that cover the Puna surface. The highest mountain in Bolivia, Sajama (rising to 21,190 feet in the northeastern corner of the country), is a deeply eroded volcano of early Pleistocene age.

Isolated embayments along the Pacific margin of Cenozoic South America accumulated great thicknesses of marine clastic sediments. The major oil fields along this coast are in northern Peru, where petroleum is obtained from Cenozoic formations. Most of this coastal belt of Peru and Chile was marked by extensive block faulting throughout the Cenozoic. The deep linear trenches off the coast of these two countries are thought to be fault troughs developed in the very late Cenozoic. From the bottom of the Milne Edwards trench off the coast of central Peru (20,000 feet deep) to the top of the Nevado Huascaran (elevation 22,000 feet), one of the highest peaks in the Peruvian Andes, there is a difference in elevation of 42,000 feet in a distance of 200 miles. Great rift valleys, some as much as 1,000 kilometers long, developed along the axial area of Chile in the late Cenozoic. The major volcanoes and volcanic activity of Chile are aligned along the eastern border of these rift valleys; the valleys themselves are floored by late Cenozoic and Quaternary terrestrial deposits.

The tectonic history and uplift of the Andes are nowhere better shown than in the

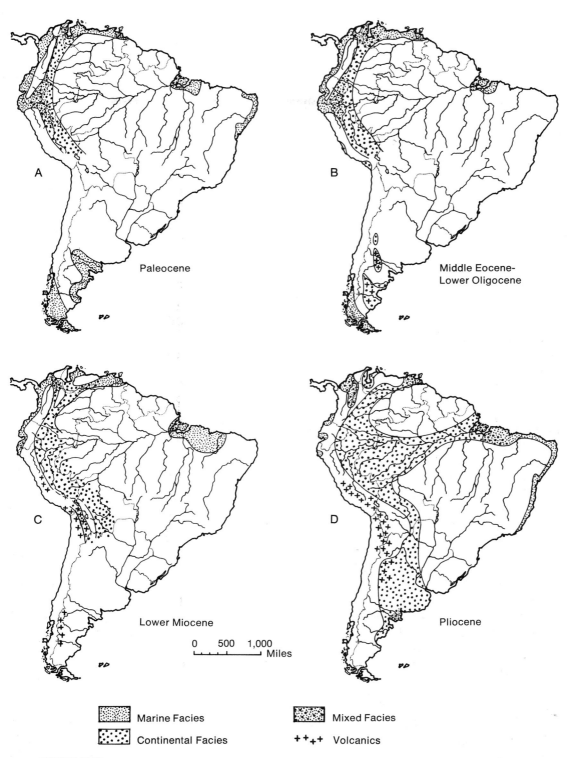

Marine Facies   Mixed Facies

Continental Facies   + + + + Volcanics

FIGURE 13-35

Paleogeographic maps of South America for various episodes of the Cenozoic: (A) Paleocene,
(B) middle Eocene-lower Oligocene, (C) lower Miocene, (D) Pliocene. (After Harrington, 1962.)

vast linear belt of Cenozoic redbeds east of the Andes. In eastern Peru thicknesses as great as 20,000 feet of Cenozoic redbeds represent fluvial and lacustrine deposits between the eastern front of the Andes and the Amazonian shield. The only marine or brackish-water strata in this sequence are of earliest Cenozoic age; Eocene and Oligocene strata were locally deposited in an eastern extension of the Pacific embayment across northern Peru. In Peru, at least, these redbeds range up to Miocene in age. They were folded during mid-Cenozoic orogenic phases that affected the Andes, after which

there was erosion, followed by the deposition of younger, flat, alluvial sediments. The Amazon region of eastern Ecuador and Peru is a vast, low, undulating plain through which anticlinal highlands rise, exposing various Mesozoic and Paleozoic formations. The eastern front of the Andes in much of this area is in fault contact with the Amazon plains (Fig. 13-36).

General stability characterized central and eastern South America during the Cenozoic. Deposition was mainly along the Amazon valley and in Argentina. Cenozoic strata are exposed in the Amazon valley in the

FIGURE 13-36

Eastern front of the Andes Cordillera near the Huallaga River in the Amazon country of eastern Peru. Jurassic (*Js*) and Cretaceous (*Ko, Kc, Kv*) formations are faulted against gently folded Cenozoic redbed formations (*C*) that form the undulating plains to the left. (Photograph by Servicio Aerofotográfico Nacional, Lima, Peru.)

banks of the main river and its principal tributaries. Wells drilled on the largest island in the mouth of the Amazon have encountered Cenozoic strata to a depth of 8,500 feet. From 8,500 to 6,100 feet there are black argillaceous shales, rich in plant remains and coal; dark siltstones and fine-to-medium-grained sandstones are also found. The few reworked Cretaceous and Eocene foraminifers in these beds appear to indicate an Eocene age. From 6,100 to 5,200 feet there occur only spores and fragments of plants of uncertain age in grayish-red silts and clay. From 5,200 to 2,400 feet plant remains are very abundant, and a few badly preserved foraminifers have been recorded. This section may have a marine origin, and its age could be Miocene. In the interval between 2,400 and 900 feet plant remains and coal abound. Foraminifers appear in larger numbers than in the preceding interval and include Miocene index fossils. From 900 feet to the surface, the evidence indicates that the sediments are Pleistocene and Holocene.

The Patagonian and Pampian regions of Argentina were the sites of thick accumulation of terrestrial deposits derived from the rising Andes to the west. At various times during the Cenozoic marine embayments invaded the Argentine coast, especially in the region of the Plata estuary and in Patagonia. Most of the marine invasions were of limited extent and duration.

The northern part of South America, taking in Venezuela, will be discussed in the next section.

13-15 THE CARIBBEAN REGION

Within the West Indies the oldest dated rocks are the Upper Jurassic of Cuba. Cretaceous formations are fairly widespread in the Greater Antilles, but not until the Cenozoic is there a reasonably complete record of deposition throughout the West Indies. The rocks exposed are only a fraction of the total record, however, since 90 percent of the region is under water.

The major geographic and structural features of the Caribbean region are summarized in Figure 13-37. The northern part of this geologic province consists of the Gulf of Mexico, the Florida peninsula, and the Bahama Islands. Between the Greater Antilles, Central America, and northern South America is a series of deep basins with intervening swells and ridges. North of Haiti, the Dominican Republic, and Puerto Rico is a narrow, linear, east-west deep, the Puerto Rico deep. The only Precambrian rocks are in northern South America, where portions of the Guianan shield crop out. Some Precambrian rocks may be exposed in the folded Venezuelan Andes. The vast area of volcanic rocks in Central America is of late Cenozoic age.

Tectonic instability, volcanism, and orogeny, which characterized the Greater Antilles during the Late Cretaceous, continued through the early Cenozoic. Orogeny commenced in Cuba during the Late Cretaceous; from late Cretaceous into Eocene time the focus of orogenic movements migrated eastward into Hispaniola and Puerto Rico. This episode of orogeny was accompanied by uplift and rapid erosion. In troughs bordering the uplifted areas, thick sequences of conglomerate, sandstone, and shales were deposited. By middle Eocene time volcanic activity in this region was almost entirely confined to southeastern Cuba; elsewhere carbonate deposits were accumulating. The volcanic foundation of the outer sedimentary arc of the Lesser Antilles (the so-called Limestone Caribbees) was built up during the middle Eocene. The oldest rocks on these islands are interbedded complexes of volcanics and limestones.

The early Cenozoic history of northern Venezuela is exceedingly complex. The Paleocene deposits are transitional to the underlying Cretaceous formations in many places. Toward the close of the Paleocene

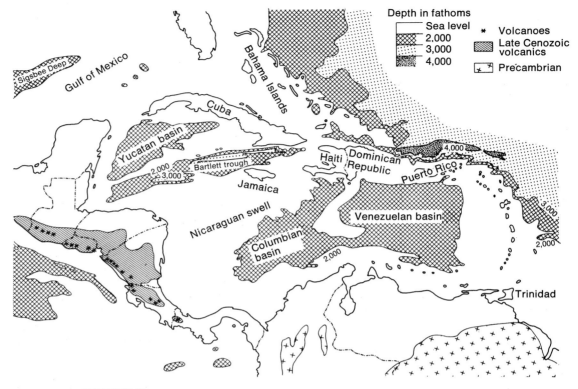

FIGURE 13-37
Geographic and structural setting of the Caribbean region. (After Woodring, 1954.)

the seas had regressed from most of Venezuela, but two areas of marine embayments persisted, one around Lake Maracaibo and the other in southern Trinidad and a small part of adjacent Venezuela. In the basin around Lake Maracaibo the deposits of this age consist of marine, brackish, and freshwater facies, largely clastic. These deposits grade downward into the Cretaceous formations with no perceptible break. At this time the Andes still had not risen, and the low central Venezuelan land sloped gradually westward and northwestward into swamps and shallow water. By the late Eocene marine conditions prevailed through a greatly enlarged Maracaibo Basin, and the coastal cordillera had subsided and was inundated by the seas, as was most of Trinidad (Fig. 13-38). This transgression was short, for at the close of the Eocene the

coastal cordillera was again uplifted, along with western Venezuela. The region around Lake Maracaibo was isolated and became a site of non-marine deposition. In eastern Venezuela a large embayment became established south of the coastal cordillera and opened to the sea through Trinidad (Fig. 13-38). In this basin were deposited the sediments of the prolific oil fields of eastern Venezuela. In some places within this basin nearly 20,000 feet of Cenozoic strata accumulated.

The Costa Rica–Panama region continued to have volcanic islands through the early Cenozoic.

The early and the middle Oligocene appear to have been times of fairly widespread emergence, for deposits of this age are not common. A wisespread transgressive phase began in the middle Oligocene, however, and continued through the early Miocene. At this time the Greater Antilles were largely submerged, and carbonates were the principal deposits (Fig. 13-39). Volcanic activity appears to have begun at the same time on the islands forming the inner arc of the Lesser Antilles (the so-called Volcanic Caribbees) and has continued intermittently to the present. On the outer arc of the Lesser Antilles there was no volcanism, and carbonate deposition prevailed.

Volcanic activity continued in the Costa Rica—Panama region. In Venezuela the seas again invaded the Maracaibo Basin, and deposition continued in the basin south of the coastal cordillera.

Early in the middle Miocene tectonic movements, mainly faulting, affected a large part of the Caribbean region, and they continued intermittently through the Pliocene. Most of the land areas within and around the Caribbean took shape during this time, and probably the deep basins within the Caribbean were also formed then. In northern South America, the Andean chains of Colombia and Venezuela and the coastal range of Venezuela were rising. The large basin south of the coastal cordillera gradually became a region of fresh-to-brackish-water deposition as the seas retreated eastward. It is of particular interest to note that every structure on which oil is found in Venezuela either originated or received its principal configuration during the Miocene-Pliocene phase of deformation.

13-16 THE LAST 70 MILLION YEARS— A RECAPITULATION

The Cenozoic Era was a time of widespread tectonic activity—activity that shaped and molded our modern earth. As the Cenozoic was the shortest of the Phanerozoic eras, it is often suggested that the degree of tectonic activity must have been greater in the Cenozoic than in previous eras. This suggestion, however, does not square with the facts; it is the newness of the mountain systems, their topographic grandeur, and the completeness of the geologic record that deceive us.

Continental and geosynclinal seas were small throughout the Cenozoic; they were greatest in the Eocene, and since then they have gradually been reduced. The geosynclines of the Cenozoic were the Tethys and

FIGURE 13-38

Paleogeographic map of Venezuela (A) during late Eocene time and (B) during early Oligocene time. (After E. Mencher *et al.*, *Bull. Am. Assoc. Petrol. Geol.*, 1953.)

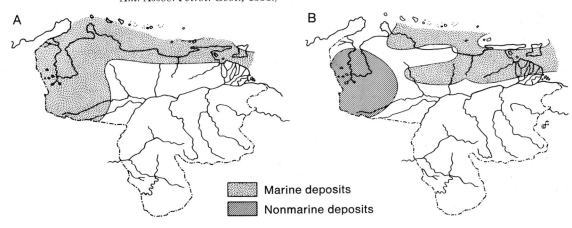

☐ Marine deposits
▨ Nonmarine deposits

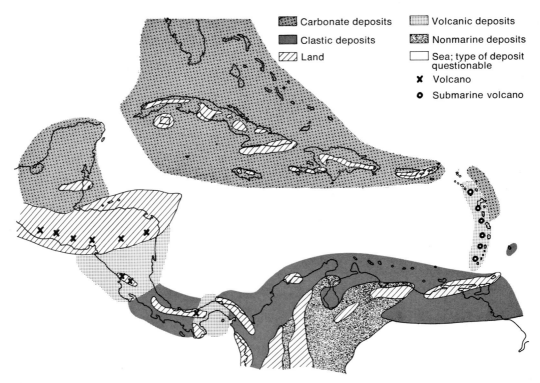

Carbonate deposits   Volcanic deposits

Clastic deposits   Nonmarine deposits

Land   Sea; type of deposit questionable

X Volcano

O Submarine volcano

FIGURE 13-39

Caribbean region during late Oligocene and early Miocene time. (After Woodring, 1954.)

the great circum-Pacific belt. The continental seas upon the stabler continental blocks were of limited extent and duration. On most of the continents, however, the pattern and location of these Cenozoic seas were closely similar to those of the later Mesozoic seas.

The distribution of Cenozoic folding is shown by Figure 13-40. This map is like that of Figure 3-19 (p. 44) except that the main periods of Phanerozoic folding have been differentiated. The map also presents a summary of the tectonic evolution of the continents. Note that the older folded rocks lie close to the margins of the shields and that younger folded rocks are encountered farther from the shields. The regions outside the belts of Cenozoic folding were characterized by shallow continental seas, terrestrial basin deposition, and epeirogenic movements.

Most of the continental areas of the world are now composed of rather stable components — that is, one or more shields surrounded by folded and rigid former mobile belts. The Eurasian and African continents are now each a rigid, unified mass. With the final orogenic phase, which broke up and solidified the Tethyan geosyncline, all of the region between the Baltic and Angaran shields on the north and the Ethiopian and Indian shields on the south was occupied by folded mountain systems, which represent former geosynclines and platforms. The geosynclinal history of Australia ended in the late Paleozoic. In North America the eastern geosyncline terminated in the late Paleozoic with the Allegheny orogeny. The area of marked instability and tectonic activity today is the circum-Pacific belt.

Is the geological evolution of the

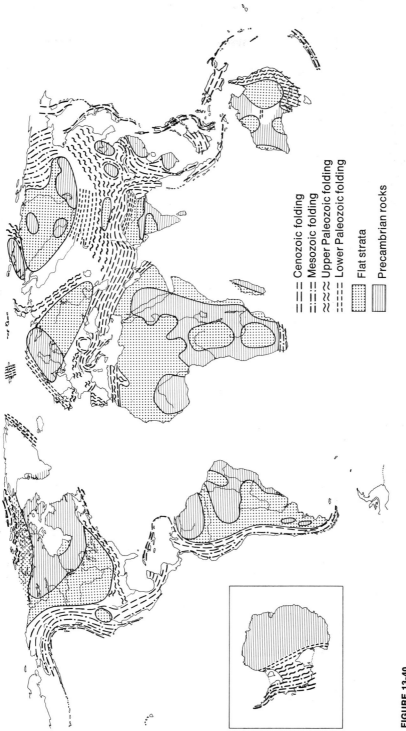

FIGURE 13-40

Distribution of the principal folded belts formed since the Precambrian. (After Umbgrove, 1947.)

Cenozoic folding

Mesozoic folding

Upper Paleozoic folding

Lower Paleozoic folding

Flat strata

Precambrian rocks

continents at or near completion? It is much easier to decipher the earth's past than its future, but the answer to that question is definitely no. Without trying to predict the future, we can, on the basis of the history of the earth, make some speculations. To base our guesses on what we have learned about the earth's history, we must look for areas in the world that are unstable today and therefore likely to move and create new geographic features. Are there any persisting geosynclines? Two such regions should be evident in Figure 13-40. The first is the western margin of the Pacific Ocean, from the Kamchatka Peninsula through Japan, the Philippines, and Indonesia to New Guinea. The second is the Caribbean region. Both these regions have either no Paleozoic record or a very limited one; the Mesozoic record is substantial but fragmentary; the Cenozoic record is a great deal better, but still, because of the island nature of the regions, fragmentary and incomplete. The geologic history, as far as we know it, clearly indicates that these regions have had complicated geosynclinal patterns, have seen much volcanic activity, and have undergone great tectonic instability. Their geologic and geographic setting, in view of the past history of the earth, supports the suggestion that these are active geosynclinal areas. On the continents, however, instead of expecting geosynclines to develop, we should rather look for erosion, terrestrial deposition, faulting, and epeirogenic movements such as characterized Eurasia north of the Tethys during the Mesozoic and Cenozoic.

SUGGESTED READINGS

See the list at the end of Chapter 4.

CENOZOIC LIFE

The last era in the evolution of life, the Cenozoic, is in many respects the most dramatic, for it is in the record of this era that we can trace the development of our modern faunas and floras. The fossil record of the Cenozoic is much richer than that of the Paleozoic and Mesozoic: it is now possible to make a much more detailed analysis of the faunas and floras and so arrive at a fairly reasonable correlation between them and the changing physical patterns and relations of the continents. The most interesting and instructive representatives of Cenozoic life are the mammals, and most of this chapter deals with this group. Marine invertebrate faunas had become almost completely modern by the earliest Cenozoic and are merely briefly mentioned, with emphasis on the origin and significance of the modern zoogeographic provinces. The unfolding of the modern flora began back in the Cretaceous, and needs no further comment here. The surviving reptiles were briefly discussed in Chapter 10. Certain birds deserve a few comments.

14-1 MARINE INVERTEBRATES

Cenozoic marine faunas were largely molluscan, composed of pelecypods and gastropods; echinoids also are abundant in certain facies, as well as bryozoans. The most striking difference between the Cenozoic and Mesozoic faunas was the absence from the former of the ammonoids. Foraminifers underwent an extensive evolutionary radiation, becoming extremely diverse and abundant. These organisms have been intensively studied by oil geologists. Their small size, rapid evolution, and abundance make them excellent guide fossils for purposes of correlation. In recent years they have become extremely useful for paleoecological studies also. Among the most characteristic early and middle Cenozoic foraminifers were the nummulites (Fig. 14-1). Throughout the Tethyan realm vast limestone formations of Eocene age are made up of the remains of these large foraminifers.

During the early Cenozoic the Tethyan geosyncline had a very diverse and abundant fauna that was homogeneous throughout its whole extent from Europe to Indonesia. A somewhat similar fauna existed in the western hemisphere in the Caribbean and adjacent regions. Though these two faunas differed, they were more closely related to each other than to faunas farther north or south.

At present the warm-water shelf faunas of the world are distinctly separated into four provinces, indicated in Figure 2-19 (p. 38). The largest province is the tropical Indo–West Pacific, which stretches from the eastern coast of Africa eastward through Indonesia, north of Australia, and on to the Hawaiian Islands; it ranks first in number and diversity of marine organisms. The ancestral forms of this fauna are found in the early Cenozoic strata of the Tethyan geosyncline. The tropical West African fauna is poorer than any other tropical coastal fauna. Many of the species and genera found there, however, show affinities with the Indo–West Pacific and American tropical faunas. There are now two faunal provinces in tropical America, one on each side of the isthmus linking North and South America (see p. 38), yet the tropical Pacific American fauna has more relationship with the tropical Atlantic American fauna than it does with Indo–West Pacific fauna. This relationship is due to the East Pacific Oceanic Barrier, a feature as significant in preventing faunal migration as any land barrier. This oceanic barrier is too wide for pelagic larvae to cross, and the water is too deep for shelf faunas.

The development of four instead of two faunal provinces in the circum-tropical regions is due entirely to earth movements during the middle and late Cenozoic. The breakup of the Tethys began in the Oligocene and was well developed by the Miocene, and from that time on the Indo–West Pacific realm was isolated from the western end of the former Tethys. In the Pliocene, North and South America became connected by an isthmus, and from that time on the faunas on the two sides of the isthmus evolved separately. Since all the faunas of these tropical provinces were adapted to warm shelf waters, it became almost impossible for any to migrate from one province to another, as the only routes available, such as that round Cape Horn, passed through colder zones. In addition to the geographic changes in the middle and late Cenozoic, climates also grew colder and thus further restricted the distribution of tropical faunas. The impoverishment of the Atlantic faunas was due to this drastic cooling during the Pleistocene. At the same time in the Indo–West Pacific province, where evidence shows that faunas were very diverse and abundant, climatic changes were slight.

14-2 BIRDS

The fossil record of the birds is a poor one. There are only three specimens of Jurassic birds (§ 10-14); from the Cretaceous there are considerably more, but the record is still scant; for the Cenozoic the record improves a little. The few data suggest that the birds were completely modernized by the Cenozoic; their skeletons were highly developed, and the Cenozoic radiations and evolutions were along lines of more subtle features not reflected in the skeleton. There is, however, one aspect of the Cenozoic evolution of birds worth brief comment, and this was the development of very large flightless birds. One of these, *Diatryma* (Fig. 14-2), attained a height of seven feet, had a large skull one

FIGURE 14-1

Nummulites from an Eocene formation of eastern Libya. Specimens average one inch in diameter. Lower figure is diagrammatic cutaway, showing complex internal structures.

and a half feet long, a very strong beak, and strong legs with four large claws. Such a creature must have been a serious menace to the small mammals of that time. At one place or another birds like *Diatryma* are found throughout the Cenozoic.

14-3 MAMMALS

The mammals had been in existence for nearly 100 million years before the Ceno-

zoic. Throughout that time they had remained small, inconspicuous creatures, completely overshadowed by the reptiles, which were reaching the climax of their evolution. With the extinction of the dinosaurs at the close of the Mesozoic, a wide range of ecologic niches became vacant. The Cenozoic history of mammals is, in essence, that of invasion of all the ecologic niches formerly occupied by reptiles—and of many more besides. Because of the richness of the geologic and fossil record of the Cenozoic, much is known about mammalian evolution.

We have a strange situation: a class of animals that had existed for many millions of years, remaining small, insect-eating creatures and undergoing no considerable evolutionary radiation, suddenly underwent an evolutionary radiation second to that of no other class of animals. In the Mesozoic they were dominated by the reptiles. Their great success in the Cenozoic was due to

FIGURE 14-2

Diatryma, a large flightless bird that lived in North America during the Eocene. (Drawing by Z. Burian under the supervision of Prof. J. Augusta.)

significant improvements in their anatomy, physiology, and methods of reproduction over those of the reptiles.

Some of the significant differences between mammalian and reptilian characteristics are summarized in Figure 14-3. As can be seen, nearly all of the characteristics that mammals possess render them more efficient in obtaining food, protecting themselves, and reproducing. In the first place, the mammals' larger brain was an obvious advantage. Other changes in the head, such as differentiated teeth and more complex cheek teeth, resulted in greater efficiency in the utilization of food and led to a greater diversity of food among later mammals. Another extremely interesting change is in the bone structure of the lower jaw. Reptiles have several bones in the lower jaw, but mammals have only one. During evolution two of the reptilian bones became incorporated to form two extra ear ossicles, which make a chain that transmits vibrations from the eardrum to the inner ear. The development of a four-chambered heart, the acquisition of a diaphragm, and the more nearly complete separation of respiratory and alimentary passages improved the continuity and efficiency of respiration. These characteristics, and others illustrated in Figure 14-3, all help to account for the great success of the mammals.

Modern mammals belong to three major categories, the monotremes, the marsupials, and the placentals. The monotremes include the duckbilled platypus (*Ornithorhynchus*) and the spiny anteaters (*Echidna*), both endemic to Australia and New Guinea. This group combines some reptilian and some mammalian characteristics; for example, monotremes lay eggs! The fossil record of monotremes extends back only to the Pleistocene, but they certainly represent, on anatomical grounds, a distinct line of descent from the mammal-like reptiles. The marsupial and placental mammals are also distinct groups, but they probably diverged from a common ancestry in the Cretaceous (Fig. 10-40, p. 380).

14-4 MARSUPIAL MAMMALS

Marsupials are characterized by their peculiar method of reproduction. The young are born after only a short period of gestation and are kept by the mother in a pouch, where they remain until large enough for independent living. The opossum and the kangaroo are familiar enough animals so that no extended description is needed.

As mentioned above, the marsupials and the placentals probably arose about the same time (most likely in the mid-Cretaceous) from a common ancestry, and by the Late Cretaceous both groups were probably widespread and about equally abundant. With the advent of the Cenozoic, however, the placental mammals underwent a radiation far greater than that of the marsupials, and this expansion brought about, in many parts of the world, a severe reduction, and in some places, the extinction of the marsupials. Marsupials fared best during the Cenozoic on two island continents, Australia and South America.

Australia is traditionally the home of the marsupials and monotremes, which (except for some rat-like placentals, which drifted in late) are the only mammals native to that island continent. With no competition from any other vertebrate group, the evolutionary radiation of the marsupials was almost without bounds. Marsupials occupied a wide range of ecologic niches and developed forms outwardly similar to placental species occupying similar niches on other continents. There are, among others, several varieties of carnivores, squirrel-like forms, the koala, marsupial moles and mice, and, of course, the kangaroo.

Unfortunately, there is practically no fossil record of Australian marsupials older than the Pleistocene. All our conclusions are based on modern forms and the scanty fossil record.

South American marsupials also developed in the isolation of an island continent but under somewhat different circumstances. Some placental groups had got into

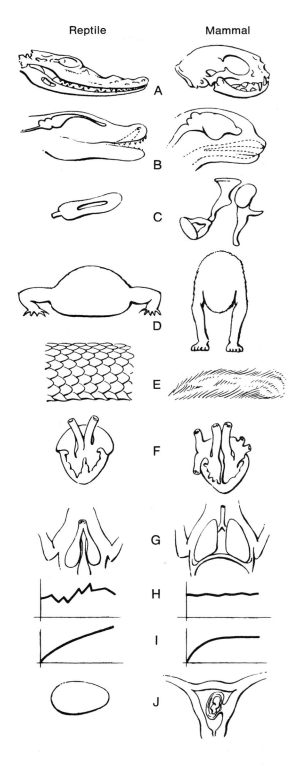

Reptile   **Mammal**

FIGURE 14-3

Mammal and reptile compared.

A. Skull: the mammal has a larger brain case and complex cheek teeth; its jaw consists of only one bone (the dentary), whereas the reptile's consists of three.

B. Brain: the mammal's forebrain is much more developed than that of the reptile; palate: the respiratory and alimentary passages are separated by a bony palate (heavy outline) in the mammal, not in the reptile.

C. Ear ossicles: the mammal's two additional ossicles are transformations of the reptile's two extra jaw bones.

D. Posture: the mammal's limbs are rotated under the body, which is thus kept off the ground—a contribution to the control of body temperature.

E. Skin: the mammal's skin is covered with hair, not scales—another aid in the control of body temperature.

F. Circulation: the lung circulation and the body circulation of the mammal are kept separate by a complete partition of the ventricle—an improvement in efficiency of respiration.

G. Diaphragm: the thorax of the mammal is completely separated from the abdomen by a diaphragm—an improvement in efficiency of lung ventilation.

H. Body temperature: uncontrolled in the reptile, controlled in the mammal.

I. Growth: continuing throughout life in the reptile, limited in the mammal.

J. Reproduction: eggs, no care of young, no milk, in the reptile; placental reproduction, parental care of young, milk, in the mammal.

(Redrawn from G. G. Simpson, C. S. Pittendrigh, and L. H. Tiffany, *Life: An Introduction to Biology*, Harcourt, Brace, New York, 1957.)

South America, along with the marsupials, before the continent was isolated. These placental mammals were plant-eaters and insect-eaters; the meat-eating adaptation was taken over by some of the marsupials. In this evolution a few aggressive carnivores evolved, one form even converging with the famous placental saber-tooth cat of Pleistocene age (§ 14-13).

Outside these two island continents the marsupials did not fare very well during the Cenozoic. In Europe they apparently disappeared by the mid-Cenozoic. In North America a few groups held on through the era. One of the famous "living fossils" is the common American opossum, which has changed little since the Late Cretaceous. The opossum, in fact, is fairly representative of the ancestral marsupials. The opossum, however, did not enter North America until late in the Cenozoic.

14-5 PLACENTAL MAMMALS

The dominance of the placentals over the marsupials and monotremes during the Cenozoic is clear: 95 percent of the known mammals are placentals. The group is characterized by its mode of reproduction and care of the young. It differs from the marsupials in various anatomical features, most important of which is the superior brain.

Of all the anatomical features of a placental mammal the teeth are probably the most important to the student of evolution and of the earth's history. The teeth, more than any other part of the anatomy, reflect evolution, environment, and adaptation. This is fortunate, for the teeth, being hard, are generally well preserved in the fossil record. In fact, a fairly large number of fossil species and genera are known only from isolated teeth and from jaw fragments. Twenty-eight orders of mammals are now recognized. Some authorities believe that if all mammals were extinct except Man, and the fossil record consisted only of teeth, the

resulting classification would be essentially the same as that now used, which is based on knowledge of the complete anatomy.

Placental mammals have a basic tooth pattern for each side of each jaw of three incisors, one canine, four premolars, and three molars. The molar teeth, more than any of the others, reflect evolution and adaptation to particular food habits.

Placental mammals are considered to have developed parallel with the marsupials, from a pantothere ancestry in the Cretaceous (Fig. 10-40). In the Late Cretaceous placental mammals were still small creatures belonging to the order Insectivora.

14-6 HOOFED MAMMALS— THE UNGULATES

One of the more important radiations from an insectivorous ancestry at the Cretaceous-Cenozoic transition was the group of plant-eating hoofed mammals—a radiation that encompassed during its history an extremely wide range of adaptive types and specializations and accounts for the majority of living mammals. The oldest fossil ungulates, whose small size and clawed feet show that they are not far removed from their insectivorous ancestry, appear in Late Cretaceous strata. In the early Cenozoic there evolved a large number of primitive ungulates, many of which attained a very large size. These represent the initial radiation of the group, but in the Eocene there appeared ungulates of more modern aspect, which underwent a very rapid and wide radiation, completely replacing the primitive forms by the Oligocene. There were, in effect, two phases in the evolutionary history of the ungulates: a first radiation, and then a second radiation of more modern forms. Parallel to this development there evolved in South America a group of ungulates distinct from those of the other continents; these will be discussed in § 14-11.

14-7 PRIMITIVE UNGULATES

The Paleocene fossil record of primitive ungulates clearly shows that their adaptive radiation was well underway and that several distinct evolutionary lines existed. One of the best-known of these early ungulates is *Phenacodus*, of which several complete skeletons have been found in the Big Horn basin of Wyoming. A restoration of *Phenacodus* is illustrated in Figure 14-4. This animal, as big as a sheep, had a long tail, five toes, and short, somewhat heavy legs. Its skull was long, narrow, and low; the toes had hoofs, and the molar teeth were clearly adapted for eating plants.

One of the most unusual evolutionary trends in this early radiation of ungulates was toward great size. By the middle Paleocene, large, massively built ungulates had already begun to appear. Probably the best-known of these are *Coryphodon*, which lived in the early Eocene and *Barylambda*, which lived in the late Paleozoic (see Figs. 14-5, 14-6). In size and appearance *Coryphodon* strongly resembled a hippopotamus. The limbs were short but very massive, obviously not adapted for fast running. Peculiarly, *Coryphodon* had elongated, saber-like canine teeth, probably used for protection. Another group, represented by *Uintatherium*, eventually attained the size of a large rhinoceros (Fig. 14-7). This grotesque animal was distinguished by the three pairs of horn-like structures on the top of the head. As in many of these large primitive ungulates the upper canines were greatly enlarged and must have served as effective defensive weapons. *Uintatherium* was the largest land animal of its time (late Eocene), but by the Oligocene it had become extinct, as had most of the primitive ungulates.

FIGURE 14-4

Phenacodus, one of the better-known early Cenozoic primitive ungulates. (C. R. Knight, American Museum of Natural History.)

FIGURE 14-5

Reconstruction of a scene in southeastern England during the Eocene. The climate was warm, and vegetation included palm trees (left center), swamp cypress (the tall tree on the right), magnolias (seen in flower on the left), and species of ginkgo, oak, walnut, breadfruit, and other trees. Animals included a large primitive ungulate, *Coryphodon*, and the five-toed ancestor of the modern horse, about as big as a fox. Crocodiles and ostrich-like birds are also shown. (Courtesy of Her Majesty's Geol. Surv., British Crown copyright.)

14-8 ODD-TOED MAMMALS — THE PERISSODACTYLS

This order, which includes the horses, the tapirs, the rhinoceroses, and two extinct groups, the chalicotheres and the titanotheres, was one of the progressive ungulate groups that appeared in the Eocene and became extremely abundant and diverse during the remainder of the Cenozoic (Fig. 14-8). In the late Cenozoic the group underwent a severe decline and is now completely overshadowed by the even-toed ungulates, the artiodactyls.

The main characteristic of the order is that the axis of the foot runs through the middle toe. There is a tendency for toe reduction from five to four or three and finally to only one, as in the modern horse. The forefeet of several perissodactyls are illustrated in Figure 14-9.

The molar teeth of perissodactyls underwent considerable modification, reflecting the food habits of each particular group. The modern horse, for example, has a high-crowned tooth with complexly folded enamel, adapted for grazing on hard grasses (Fig. 14-10, E). In contrast, the tapir, a browsing jungle animal, has cross-crested teeth (Fig. 14-10, F). The molar teeth of other perissodactyls are also illustrated in Figure 14-10. One interesting trend among the perissodactyls was that the premolar teeth became like the molars in shape and form. This adaptation, of course, increased the efficiency of the teeth for grinding plant foods.

The Horses

Probably no other group of mammals has as rich a fossil record or has been studied so thoroughly as the horse. The center of horse evolution was North America, where there is fortunately a splendid sequence

FIGURE 14-6

Coryphodon and *Barylambda,* two of the larger primitive ungulates that lived during the early Cenozoic. (Denver Museum of Natural History.)

of continental deposits from the early Eocene to the Recent.

The earliest ancestral horse is eohippus of Paleocene and early Eocene age. (According to the rules of zoological nomenclature, the proper name is *Hyracotherium;* but, because "eohippus" is well established, we continue its use.) Eohippus, as big as a fox, had an arched back, a slender body, and long slender legs obviously adapted for running (Fig. 14-7). The forefeet had four toes, and the hind feet three. The molar

FIGURE 14-7

The uintatheres, the largest of the primitive ungulates that lived during the Eocene, are illustrated here with an early primitive horse, *Eohippus.* In the foreground are contemporaneous reptiles. (Denver Museum of Natural History.)

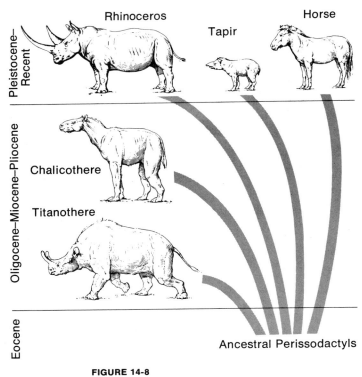

FIGURE 14-8

Simplified evolutionary tree of the perissodactyls.
(Drawings of animals from H. F. Osborn, 1929.)

FIGURE 14-9

Front feet of various odd-toed ungulates (perissodactyls).

A. The chalicothere *Moropus*.
B. The titanothere *Brontotherium*.
C. Modern tapir.
D. Modern horse, *Equus*.
E. Running rhinoceros of Oligocene age, *Hyracodon*.

(After figures by W. B. Scott, 1937.)

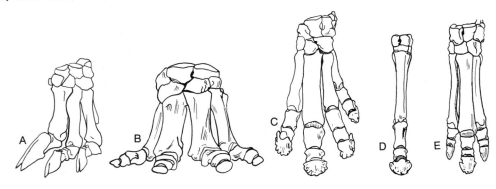

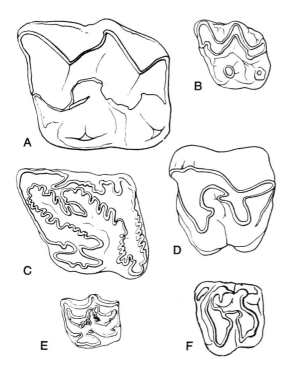

FIGURE 14-10

Pattern of upper molars of six perissodactyls, showing the widely different extent of the enamel edges and surfaces. The simpler molar crowns, with short enamel edges, are adapted to crushing coarse vegetation; the more complex crowns, with greatly elongated enamel edges, are adapted to the grinding of hard siliceous shrubs and grasses.

A. Oligocene titanothere.
B. Middle Eocene titanothere.
C. Pleistocene grazing rhinoceros.
D. Miocene browsing rhinoceros.
E. Modern horse.
F. Modern tapir.

(From H. F. Osborn, 1929.)

teeth were low-crowned, with cone-like cusps, and the premolars were not yet molariform. These teeth were adapted for browsing rather than grazing; eohippus was probably a forest dweller rather than a plains dweller.

The general anatomical features of eohippus show that it was very primitive and must have closely resembled the ancestral perissodactyl. No earlier fossils have yet been found to indicate definitely the ancestory

of the perissodactyls, but the group was probably derived from primitive ungulates of Paleocene age.

The evolutionary history of the horse from eohippus to the modern *Equus* is a complex story. The major trends were as follows: (1) they increased in size; (2) the molar teeth became more complex in crown pattern and changed from low-crowned to high-crowned; (3) the premolars became molariform; (4) the face grew longer to accommodate the high-crowned teeth; (5) the jaws grew deeper to accommodate the high-crowned teeth; (6) the arched back straightened; (7) the legs grew longer; (8) the toes became fewer; and (9) the brain increased in size and complexity.

From the small eohippus there evolved a radiation that invaded the Old World but soon became extinct. But from them evolved giant horse-like mammals called palaeotheres, which survived into the early Oligocene and reached the size of a rhinoceros.

In North America a series of changes, in the Eocene and Oligocene, led to *Miohippus*. All the feet of this horse had three toes, the middle toe being much the larger. The molar teeth had become strongly crested, and some premolars had become molariform. The teeth were still low-crowned and adapted for browsing.

With the beginning of the Miocene several distinct adaptive lines developed from *Miohippus*. One of these lines maintained the three-toed condition of *Miohippus* and the low-crowned teeth adapted for browsing, but increased in size. This trend is represented by *Hypohippus*. Another line evolved in the direction of the modern horse, *Equus*. The first genus in this trend was *Merychippus*, which was about as big as a pony, had three toes, of which only the central was functional, and—most significantly—had high-crowned teeth, capable of eating the hard grasses of the plains. In the early Pliocene two lines evolved from *Merychippus*. One, illustrated by *Hipparion*, maintained the three-toed condition but showed progressive development of the teeth and skull. They were abundant and widely distributed in the Pliocene but became extinct in the Pleistocene. The other line from *Merychippus* progressively changed in teeth, skull, and feet toward horses with a single toe, the side toes becoming mere splints concealed beneath the skin of the upper part of the foot. This line is represented by *Pliohippus,* from which *Equus,* the modern horse, evolved in the Pleistocene. Also from *Pliohippus* evolved another distinct horse (*Hippidium*), which inhabited South America during the Pleistocene but became extinct before the close of the epoch.

This discussion of the horse's evolution is greatly simplified, but it does outline the major features. Figures 2-23 and 14-11 clearly show that many adaptive trends are involved in the progression from *Eohippus* to *Equus* and that not all of them developed at the same time or rate. A strange fact of the horse's evolution is that the main scenes were played in North America, where the horse became extinct in the Pleistocene, although it survived in Eurasia. The horse was first reintroduced to North America by the early Spanish conquerors.

The Tapirs

These strange animals, now confined to Malaya and South America, are in many respects the most primitive of living perissodactyls. The primitive character is seen in the persistence of four toes on the front feet and three toes on the hind feet and in the arched back. The teeth are low-crowned, the cross-crested molars being adapted for browsing in the jungle (Fig. 14-10, F). The last three premolars are molariform. A peculiar specialization of the tapirs is the proboscis. Except for this enlarged nose the basic features of the tapirs are like those of the earliest horses.

The present distribution of the tapirs came about during the Pleistocene, when the cold of the northern hemisphere probably led to the group's extinction there. During the earlier Cenozoic the tapirs had been widely distributed in North America and Eurasia.

The Rhinoceroses

The rhinoceroses are represented today by only two species in Africa and three in Asia, mere remnants of a once flourishing group. The Cenozoic evolution of these animals was extremely complex and is not yet well understood, but by concentrating on a few of the main adaptive types we can get some idea of their diversity and history.

The earliest rhinoceroses, animals slightly larger than eohippus, showed all the basic primitive perissodactyl characteristics: they had four toes on the forefeet and three on the hind feet, and the teeth of most forms were low-crowned and adapted for browsing (Fig. 14-10, D). The early rhinoceroses were rather graceful and had long legs adapted for running. Several groups of running rhinoceroses are known from Eocene and Oligocene deposits, but by the end of the latter epoch they had become extinct. Another group of rhinoceroses living at the same time appear to have adopted habits like those of the modern hippopotamus (Fig. 14-18); they had large, heavy bodies, with short but strong legs and broad feet.

One evolutionary line of the rhinoceroses led to the development of one of the largest land mammals that ever lived. In various parts of Asia (Mongolia, Baluchistan, southern Russia) Oligocene and lower Miocene formations have yielded fossils of rhinoceroses that were seventeen feet high and built somewhat like a heavy giraffe. One of these forms, *Indricotherium,* is illustrated in Figure 14-12. It had no horns and lived mainly off the leaves of trees whose crowns it could easily reach. The fossil remains of *Indricotherium* are found in southern Russia on the shores of the Aral Sea.

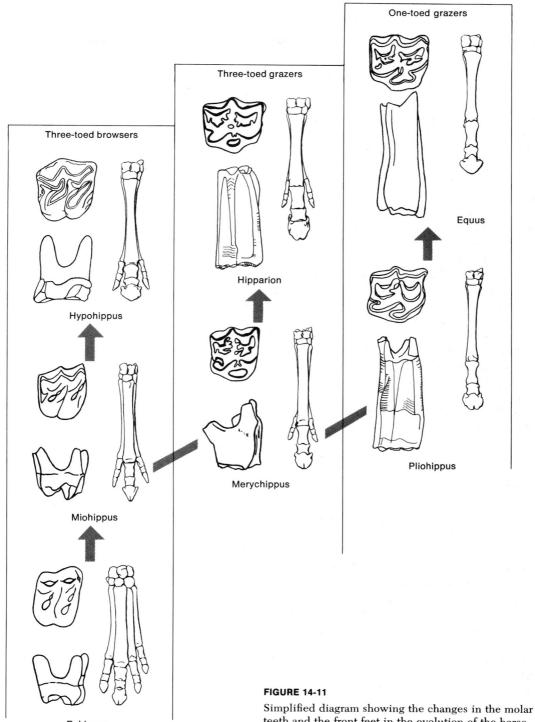

One-toed grazers

Three-toed grazers

Three-toed browsers

Equus

Hipparion

Hypohippus

Pliohippus

Merychippus

Miohippus

Eohippus

FIGURE 14-11

Simplified diagram showing the changes in the molar teeth and the front feet in the evolution of the horse. Compare with Fig. 2-23 (p. 83).

FIGURE 14-12

Indricotherium, a hornless rhinoceros, one of the largest land mammals that have ever roamed the earth. It lived in Asia during the Oligocene and Early Miocene. (Drawing by Z. Burian under the supervision of Prof. J. Augusta.)

Various lines of one- and two-horned rhinoceroses evolved during the Miocene and later epochs (Fig. 14-13). The rhinoceros horn is unique among mammals in that it is formed of coalesced hair rather than bone and is therefore rarely fossilized. In the Pleistocene the famous woolly rhinoceros was common in the northern fauna. During the Pliocene and Pleistocene the rhinoceroses began to die out.

The Titanotheres

The titanotheres, a group of perissodactyls, underwent a most phenomenal evolution during the Eocene and Oligocene and then became extinct. The earliest forms were much like eohippus in size and anatomical structure. They maintained primitive teeth and feet while evolving into giants, and later they developed large bony horns on their skulls. The culmination of this evolution was *Brontotherium*, which measured eight feet in height. The increase in size in the history of the titanotheres is illustrated in Figure 14-14. Note that the increase in size preceded the appearance of the horns.

North America was the main center for the evolution of the titanotheres, which during the Eocene and Oligocene were among the commonest animals of western North America. One group invaded Eurasia during the late Eocene and early Oligocene

FIGURE 14-13

The large animals with clawed feet are chalicotheres of the genus *Moropus;* in the right foreground is a large entelodont, *Dinohyus;* in the center are rhinoceroses of the genus *Diceratherium,* peculiar for the pair of horns on the nose. (American Museum of Natural History.)

and died out there, just as they did in North America. It seems strange that an animal as large and formidable and apparently successful as the titanothere should undergo such a rapid evolution and then become extinct right after reaching its culmination. One clue to a solution of this problem may be the lack of development of the teeth, which remained low-crowned, with simple crests and cones for feeding upon soft vegetation (Fig. 14-10, A, B). A slight change in the kind of vegetation available may have been enough to lead to their extinction.

The Chalicotheres

Probably the strangest of the odd-toed mammals were the chalicotheres, which were closely related to the titanotheres. The group had a long and successful history, first appearing in the Eocene and not becoming extinct until the Pleistocene. The chalicotheres were unique among the perissodactyls in having claws instead of hoofs on their feet. A very common Miocene form was *Moropus,* illustrated in Figure 14-13. *Moropus* looked somewhat like a horse but

FIGURE 14-14

Outlines of the bodies of titanotheres at different stages of evolution. (From H. F. Osborn, 1929.)

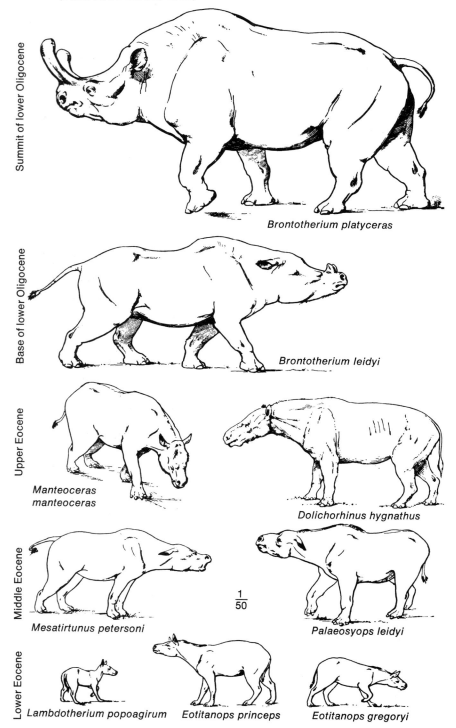

Summit of lower Oligocene — *Brontotherium platyceras*

Base of lower Oligocene — *Brontotherium leidyi*

Upper Eocene — *Manteoceras manteoceras* — *Dolichorhinus hygnathus*

Middle Eocene — *Mesatirtunus petersoni* — $\frac{1}{50}$ — *Palaeosyops leidyi*

Lower Eocene — *Lambdotherium popoagirum* — *Eotitanops princeps* — *Eotitanops gregoryi*

had more massive legs. The teeth were like those of the titanotheres. Apparently the chalicotheres were adapted to feeding on roots and lived along streams where they could dig up roots with their clawed limbs.

14-9 EVEN-TOED MAMMALS— THE ARTIODACTYLS

The even-toed ungulates are the commonest and most widespread mammals living today. The order includes both nonruminant forms, such as pigs, peccaries, and hippopotamuses, and ruminant forms, such as camels, llamas, mouse-deer (chevrotain), deer, giraffes, pronghorns, sheep, goats, musk oxen, antelopes, and cattle (Fig. 14-15). This list alone is ample evidence of their great diversity and evolutionary success. The artiodactyls have either four or two toes (Fig. 14-16), and two basic molar types: the pigs and their relatives have low-crowned teeth with rounded cusps, but the deer and related forms have molar teeth with crescentic cusps (Fig. 14-17).

In the early and middle Cenozoic the perissodactyls were much more abundant and widespread than the artiodactyls; not until the mid-Cenozoic did the artiodactyls

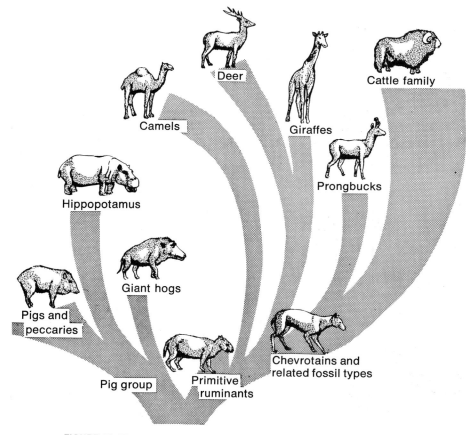

FIGURE 14-15

Simplified evolutionary tree of the even-toed ungulates (artiodactyls). (After A. S. Romer, 1933.)

become significant in mammalian faunas. During the later Cenozoic the perissodactyls declined greatly, but the artiodactyls have continued to increase in number and diversity. The most significant adaptive advantage of artiodactyls is apparently the complex digestive system developed by the ruminants, or cud-chewers. The stomach of a ruminant is divided into four chambers; newly cropped food passes into two of these chambers and is partly digested, is then regurgitated into the mouth and chewed again, is swallowed again and passed to the other chambers. This strange adaptation has the distinct advantage that food can be eaten in a hurry and then chewed and digested at leisure in a safe place, free from attack by enemies.

The classification of the artiodactyls is extremely complex, but they can be reduced to two main groups. One group includes the pig and his relatives, and the other includes all the ruminants.

Pig-like Artiodactyls

A number of primitive artiodactyls with piglike molar teeth are known from the Eocene, indicating an initial wide adaptive radiation. One trend evolved a group of grotesque, giant, pig-like animals known as entelodonts, illustrated in Figure 14-13. Oligocene and early Miocene entelodonts attained the size of a bison, with a skull fully a yard long. The skull had enlarged knobs on the cheekbone and on the lower jaw, giving the face a rather strange appearance. These animals, though often referred to as giant hogs or pigs, are not pigs at all but a separate evolutionary line. The entelodonts developed long slender legs and a straight back adapted for running. Early in their history the toes became reduced to two.

The first pigs and peccaries appeared in the Oligocene, the former in Eurasia and the latter in North America. Throughout their history the pigs and peccaries have

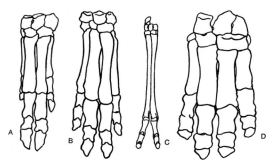

FIGURE 14-16

Front feet of various even-toed ungulates (artiodactyls).

A. Modern pig, *Sus.*
B. Mid-Cenozoic pig-like mammal, *Bothriodon.*
C. Miocene camel, *Procamelus.*
D. Modern hippopotamus.

(*A*, *B*, and *D* from E. H. Colbert, 1935; *C* from W. B. Scott, 1937.)

probably been forest-dwelling animals rooting in the ground for the food. During the middle and late Cenozoic several adaptive lines evolved. The modern domestic pig, *Sus*, is descended from Asiatic pigs.

The hippopotamuses did not appear until the Pliocene. They evolved from an extinct line of pig-like artiodactyls. During the Pliocene and Pleistocene they were widespread in Eurasia and Africa, but are now greatly restricted in distribution.

FIGURE 14-17

Crowns of the two main types of artiodactyl molar teeth.

A. *Paleochoerus*, a primitive pig-like form. (After E. H. Colbert, 1935.)
B. *Procamelus*, a Miocene camel. (After W. B. Scott, 1937.)

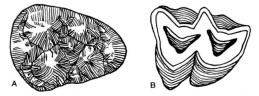

FIGURE 14-18

An Oligocene landscape in western North America. The largest animals
are aquatic rhinoceroses, *Metamynodon;* in the center foreground is
the earliest of the saber-tooth cats, *Hoplophoneus;* at the left are
Mesohippus, an early horse, and *Merycoidodon,* an oreodont.
(Denver Museum of Natural History.)

Primitive Ruminants

During the early and middle Cenozoic the
earliest and most primitive ruminants were
a North American group, the oreodonts (Fig.
14-18). The oreodonts are the commonest
fossils in the Badlands of South Dakota.
Many skeletons are in huddled groups, suggesting that they had been overcome by
some sudden disaster such as a sandstorm.

FIGURE 14-19

Miocene landscape in western North America. In the center are camels of
the genus *Stenomylus;* to the left are ruminants of the genus *Syndyoceras.*
(Denver Museum of Natural History.)

The oreodonts were between a sheep and a large pig in size, with four toes on each foot. The group was highly successful during the mid-Cenozoic, evolving a wide range of species, but by the Pliocene it had become extinct.

The Camels

One of the oldest and most distinctive ruminants is the camel, which, like the horse, underwent most of its evolution in North America. The present distribution of the camels in North Africa and Asia and of the related llamas in South America were Pleistocene developments. The first camels appeared in the late Eocene and were small artiodactyls with four toes. During the remainder of the Cenozoic many different adaptive types evolved. Some forms remained small (Fig. 14-19); others grew to giant size, such as the Pliocene *Alticamelus*, whose head was fully ten feet above the ground. By the Oligocene the two-toed pattern of the feet had become established. In the more advanced camels the upper incisors were lost and replaced by a horny pad. Most of the Cenozoic camels appear to have been plains dwellers rather than desert dwellers. The hump of the modern camel is probably a recent development.

Modern Ruminants

The mouse-deer, deer, giraffes, pronghorns, bison, sheep, goats, antelopes, and cattle comprise the more highly developed ruminants. The group originated in the late Eocene, but most modern representatives are of more recent (for example, Pliocene) origin. A modern animal that is closely related to the ancestors of the higher ruminants is the oriental mouse-deer (*Tragulus*), which is about the size of a jackrabbit, has retained four complete toes (unlike other higher ruminants), and has no horns but possesses sharp canine teeth. During the mid-Cenozoic the males of one group of this evolutionary line developed spectacular horns. Some forms had six horns—two on the nose, two above the eyes, and two on the back of the head (Fig. 14-19). One form had a long horn that grew from just above the nose and diverged like a Y near its tip (Fig. 14-20). These peculiar types were common from the Oligocene to the Pliocene and then became extinct.

The deer evolved during the Oligocene from a mouse-deer ancestry. Their most striking specialization was the development of antlers on the skulls of the males. Throughout their history the deer and related forms (elk, moose, reindeer, etc.) have maintained their forest-dwelling, browsing habits and the lower-crowned type of molar teeth. The deer never invaded Africa south of the Atlas Mountains but remained adapted to temperate regions.

Closely related to the deer are the giraffes, which first appeared in the Miocene. The early giraffes had short necks and must have been similar to the modern okapi of the Belgian Congo. The modern giraffe evolved from these short-necked forms in the Pliocene or Pleistocene.

By far the most familiar ruminants are the sheep, goats, musk oxen, antelopes, and cattle. The group to which these animals belong first appeared in the Miocene, but not until the Pliocene did it begin to evolve and radiate at a high rate. Now, throughout the world, the ruminants are the most abundant and diverse of all mammalian orders. Most of their evolution took place in Eurasia and Africa.

14-10   THE ELEPHANTS

The elephants are represented today by a single African and a single Asiatic species, a mere remnant of fantastically abundant, adaptive types of the middle and late Cenozoic, when they were nearly world-wide in distribution.

The first true proboscidians, *Palaeomastodon*, are known from early Oligocene deposits in the Fayum oasis of Egypt, sixty miles southwest of Cairo. They were already of impressive size, measuring seven or eight feet at the shoulders. The body and legs were massive, and a rather large head supported a proboscis, as shown by the position of the nasal bones far back on the skull.

jawed mastodonts evolved more complex cusp patterns, increasing the effectiveness of the teeth as grinding mills. There also evolved many different sizes and shapes of tusks. Among the most unusual forms were the so-called shovel-tuskers of Pliocene age: the lower tusks were flattened, instead of round, and joined together to form an effective scoop or shovel for digging plants from the bottom of shallow waters (Fig. 14-20).

Developing parallel to the long-jawed

FIGURE 14-20

Pliocene landscape in western North America. In the center is the shovel-tusk mastodont, *Amebelodon;* to the right is a rhinoceros, *Teleoceras;* on the left is the unique horned ruminant, *Synthetoceras*. (Denver Museum of Natural History.)

The second upper incisors were enlarged into tusks that extended forward and downward. The lower jaw was long and also bore two tusks that extended forward. From such an ancestral form all later elephantoids evolved. The Miocene mastodonts, as illustrated by *Trilophodon* (Fig. 14-21), were much like their early Oligocene ancestor. The same four-tusked condition and the long upper and lower jaws persisted. The molar teeth developed three pairs of transversely arranged conical cusps. Later, long-

mastodonts was a line that had a short, virtually tuskless lower jaw. The well-known Pleistocene *Mastodon* represents this line. The group first appeared in the Miocene. The upper tusks were often strongly curved. The American *Mastodon americanus* is probably one of the best-known fossil vertebrates because of the abundance and wide distribution of its remains. It survived in North America well after the first appearance of man but became extinct several thousand years ago.

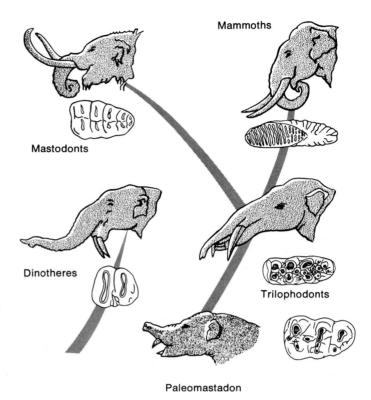

Mammoths

Mastodonts

Dinotheres

Trilophodonts

Paleomastadon

FIGURE 14-21

Simplified evolutionary tree of the proboscidians.
(After H. F. Osborn, 1936–42.)

During the late Miocene or early Plio-cene there evolved from the long-jawed mastodonts a group whose molar teeth be-came longer, strongly cross-crested in con-trast to the large cones of the mastodonts, and had more crests. This was an intermedi-ate step in the evolution of the first true elephants, which appeared in the late Plio-cene. The early true elephants had very large, curved upper tusks, short, tuskless lower jaws, and greatly enlarged molar teeth with many low cross-crests. The great enlargement of the molar teeth created a problem in jaw dimension. In all masto-donts all cheek teeth were emplaced in the jaw at the same time, but with the greatly enlarged cheek teeth of the elephants, only one such tooth was present on each side of

the jaw, and some change in the method of emplacement was necessary. The jaws be-came greatly deepened, the upper teeth were recessed in the skull, and the teeth came into use one at a time. As an emplaced tooth was worn down, it was gradually pushed forward by the erupting new teeth and finally broke away at the front and dis-appeared. The molar teeth in fossil and Holocene elephants became strongly cross-crested in the form of parallel plates (Fig. 14-21). The early Pleistocene true elephants still had low-crowned teeth but had evolved high-crowned teeth before the end of this epoch, this in a time span of approximately three million years—a most rapid evolu-tionary rate. The Pleistocene was the hey-day of the elephants. The fossil elephants,

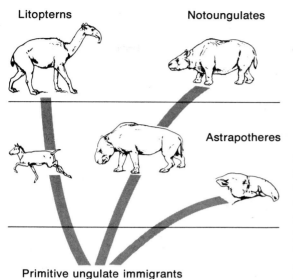

Litopterns   Notoungulates

Astrapotheres

Primitive ungulate immigrants

FIGURE 14-22

Simplified evolutionary tree of the principal orders of South American ungulates. (Drawings of animals from W. B. Scott, 1937.)

the mammoths, evolved many species and were nearly world-wide in distribution. The woolly mammoth was familiar to prehistoric European man, who drew its picture on the walls of caves. The mammoths became extinct in the late Pleistocene, and only the African and Asian elephants survived.

A very strange group, probably far from the main evolutionary path of the proboscidians, is represented by *Deinotherium* (Fig. 14-21). The deinotheres first appeared in the Miocene and died out during the Pleistocene. They were a highly successful group but evolved little during their long history. Most deinotheres were large, and some developed into the largest of the proboscidians, measuring ten feet high at the shoulders. Their molar teeth, quite distinct from those of all other proboscidians, had two cross-crests that changed little throughout the history of the group. They had no tusk in the upper jaw but had two large recurved

tusks in the lower jaw. The group was widespread in Africa and Eurasia but never entered the western hemisphere.

14-11 SOUTH AMERICAN UNGULATES

South America was an island continent during most of the Cenozoic. Because of this isolation the mammal groups there evolved completely independently of the northern continents. Late in the Cretaceous or early in the Cenozoic marsupials, edentates, and primitive ungulates entered South America either across the isthmian link of Central America or possibly by rafting. Whatever the method, these forms did get into South America, but subsequent entry was greatly reduced, being confined to some small rodents, primates, and relatives of raccoons. The primitive mammals did not survive beyond the early Oligocene, but early developed five new orders; only one of which is known outside South America. The evolution of these orders is one of the most remarkable and instructive stories in mammalian history. Some orders underwent a broad adaptive radiation and evolved many types that superficially are astoundingly close to those of the northern continents. This strong likeness to mammals elsewhere illustrates the principle that, under similar environments, genetic processes can evolve remarkably similar animals. Of the five orders of South American ungulates only three are fairly common and well documented, the other two being known from few forms. The general relations of the principal orders, with illustrations of typical forms, are shown in Figure 14-22.

The commonest and most diverse order, the notoungulates, ranged from the Paleocene to the Pleistocene. Within this order a fantastic array of different adaptive types evolved, ranging from large rhinoceros-like to small rodent-like forms. There was even a group with claws, remarkably similar to the chalicotheres. A trend toward great size was noticeable in many of the notoungulate groups. One of the largest was the genus *Toxodon*, of Pleistocene age, a heavy, massive animal nearly six feet high at the shoulders (Fig. 14-23).

Another group, the litopterns, also ranged from the Paleocene to the Pleistocene. The litopterns, soon after their appearance, evolved in two distinct trends, one developing animals very similar to the horse and the other camel-like animals. The horse-like forms were small but agile running animals that evolved feet remarkably similar to those of the real horses. One group evolved a three-toed foot with the central toe greatly enlarged; another developed a one-toed foot like that of the modern horse. The premolars became molariform, as in the true horses. This group became extinct in the Pliocene. The camel-like line displayed equally amazing parallel evolution. Two are illustrated in Figure 14-22. The large Pleistocene *Macrauchenia* (Fig. 14-23) had the nasal bone high on the head, suggesting a proboscis, probably like that of the tapir. This whole group survived well into the Pleistocene, even after the continent had been invaded by modern ungulates and carnivores from the north, but then became extinct.

The final group to discuss is the astrapotheres, which lived from Eocene to Miocene time. The later forms were large, heavy animals reminiscent of ungulates such as *Uintatherium* or the earliest elephants. The skull was remarkably modified. The upper canines were greatly enlarged into tusk-like daggers, but the incisors were lost. The lower jaw was longer and had both enlarged canines and well-developed incisors. There may have been a short proboscis.

FIGURE 14-23

The pampas of Argentina during the Pleistocene. The large animals in the center are ground sloths; to the right are armored glyptodons; behind the glyptodons is the strange, camel-like litoptern *Macrauchenia;* in the left background are two toxodonts (notoungulates). (C. R. Knight, American Museum of Natural History.)

14-12 MORE SOUTH AMERICAN MAMMALS—THE EDENTATES

The long Cenozoic isolation of South America was responsible for the evolutionary success and development of another group of mammals, the edentates. This order, of insectivore ancestry, had greatly simplified cylindrical or columnar teeth, without roots or enamel; or lacked teeth entirely. The order includes the armadillos, glyptodons, tree sloths, ground sloths, and anteaters. The edentates were present in North America during the early Cenozoic but had become extinct there by the end of the Oligocene. In the earliest Cenozoic they were among the first immigrants into South America, where they underwent a tremendously successful evolutionary radiation that continues today. A remarkable fact of their history is that rather than dying out when North and South America were again joined in the late Cenozoic, they successfully migrated northward into North America.

The armadillos have been highly successful since early Cenozoic time. They are distinctive in their armored body, consisting of many rows of bony plates over the back and sides. The armadillos are the only living mammals with a bony armor. Related to the armadillo was an extinct group, the glyptodons (Fig. 14-23). These were remarkable in that their armor consisted of a solid domed carapace, even the head and tail being similarly covered. Some species measured ten or more feet in length, and in some the end of the tail was equipped with a formidable spiked knob of bone. The glyptodon is reminiscent of the armored Cretaceous dinosaurs.

Another adaptive trend among the edentates evolved the tree sloth, the ground sloth, and the anteaters. Of these only the ground sloth is now extinct, but it had the most impressive history. This line tended toward huge size, developing animals as big as elephants and weighing several tons (Fig. 14-23). The feet were equipped with huge claws probably used for digging. The teeth were confined to the sides of the jaw. The ground sloths were plant-eaters that lived largely upon the leaves of trees and bushes, as shown by the plant residues in coprolites.

Both the glyptodon and the ground sloths successfully invaded North America in the Pleistocene and survived there until just a few thousand years ago. Partial mummies of ground sloths, with portions of skin and hair intact, have been found in the southwestern United States.

14-13 THE CARNIVORES

Most mammals we have discussed so far were adapted for plant-eating, but from the ancient insectivores there developed another group that was adapted for flesh-eating—the carnivores. A flesh-eater required agility and intelligence to catch his dinner, whereas the food of the plant-eater was always easier to come by. The plant-eaters evolved many different sizes and shapes with much modification of the limbs and teeth. The carnivores underwent nowhere near so much structural modification of their skeletons; their feet, for instance, were equipped with claws, which generally remained at the primitive number of five. The basic pattern of the skeleton and limbs generally remained rather primitive. As there was a premium on agility and mental alertness, however, carnivores had to develop these traits along with a keen sense of smell.

The most significant specialization among the carnivores was in the development of their teeth. Because the carnivore generally kills his prey with his teeth, and because meat is comparatively easy to digest and does not need elaborate chewing, the incisors were usually well developed and the canines large and pointed. Most carnivores have developed a specialized pair of teeth called the carnassials, which are large and elongated and serve to cut hard pieces of food. In living carnivores the carnassials are always the last premolar above and the first molar below (Fig. 14-24, A). In some early carnivores, however, the carnassials consisted of the first molar above and the second molar below, in others of the second molar above and the third molar below (Fig. 14-24, B). The carnassial teeth do not meet directly but pass each other, with the upper one on the outside, the two acting as a pair of shears. As the carnassials developed, the teeth behind them became greatly reduced or even disappeared. Animals that exist on a more varied diet, such as the bears, maintain molars with large grinding surfaces that enable them to cope with vegetation as well as meat.

The dominant flesh eaters of the early

FIGURE 14-24

Carnassials, or flesh-cutting teeth, of several carnivores.

A. Wolf: the only carnassials are the fourth premolar in the upper jaw and the first molar in the lower jaw.
B. *Hyaenodon*, a creodont: the main carnassials are the second molar in the upper jaw and the third molar in the lower jaw.

(After W. D. Mathew, 1905.)

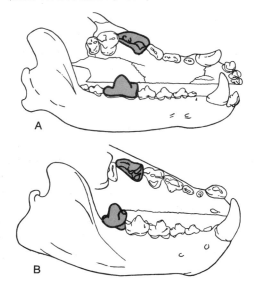

Cenozoic were the creodonts, a primitive group that was related to the order Condylarthra—an order completely unrelated to the Carnivora (Fig. 14-25). Most creodonts were small animals with slender, agile bodies, clawed feet, and carnassial teeth (Fig. 14-26). The carnassial cheek teeth were not the same in all the groups of primitive carnivores. In all these early carnivores the brain was usually small. Although most of the creodonts were small, some did develop great size. An Oligocene form found in Mongolia, whose skull was a yard long, was the largest flesh-eating mammal that ever lived.

These primitive carnivores were dominant during the Paleocene and Eocene, but most had become extinct by the close of the latter epoch. Only one group persisted into the early Pliocene. This decline coincided with the appearance of modern carnivores, ancestral to the modern cats, dogs, bears, etc., who were descended from an early carnivorous group, the miacids, that had larger and better-developed brains than the other primitive carnivores. This group also had its carnassial teeth in a more forward position, they being composed of the upper fourth premolar and the lower first molar, just as in the modern carnivores. Early in the Oligocene a new radiation occurred, one trend evolving the cat-like, another the dog-like, and still another the marine carnivores—the walrus, seals, and sea lions. This change from primitive to modern carnivores coincides approximately with the similar change in the ungulates. As the ungulates became more versatile in their adaptations, the carnivores likewise had to improve their brain power and agility in order to survive.

The wolves and dogs can trace their ancestry back to the late Eocene. Aside from specializations in the carnassial teeth and lengthening of the feet, the body plan remained rather primitive; but the brain grew larger. The fossil record of the dogs' evolution is rather good.

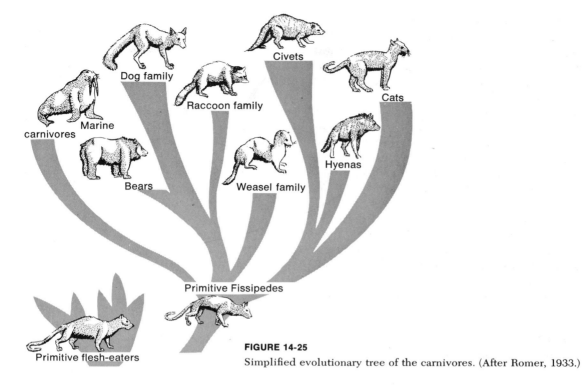

FIGURE 14-25

Simplified evolutionary tree of the carnivores. (After Romer, 1933.)

In the Miocene dogs began to evolve with heavier, more massive bodies and blunt, square molar teeth, the carnassial teeth being lost. These forms evolved into the bears, which have tended more and more away from a fleshy diet. The bears have only two molar teeth, but they are greatly elongated and crowned with blunt cusps. The bears today are among the largest land-living carnivores. Other evolutionary offshoots of the dogs are the raccoons and pandas.

The weasels, martens, badgers, skunks, wolverines, and otters belong to a separate group descended from miacids through a very complicated evolutionary history involving a wide range of adaptations.

In the evolutionary trend involving the cat and its relatives, the most primitive forms are the Old World civets, mongooses, genets, etc. Most of these are structurally quite close to the late Eocene and Oligocene carnivores, having changed very little.

They are mostly small with short limbs and arboreal habit. Late in the Miocene the hyenas branched off from this group. The hyenas are actually large descendants of the civets; they have elongated legs for running and a massive jaw with strong teeth. By the late Miocene the hyenas had already evolved to something very similar to the modern types; an example of an animal entering a certain ecologic niche and becoming perfectly adapted to it in a very short time.

The evolutionary history of the cats is similar to that of the hyenas in that soon after their appearance (early Oligocene) they were already structurally so well adapted to their habitat that little subsequent change took place. The cats are the most carnivorous of all carnivores and have the most highly specialized teeth for catching prey and eating meat. Along with high intelligence, they are extremely agile animals with strong legs and retractable claws. From their first appearance the cats evolved in two distinct lines, the normal cats we are familiar with and the famous saber-tooth cats with greatly enlarged canines. The saber-tooth cats evolved at several different times from the normal cat line. The skulls of an early and a late representative of normal cats and saber-tooth types are illustrated in Figure 14-27. An early normal cat is illustrated by the genus *Dinictis* and the modern cat by the genus *Felis*. Both had well-developed carnassial teeth, and the molars behind these were greatly reduced. In *Dinictis* the canines were fairly large, but in the modern cats they are smaller; otherwise the dentition remains much the same. The members of this evolutionary line were active, agile predators relying on fast movement and stealth to capture their prey.

The saber-tooth cats, in contrast, were generally larger, slower animals, which relied on their large canines to stab their victims. Figure 14-27, A, B shows there is not much difference in skull structure or teeth between the Oligocene saber-tooth and the Pleistocene *Smilodon*. The saber-tooth cats could open their jaws very wide so as to use most effectively their large canine sabers. Probably the saber-tooth specialized in preying on large, thick-skinned plant-eaters, such as the proboscidians (Fig. 14-28). The extinction of the saber-tooth cats, in fact, is thought by some to be due to the extinction of the large, thick-skinned animals. Growing evidence suggests that this extinction may have been due to hunting by early man. It is notable that the last saber-tooth cats in North America are of late Pleistocene age, about the time that man entered this continent.

FIGURE 14-26

The creodont *Oxyaena* making a meal of the primitive horse *Eohippus*.
(C. R. Knight, American Museum of Natural History.)

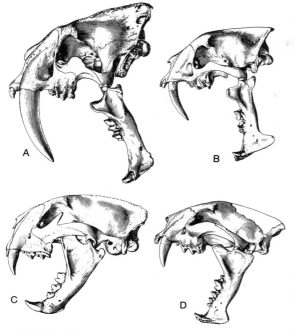

FIGURE 14-27

Skulls of saber-tooths and true cats.

A. *Smilodon*, a Pleistocene saber-tooth.
B. *Hoplophoneus*, an early saber-tooth, of Oligocene age.
C. *Felis*, a modern cat.
D. *Dinictis*, an early true cat, of Oligocene age.

(From W. D. Mathew, *Bull. Am. Mus. Nat. Hist.*, 1910.)

14-14 THE PRIMATES

Probably no other mammalian order stimulates a greater interest than the primates, for this is the order to which man belongs. Besides man the primates include the tree shrews, the tarsiers, the lemurs, the monkeys, and the apes. A concise definition of the order is difficult because of its lack of specialization. The primates have maintained many generalized and rather primitive characteristics; for instance, the primitive number of fingers and toes (five), and teeth less differentiated than in many ungulate and carnivorous groups. Most primates are arboreal, and in this habitat maintained

many primitive characteristics on the one hand and acquired specialized characters on the other. The five-fingered hand is perfectly suited for movement and grasping in a life in the trees. The food available in this habitat, insects, leaves, fruits, etc., did not necessitate specialized teeth. But life in the trees would be precarious without improved vision. In the primates the eyes came closer together, giving them binocular vision. In the monkeys and higher primates a further improvement, stereoscopic vision, gives the depth perception essential for arboreal life. Along with their improved vision came a decline in their sense of smell, which is of little use to an arboreal animal. With this decline the olfactory region of the brain also decreased in size, but the remaining parts of the brain, especially the cerebrum, increased, and the snout shortened. Grasping hands took over many of the functions performed by the front teeth of other mammals; so the jaws and teeth of primates became smaller and less obtrusive.

The classification of the primates now generally used divides the order into two suborders: the Prosimii, which include the tree shrews, lemurs, and tarsiers; and the Anthropoidea, which include the monkeys, apes, and man (Fig. 14-29).

The Tree Shrews

The most primitive primate in the opinion of some is the tree shrew; it is so primitive in many features, in fact, that some zoologists question its classification as a primate. But the tree shrews are reasonably close to the earliest primates evolved from generalized mammalian ancestors. They are about as big as squirrels, with long bushy tails, grasping hands with five claw-bearing fingers, and pointed snouts. The eyes are relatively large, and the brain is definitely advanced over that of the true insectivores. The modern tree shrews are widely distributed in India, Burma, the Malay peninsula, Sumatra, Java, and Borneo. They are

FIGURE 14-28
Typical scene at the Rancho La Brea tar pits near Los Angeles, California, during the Pleistocene. The saber-tooth cat is himself caught in the tar along with his prey. (Drawing by Z. Burian under the supervision of Prof. J. Augusta.)

FIGURE 14-29

Representative modern primates.

A. Tree shrew
  (*Tupaia*).
B. Lemur (*Galago*).
C. Tarsier (*Tarsius*).
D. Macaque monkey
  (*Macaca*).
E. Gibbon (*Hylobotes*).
F. Chimpanzee (*Pan*).

(British Museum, Natural History.)

largely arboreal animals, but some large species inhabit the undergrowth of the forest rather than the higher parts of trees. Their diet is largely insects.

The Lemurs

The lemurs are characteristic of Madagascar, where they have evolved successfully for a long time. They are almost all entirely arboreal, and many are nocturnal. Next to the tree shrews they are the most primitive of the primates. The usually pointed snout, large ears, and rather immobile face are a few of the characteristics that indicate their primitive state. Lemurs' hands and feet look like hands rather than paws and are well adapted for grasping with mobile thumbs and big toes. In contrast to the more advanced primates, the lemurs have a heavy covering of fur. They vary considerably in size, from that of a cat to much smaller forms.

The Tarsiers

These remarkable animals are the most highly advanced of the prosimians. The only living genus known in Borneo, the Philippines, and Celebes. The tarsier is tiny, about as big as a two-week-old kitten; it is nocturnal and entirely arboreal. Its most striking feature is the very large eyes. It has a long, hairless tail and long hind legs adapted for leaping from branch to branch. In most anatomical features it is much advanced over the lemurs and is, in fact, closer to the monkey grade of development. It is not a monkey, however, and it is far too specialized to be a simple intermediate form between monkeys and lemurs. The early Cenozoic fossil tarsiers, however, clearly demonstrate the intermediate position of the group.

The Monkeys

The monkeys represent distinct advances over the prosimians, having superior brain development and stereoscopic vision. Most monkeys are thoroughly arboreal, though some Old World monkeys do live on the ground. Two major groups of monkeys are anatomically and geographically distinct. One, the Ceboidea, includes all the New World monkeys, and the other, the Cercopithecoidea, includes all the Old World monkeys. The anatomical features distinguishing these groups are the shapes and positions of the nostrils and the number of premolar teeth.

The Ceboidea comprise many genera, which vary greatly in size. Some are remarkable for their "fifth hand" — the prehensile tail by which they can hang and swing from branches and can even grasp objects. Other forms include the howler monkeys and the typical organ-grinder monkeys.

The Cercopithecoidea are widely distributed in Asia and Africa and even reach southern Europe. One of the commonest is the macaque (Fig. 14-29, D), which ranges from Gibraltar east to Japan. This form belongs to a large group of Old World monkeys that have cheek pouches for the storage of food. Another group have complicated stomachs, somewhat comparable to those of ruminants, for the digestion of vegetable food. These particular monkeys are all completely arboreal, but many of the cheek-pouched forms, such as the baboons, have become adapted for ground living.

The Anthropoid Apes

The four living types of anthropoid apes are the gibbon, the orangutan, the chimpanzee, and the gorilla. They are clearly more closely related to man than any of the primate groups discussed so far. The forms living today are mostly arboreal, each group having undergone certain adaptive specializations. The skeleton of an ape is somewhat similar to that of man but differs sharply in the longer arms and the shorter legs. This difference is related to the unique mode of movement of most of these apes, which is swinging from branch to branch, a mode that favors a great lengthening of the arms and a shortening of the thumbs. The hand functions more as a hook than as a grasping mechanism. Another great difference from man is, of course, the size of the brain.

The smallest and most primitive anthropoid apes are the gibbons, which today are fairly common in southeastern Asia. Among all the apes the gibbon is the most adept at walking on two legs; but, it uses its long arms as balancers. The orangutan, confined today to Borneo and Sumatra, is much larger and thoroughly arboreal in habit.

In contrast to the gibbon and the orangutan, which are almost exclusively arboreal, the gorilla and chimpanzee are quite well adjusted to life on the ground. Anatomical evidence indicates clearly that the gorilla has only recently in its evolutionary history abandoned an exclusively arboreal mode of life. The gorillas and chimpanzees are confined to equatorial Africa and are the largest living anthropoid apes. Male gorillas attain six feet in height and a weight of about 600 pounds. The gorilla's brain, more highly developed than that of other apes, may attain a volume of 630 cubic centimeters, in contrast to about 500 cubic centimeters for the orangutan and only 90 cubic centimeters for the gibbon. The gorilla and the chimpanzee both spend most of their time on all fours.

14-15 FOSSIL PRIMATES

Among the rarest fossils are those of the primates. Their arboreal existence in wooded regions and their general high level of intelligence were unfavorable for fossilization. Wooded regions, especially in the tropics, where constantly changing stream patterns and high rates of decay prevailed, are not good environments for the preservation of fossils. Most of the evolution of the primates took place in the Old World. Fossil primate finds are quite fragmentary, consisting generally of teeth and pieces of jaws. Skulls and post-cranial skeletons are rare, even though many discoveries have been made since 1960.

The earliest fossil primates were very small creatures, not far removed from their insectivorous ancestors. Slightly later in the Paleocene several distinct evolutionary lines were already established, representing the initial prosimian radiation. The best-known Paleocene primate is *Plesiadapis* (Fig. 14-30), known from a remarkably complete skull and a relatively complete series of limb and foot bones near Cerney-les-Reims, France. Fragmentary remains of the same genus are known from several Paleocene localities in the western United States—this is the only genus of primate, other than *Homo* (man), that is surely known to have inhabited both the Old and the New World. Species of *Plesiadapis* varied in size from that of a squirrel to the size of a housecat. The pattern of the crowns of the cheek teeth and structure of the limb bones link *Plesiadapis* with modern lemurs, such as those in Madagascar. The incisors, however, were very rodent-like, being enlarged,

forward slanting, and widely separated from the cheek teeth (Fig. 14-30). Another striking feature was that the fingers and toes ended in long claws flattened on the sides. Among living primates only the tree shrews have a claw on each digit; all other species have either a combination of nails and claws or nails exclusively. Many details of its skeleton link *Plesiadapis* with the even earlier placental mammals—the Insectivora, from which the primates arose.

The next well-preserved fossil primates come from the middle Eocene of North America (Fig. 14-31). The best-known reconstruction of *Notharctus* does look

FIGURE 14-30

Plesiadapis, the best-known Paleocene Prosimian. (Above) Restoration based on analogy with living tree shrew. (Below) Skull details. (From Elwyn L. Simons, "The Early Relatives of Man." Copyright © 1964 by Scientific American, Inc. All rights reserved.)

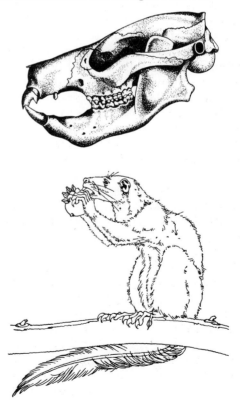

surprisingly like a modern lemur, but this is at least partly the artist's intent. The character of the skull and teeth and the long limbs clearly indicate that *Notharctus* was well adapted to the arboreal life that has been successful for the group throughout its history. *Notharctus* and its relatives lived in the tropical and subtropical forests of North America and Eurasia during the Eocene, but at its close they disappeared from the northern hemisphere. They have persisted since that time in the Old World tropics. A group of lemurs reached Madagascar and became isolated there, where they underwent a greatly expanded evolutionary radiation. One of the most unusual was a now extinct form that attained the size of a chimpanzee. Others evolved astonishingly close to monkeys in appearance.

The tarsioids likewise had become remarkably specialized by the Eocene and extremely diverse in details of their anatomy. As a group they were already well established in their characters and continued to evolve along this adaptive line. Some of them, however, show progressive features in the teeth, details of the skull, and size of the brain; further evolutionary advances later led to the appearance of true monkeys.

The evolutionary radiation of the prosimians in the Paleocene and Eocene of North America and Europe was great and the fossil record not insignificant. Yet not a single fossil from either continent appears to be ancestral to the living, higher, Old World primates, man included. Some fragmentary fossils from the upper Eocene formations of Burma may ultimately contribute to this evolutionary transition.

The only Oligocene fossil primates are from the Fayum oasis of Egypt about 60 miles southwest of Cairo. Professor E. L. Simons of Yale University has collected more than 100 individual primate specimens, most only single teeth. Among them several ape-like primates appear to be ancestral to modern apes and possibly Old

World monkeys. One, named *Aegyptopithecus*, known from a nearly complete skull and other fragmentary remains (Fig. 14-32), was about the size of a gibbon and had teeth much like a gorilla's. The canines were large and the lower front premolars elongated. *Aegyptopithecus* may be near the ancestral form from which both the

FIGURE 14-31
Notharctus, a lemuroid that lived in North America during the Eocene. (F. L. Jaques, American Museum of Natural History.)

higher apes and man arose, but it is premature to make definitive statements.

The primate record of the Miocene is much better than that of the Oligocene, for fossils come from many localities in Europe, the Middle East, Africa, India, and China. A Miocene primate, *Pliopithecus* (Fig. 14-33), is considered an ancestor of the modern Gibbon. The arms of today's gibbons are considerably longer than their legs; *Pliopithecus*, in contrast, had hind limbs and forelimbs of nearly equal length. In other words, these apes were still generalized and had not yet specialized like modern gibbons.

Another fairly common fossil primate in the Miocene and Pliocene of Europe is *Dry-opithecus*, whose fossil record consists

FIGURE 14-32

Aegyptopithecus, the earliest known fossil ape, from Oligocene strata of the Fayum region of Egypt. The lower jaw is a restoration based on jaw fragments not found in association with the cranium; the four incisor teeth of the upper jaw are also restorations. (Photographs furnished by Elwyn L. Simons.)

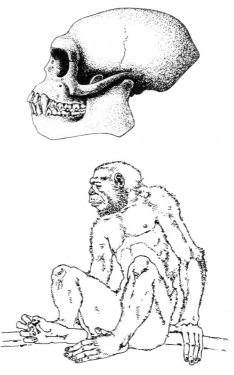

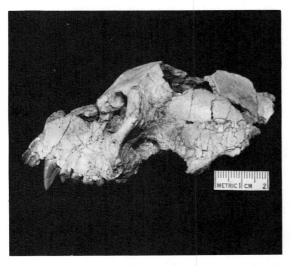

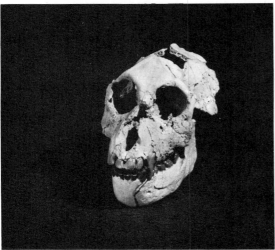

FIGURE 14-33

Skull and restoration of *Pliopethicus*, a Miocene anthropoid that is probably the ancestor of the modern gibbons. (From Elwyn L. Simons, "The Early Relatives of Man." Copyright © 1964 by Scientific American, Inc. All rights reserved.)

mainly of individual teeth and teeth in incomplete jawbones. Fossil teeth assignable to *Dryopithecus* have also been found in brown coal in southwestern China. In spite of the incompleteness of the record, the dryopithecines were an extremely important fossil group because their molar teeth had a cusp pattern like that of early man. Recently many remarkable fossil discoveries around Lake Victoria in East Africa have greatly added to our knowledge of the dryopithecines. One of the most complete of these fossils, *Proconsul africanus*, is known from two skulls and some limb bones, including parts of a foot and a forelimb with a hand (Fig. 14-34). These fossils show some monkey-like traits of hand, skull, and brain but hominoid and even partially hominoid characteristics of face, jaw, and dentition. The foot and forelimb also suggest some ape-like adaptations, including an incipient ability to swing by the arms from tree branch to tree branch.

It is now generally agreed that the East African *Proconsul* belongs to the same group as the European and Asiatic *Dryopithecus,* at approximately the same grade of evolutionary development. Detailed relations of the dryopithecines are by no means known at present. They possibly include ancestors of both apes and man, suggesting that during the early Miocene the hominoids were a single cosmopolitan group (Fig. 14-35).

The Siwalik deposits of northwestern India and adjacent Pakistan have yielded important mammal faunas. Among the primate remains, *Ramapithecus* is of particular interest. The first find was part of one side of the jaw, including the first two molars, both premolars, the socket of the canine tooth, the root of the lateral incisor, and the socket of the central incisor. When it and other specimens of *Ramapithecus* are used to reconstruct an entire upper jaw, complete with palate, the result is surprisingly human in appearance (Fig. 14-36). The proportions of the jaw indicate a foreshortened face. The size ratio between front teeth and cheek teeth is about the same as in man. The size of its sockets suggest that the canine tooth was not much larger than the first premolar—another hominoid characteristic. The arc formed by the teeth is curved as in man, rather than being parabolic, or U-shaped, as in the apes.

Recently a very similar form, *Kenyapithecus,* was found near Fort Ternan in southwestern Kenya. Potassium-argon dating yields a radiometric age of about 11 million years, near the boundary between the Miocene and Pliocene. This discovery strengthens the conclusion that sometime between late Miocene and early Pliocene time, both in Africa and India, an advanced

hominoid species was differentiating from more conservative ape-like stocks and developing hominoid characteristics.

14-16 PRIMATE EVOLUTION— A RECAPITULATION

The evolution and relationships of the primates is diagrammatically shown in Figure 14-37. The first primates evolved from insectivorous ancestry possibly in Late Cretaceous time. During the Paleocene and Eocene the primates (prosimians) under-

FIGURE 14-34

Skull of an East African Miocene ape (*Proconsul*), × ½. (From W. E. Le Gros Clark, 1950, by courtesy of the Trustees of the British Museum.)

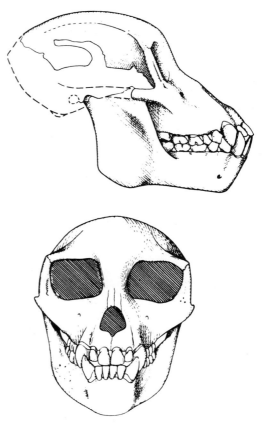

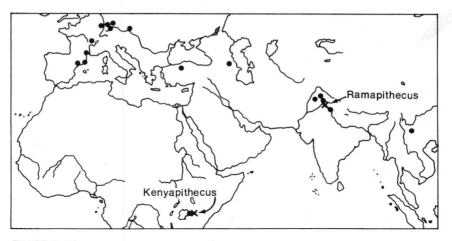

FIGURE 14-35

Geographic distribution of dryopithecines during the Miocene and early Pliocene. Also shown are the localities where the advanced homonoid genus *Ramapithecus* and *Kenyapithecus* have been found. (After Elwyn L. Simons, "The Early Relatives of Man." Copyright © 1964 by Scientific American, Inc. All rights reserved.)

went rapid and widespread evolutionary radiation along several lines. From within this early radiation the surviving stocks arose by the close of the Eocene, many of the earlier lines having become extinct. The surviving prosimians were restricted to Africa, Madagascar, and Asia. The New World monkeys on the one hand and the Old World monkeys and apes on the other, appear to have risen independently from a tarsoid stock in the early mid-Cenozoic. The Old and New World monkeys continued to evolve separately in their own structural and ecologic frameworks. The apes first appeared in the Oligocene in a variety of forms. Present evidence indicates that in the Miocene the early apes were generalized, lacking the specialized features of modern apes. Through the remainder of the Cenozoic, apes acquired the specialized limb features, fitting them for an arboreal life. Another branch, that leads toward Man, maintained rather primitive limb structures more adapted for life on the ground.

FIGURE 14-36

Upper jaw of *Ramapithecus* (A) compared with that of an orangutan (B) and a man (C). (From Elwyn L. Simons, "The Early Relatives of Man." Copyright © 1964 by Scientific American, Inc. All rights reserved.)

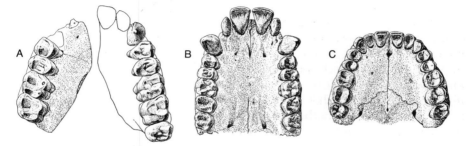

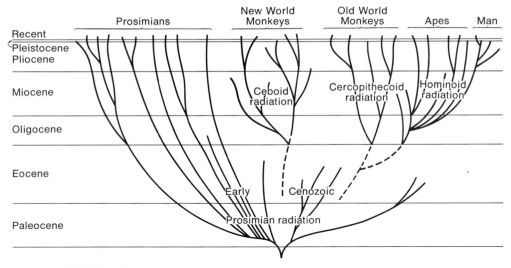

FIGURE 14-37

Diagrammatic representation of the evolutionary radiation of the primates. (After Simpson, 1949, with suggestions from B. Patterson.)

14-17 MAMMALIAN EVOLUTION AND GEOGRAPHY

One of the basic themes in the evolution of several mammalian orders was an early radiation of primitive forms, beginning probably in the mid-Cretaceous and continuing into the Paleocene and Eocene and their replacement by more modern types in the mid-Cenozoic. This type of evolutionary picture characterized the ungulates and carnivores especially and can be well illustrated from a study of the composition of Cenozoic mammalian faunas of North America. The basic data on that composition are summarized in Figure 14-38. Note how the Paleocene and Eocene faunas are largely made up of primitive ungulates, primitive carnivores, early primates (prosimians), insectivores, and multituberculates. By the late Eocene and Oligocene, however, the modern ungulates (perissodactyls and artiodactyls), modern carnivores, and rodents comprised the bulk of the fauna. The primitive types were on the verge of extinction. This general pattern continued through the remainder of the Cenozoic into the Pleistocene, and then the perissodac-

tyls, artiodactyls, and carnivores underwent a wave of extinction. The perissodactyls had been declining rapidly since the mid-Cenozoic, at the time when the artiodactyls were expanding. The proboscidians became extinct in the Pleistocene, but the rodents greatly expanded. The rodents, in fact, now comprise fully half of the North American mammalian fauna.

A similar type of graph, plotting the composition of Eurasian mammalian faunas, would not be very different, but such a graph of Australian or South American faunas would be completely different. This difference reflects the high degree of intermigration of faunas between North America and Eurasia during the Cenozoic; North American fossil faunas are more clearly related to those of Eurasia than to those of South America or Australia. There is a great deal of evidence to demonstrate that connections between North America and Eurasia were by way of the Bering Strait. It is also possible that some sort of connection may have existed from East Greenland to Iceland to the British Isles in the early Eocene. Detailed analysis of Eurasian and North American fossil mammalian faunas has shown that at several times during the Cenozoic the two continents were joined across the Bering Strait. A sort of filter bridge was formed; that is, animals that were acclimated to the region of the bridge could and did pass, and those that were not so acclimated did not pass. The times of exceptionally intense intermigration were the early Eocene, the early Oligocene, the late Miocene, and the Pleistocene. The middle Eocene, the middle and late Oligocene, the Early Pliocene, and the Holocene were times of little or no intermigration.

The composition of the mammalian fauna in South America during the Cenozoic was completely different from that in North America because of the isolation of South America during most of this era. Early in the Cenozoic, you will recall, South America was an island continent. Early ungulates, marsupials, and edentates were present. In isolation and without serious outside competition a unique evolutionary drama was unfolded. The earliest phase of the fossil record already has a composition very unlike that of North America. The greater part of the fauna is composed of unique orders of ungulates endemic to this continent, and the remaining members include the primi-

tive ungulates, edentates, and marsupials. This general pattern continued through the Cenozoic to the Pliocene except for the addition of rodents, which entered South America after it had already been separated from North America. The great majority of the mammals are insect-eaters and plant-eaters. The carnivorous adaptation was taken over by the marsupials early in the Cenozoic. Some small raccoon-like carnivores did enter South America, doubtless by rafting, in the mid-Cenozoic, but these never became a large component of the fauna.

In the late Cenozoic, probably in late Pliocene time, North and South America were again connected by an isthmus, and immediately there was a migration southward of modern ungulates and carnivores. The arrival of these more advanced types eventually led to the extinction of the endemic South American ungulates (notoungulates, litopterns, astrapotheres), though some of these survived well into the Pleistocene. Of the typical South American fauna only the edentates survived. They also successfully invaded the north in the form of ground sloths, armadillos, and glyptodons. Thus the present mammalian fauna of South

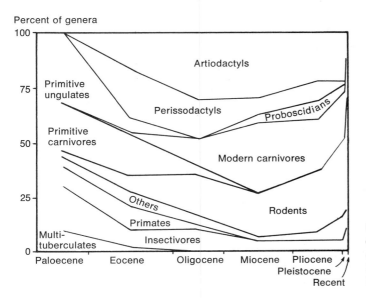

FIGURE 14-38

Changes in the known composition of the mammalian land fauna of North America during the Cenozoic. (After Simpson, 1953.)

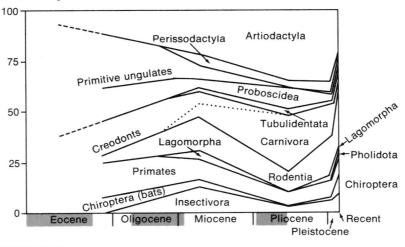

Percent of genera

FIGURE 14-39

Changes in the known composition of the mammalian land fauna of Africa during the Cenozoic. The shaded bars at the bottom of the time scale indicate the time spans for which little or no date is available. (After Cooke, 1968, with suggestions from B. Patterson.)

America is composed largely of late immigrants and a minor element of endemic survivors.

The Cenozoic history of African mammals has long been thought to parallel that of Eurasian mammalian history except that Africa was regarded as a refuge for the survival of archaic forms. Many new data demonstrate that African mammalian faunas are essentially indigenous, reflecting an important center of evolution and source for diffusion into Eurasia. The faunal and geologic evidence suggests that intermigration of mammalian faunas between Africa and Eurasia was not free throughout the Cenozoic. At the close of the Cretaceous, the general uplift of Africa created island connections across the Tethys, and primitive ungulates, carnivores, insectivores, and primates entered the continent. The intense diastrophism during late Oligocene again created conditions suitable for faunal exchange in the Tethyan region; other times of somewhat lesser interchange came in the late Miocene and, mainly from the Arabian region, in the Pliocene. Figure 14-39, like Figure 14-38, shows the variation in the

Cenozoic mammalian fauna of Africa. The most conspicuous features of the diagram are the early dominance of the primitive ungulates and proboscideans and their later declines and the great growth of the horned artiodactyls, an immigrant group that underwent great evolutionary radiation.

This interpretation of African mammalian history is a distinct break from previously held ideas and many additional data are needed for a final evaluation. Nevertheless, the evidence suggests that this continent was an important center of evolution with a long history of indigenous evolution for its wide variety of mammals. Man himself may perhaps have been among the important exports from the African continent.

Australia, another island continent, is unique in that nearly the whole of its mammalian fauna is composed of marsupials, with a few monotremes. The fossil record is extremely poor, mostly from the Pleistocene and Holocene. In no other region did the marsupials undergo such an extensive evolutionary radiation as in Australia, evolving into carnivores, insectivores, and herbivores. Many genera converged toward

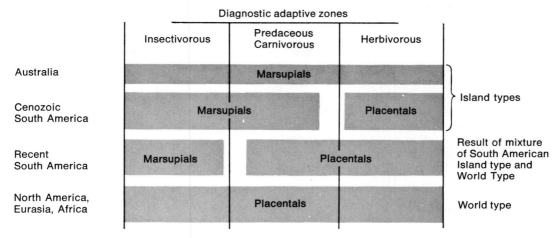

FIGURE 14-40

Adaptive range of marsupial and placental mammals during the Cenozoic.
(After Simpson, 1953.)

placental types on other continents. One puzzling problem is: Why did the marsupials take over Australia with no interference from placental mammals? In the late Mesozoic both placental and marsupial mammals appear to have been equally abundant and widespread. If there was a land connection between Australia and Asia, it is difficult to explain why one group and not the other made the trip. Here, again, it seems that chance island hopping by rafting is the most plausible explanation, and the marsupials were the first to get there. The probability of this theory is enhanced by

the evidence that certain placentals (members of the rat family) reached Australia later by island hopping from Asia.

Thus the composition of the fauna on any continent is determined by evolution and continental geography. A summary of the basic continental faunal types in terms of the main adaptive zones is given in Figure 14-40.

The geographic relations of the continents during the Cenozoic are diagrammatically illustrated in Figure 14-41. Compare this diagram with the map of the modern zoogeographic provinces (Fig. 2-19, p. 38).

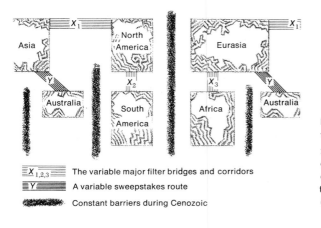

$X_{1,2,3}$ The variable major filter bridges and corridors
Y A variable sweepstakes route
Constant barriers during Cenozoic

FIGURE 14-41

We can best account for the geographic history of mammals during the Cenozoic by considering the continental blocks and the main sea barriers as constants and the three main filter bridges and the main sweepstakes route as variables.
(After Simpson, 1953.)

SUGGESTED READINGS

Colbert, E. H., *Evolution of the Vertebrates* (Wiley, New York, 1955).

Ekman, Sven, *Zoogeography of the Sea* (Sidgwick & Jackson, London, 1953).

Romer, A. S., *Vertebrate Paleontology*, 3rd ed. (Univ. Chicago Press, Chicago, 1966).

Simpson, G. G., *Evolution and Geography, An Essay on Historical Biogeography with Special Reference to Mammals* (Condon Lectures, Oregon State System of Higher Education, 1953).

Symposium: Evolution of Mammals on Southern Continents, Quart. Rev. Biol., 43 (1968).

A. Keast, "Introduction: The Southern Continents as Backgrounds for Mammalian Evolution."

H. B. S. Cooke, "The Fossil Mammal Fauna of Africa."

R. C. Bigalke, "The Contemporary Mammal Fauna of Africa."

A. Keast, "Australian Mammals: Zoogeography and Evolution."

B. Patterson and R. Pascual, "The Fossil Mammal Fauna of South America."

——44 (1969).

P. Hershkovitz, "The Recent Mammals of the Neotropical Region. A Zoogeographic and Ecologic Review."

Scientific American Offprints

622 Elwyn L. Simons, *The Early Relatives of Man* (July, 1964).

636 Elwyn L. Simons, *The Earliest Apes* (December, 1967).

655 Elwyn L. Simons and Peter C. Ettel, *Gigantopithecus* (January, 1970).

15

THE PLEISTOCENE EPOCH

The doctrine of uniformitarianism has been such a strong guiding principle in the reconstruction of the earth's history that it may come as a surprise to state that the world today is in a very abnormal condition. For one thing, the area and general elevation of the continents are exceptionally great in comparison with conditions that prevailed during most of the earth's history. The world's climate at the moment, with its clear-cut belts and extremes of temperature, is quite exceptional, for the "normal" climate of the geological past appears to have varied from moderate to warm, with little significant difference in temperature between the equatorial and high latitudes.

In the past one or two million years the world has undergone a glacial period during which vast ice sheets formed on the northern continents and on the highlands of the southern continents. Great variation in climate caused a series of advances and retreats of the ice sheets during the Pleistocene. Glacial conditions had existed previously in the southern hemisphere during the Carboniferous and the early Permian (§ 11-1), and there were glacial episodes during the Precambrian (§ 4-6). But it was during the last glacial episode, during the Pleistocene, that man evolved. The evolution of man is intimately involved with the physical and climatic changes of the earth during the past 2 or 3 million years.

The interpretation of past climates is particularly complex and rests mainly on deductions from the fossil and rock record. Here, as in many other geological problems, the doctrine of uniformitarianism provides the framework within which the historical geologist operates. Also it is clear that the younger the fauna or flora, the more confidence can be had in the climatic interpretations, whereas those based on extinct organisms are fraught with difficulties and uncertainties. The older forms, however, can give some clues to the climate in which they thrived. Conclusions cannot be based on individual genera or species; the whole fauna or flora of any era must be considered. The time and space distribution of faunas and floras make climatic interpretations possible. The basic assumption is that the animals of the past resembled their modern relatives in physiologic requirements respecting heat and cold, water and the lack of it, and hence in ecological tolerances, although it is possible that closely related animals might have lived in strikingly different environments. Modern elephants are limited to tropical and subtropical climates, but some of their cousins, the extinct mammoths, flourished in arctic realms.

The "normal" climate during the earth's history has varied from warm to moderate and the glacial episodes were short interruptions of this basic pattern. Figure 15-1 illustrates in a general way the probable temperature variations in middle latitudes since the late Precambrian. Because climate is important to the understanding of Pleistocene history, let us first discuss past geologic climates.

15-1 CLIMATIC EVIDENCE FROM VERTEBRATES

The Devonian is the first period that has enough of a fossil vertebrate record to furnish a basis for climatic interpretation. Primitive jawless fish (ostracoderms) and early jawed fish (placoderms) of Devonian age are known in northern Europe, eastern Greenland, North America (Quebec, New York, Ohio, Wyoming, and Arizona), New South Wales in Australia, and Antarctica. Even though this is a spotty distribution, it suffices to show a nearly world-wide

distribution. As all of the known faunas show very close relationships, environmental conditions were probably roughly the same in all these regions.

Among the Devonian vertebrate faunas a large group of air-breathing fishes, represented by both lungfishes and crossopterygians, allow certain climatic deductions.

FIGURE 15-1

Generalized climatic trends in middle latitudes. (After J. Wyatt Durham, "Palaeoclimates," *Physics and Chemistry of the Earth* (Vol. 3), L. H. Ahrens et al., eds., Pergamon Press, New York, 1959.)

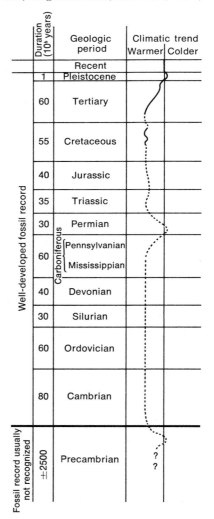

They were probable ancestors of the first land-living animals. Analogy to modern lungfishes suggests that the waters in which they lived were subject to periods of great reduction or even drying up. As was pointed out in § 7-6, alternating wet and dry seasons were probably closely related to the origin of the first land-living animals. The movement of these fishes from one pool of water to another during dry seasons was perhaps the first step in the invasion of land by the vertebrates and the origin of the amphibians.

Modern amphibians are restricted by narrow temperature tolerance, and presumably this limitation also applied to earlier amphibians. The earliest amphibians are from the Upper Devonian of eastern Greenland, thus suggesting a temperate climate in this region.

The Carboniferous was the period of amphibian predominance. The fossil evidence of these vertebrates indicates that the climate of the Carboniferous continued without much change from that of the Devonian. Even the orogenic activity that affected vast areas in the northern hemisphere during the Late Carboniferous appears to have had no appreciable effect on Carboniferous faunas and floras.

Significant changes in the physical character of the continents took place in the Permian. Hercynian orogenic phases continued into the Permian in many places, expanding the emergent portions of the continents and causing regression of the continental seas and geosynclines. In the southern hemisphere, during the mid-Carboniferous and early Permian, vast ice sheets covered portions of east-central South America, Antarctica, South Africa, peninsular India, and Australia. The initial radiation of the reptiles was well under way by the Permian, and during this period they became the dominant land-living animals. Because of their method of reproduction the reptiles were particularly suited to cope with the rigorous changes taking place at this time. Permian terrestrial deposits with fossil reptiles in North America and Europe are mainly in redbed facies. Besides reptiles, however, these deposits contain fish and amphibians. Redbed strata are often interpreted as desert deposits, but these particular formations do not fit this interpretation, for the presence of fish and amphibians implies the existence of streams and lakes. The climate, therefore, was probably warm, and there were probably alternate wet and dry spells.

On the other hand, the reptile faunas of the Karroo deposits of South Africa suggest quite different ecological conditions. The Karroo deposits are particularly famous for their abundant faunas of mammal-like reptiles § 11-8. These faunas include numerous large herbivorous types associated with many carnivorous forms. This association of herbivores and carnivores is common among later land-living vertebrates, and it has been suggested that these Karroo reptiles were plains-dwelling animals under conditions not unlike those of the High Plains of North America, during the Cenozoic.

The distribution of late Permian reptiles is shown by Figure 11-13 (p. 399). The wide latitudinal range of these reptiles suggests that there was not much difference in climatic conditions between the high and low latitudes. Modern reptiles have varying temperatures that correspond roughly with the temperatures of the environments in which they live. But their temperature tolerance is rather narrow, and they cannot survive body temperatures that vary widely. In this modern world of definite climatic belts all reptiles find the environment most suitable to them in the tropical, subtropical, and temperate regions of the earth, the large reptiles inhabiting only the tropics and subtropics. If Permian reptiles had similar physiological requirements, the presence of reptiles in northern Russia and southern Africa indicates that the climate in those widely separated regions was moderate or warm but not cold.

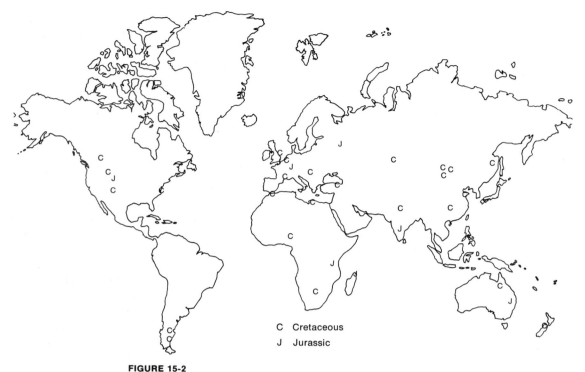

C Cretaceous
J Jurassic

FIGURE 15-2

Distribution of Jurassic and Cretaceous amphibians and reptiles.
(After E. H. Colbert, 1954.)

The general emergence of the continents and the wide distribution of terrestrial red-bed sedimentation continued through Triassic time. The reptiles and amphibians of this period represented continuations and elaborations of evolutionary lines already well established in the Permian. The wide distribution of the reptiles (Fig. 11-13) and the types of facies developed are reminiscent of the Permian, and Triassic climates are, therefore, thought to have been moderate, like those of much of the Permian.

The Jurassic and Cretaceous land vertebrate faunas were dominated by the dinosaurs. Their large size and their nearly world-wide distribution suggest mild and uniform conditions, especially during the Jurassic (Fig. 15-2). Slightly more varied climatic conditions are indicated for the Late Cretaceous by the fact that the gigantic herbivorous dinosaurs were generally smaller than those in the Jurassic. Most of the other dinosaurs of the Cretaceous, especially the dominant ornithischians, were also smaller than the Jurassic giants. This change probably indicates that environmental conditions were not so favorable to giantism as in the Jurassic, that the climate was perhaps not so uniformly tropical or subtropical, and that plant life was not so luxuriant as it had been in Jurassic time.

Beginning in the Late Cretaceous and continuing into the Eocene, widespread diastrophism caused mountain-building, general emergence of the continents, and regression of the continental and geosynclinal seas. During this span of time came the gradual extinction of the dinosaurs and the explosive evolutionary radiation of the mammals. The persisting reptiles — crocodiles, turtles, etc. — remained as good temperature indicators. The mammals, however, since they tolerate a greater temperature range, are not such effective indicators of the conditions under which they lived. Nevertheless, the general distribution and association of early Cenozoic mammal faunas indicate moderate climates in the higher latitudes and a continuation of the moderate conditions of the Late Cretaceous. In the later Cenozoic, however, there are indications that climatic zonation had begun. The appearance of steppe forms—animals that lived on the high plains where temperatures at times were certainly severe—is among the data that support this conclusion. Wide climatic variation during the Pleistocene resulted in a great wave of extinction within many orders of mammals and in the establishment of the modern zoogeographic provinces (Fig. 2-19, p. 38).

15-2 CLIMATIC EVIDENCE FROM MARINE INVERTEBRATES

The distribution of modern shelf marine faunas is largely controlled by temperature, and the present faunal provinces (shown in Fig. 2-19 and discussed in § 2-3) developed during the latter half of the Cenozoic as a result of the break-up of the Tethys, the rise of the isthmian link between North and South America, and the deterioration of the climate during these periods.

In the interpretation of climate based upon marine invertebrates, the following general rule is used: organisms with a narrow temperature tolerance are usually much more critically affected by temperatures below their tolerated minimum than by temperatures above their tolerated maximum. For instance, numerous tropical shallow-water genera of gastropods drop out of the faunas as we leave the tropics, whereas equally numerous boreal shallow-water genera of mollusks are found in deep water (lower temperature) nearer the tropics. So far as Cenozoic faunas are concerned, a

working rule is that, in general, the past ecological requirements of associations of organisms are similar to those of similar Holocene associations.

Paleozoic faunas present the greatest difficulty in the interpretation of past climates by marine invertebrates because the faunas of this era are composed largely of extinct groups and, in addition, we have many fewer data on Paleozoic faunas (especially the older ones) than on younger faunas. The rock and fossil record is so incomplete that establishment of correlation is generally on a much broader scale than for the Mesozoic or Cenozoic. Finally, paleogeographic reconstructions of the continents are still rather crude. If, for example, the distributional pattern of marine faunas shows the existence of faunal provinces (as it does in many levels of the Paleozoic), we have difficulty in deciding whether the pattern is due to climatic factors or to geographic barriers of which we are ignorant.

These remarks apply more to the earlier Paleozoic than to the later. In § 11-9 we discussed the relative diversity of nine Permian faunas from North America. These data show a clear increase in faunal diversity from the Arctic Islands of Canada to central Mexico (Fig. 11-23). More qualitatively, Permian brachiopods and fusulinids suggest that a boundary between temperate and subtropical waters in the northern hemisphere lay between 50° and 60° north latitude (Fig. 15-3). For one zone of the overlying Lower Triassic faunas there is likewise a clear indication of a faunal gradient. In this particular zone the Tethys has an ammonoid fauna of 57 genera, of a total for the zone of 65 genera, and 24 of these genera are endemic to this region. The fauna of this zone in the circum-Arctic region consists of only 16 genera of which only 2 are endemic. The difference in temperature between Tethys and the circum-Arctic region, however, may not have been great.

Of all the invertebrate groups of the Mesozoic and Cenozoic, the corals, more than any

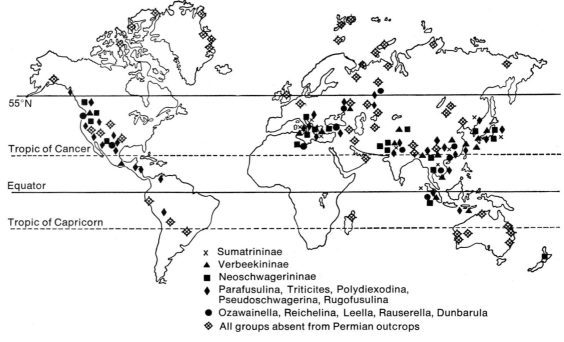

55°N

Tropic of Cancer

Equator

Tropic of Capricorn

x Sumatrininae
▲ Verbeekininae
■ Neoschwagerininae
♦ Parafusulina, Triticites, Polydiexodina,
   Pseudoschwagerina, Rugofusulina
● Ozawainella, Reichelina, Leella, Rauserella, Dunbarula
⬦ All groups absent from Permian outcrops

FIGURE 15-3

Distribution of various fusulinid groups in Permian rocks.
(From F. G. Stehli, *Am. J. Sci.*, 1957.)

others, suggest climatic implications by their distribution. The corals of these eras are of the same order (Scleractinia) as those that make up modern coral reefs in the tropical seas. The time and space distribution of scleractinian corals were discussed in § 10-6, and the important feature is the exceedingly widespread latitudinal distribution of Late Triassic corals (and their general homogeneity), which was followed by a gradual reduction through the remainder of the Mesozoic and Cenozoic. This reduction was by no means constant but fluctuated frequently. Another group of organisms that are considered to have inhabited only warm waters are the peculiar sedentary pelecypods — the rudistids. Their distribution during the Late Cretaceous is shown by Figure 15-4. The most northerly and southerly faunas are dwarfed. Some attribute this dwarfing to the influence of waters colder than those of the Tethyan realm.

In recent years isotope chemistry, by giving a means of determining paleotemperatures, has helped to solve the problem of climate. The technique consists in measuring the proportions of oxygen 16 and oxygen 18 in a calcareous shell. These two isotopes have an equilibrium relation that is temperature-dependent: at higher temperatures the proportion of oxygen 18 is greater. A difference of one degree centigrade in water temperature produces a difference of 0.02 percent in the ratio of oxygen 18 to oxygen 16 in the carbonate. Extremely sensitive mass spectrometers are needed to measure the ratio of these two oxygen isotopes in a carbonate shell, but interesting results have already been published. For instance, a recent study, using this method, of a fairly large number of Late Cretaceous belemnites, oysters, *Inoceramus* (a pelecypod), and brachiopods, and of a fair quantity of broken shell material, indicates that the temperature of the ocean rose to about the middle of the Late Cretaceous and then declined to the end of that period. The belemnite data adduced by this method were particularly useful in establishing this climatic history, but data from the associated fossils corroborate the findings (Fig. 15-5).

Cenozoic marine faunas were modern in aspect even in the early part of that era, and their close relationship to modern faunas enables us to make much more refined climatic interpretations. Cenozoic faunas of the Pacific coast of North America show the climatic implications. The association of organisms at a particular place is the unit of study, and most of these fossil faunas contain a large number of genera that are living today and for which temperature data are available. Assuming that (1) organisms are more limited by minimum temperatures than by maximum temperatures and (2) that past requirements of associations of organ-

isms are the same as those of similar Holocene associations, Figure 15-6 shows the inferred centigrade temperatures of all the known Cenozoic faunas of the Pacific coast. The broken lines show the past positions of the February marine isotherms. The graph clearly shows that the 20° isotherm during the Paleocene was north of 49° north latitude. Beginning in the Oligocene, the 20° isotherm shifted southward, with minor fluctuations, to its present position at approximately 25° north latitude, off the coast of Baja California.

15-3 CLIMATIC EVIDENCE FROM PLANTS

Many land plants are even better climate indicators than animals. As with vertebrate animals, the first appreciable fossil record of plants becomes available in the Devonian.

FIGURE 15-4

Distribution of Late Cretaceous rudistid pelecypods: x's, normal types; round spots, dwarfed types. (After E. Dacqué, 1915.)

The emergence of plants from the sea and the establishment of a true land flora took place in the Silurian. Early Devonian land plants were small and simple in structure but distributed very widely. The first land plants are known from such widely separated areas as the Falkland Islands, Spitsbergen, and the interior of Asia and North America. Such a distribution can have one of only two meanings: either these primitive plants had a wide temperature tolerance, or temperature ranges in the Devonian were far narrower than they are today. Middle and Late Devonian floras indicate similar conditions. Early Carboniferous (Mississippian) floras are widely distributed. The Late Carboniferous (Pennsylvanian) was truly exceptional in the evolutionary history of plants. The accumulation of thick, widespread coal deposits in the northern hemisphere indicates rapid and prodigious growth. The factors involved in coal deposition were discussed in § 7-11. One impressive indicator of uniform climate over great areas of the Carboniferous continents is the general absence of annual growth rings in coal-swamp trees. In climates with distinct seasons, seasonal change is almost invariably reflected in pronounced annual growth rings.

The Late Carboniferous and early Permian, however, saw a complete breakup of uniform climatic conditions. This was the period of glaciation in the southern hemisphere and of the widespread distribution of the *Glossopteris* flora (see § 11-7). But in the northern hemisphere Permian floras were still like those of the Carboniferous.

Triassic floras, though not as well known or diverse as those of the Carboniferous, indicate a return to a more nearly uniform worldwide climate. Jurassic floras were extremely varied and remarkably wide-spread. They are known from some of the most northerly land areas of the Arctic, such as northern Greenland and Franz Josef

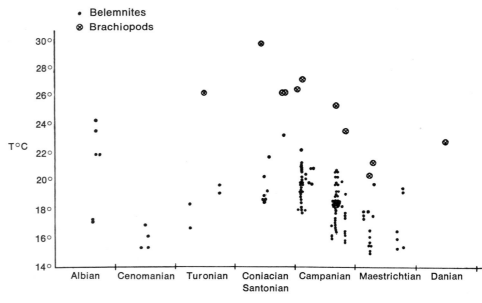

FIGURE 15-5

Average temperatures derived from oxygen-isotope analysis of belemnites and brachiopods from Albian (late Early Cretaceous) to Danian (earliest Cenozoic). (From H. A. Lowenstam and S. Epstein, *J. Geol.*, 1954.)

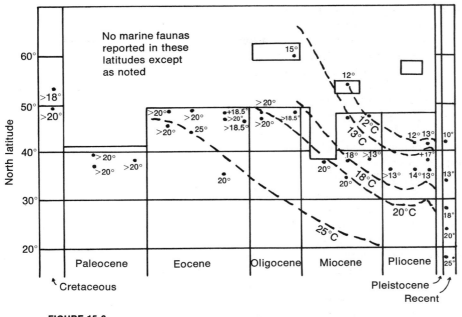

FIGURE 15-6

Past positions of the February marine isotherms along the Pacific coast of North America. (After J. Wyatt Durham, *Geol. Soc. Am. Bull.*, 1950.)

Land, north of the eightieth parallel. The most remarkable evidence of the nonglacial climate of the polar regions in Jurassic time is the presence of fossil plants and coal on Mount Weaver, at an elevation of 10,000 feet, at latitude 80°58′ south, in Antarctica.

The appearance and evolutionary radiation of the angiosperms in the Cretaceous make it possible to refine greatly the paleoecologic interpretations of climate. By the Late Cretaceous and early Cenozoic a large number of genera of trees and shrubs that are still living had already appeared. The distribution of early Cenozoic plants clearly shows that moderate-to-warm conditions still prevailed in both the Arctic and the Antarctic. These floras are composed of types now confined to subtropical or even tropical regions. The flora illustrated in Figure 14-5 (p. 000), an Eocene landscape of southeastern England, has its present equivalent in the Austro-Malaysian tropics. Figure 15-7 shows the location of most of the better-known Eocene fossil floras; the con-

necting lines (called **isoflors**) join fossil floras that are considered to represent similar climatic conditions. We believe that these isoflors have a general correspondence to the position of the Eocene isotherms. Modern vegetation shows a similar northwest-southeast trend across the northern continents, and such trends are roughly parallel to existing isotherms (Fig. 15-8). Wherever present isotherms do not coincide with parallels of latitude, one or more of several factors may account for the variations: the general emergent state of the continents, the tempering effect of ocean currents on the western sides of the continents in the middle and high latitudes, and the increasing continentality toward the east. We therefore conclude that the similar distribution of vegetation and temperature during the Eocene resulted from the same controlling factors we have today.

A shift in vegetation belts initiated by gradual climatic change and manifested by migration of forest associations apparently

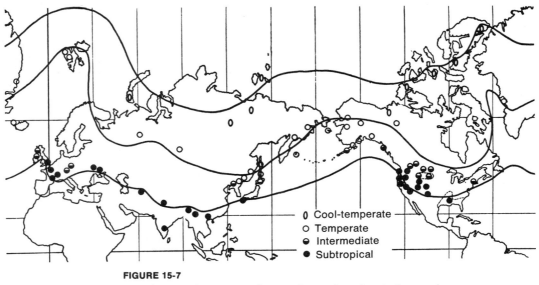

0 Cool-temperate
O Temperate
◒ Intermediate
● Subtropical

FIGURE 15-7

Distribution of Eocene isoflors in the northern hemisphere.
(After R. W. Cheney, *Geol. Soc. Am. Bull.*, 1940.)

began in the late Eocene. By the mid-Cenozoic temperate forms began to appear in increasing numbers in mid-latitude floras, and with them a sudden influx of herbaceous plants. Climatic deterioration probably went on at an accelerated rate during the late Cenozoic. The great change in Cenozoic floras is illustrated graphically in Figures 15-9 and 15-10, the results of an analysis of thirty-two of the larger Upper Cretaceous and Cenozoic fossil floras of North America. In Figure 15-9, the higher a point is on the curve, the greater the similarity in generic composition to the existing flora. The slope of the curve from the middle Eocene to the Pleistocene indicates the rate of trend toward the modern flora, and so can be regarded as a curve of climatic change. In Figure 15-10, the higher a point is on the curve, the greater the difference in generic composition from the existing flora. The rare or absent element is composed largely of genera now limited to warmer temperate or subtropical regions. The number of rare or absent genera rapidly declines in the later Cenozoic, because of the retreat south-

ward of the elements of the various floras that required a warmer climate. Both of these graphs show that the cooling trend in the Cenozoic climate, which culminated in the Pleistocene glaciation, probably began in the mid-Cenozoic, more than 20 million years ago.

15-4 THE PLEISTOCENE RECORD

The deterioration of the Cenozoic climate culminated in the Pleistocene, when temperatures in the higher latitudes were lowered sufficiently to allow the accumulation and spread of immense ice sheets. The ultimate cause of this (or any other) glacial episode is not known, though many theories have been proposed. But, whatever the cause, the last epoch of the geologic time scale is dominated by two features, glaciation and the evolution of man. The temporal and spatial distribution of both the glacial deposits and the material evidence of man's evolution will be the focal points of this discussion.

Area of Pleistocene Glacial Deposits

At the present time ice caps cover Greenland and Antarctica, and small isolated glaciers exist in many of the higher mountain ranges of the world; during the maximum extent of Pleistocene glaciation, however, vast areas of the continents were completely covered by ice (Fig. 15-11). Ice sheets occupied Greenland and much of Canada, where they merged, in the west, with glaciers formed in the Canadian Rockies. Glaciers also occupied the mountain ranges of southern Alaska and a portion of northern Alaska. The southern part of this ice field occupied the United States to the approximate positions of the Ohio and Missouri rivers. In Europe a great ice field spread out from the Scandinavian Mountains and extended south to cover the Baltic and North seas, eastern Britain, a large part of Germany, Poland, and Russia to about the Ural Mountains. A vast complex of glaciers also spread out from the Alps. A similar pattern of extensive glaciers developed in the Himalayan and associated mountain systems, and other high regions of central and eastern Asia were glaciated. In the southern hemisphere an ice field occupied all of Antarctica and great areas in the southern Andean Cordillera of South America. The highlands of New Zealand and Tasmania were occupied by glaciers. Even Mount Kilimanjaro, near the equator in Kenya shows evidence of important glaciation, and Mount Ruwenzori, on the equator, still has glaciers.

Nature of Pleistocene Glacial Deposits

Soon after the glacial origin of the Pleistocene deposits was established, it was recognized that there was not just a single advance but as many as six major advances of the ice sheets over the northern hemisphere. The major advances were separated by extensive retreats of the ice sheets and interglacial

FIGURE 15-8

Distribution of January isotherms and modern vegetation in the northern hemisphere. (After R. W. Cheney, *Geol. Soc. Am. Bull.*, 1940.)

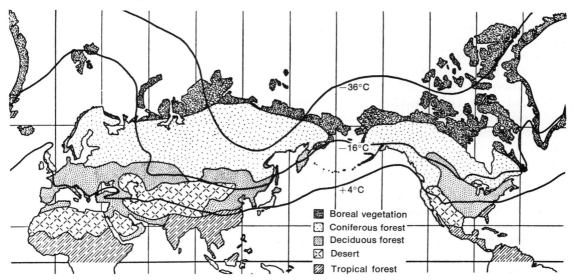

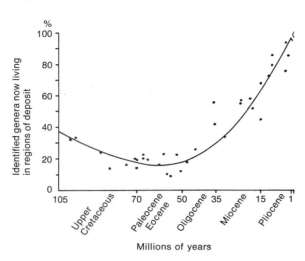

FIGURE 15-9

Graph showing the relation between geologic time and the proportion, in fossil floras, of genera that still live in the region of the deposits. (After E. S. Barghoorn, 1953.)

often contain fossil plants and animals, which clearly indicate, among other things, more moderate climatic conditions.

Chronology of Pleistocene Deposits

The methods of stratigraphic analysis and correlation of Pleistocene deposits are essentially the same as those used in the study of older rocks. Superposition is the primary criterion, but seldom is it possible to see more than two glacial tills in a single sequence. Considering the nature of the deposits and the topography of much of the area where they are present, we have seldom found exposed a sequence through the whole glacial deposit. As younger glacial advances came along, they frequently destroyed the older tills. The fauna and flora of interglacial deposits are also an important means of establishing chronology and correlations, but the record is obscure because of multiple migrations resulting from

episodes during which the climate was much warmer. The sequence of glacial and interglacial stages most generally recognized in North American and Europe is shown in Table 15-1. Successive layers of till are recognized by the soil profiles, by the degree of weathering, and especially by the nature of the contained fossil faunas and floras. At many places in North America and Europe, roadcuts or natural exposures reveal two tills, one on top of the other. The upper till may contain boulders of almost fresh granite and gneiss, some with polished facets and striae. In the lower till, however, beneath a layer of thoroughly weathered clayey soil, though the outlines of similar boulders can be made out, they have weathered to the consistency of cheese—the feldspars are thoroughly rotted to clay, and the ferromagnesian minerals have decomposed to limonite and other materials. Only chemically resistant rocks like quartzites are found intact in the lower till. Deeply weathered tills are called gumbotils.

The interglacial episodes are represented by stream, lake, or cave deposits, which

FIGURE 15-10

Graph showing the relation between geologic time and the proportion, in fossil floras, of genera that no longer live in the region of the deposits. Because of the unidentified or extinct elements in the fossil floras, the data are not the direct inverse of those in Fig. 15-9. (After E. S. Barghoorn, 1953.)

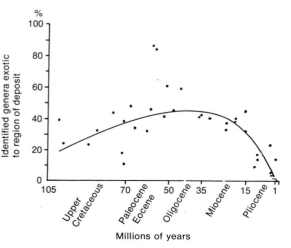

TABLE 15-1

Pleistocene chronology

North America	Europe (Alps)
Wisconsin Glacial	Würm Glacial
Sangamon Interglacial	Third Interglacial
Illinoian Glacial	Riss Glacial
Yarmouth Interglacial	Second Interglacial
Kansan Glacial	Mindel Glacial
Aftonian Interglacial	First Interglacial
Nebraskan Glacial	Günz Glacial

climatic changes. Owing to the shortness of the Pleistocene Epoch, little evolutionary change is noticeable in most animal or plant groups. In northeastern France, near Strasbourg, there are exposures of alluvium and loess about 100 feet thick. At the bottom of this sequence are sands with a molluscan fauna of interglacial character, deer, hippopotamus, and rhinoceros—a warm-weather fauna. Above this are gravels and sandy loess with a cold-weather fauna of reindeer and mammoth. Then comes a loess with several old soil surfaces indicating interruptions in the deposition of the loess. At the base of this unit is a warm-weather fauna containing elephants (*Elephas antiquus*), *Rhinoceros mercki*, *Felis spelaeus* (a cat), and *Hyaena*. At the top a cold-weather fauna is developed with reindeer and mammoth. This, of course, is an ideal circumstance: successive faunas occur in one sequence and show clearcut changes of climate. Such occurrences, however, are unfortunately not common. The remains of prehistoric man and his tools are, of course, an additional paleontological clue of immense importance, and these we shall discuss in a later section.

FIGURE 15-11

Maximum distribution of ice sheets and glaciers in the northern hemisphere during the Pleistocene. (After Flint, 1957.)

In the past decade a new chronologic tool has already had a tremendous influence on our knowledge of Pleistocene chronology. This is radiocarbon dating, a method useful for dating events back to approximately 40,000 years ago. The bombardment of nitrogen by cosmic rays in our atmosphere creates radioactive carbon (C^{14}), which by oxidation becomes part of the carbon dioxide of our atmosphere. Atmospheric carbon dioxide is metabolized by plants, which in turn are eaten by animals. As long as the plants or animals are living, there is an equilibrium in the concentration of C^{14} in the atmosphere, in animals, in plants, and, of course, in water. As soon as an animal or plant dies, however, this equilibrium is terminated, and the C^{14} disintegrates to nitrogen. The half-life of C^{14} is about 5,570 years; theoretically this limits the use for dating to about eight half-lives, or a maximum of about 50,000 years. The technique has been checked on archaeologically dated woods, and the dates derived demonstrate the approximate validity of the method for at least the last 5,000 years. No accurate checks are available on the older dates; there appears to be fair agreement, however, among several determinations on the same or comparable samples, from set horizons. The method has to be used with great caution, since there are numerous factors that can lead to contamination of the sample and thus give erroneous dates.

Radioactive carbon dates have been particularly significant in clarifying our understanding of the last glacial stage (the Wisconsin in North America and the Würm in Europe). Establishment of an absolute chronology beyond 40,000 years ago is still, however, an extremely difficult problem. Before the development of the radioactive carbon dating method, chronologic scales had been based on rates of various earth processes, such as weathering and soil development, stream development, recession of waterfalls, and rate of delta building.

The duration of the glacial phase of the Pleistocene, arrived at through one or another of these methods, has been estimated at from a million to about 400,000 years. In recent years radioactive carbon dates have enabled us to recheck the rates of earth processes (weathering, recession of waterfalls, etc.) and have clearly shown that the original assumption of the rate of a process was usually much too low.

It now appears that deep ocean sediments can assist greatly in deciphering Pleistocene chronology. There have been three primary approaches to the use of deep ocean cores for this purpose, resulting in widely divergent conclusions. The first of these methods relies on oxygen-isotope ratios of calcareous foraminifers to obtain data on the temperature of the water at the time these creatures were living. It is generally assumed (though probably somewhat incorrectly) that deposition of these deep-sea sediments was continuous and uniform, so that the isotopic temperature plotted against depth in the core (stratigraphic position) shows the variation of temperature with time. A generalized temperature curve based on several cores for the Caribbean region is shown in Figure 15-12. This curve shows eight temperature cycles over a span of time from the present back to 425,000 years ago. The dating is derived from carbon-14 and protactinium-231/thorium-230 age measurements and extrapolation therefrom to the more ancient ages. Recent K-Ar measurements on deposits of an early glacial stage in Europe (Günz) give an age of 350,000 to 400,000 years, which leads the authors of the above interpretations to believe the curve of Figure 15-12 represents the entire Pleistocene as classically understood. If this interpretation of Pleistocene climatic fluctuations is indeed reliable, it implies a picture of many alternating high and low temperature stages rather than the four or six that have generally been accepted.

The second approach in using deep ocean sediments for Pleistocene chronology is primarily biological. Many species of living planktonic foraminifers are temperature sensitive. The cores show variations, from level to level, in the relative abundance of species sensitive to temperature. These variations are interpreted as records of shifts in the geographical ranges of the species — shifts presumably due to climatic changes in the late Pleistocene. Radiometric age measurements by the carbon-14, protactinium-ionium, and protactinium methods provide a time scale from the present back to about 175,000 years ago. Careful examination of cores whose homogeneous texture is thought to reflect a constant rate of sediment accumulation, gives a deposition rate of about 2.5 cm per 1,000 years. Analysis of 26 homogeneous cores raised by the Lamont Geological Observatory indicates that the Pleistocene deposits are about 38 meters thick. If these figures are correct the glacial cycles that mark the Pleistocene began about 1.5 million years ago.

The recent discovery of reversals in the earth's magnetic field has provided a new means of correlation. Paleomagnetic studies of volcanic rocks on land have established a series of reversals of polarity of the earth's magnetic field and have provided K-Ar dates for these events. The earth's field has had its present, or normal, polarity for the last 700,000 years, the Brunhes normal epoch. For approximately the preceding 1.7 million years the field had an opposite or reversed polarity; this is the Matuyama epoch. Within the Matuyama epoch, two short periods of normal polarity occurred, at about 900,000 and 1.9 million years ago; these are the Juramillo and Olduvai events, respectively. Remanent magnetism of deep-sea sediments is sufficiently strong and stable that these polarity reversals can be used to date and correlate geological events recorded in these sediments throughout the world during the past five million years. Figure 15-13 summarizes the data on relative abundance of certain temperature-sensitive foraminifers and magnetic determinations for four cores from the mid and south Atlantic Ocean. The warm and cold cycles as indicated by the foraminifers are then matched with the generally recognized glacial and interglacial cycles as determined from Pleistocene deposits in Europe and North America. The Pliocene-Pleistocene boundary is determined on paleontological criteria. This approach to deciphering the chronology and duration of the Pleistocene suggests that the glacial episode of the epoch began about 2 million years ago.

The striking contrast in the results of these two methods emphasizes the complexity of the problem. This whole area of study is new, and, needless to say, much more work is needed before we can explain completely the discrepancies in these conclusions.

FIGURE 15-12

Generalized temperature curve based on isotopic analysis of sediment cores from the central Caribbean. (After C. Emiliani, *Science*, **154**, Nov. 18, 1966. Copyright 1966 by the American Association for the Advancement of Science.)

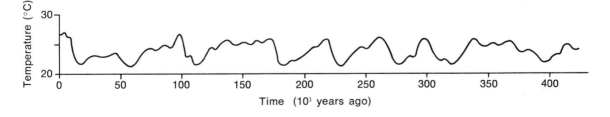

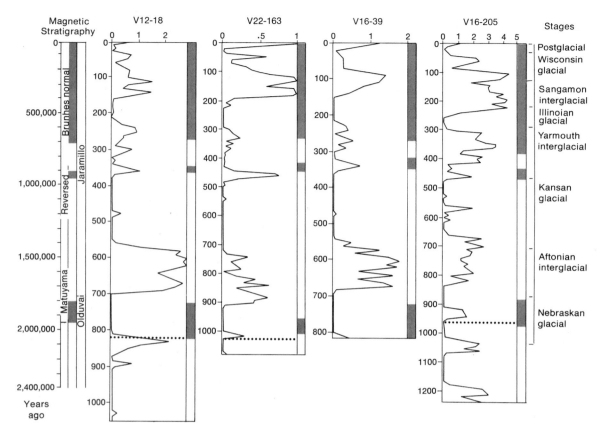

FIGURE 15-13

Magnetic and paleontological data for four cores from the mid and south Atlantic ocean. The frequency curves are for the foraminifer *Globorotalia menardii;* warm climatic conditions are indicated by a shift of this curve to the right and cold with a shift to the left. The climatic zones are correlated with the glacial and interglacial stages. The time scale for the magnetic stratigraphy is indicated on the left; the magnetic stratigraphy is indicated on the left; the magnetic stratigraphy for each core is designated by black (normal) and white (reversed) areas to the right of the columns. The x-lines mark the first appearance of abundant *Globorotalia truncatulinoides,* the criterion used for defining the Pliocene-Pleistocene boundary. Numbers to the left of the columns are depths in cores, in centimeters. (After D. B. Ericson and G. Wollin, *Science,* **162,** Dec. 13, 1968. Copyright 1968 by the American Association for the Advancement of Science.)

15-5 PREHISTORIC MAN

One reason why there is widespread interest in the Pleistocene is that in the record of the epoch we find the fossils of prehistoric man and his tools. The chronology and correlation of these fossils are almost entirely dependent on an understanding of the sequence and distribution of Pleistocene deposits. At the same time the fossils and cultural remains of prehistoric man are extremely helpful to our understanding of the geological history of the Pleistocene Epoch. The fossil remains of prehistoric man are not common.

Biologically, man is a product of evolution just as any other species of animal is; but man has, in addition, a cultural heritage, and in this he differs from all other animal species. The history of man during the Pleistocene is expressed in two independent ways, through his skeletal remains and through his implements, mostly of stone, which are an expression of his culture. The stone implements and other artifacts of early man are much commoner than his fossil remains; and archaeologists have been able to establish a useful chronology from the temporal and spatial distribution of the various types of implements.

Man the Tool-maker

The broad stages of human progress are closely related to the materials primitive man learned to use in making his cutting tools. He began his slow development as a tool-maker by using and shaping stone as he found it in its natural state. He next learned to smelt copper and tin, both soft metals, which he learned to combine into a useful material for tools, bronze. With the discovery of iron, and the development of skill in using it, primitive man took a long step toward becoming modern; and he finally achieved that state with his development of steel. The history of man is therefore broadly divisible into the Stone Age, the Bronze Age, the Iron Age, and the Steel Age. These are stages of technical development rather than chronological periods, and are illustrated in the development of the hand ax (Fig. 15-14).

The Stone Age, which dawned with man early in the Pleistocene, ended in the Middle East with the discovery of metals about 4000 B.C., but the Stone Age culture persisted until after 2000 B.C. in Britain, and in parts of Australia has lasted until today. The early Stone Age cultures are generally known as the Paleolithic and the later stages as the Neolithic. In the Paleolithic stone implements were made by chipping or flaking of flints or similar siliceous rocks; in the Neolithic stage man learned to shape stone tools by grinding instead of chipping. An intermediate stage, the Mesolithic, is also generally recognized. Since modern man was already established in Europe, Africa, and Asia during the Late Paleolithic, we shall not discuss any further aspects of

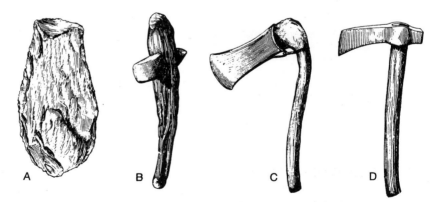

FIGURE 15-14

Development of hand axes. (A) chipped stone ax for use in the hand, Paleolithic, from Kenya. (B) Polished stone ax head in wooden haft, Neolithic, from Irish peat bog. (C) Socketed bronze ax head on wooden haft, from salt mine near Salzburg, Austria. (D) Iron ax head, Roman. (From K. P. Oakley, 1950, by courtesy of the Trustees of the British Museum.)

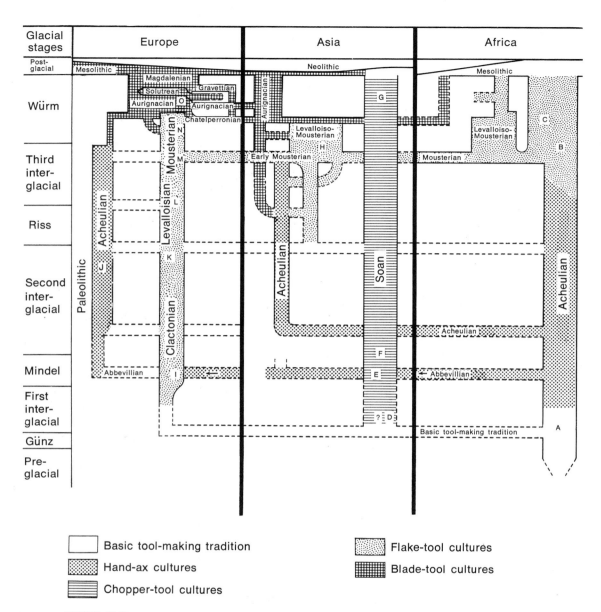

Glacial stages	Europe	Asia	Africa

FIGURE 15-15

Chart showing the cultural traditions of early man in Europe, Asia, and Africa and the probable connections between the various streams of cultural tradition. The relation of the earlier cultures to glacial stages is far from certain; in this (as in other respects) the chart is merely provisional. A letter plotted in relation to the name of a culture shows that fossil remains (discussed in § 15-5) of a bearer of the tradition have been found in the deposits of the age and location indicated. The following is the key to the localities of fossil hominids listed on the chart and discussed in § 15-5.

A. Transvaal, S. Africa.
B. Saldanha, S. Africa.
C. Broken Hill, N. Rhodesia.
D. Sangiran, Java.
E. Choukoutien, China.

F. Trinil, Java.
G. Ngandong, Solo River, Java.
H. Mount Carmel, Israel.
I. Heidelberg, Germany.
J. Swanscombe, England.

K. Steinheim, Germany.
L. Fontéchevade, France.
M. Ehringsdorf, Germany.
N. La Chapelle-aux-Saints, France.
O. Cro-Magnon, France.

(After K. P. Oakley, 1958, by courtesy of the Trustees of the British Museum.)

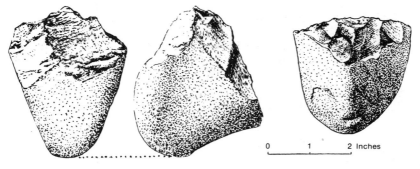

FIGURE 15-16

Some of the earliest known stone implements, from Bed I (lower Pleistocene), Olduvai Gorge, Tanzania. (From K. P. Oakley, 1950, by courtesy of the Trustees of the British Museum.)

the Mesolithic or later stages. A summary of the distribution and tentative relations of the cultural traditions of early man is shown by Figure 15-15.

The oldest known implements are crudely chipped pebbles of lava, quartz, and quartzite found in the lower Pleistocene of eastern Africa. These consist of rounded pebbles that have been chipped only on one end (Fig. 15-16). Stratigraphically above these crude, partially chipped implements are found stone tools that have been chipped more extensively, that represent the dawn of the hand-ax culture, usually known as Abbevillian (Fig. 15-17). Hand axes of this type are found in many places in western Europe, but their place of origin is thought to be central Africa. With the advent of the second glaciation the Abbevillian hand-ax makers disappeared from parts of northwestern Europe but persisted in Africa. Hand axes again made their appearance in Europe during the second interglacial interval, but they were pointed or almond-shaped. These hand axes, products of the Acheulian stage (Fig. 15-18), had fairly

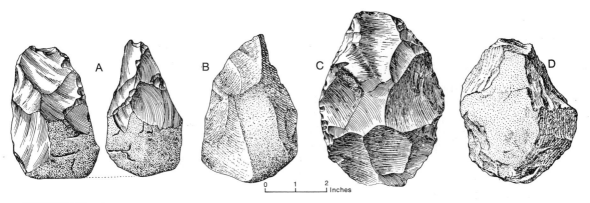

FIGURE 15-17

Abbevillian tools.

A. Hand ax, 150-foot terrace of Thames, near Caversham, Berks, England.
B. Quartzite hand ax, raised beach, Morocco.

C. Hand ax, Chelles-sur-Marnes, France.
D. Lava hand ax, Bed II, Olduvai Gorge, Tanzania.

(From K. P. Oakley, 1950, by courtesy of the Trustees of the British Museum.)

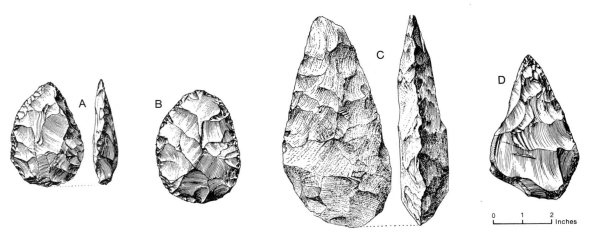

FIGURE 15-18

Acheulian tools.

A. Twisted, ovate *argile rouge*, 30-meter terrace, St. Acheul, near Amiens, France.
B. Ovate hand ax of flint, south of Wadi Sidr, Israel.

C. Lava hand ax, O1 Orgesailie, Kenya.
D. Hand ax, Hoxne, Suffolk, England.

(From K. P. Oakley, 1950, by courtesy of the Trustees of the British Museum.)

straight edges and shallow flake scars, in contrast to the irregular edges and deep flake scars of the Abbevillian tools. Acheulian hand axes became a standardized implement by the third interglacial interval and are found in western Europe, Africa, and southern Asia. Specimens from these widely separated regions are remarkably similar except that they are made of different kinds of rocks.

Parallel to the development of the hand-ax cultures (Abbevillian and Acheulian) there was another cultural trend represented by flake tools (Fig. 15-19). In Europe east of the Rhine flake cultures predominated throughout Early Paleolithic times. In western Europe, especially in France and southern Britain during the glacial stages, the flake cultures are represented almost to the exclusion of the hand-ax cultures that flourished in these areas during the interglacial periods. The early phase of the flake culture is known as Clactonian and the later phase as Levalloisian. There are numerous sites where flake tools intermingle with hand axes.

A somewhat different stone-tool culture, called Soan, prevailed in eastern Asia. The most characteristic implement was a chopper, but flake tools also were used. The tools are known from the Choukoutien caves of Peking, China, where the remains of the famous *"Sinanthropus"* were excavated, and also from India, Burma, and Java (Fig. 15-20). This flake-and-chopper culture predominated in southeastern Asia throughout the Pleistocene. The Clactonian flake culture of Europe was possibly an early offshoot.

A different stone culture prevailed in Europe, Africa, and western Asia during the latter part of the third interglacial interval and the first part of the last glacial stage; this culture was the Mousterian, which developed out of the Clactonian (flake-culture) group and in some regions was influenced by the hand-ax culture groups. Typical implements of Mousterian industry are illustrated in Figure 15-21. The implements consist of a variety of differently shaped scrapers and hand axes. This cultural group makes up the Middle Paleolithic.

During the later phases of the fourth

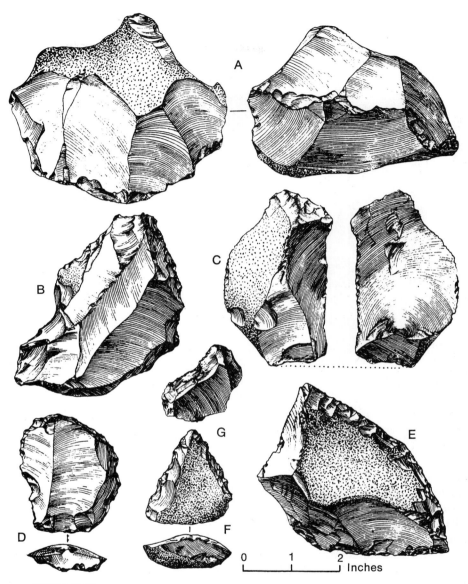

FIGURE 15-19

Clactonian artifacts.

A. Flint core.
B. Flake tool, lower gravel, 100-foot terrace, Swanscombe, Kent, England.
C. Flake tool, lower gravel, 100-foot terrace, Swanscombe, Kent, England.
D. Flake tool, *Elephas antiquus* gravel, Clacton-on-Sea, Essex, England.
E. Acheulo-Clactonian scraper, High Lodge, Mildenhall, Suffolk, England.
F. Proto-Mousterian flake tool, Combe-Capelle, Montferrand, Dordogne, France.
G. Flake from interglacial river gravel, La Micoque, Tayac, Dordogne, France.

(From K. P. Oakley, 1950, by courtesy of the Trustees of the British Museum.)

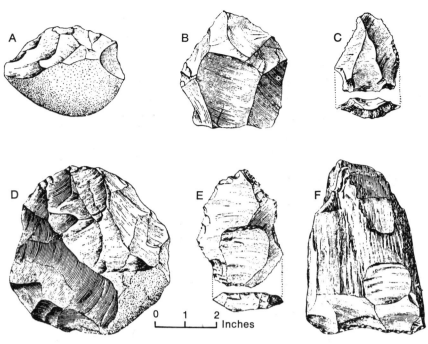

FIGURE 15-20

Artifacts of the Soan culture group.

A. Pebble chopper, northwestern India.
B. Flake tool, northwestern India.
C. Late Choukoutien flake tool of chert, Choukoutien, China.

D. Chopper of silicified tuff, Patjitanian, Java.
E. Flake tool of silicified tuff, Patjitanian, Java.
F. Chopper of fossil wood, Anyathian, upper Burma.

(From K. P. Oakley, 1950, by courtesy of the Trustees of the British Museum.)

glaciation (Würm), cultural development and evolution accelerated rapidly. At least five successive cultural levels are recognized in Europe, and these make up the Late Paleolithic. All have blade-tool traditions (Fig. 15-22) and are especially remarkable for the great variety of specialized tools and weapons. Members of these groups had also mastered the working of bone and other substances. The Late Paleolithic peoples appear to have been nurtured in some part of southwestern Asia, perhaps on the Iranian plateau, whence they, or at least their traditions, spread, in a series of waves, eastward as far as China and westward to the Mediterranean and Atlantic seaboards.

The Late Paleolithic peoples were artistic and produced the remarkable cave paintings of France and Spain (Fig. 15-23). The principal subjects of most of the paintings were food animals (reindeer, bison, mammoth, fish, and an occasional reptile or fowl) or animals dangerous to a hunting community (bear, lion). The artistic versatility of the cave painters included modeling in clay, making jewelry of shells and teeth, carving friezes, and decorating and carving bones.

The Fossil Evidence of Hominids

Fossil hominids of early and middle Pleistocene age are extremely rare. The most important finds to date have been made in Africa, Java, China, and a few places in Europe. The record improves greatly for the

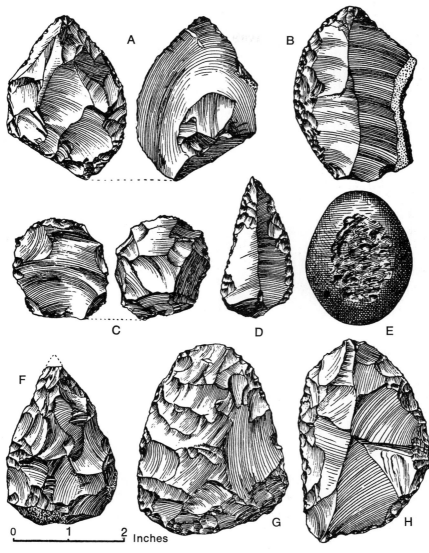

FIGURE 15-21

Mousterian implements. (A–D are typically Mousterian; F, G, and H are Mousterians of Acheulian tradition.)

A. Side scraper, Le Moustier, near Peyzac, Dordogne, France.
B. Side scraper, Le Moustier.
C. Disk core, Le Moustier.
D. Point, Le Moustier.

E. Small anvil or hammerstone, Gibraltar caves.
F. Hand ax, Le Moustier.
G. Hand ax (chert), Kent's Cavern, Torquay, England.
H. Oval flake tool (flint), Kent's Cavern.

(From K. P. Oakley, 1950, by courtesy of the Trustees of the British Museum.)

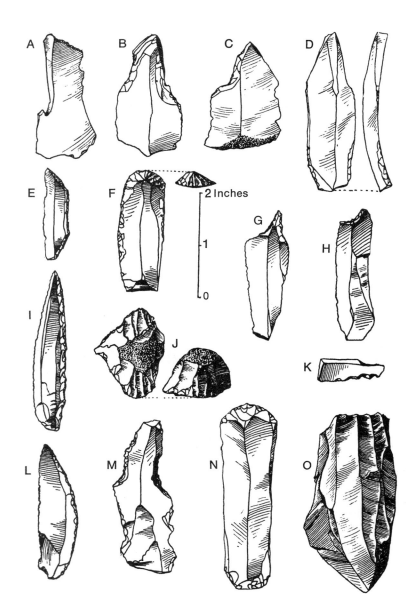

FIGURE 15-22

Late Paleolithic flint tools.

A. Gravettian graver, Laugerie Haute, Dordogne.
B. Aurignacian nosed graver, Ffynnon Bueno,
   Vale of Clwyd, Wales.
C. Aurignacian graver, Cro-Magnon, Les Eyzies,
   Dordogne.
D. Magdalenian graver, La Madeleine rock shelter,
   Tursac, Dordogne.
E. Trapezoid blade, Kent's Cavern, Torquay,
   England.
F. End scraper, Cae Gywn, Vale of Clwyd.
G. Solutrean piercer, Laugerie Haute.

H. Magdalenian concave end scraper, Limeuil,
   Dordogne.
I. Knife point, Laussel, Dordogne, France.
J. Aurignacian nosed scraper, Laugerie Haute.
K. Magdalenian fragment of saw blade,
   Laugerie Haute.
L. Knife point, Châtelperron, Allier, France.
M. Aurignacian double spokeshave, Laugerie Haute.
N. Magdalenian double-ended scraper,
   Grotte des Euzies, Dordogne.
O. Magdalenian blade core, Grotte des Eyzies.

(From K. P. Oakley, 1950, by courtesy of the Trustees of the British Museum.)

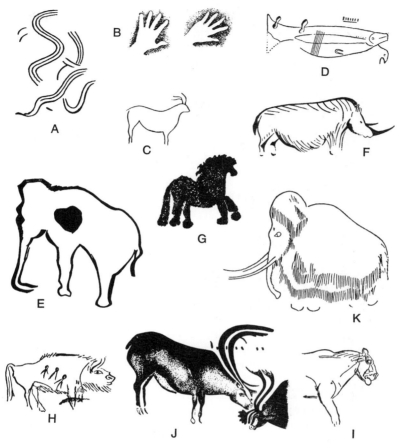

FIGURE 15-23

Late Paleolithic cave art. (A–C and E are early-middle Aurignacian;
D, F, and G are late Aurignacian; H–K are Magdalenian.)

A. Snake-like scribbles in yellow ocher,
   La Pileta, Málaga, Spain.
B. Hands stenciled in red ocher, Castillo,
   Santander, Spain.
C. Engraving of ruminant, Pair-non-Pair,
   Gironde, France.
D. Salmon engraved on roof of rock shelter,
   Gorge d'Enfer, near Les Eyzies, Dordogne,
   France. Note associated bird's head, tally
   marks (?), and dumbbell-shaped holes.
E. Elephant painted in red ocher, Pindal,
   Oviedo, Spain. Note the heart.

F. Woolly rhinoceros painted in red ocher,
   Font-de-Gaume, near Les Eyzies.
G. Horse painted in black oxide of manganese,
   Lascaux, near Montignac, Dordogne.
H. Wounded bison, engraved on floor of cave,
   Niaux, Ariège, France.
I. Engraving of cave lion, Combarelles,
   Dordogne.
J. Reindeer painted in black (male) and
   in red (female), Font-de-Gaume.
K. Engraving of mammoth, Font-de-Gaume.

(From K. P. Oakley, 1950, by courtesy of the Trustees of the British Museum.)

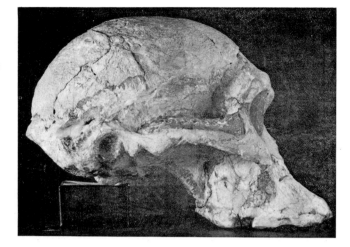

FIGURE 15-24

Side view of the skull of *Australopithecus*. (Photograph by J. T. Robinson.)

late Pleistocene, from which a number of remains are known in Europe, Africa, and Asia. For the moment let us concentrate our attention on the nature of the fossil evidence, the dating problem, and some tentative interpretations. In so doing we will use the names originally proposed for most of these fossils. With this background we will make a summary of the phylogenetic history of man.

In 1925 Professor Raymond Dart of Johannesburg reported the discovery of part of the skull and jaws and the impression of the braincase of an immature creature of apelike appearance. The specimen was found in a cave deposit embedded in a limestone matrix at Taungs in Bechuanaland, South Africa. Dart named his fossil *Australopithecus africanus*, or the "Southern Ape." Since that time many other skulls and parts of limbs have been found in other localities in the Transvaal of South Africa (Fig. 15-24). There has been considerable debate as to the age of the deposits containing australopithecine remains, for, since the formations are cave deposits of precipitated calcium carbonate, there is no normal stratification. The associated animal bones include many extinct types; in fact, as some of them became extinct in Europe in the Pliocene, some authorities would date them as Pliocene. We know, however, that many animal groups

found Africa a haven and persisted there long after becoming extinct on other continents. Among the associated bones are remains of *Equus*, the modern horse, which is Pleistocene in age. Thus an early Pleistocene age is generally accepted for these cave deposits.

The skull of *Australopithecus* has a superficial resemblance to that of one of the large apes, but it is actually quite different in significant features. The brain size of *Australopithecus* was about 600 cubic centimeters, about half of the average brain size of modern man. In modern apes there is a ridge of bone, called the nuchal crest, high on the back of the skull, for the attachment of strong neck muscles. In the australopith-

FIGURE 15-25

Palate and upper teeth (A) of a male gorilla, (B) of *Australopithecus*, and (C) of an Australian aboriginal. (From W. E. Le Gros Clark, 1956, by courtesy of the Trustees of the British Museum.)

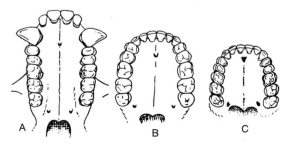

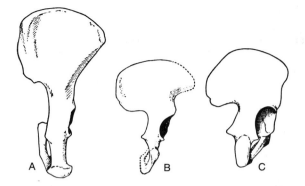

ecines the nuchal crest is much lower—is, in fact, in much the same position as in man. The massive, protruding jaw is, of course, an ape-like characteristic, but the occipital condyle (by which the skull articulates with the top of the spinal column) is situated much farther forward in relation to other structures of the skull base, approaching the condition in man. This condition suggests a body posture somewhat like that of modern man. The teeth are especially significant for the interpretation of the australopithecines. The palates and upper teeth of a male gorilla, an australopithecine, and an Australian aboriginal are illustrated in Figure 15-25. Note, in the gorilla, that the rows of teeth are sub-parallel, that there is a space between the canines and the incisors, and that the canines are greatly enlarged. In *Australopithecus*, however, there is a continuous, evenly curved arch of teeth, and the canines are not enlarged. These features are remarkably similar to man's.

The limb bones associated with some of the skulls are of special interest because of their strong similarities to those of modern man and their contrast to those of the apes. The similarities are so striking, in fact, that, if found alone, without the associated skulls, the limb bones would probably have been thought to be human bones. The leg and arm bones show none of the arboreal specializations found in modern apes. The shape of the pelvis is especially significant; we can readily see (Fig. 15-26) that an australopithecine pelvis is very close to

a human pelvis, but quite unlike that of an ape. This additional bit of evidence suggests that these creatures stood and walked almost as man does today. This conclusion is entirely consistent with the climatic evidence, which indicates that the australopithecines lived, not in a tropical forest, but rather in a more arid region (Fig. 15-27).

Australopithecus, we see, was an animal that appears to have combined ape and human characteristics: the limb structures and teeth were distinctly human in aspect, but the brain size was more like that of the gorilla and chimpanzee. Just where does *Australopithecus* belong in a scheme of classification, to the ape family or to the family that includes man? At the moment, at least, no definite evidence has been found to indicate that *Australopithecus* made tools or knew the use of fire. In one of the sites that yielded australopithecine skulls, however, there were also a number of baboon skulls showing depressed fractures on the top; this fact suggests that the baboons were killed by well-aimed blows with a weapon of some sort, and we may surmise from this evidence that baboons were systematically hunted for food by *Australopithecus*. The whole complex of anatomical features, however, leads most authorities to conclude that *Australopithecus* is a very primitive representative of the family that includes modern and extinct man.

Australopithecus is now known from three localities (Taung, Sterkfontein, and Makapansgat) in South Africa (Fig. 15-28). There are in addition two localities (Kromdraai and Swartkrans) that have yielded a quite different form originally named *Australopithecus robustus* but regarded by some as a distinct genus, *Paranthropus*. In this form the incisors and canines are very small, but the molars are huge, and there are pronounced bony crests on the skull, particularly in the males. The cranial capacity appears to have been about the same as that of *Australopithecus*. The character of the teeth suggests that *Paranthropus* was more of a vegetarian; the character of the teeth of *Australopithecus* indicates a more diversified diet, including a liberal portion of animal prey. Dating of the two sites that have yielded the specimens of *Paranthropus* is difficult and uncertain, but the best available determination at present is basal middle Pleistocene.

Important fossil remains of early man have been found at three localities in Tanzania, of which the most important is Olduvai (Fig. 15-28). Pleistocene beds, up to 350 feet thick, are exposed along a 15-mile length of Olduvai Gorge and have yielded several fossil hominids, artifacts, and occupation sites. In addition to having a stratigraphic sequence of strata, K-Ar dates on volcanic rocks from the base of the sequence give a radiometric date of 1.75 m.y. Thus at Olduvai we have two important geologic bases for determining age, neither of which can be used in the South African localities mentioned above.

The lowest unit of the sequence exposed

FIGURE 15-27

Reconstruction of *Australopithecus* in an early Pleistocene landscape in South Africa. (Drawing by Z. Burian under the supervision of Prof. J. Augusta.)

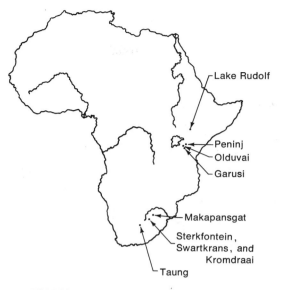

FIGURE 15-28

The African sites which have yielded fossilized remains of *Australopithecus*. The Lake Rudolf site is in Kenya, the next three sites are in the Republic of Tanzania, and the five southern sites are in the Republic of South Africa. (After P. V. Tobias, *Science*, **149**, July 2, 1965. Copyright 1965 by the American Association for the Advancement of Science.)

in Olduvai Gorge, known as Bed I, is considered to be of lower Pleistocene age. Above this is a unit known as Bed II, of middle Pleistocene age. Bed I has yielded a nearly complete cranium, including all 16 upper teeth, that has been named *Zinjanthropus boisei*. This form, however, is considered to be essentially identical to *Paranthropus* of South Africa. In addition, four levels within Bed I have yielded another type of hominid that has been given the name of *Homo habilis*. What is of particular interest is that at each level where remains of *Homo habilis* have been found, primitive stone implements have been found, such as those illustrated in Figure 15-16. The describers of this new fossil hominid concluded that it is clearly a primitive member of the genus *Homo*, to which modern man

belongs, and quite distinct from *Australopithecus*. This conclusion has aroused controversy. The cranial capacity of *Homo habilis* is about 680 cubic centimeters, less than 100 cubic centimeters greater than the current maximum estimate for *Australopithecus* of South Africa. The opponents of the above conclusion claim that *Homo habilis* is nothing more than a slightly advanced *Australopithecus*. The controversy continues. In Bed II two distinct hominids have also been discovered. The first of these is clearly assignable to *Paranthropus*, thus confirming the existence of this form at least into the early middle Pleistocene. The second is a form that was originally placed in *Homo habilis* but is now thought to be a primitive representative of *Homo erectus* ("*Pithecanthropus*"), which we shall discuss below. At Swartkrans, in the Transvaal of South Africa, a number of fragments were found in association with *Paranthropus* (*Australopithecus robustus*) that were given the name *Telanthropus*. This form is now considered to be a representative of *Homo erectus* ("*Pithecanthropus*").

Recent discoveries in the area of Lake Rudolf, by F. Clark Howell of the University of Chicago and Bryan Patterson of Harvard University, are of special importance, as they add considerably to the data on the antiquity of fossil man. Professor Howell's discovery, at the north end of Lake Rudolf in southern Ethiopia, consists of approximately 40 teeth and two lower jaws. Among these specimens are broad-toothed forms representing *Paranthropus* (*Australopithecus robustus*) and small-toothed forms representing *Australopithecus* (*A. africanus*). Volcanic ash in the deposits that have yielded these fossils give a K-Ar date of 4 m.y. Professor Patterson's discovery consists of the distal end of a hominoid humerus from early Pleistocene sediments at the southern end of Lake Rudolf. This fragmentary fossil comes from sediments that are capped by lavas that have yielded

FIGURE 15-29

Pithecanthropus. (Drawing by Z. Burian under the supervision of Prof. J. Augusta, Prof. J. Filipa, and Dr. V. Fettra.)

K-Ar dates of 2.5 m.y. and 2.9 m.y. Exhaustive analysis of this fossil fragment has led to the conclusion that it is most probably an *Australopithecus.*

The importance of these two new discoveries is that they extend the existence of *Australopithecus* back in time and show that two distinct species were already well established 4 million years ago—that is, in the late Pliocene.

Probably the best-known fossil man is the Java ape man, "*Pithecanthropus.*" A Dutch anatomist, Eugène Dubois, had been so stirred by the idea of a "missing link" that he accepted a position as an army doctor in Indonesia for the specific purpose of searching for that link. After several years of arduous work, much to everyone's surprise (and probably his own), Dubois actu-

ally found some remains in 1891, at Trinil, Java. The find consisted of a skull cap, a thigh bone, and a few teeth. These fossils came from alluvial deposits exposed along the banks of the Solo River. The thigh bone was found some 40–50 feet away from the skull cap. The skull cap is very flattened, with large eyebrow prominences, and there is no forehead. This indeed is a far more primitive and ape-like skull cap than that of any existing human being. The thigh bone, on the other hand, in its shape and proportions, was obviously adapted mechanically for the same kind of stress and strain for which a modern human thigh bone is adapted; its owner evidently stood erect and walked much as we do today. The teeth were human in character, but this evidence is inconclusive, for some of the higher apes

FIGURE 15-30

Replica of the original *Pithecanthropus erectus* skull found by Dubois in 1891 (left) compared with more nearly complete specimen discovered in 1937. (American Museum of Natural History.)

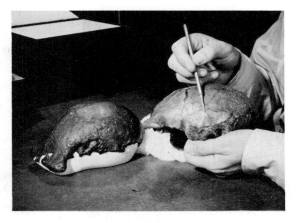

have teeth not very different from those of some primitive human beings.

Dubois named his fossils *Pithecanthropus erectus,* or the "erect ape man," and considered this creature to be an intermediate form between man and apes, and thus a true "missing link" (Fig. 15-29). Publication of Dubois' discovery precipitated a heated scientific controversy that lasted several decades. Many people firmly denied any human aspect of *Pithecanthropus*. Then, during the four years immediately before World War II, further highly important fossil discoveries were made in Java. Most of these new fossils came from a site about ninety miles from Trinil. They consisted of a skull cap much more nearly complete than the specimen found by Dubois, a portion of a lower jaw with several teeth in place, the back part of a skull with a considerable part of the upper jaw and teeth, and a portion of the roof of a third skull. At another site in eastern Java, not far from the site of the original discovery by Dubois, the skull of a child was discovered.

With these new materials, at least the skull of *Pithecanthropus* is fairly well known (Fig. 15-30). The most significant features are the flattened skull roof, the prominent eyebrow ridges, the lack of a forehead, the protruding jaw, the size of the brain, and the unusual thickness of the cranial wall. The mean cranial capacity of the three skulls available is 860 cubic centimeters. It is doubtful if the cranial capacity of *Pithecanthropus* ever exceeded 1,000 cubic centimeters.

There has been much controversy over the age of the strata that yielded the *Pithecanthropus* remains. It is now generally agreed that the beds that yielded the original fossils at Trinil are of middle Pleistocene

age and probably correspond to the second glaciation in other parts of the world. The new fossils found just before the last war apparently came from two horizons. The upper corresponds to the Trinil deposits, but fossils were also found in an underlying stratum of black clay, called the Djetis layer, which the associated mammal fauna indicates to be of early Pleistocene age. Beneath the Djetis layer is a still older formation with fossils that suggest a horizon at the base of the Pleistocene. There is some doubt whether the fossil hominids of the Djetis layer (the lower level) are actually contemporaneous with the associated mammal remains of that bed. Other doubts arise because these are alluvial deposits, most of the collecting was done by native assistants, and there is always the possibility of slumping of fossils along steep river banks.

The second important site where fossils of prehistoric man have been found in Asia is near the Chinese village of Choukoutien, not far from Peking. Excavations of caves at Choukoutien yielded in 1927 a single molar tooth of a primitive human type. On the basis of this tooth its discoverer postulated an extinct type of man and gave it the name *Sinanthropus* (man of China). In the following years continued excavation yielded a large number of fossils and made available for study skulls, or portions of skulls, of at least fourteen individuals, young and old, with teeth and fragments of jaws belonging to more than forty individuals, as well as the shafts of two thigh bones and an upper arm bone, a collar bone, and one of the wrist bones. This cave deposit is middle Pleistocene in age and is thought by some authorities to correspond to the second glacial interval. One of the tragedies of World War II is that all these fossils disappeared. It is suspected that in an attempt to save them they were lost at sea in a ship sunk during the early part of the war. Fortunately, excellent casts, photographs, and drawings of all the material are still available. (The new finds of the Java *Pithecanthropus* might have suffered the same fate if it had not been for the ingenuity of their discoverer. On the eve of the Japanese invasion he distributed the fossils among trusted friends. Upon being released from internment after the war, he was able to bring together all of the specimens.)

The Peking skulls are remarkably similar in structure to those from Java. The top of the skull is not quite so flattened, and there is a slight forehead, but the prominent eyebrow ridges and protruding jaws are similar. Most significant is the greater cranial capacity of the Peking Man, which ranges from 850 to 1,300 cubic centimeters, with a mean of 1,075. This is roughly 100 cubic centimeters greater than the mean for the Java specimens discussed above. The mean for modern human races is 1,350 cubic centimeters. The teeth and jaws show many primitive, ape-like characteristics but, when compared with those of modern apes, are essentially human in their gross aspect. The teeth are arranged in an even curve (as in *Australopithecus*), whereas in modern apes the grinding teeth on the sides are disposed in parallel straight rows ending in front in the sharp, projecting canines (Fig. 15-25). The limb bones of Peking Man, like those of Java Man, are very human in aspect and clearly indicate he walked upright. The dissimilarity between the Chinese fossils and *Pithecanthropus* is not great enough to warrant a separate generic name

for each. Most students of fossil man now place both in the same genus, and we are justified in assuming that they belong to the same species, and that their differences warrant only subspecific rank. We will discuss this whole problem in § 15-12.

Even though the intelligence of Peking Man was probably low, as deduced from brain size, he was able to fabricate stone implements of the flake and chopper types (Fig. 15-20). Numerous hearths found in the caves at Choukoutien indicate that Peking Man knew how to use fire. He must have been a fairly skilled hunter, for 70 percent of the animal remains found at one site belonged to deer. No stone implements have been found in the beds that yielded the specimens of Java Man. Chopper and flake tools, however, have been found at slightly higher horizons in Java (Fig. 15-20), and it is not improbable that those tools were made and used during the middle Pleistocene by men of the *Pithecanthropus* type.

In Europe, where several fragmentary fossils of early man have been discovered, the presence of prehistoric man is well documented by abundant stone implements, and at least a working chronology has been developed. The oldest European fossil of early man is a single lower jaw found in 1907 in a sand pit at Mauer, near Heidelberg, Germany (Fig. 15-31), and the specimen is known as Heidelberg Man (*Homo heidelgergensis*). The deposits that yielded this specimen are of fluvial facies and also contain bones and teeth of horse, rhinoceros, elephant, and other mammals characteristic of the early part of the Pleistocene—probably about the time of the early part of the second interglacial interval. The Heidelberg jaw is extremely massive and has a receding chin—features that are primitive and ape-like. The teeth, however, though rather primitive, are fundamentally of human type.

The specimens from Java assigned to "*Pithecanthropus*," from the caves around Peking, and from Mauer, near Heidelberg, Germany, are now considered to belong to a single species, *Homo erectus*. This is one of only two species of the genus *Homo* generally recognized—a species that evolved during the period between about one million and 500 thousand years ago. In addition to these classic discoveries, in recent years a number of new specimens have been uncovered which clearly show that *Homo erectus* was widely distributed in Europe, Africa, and Asia. We have already mentioned that specimens from the upper part of Bed II at Olduvai are members of this species, as are the specimens from Swart-

FIGURE 15-31

Heidelberg jaw (B) compared with the jaw of a chimpanzee (A) and the jaw of a modern man (C). (From W. E. Le Gros Clark, 1950, by courtesy of the Trustees of the British Museum.)

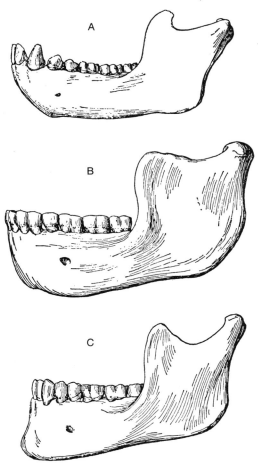

krans, South Africa, originally assigned to *Telanthropus*. At Ternifine, Algeria, three jaws and a parietal bone have been named *Atlanthropus mauritanicus*. These are now considered to be very similar to the Peking remains and thus assignable to *Homo erectus*. In the Lantian district of Shensi, China, a lower jaw and a skull cap have been uncovered from beds distinctly older than those that yielded Peking Man. These specimens are considered to be primitive, early members of *Homo erectus*. Despite the clear evidence of gradual evolutionary development of *Homo erectus* in skeletal features, the tools made by these men reveal little change.

Homo sapiens appeared in the middle Pleistocene, as shown by the discovery in 1965 at Vértesszöllös, Hungary, of an isolated occipital bone (in the back of the skull) in beds of early middle Pleistocene age (middle or late Mendel glaciation). The cranial capacity of this specimen is 1,400 cubic centimeters, close to the average for modern man and well above that of known specimens of *Homo erectus*. *H. sapiens* was at least for a while contemporary with *H. erectus*, though the two probably did not live in the same regions.

Probably the best-dated of the few fossil specimens of Early Paleolithic Man are three fragments of a skull cap found twenty-four feet below the surface in well-stratified gravels of the 100-foot terrace of the Thames River near Swanscombe, England. These are known as the Swanscombe skull bones. Associated with the three skull-cap fragments were bones of elephant, rhinoceros, and deer, as well as flint implements that are assigned to the early middle Acheulian hand-ax culture. The geological, archaeological, and paleontological data all indicate the age of the second interglacial interval. Though very fragmentary, the evidence available from these three bones of the Swanscombe skull indicates that Acheulian Man in Europe was not markedly different in anatomical features from *Homo sapiens*, modern man. It is estimated that the cranial capacity of the Swanscombe skull is 1,300 cubic centimeters.

Another early hominid fossil was found in a quarry at Steinheim on the Murr, a tributary of the Neckar, twenty miles north of Stuttgart. This specimen consists of a fairly well-preserved skull, minus the lower jaw (Fig. 15-32). It comes from an interglacial gravel deposit, but it is uncertain whether the deposit was formed during the second or third interglacial interval. Notable primitive features of the Steinheim skull are the prominent eyebrow ridges and the strongly built upper jaw. The top and back of the skull, however, are rounded. The cranial capacity is about 1,100 cubic centimeters. The combination of morphological features falls well within the range of skull variation in *Homo sapiens.*

Two fragmentary lower jaws, with several teeth and a badly broken skull cap, have been found at Ehringsdorf, near Weimar, Germany. The skull cap is somewhat similar to the Steinheim skull, but it had a much larger brain capacity (about 1,450 cubic centimeters). Associated with these fossils are remains of elephants, rhinoceroses, horses, and plants that indicate a temperate climate. Also present are stone implements of early Mousterian culture. The age of these fossils is fairly well established as the second half of the last interglacial interval.

Such is the record of Paleolithic Man in Europe up to the beginning of the last glaciation (Würm). In addition to the specimens just discussed, fossil fragments have been found near Verona and Rome in Italy, at Fontéchevade in France, at Krapina in Yugoslavia, and at Mount Carmel in Israel. The Verona specimen dates from the second interglacial interval; the remaining specimens are of third interglacial age. All these specimens of pre-Würm age in Europe are clearly similar to modern *Homo sapiens;* on anatomical grounds, in fact, they cannot be distinguished from that species. This evidence indicates that *Homo sapiens* was widespread throughout Europe during the

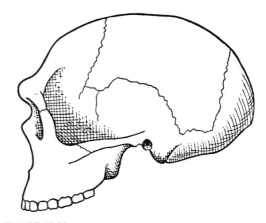

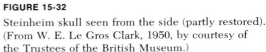

FIGURE 15-32

Steinheim skull seen from the side (partly restored). (From W. E. Le Gros Clark, 1950, by courtesy of the Trustees of the British Museum.)

second and third interglacial intervals, contemporaneously with the development of a Paleolithic culture of the Acheulian type. Of all these specimens at least the Steinheim skull shows definite traces of an ancestry of the *Pithecanthropus* type, especially in the forehead.

Archaeological evidence of early and middle Pleistocene man in eastern Africa continues to accumulate rapidly. Numerous sites of Acheulian culture are known, but as yet few hominid fossils have been found.

One of the earliest discoveries of specimens of fossil man consisted of a skull cap and portions of limb bones in a Neanderthal cave, near Düsseldorf, Germany, in 1856. The deposits that yielded these specimens are late Mousterian in age (from the early part of the last glaciation). Since that time a large number of similar fossil forms have been uncovered in Europe, Palestine, southern Russia, Siberia, and North Africa. All these specimens are representatives of the famous Neanderthal Man (*Homo neanderthalensis*). The paleontological material available includes numerous skulls and limb bones, a situation in striking contrast to the fragmentary nature of the pre-Mousterian record. The fossil material is often as-

FIGURE 15-33
Neanderthal Man. (Drawing by Z. Burian under the supervision of Prof. J. Augusta.)

sociated with late Mousterian implements and the remains of mammoth and woolly rhinoceros, which are characteristic of the last glacial period.

Anatomically Neanderthal Man differs from *Homo sapiens* in several striking features. His skull is large and has thick cranial walls, very massive brow ridges, a very low sloping forehead, and a flattened braincase. The posterior part of the skull projects back at an angle. The jaws are massive, and the teeth are relatively larger than in *Homo sapiens*. The lower jaw has a receding chin. The region of attachment of the skull to the top of the spine indicates that there was a forward tilt to the head. Indeed, many of the features about the skull suggest a rather primitive condition, although it was large. The average size of the brain (about 1,450

cubic centimeters) was greater than the average size of the brain in modern man (1,350 cubic centimeters).

The rest of the skeleton is even more unusual. The limb bones, for instance, have curved shafts, and the extremities are disproportionately large (Fig. 15-33). Though Neanderthal Man was rather short, about five feet tall, he was exceedingly strong.

Neanderthal Man was the sole hominid occupant of Europe, North Africa, and the western part of Asia during the early part of the last glacial period (the time of late Mousterian culture). He was the first hunter known to attach a flint point to a rod to form a spear. The game he hunted consisted of the woolly mammoth, rhinoceros, deer, bison, and other animals. Cave bears were a constant source of danger to him. Living as he did during a time of cold climates, he probably wore some sort of fur for protection. His favorite dwelling sites were caves. It is our good fortune that he lived in caves, for, not being a particularly tidy person, he threw his refuse about on the cave floors and thus supplied fossil material for the archaeologist. Neanderthal Man was also the first to bury his dead (often placing tools and parts of animals in the same grave), and this custom accounts for his leaving a vastly richer fossil record than other hominids (Fig. 15-34).

Since Neanderthal Man was one of the earliest types of prehistoric man discovered, it is not surprising, in spite of his rather primitive, ape-like appearance, that for a long time he was looked upon as modern man's direct ancestor. But with the discovery

of pre-Mousterian human remains, which cannot be separated anatomically from *Homo sapiens*, this interpretation is no longer tenable. We shall come back to this problem shortly.

With the close of the Mousterian cultural episode, Neanderthal Man became extinct and was replaced by people of completely European type. Before taking up this new group, however, we need to mention two other groups of prehistoric man with neanderthaloid characters. One of these groups lived in South Africa and the other in Java. The South African fossils were first found in a cave in northern Rhodesia, and in recent years a similar form was found near Capetown. The Rhodesian fossil remains consist of a nearly complete skull and part of the upper jaw of a slightly smaller skull,

FIGURE 15-34

Since he was a cave-dweller, Neanderthal Man's most dreaded enemies were huge cave bears, the skulls of which appear to have been used for some sort of cult, as illustrated in this drawing. (Drawing by Z. Burian under the supervision of Prof. J. Augusta and Prof. J. Filipa.)

portions of limb bones, part of a hip bone, and a sacrum. The skull is remarkable for the large size of the brow ridges—larger than on any other known prehistoric skull. The jaws likewise are very large. The cranial capacity was somewhat less than 1,300 cubic centimeters. In most features the skull of Rhodesian Man was similar to that of Neanderthal Man except for slight differences in features of the nose and ears. The limb bones, however, are like those of *Homo sapiens* and not at all like those of Neanderthal Man. Unfortunately the antiquity of Rhodesian Man is uncertain. Stone implements associated with the fossil bones are of Levalloisian type, but the animal bones found in the cave have a fresh appearance and belong to animals like those still living in Rhodesia. The specimen found near Capetown, known as Saldanha Man, consists of a skull cap. The archaeological, paleontological, and geological data on this find indicate that the skull probably antedates the last glacial period in Europe.

A late Pleistocene deposit at Ngandong, on the Solo River of Java, has yielded eleven skulls and a portion of a limb bone. Like the Rhodesian material, the skulls of the specimens are very much like those of Neanderthal Man, but the limb bones are more like those of modern man. The skulls indicate a medium cranial volume of 1,150-1,300 cubic centimeters.

Though we do not know why, Neanderthal Man abruptly disappeared at the close of the Mousterian, and immediately thereafter Europe was occupied by a completely modern group of men who are identified by the Aurignacian culture (Fig. 15-15). These people are known as Cro-Magnons from the locality that yielded their remains in southern France. No more need be said of their physical features than that it probably would have been impossible to distinguish them in life from some groups of people who inhabit Europe today. Cro-Magnon Man occupied Europe during the Late Paleolithic, during the culmination

of the Old Stone Age technology. The rapid development and elaboration of this technology are reflected in the five cultural divisions that are recognized for the Late Paleolithic, covering the time, approximately, from 35,000 to 800 B.C. (Fig. 15-35).

The first phase in this development was the invention of the flint blade, which marks the Early Aurignacian culture. The commonest tool of these hunters was a knife made from a blade of flint, with one edge straight and razor-like and the other curved over to the point and blunted by abrupt trimming (Fig. 15-22, A). The Late Aurignacian culture used flint blades modified into a variety of sharp, incising tools, called burins, with which other tools and weapons of bone could be made. The typical Aurignacian bone industry included polished pins or awls. The associated flint industry included fluted core-like scrapers (Fig. 15-22, I), end scrapers (Fig. 15-22, J), and various edge-trimming blades (Fig. 15-22, H). The Aurignacian peoples had some artistic talents also, for they painted and engraved outlines of animals and carved human figures in bas-relief (Fig. 15-23, A, B, C, E). The straight-backed flint knife became common (Fig. 15-22, B), and at the Solutrean cultural level the art of pressure flaking was developed. This technique enabled the making of a large number of different types of beautifully shaped tools.

The final cultural phase of the Late Paleolithic was the Magdalenian, which was centered in southwestern France. The material culture of the Magdalenians resembled that of the Eskimos, possibly because of adaptation to a partly similar environment. Their flint industry, in the blade-tool tradition, was especially elaborate. They made a considerable variety of instruments in bone, ivory, and reindeer antler, including spear heads with linkshafts, barbed points and harpoons for spearing fish, hammers, wedges, needles with eyes, and various artifacts of whose use we are uncertain. They designed these objects with great artistic skill, and on many

FIGURE 15-35

Late Paleolithic Cro-Magnon Man. (Drawing by Z. Burian under the supervision of Prof. J. Augusta, Prof. J. Filipa, and Dr. J. Molého.)

they engraved lively reproductions of animals of the chase and occasionally human beings and geometrical designs. Even more remarkable was their mural art (found chiefly in the caves of southern France and northern Spain), which appears to have been a flowering of the tradition begun by the Aurignacians (Fig. 15-23).

15-6 THE PHYLOGENY OF MAN

The reconstruction of man's phylogeny is fraught with controversy, yet considerable progress has been made in recent years. Many additional fossil hominids have been discovered, and some of the basic concepts of evolution and phylogeny have been revised. The evolution of man involves two phenomena, the branching from the ape line and the reaching of the human level within the hominid line. The earliest representatives of the hominid line were still anthropoid apes. The problem of the origin of man consists, then, of two questions: (1) When and where did the hominid line branch off from the ape line? (2) At what point on the hominid line did the truly human characters develop?

It may be well at this point to review the discussion of primates in § 14-14 through 14-16 and study closely again Figure 14-37 (p. 536). Recall that the most probable candidates as ancestors to the hominids were *Ramapithicus* of India and *"Kenyapithecus"* of Kenya, both of late Miocene or early Pliocene age. The Kenya specimen has yielded a K-Ar date of 14 m.y. The oldest known specimen that may be *Australopithecus* is 4 million years old. There is thus a gap in our record covering a period of some 10 million years, and this is critical to the story.

The australopithecines are probably the ancestors of the genus *Homo*. Even so, there is considerable difference of opinion as to the interrelationships of the various australopithecines and their origins. As the matter is by no means settled, let us discuss the two main hypotheses. The first of these interpretations, proposed by Professor P. V.

FIGURE 15-36

A provisional scheme of hominid phylogeny from upper Pliocene time to the upper Pleistocene proposed by Professor P. V. Tobias of Witwatersrand University. Increasing intensity of shading represents increasing degrees of approach toward the structure and behavior of modern man. A., the hypothetical ancestral australopithecine; A.b., *Australopithecus (Zinganthropus) boisei*; A.r., *Australopithecus robustus*; A.a., *Australopithecus africanus*. The scheme indicates the synchronic coexistence of several different hominids in the lower and middle Pleistocene, the australopithecines surviving into the middle Pleistocene alongside more advanced hominids of the genus *Homo*. (After P. V. Tobias, *Science*, **149**, July 2, 1965. Copyright 1965 by the American Association for the Advancement of Science.)

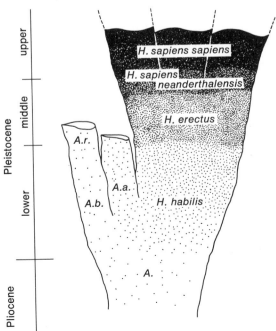

Tobias of Witwatersrand University, South Africa, is summarized in Figure 15-36. An ancestral australopithecine, unspecialized, with small teeth, and presumably omnivorous in diet is hypothesized as living in the Pliocene. In late Pliocene time this ancestral stock gave rise to three distinct evolutionary lines. The first of these is represented by *Australopithecus boisei* and *Australopithecus robustus,* both with large molar teeth reflecting an essentially herbivorous diet. The second line is merely a continuation of the ancestral stock and is represented by *Australopithecus africanus.* This line is closely linked to the presumed progressively more hominized line represented by *Homo habilis.* Soon after the establishment of the *habilis* line, the *africanus* line became extinct and shortly thereafter the *robustus* line in the early middle Pleistocene. Continued evolutionary development of the *habilis* line gave rise to *Homo erectus,* and this later gave rise to *H. sapiens.*

The essential feature of the above interpretation is that the ancestral australopithecine is a small-toothed form. From this ancestral form the broad-toothed forms, represented by *A. boisei* and *A. robustus,* evolved as an adaptation toward a primarily vegetarian diet. The small-toothed forms, presumably omniverous in diet, evolved the relatively conservative and short-lived *Australopithecus* line along with a more progressive line represented by *Homo habilis.*

A quite different interpretation of the early evolution of the hominids has been proposed by J. T. Robinson of the University of Wisconsin and is summarized in Figure 15-37. In this interpretation the ancestral australopithecines are believed to be the broad-toothed forms (e.g., *Paranthropus*). It should be noted here that for neither hypothesis is there any fossil evidence of the supposed ancestral australopithecine. According to Robinson the ancestral broad-toothed forms were plant eaters and represent a conservative line that continued until early middle Pleistocene time.

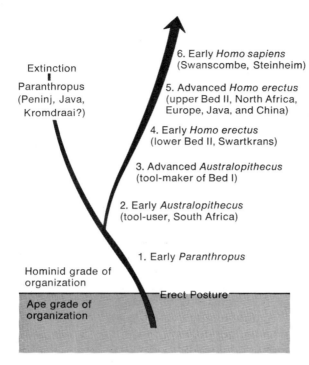

Extinction

Paranthropus
(Peninj, Java,
    Kromdraai?)

6. Early *Homo sapiens*
(Swanscombe, Steinheim)

5. Advanced *Homo erectus*
(upper Bed II, North Africa,
Europe, Java, and China)

4. Early *Homo erectus*
(lower Bed II, Swartkrans)

3. Advanced *Australopithecus*
(tool-maker of Bed I)

2. Early *Australopithecus*
(tool-user, South Africa)

1. Early *Paranthropus*

Hominid grade of
organization

Erect Posture

Ape grade of
organization

FIGURE 15-37

Scheme of hominid evolution proposed by
J. T. Robinson of the University of Wisconsin.
Stages 3, 4, and 5 are known from Olduvai
Beds I and II; 3 = Bed I *Homo habilis* material;
4 = Bed II *Homo habilis* material; 5 = the so-called
Chellean skull from the top of Bed II. Stage 2 is
known from South Africa. The early *Paranthropus*
level (Stage 1) is unknown as yet. (After
J. T. Robinson, Nat. Acad. Sci., 1967.)

On the other hand, other geologists believe that in early Pleistocene time climatic changes led to selective pressures to adapt to a more varied diet; that is, at times of aridity, when adequate vegetation was unavailable. During these times a premium would be placed on agility of movement and use of the hand to obtain whatever animal food was available. This necessarily would lead to the development and use of stone weapons for predatory purposes. The surviving broad-toothed forms (*Paranthropus*) were confined to whatever wooded areas existed. Of the two australopithecine groups, the omnivorous *Australopithecus* appears to have had a gait of walking most similar to that of modern man, whereas *Paranthropus* had a more encumbered mode of movement, one described as waddling, but at the same time appropriate for life within the forest.

Thus, carrying this hypothesis further, *Australopithecus* evolved from *Paranthropus*, reflecting a basic change in diet and mode of life. With this shift there took place the appropriate change in size of the molar teeth. The earliest austalopithecines are believed to be those of South Africa, and these were merely tool users. The next grade of evolution of *Australopithecus* is represented by *Homo habilis* of Olduvai Bed I, and these were tool makers. As with both of these hypotheses, there is a period in which there are two hominids living essentially side by side. This seems to have been possible because the basic diets of the two were so different, one solely herbivorous, the other omnivorous.

The differences between these hypotheses are of fundamental importance. So far, neither can be demonstrated to be either right or wrong. A primary handicap inherent in both interpretations is the lack of precise data on correlation of the various beds that have yielded australopithecine remains. Research in these areas, however, is particularly active and one can hope that in the near future more precise means of correlation will develop and more fossil data will become available.

All the other hominids so far discussed, from *Pithecanthropus* through Cro-Magnon Man, were human. All these forms had an essentially uniform anatomy such that a zoologist would place them in a single genus, *Homo* (Fig. 15-38).

The fundamental step in the evolution of the hominid line was the acquisition of upright posture. This preceded brain evolution and most other specializations of modern man. The evolution of the brain is a comparatively recent event; little seems to have happened after the branching off of the hominid line during the remainder of the Miocene and Pliocene. Even the australopithecines in the early Pleistocene had a brain that did not much exceed that of the anthropoids in relation to body size. But from then on brain size increased at an unprecedented rate. A rough picture of this brain evolution is given by a list of cranial capacities (in cubic centimeters):

Chimpanzee and gorilla	325– 650
Australopithecines	450– 750
Java Man	750– 900
Peking Man	900–1,200
Neanderthal Man	1,100–1,550
Modern Man	950–2,100

The increase in brain size led to a complete reconstruction of the skull, which was also affected by two other developments. One was the forward shift of the support of the skull, which resulted from upright posture; the other was the diminished need of strong jaws and big teeth as man's bigger brain and better-developed hands made possible prepared, cut-up, cooked, more easily eaten foods. All this resulted in a reduction of the jaws, the teeth, and the entire facial part of the skull while the cerebral part of the skull enlarged. These changes also resulted in a reduction of the facial muscles and of all the bony crests and ridges

to which these muscles were attached. The evolution of the hominids is an almost classical demonstration of **mosaic evolution:** each organ and each system of organs has its own rate and pattern of evolution. Mosaic evolution is the characteristic form of evolution of all types that shift into a new adaptive zone.

Java Man and Peking Man have, for some time now, been considered representative of *Homo erectus*. They are, however, separated at a subspecific level; thus the proper name of Peking Man is *Homo erectus pekinensis*. The Swanscombe skull, the Steinheim skull, the Ehringsdorf skull, and Cro-Magnon Man are considered members of *Homo sapiens*, the species to which modern man belongs. The recognition of polytypic species in the hominid line offers promise of a proper understanding of Neanderthal Man. We mentioned earlier (§ 15-5b) that this form is often looked upon as an immediate ancestor of modern man, but that the discovery of older types clearly belonging to *Homo sapiens* makes this explanation untenable. It now appears that *Homo sapiens neanderthalensis* was an offshoot localized in western Europe and Asia and more or less separated from the more easterly and southerly populations of *Homo sapiens* by water, ice, or unsuitable terrain. The characteristics of Neanderthal Man are therefore not primitive but are correlated with the climatic and other ecological conditions under which he lived. Solo Man of Java and Rhodesian Man of South Africa can be looked upon as similar peripheral isolates.

The fate of Neanderthal Man is a puzzle. Wherever found, Neanderthal bones are associated with artifacts of the so-called Mousterian culture. They are succeeded by typical members of the modern *Homo sapiens*, associated with the Aurignacian culture. There is no evidence of hybridization between the aborigines and the invaders. We do not know whether Neanderthal Man had become extinct before Cro-Magnon

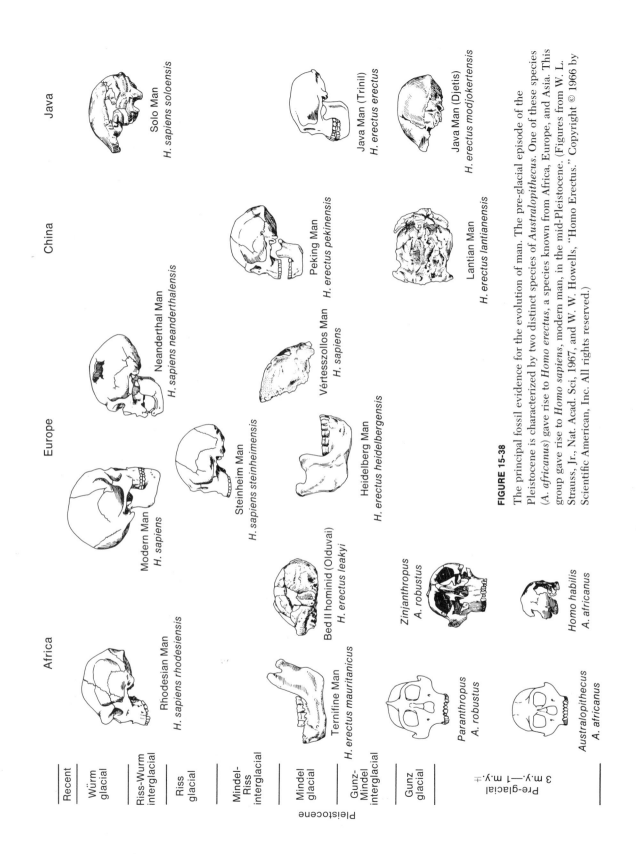

FIGURE 15-38

The principal fossil evidence for the evolution of man. The pre-glacial episode of the Pleistocene is characterized by two distinct species of *Australopithecus*. One of these species (*A. africanus*) gave rise to *Homo erectus*, a species known from Africa, Europe, and Asia. This group gave rise to *Homo sapiens*, modern man, in the mid-Pleistocene. (Figures from W. L. Strauss, Jr., Nat. Acad. Sci, 1967, and W. W. Howells, "Homo Erectus." Copyright © 1966 by Scientific American, Inc. All rights reserved.)

Man arrived or whether the Cro-Magnons exterminated the Neanderthals.

The available evidence leads to the conclusion that modern *Homo sapiens* evolved somewhere south or east of the Mediterranean and invaded central and western Europe at the end of the first stage of the Würm ice age. To understand *Homo sapiens* fully, we must think of him as a multidimensional species, polytypic in space and time. His early condition is represented by Swanscombe and Steinheim; Neanderthal is a somewhat specialized peripheral isolate in western Europe and Asia; the typical *sapiens* evolved somewhere, perhaps, in the Near East, and invaded western Europe at a later stage. Much of this is outright speculation, but it is a working hypothesis consistent with all the known facts, fossil as well as genetic and zoological.

15-7 BIOLOGICAL FACTORS IN THE EVOLUTION OF MAN*

The phylogeny and population structure of man are now largely understood in their broad outline. But, as much as we may know about the "How?" of human evolution, the "Why?" is still a great puzzle. High intelligence and harmonious social integration are undoubtedly attributes of high selective value. What has enabled man to break the chains of his animal heritage and evolve from a primate to the level of man? It is something that occurred in the past and like all historical events can never be demonstrated experimentally. Nevertheless, an analysis of the ecological conditions under which primitive hominids lived and of their social and population structure helps us to construct what is perhaps a fairly realistic model of the crucial selection pressures.

---

* The synthesis presented here is derived (almost verbatim) from the writings of Ernst Mayr of the Museum of Comparative Zoology at Harvard College.

The most astounding phenomenon of human evolution is the precipitous increase in brain size during the Pleistocene. *Australopithecus* would be nothing but a terrestrially specialized anthropoid if it were not for the subsequent evolution of man. Hominid evolution up to and including the *Australopithecus* stage was, then, merely the prologue to the crucial stage. This prologue, as we saw above, included the assumption of upright posture, the freeing of the hand for manipulation, a shift into the terrestrial niche, and the increasing use of tools. Preman was a precarious creature at this stage, for his locomotion was none too efficient and his niche none too hospitable. What the original hominids lacked in speed, in natural weapons, in nocturnal secretiveness, and in arboreal safety they had to counterbalance by the development of the various facilities that we now consider peculiarly human. These include the widespread use of tools, the use or manufacture of devices for protection against the inclemency of the weather, and the development of communication (language). Mortality from enemies, famine, exposure, and interspecific strife must have been high. The premium on the ability to cope with these difficulties must have been correspondingly high.

There were two biological factors that may have facilitated this development. One was the population structure of primitive hominids. They faced their adverse environment not as individuals but as family groups or small bands. The unit of evolution was not the individual but the population. As soon as we appreciate this fact, we can see that the conception of natural selection that is prevalent in popular writings and among post-Darwinian philosophers and sociologists is far from correct. To describe natural selection in terms of brute force — "nature red in tooth and claw," "survival of the fittest," etc. — places a misleading emphasis on only a single aspect of natural selection. Inventiveness, foresight, leadership, and, in many cases, cooperation will be far more favored by natural selection than brute force in an animal with the social population structure of primitive hominids.

Polygyny is more or less strongly developed in all the anthropoid apes. There are good reasons to believe that it was characteristic of the primitive hominids. What effect on the evolutionary rate would polygyny have? If the leader of a group has several wives (perhaps even all the mature females of the family group), he will contribute a far greater than average share to the genetic composition of the next generation of his group. Such tremendous reproductive advantage of a leader in a self-contained family group or tribe would favor the very characteristics that have made man what he is. Reproductive advantage in the primitive hominid society, we may speculate, was not so much a result of any bizarre secondary sex characters as of social position within the group. That position depended, to a considerable extent, on the genetic endowment of the individual. We thus have a situation in which reproductive advantage coincides almost completely with superior fitness.

The second biological factor responsible for the acceleration of hominid evolution is parental care. Much of the mortality among most animals, particularly at the immature level, is accidental and haphazard. The institution of parental care reduces such random mortality. Survival depends largely on the kind of care bestowed upon the child by the parents. The high selection pressure in favor of brain development is obvious. The smarter the individual, the better his chance of survival. And brain development is reinforced by a slowing down of the development of the human infant and a lengthening of the period during which parental care is required; this increases the selective value of parental care and increases the correlated selection pressure in favor of increased brain size.

This combination of selection pressures can account for the rapid increase of brain size in the hominids, but it does not tell us why this evolutionary trend has come to such a sudden halt. There has been no apparent increase in brain size since the time of Neanderthal. To be sure, there may have been an improvement of the brain without an enlargement of the cranial capacity, but there is no real evidence of this. Something must have happened to weaken the selection pressure drastically. Among several possible factors two are most important. One is an increase in the "unit of population"—that is, the family group, tribe, or nation that has a selective success as a whole, compared with similar competing units. The larger such a unit is, the less important will be the genes of its leader and the more protected (biologically) will be its average or below-average individual. Reproductive success will no longer be closely correlated with genetic superiority. If we add to this the dysgenic effects of urbanization and of density-dependent diseases, we can see why the trend that created man has not continued to produce superman. The changed social structure of the human community no longer rewards superiority with reproductive success. The second factor that has reduced selection pressure is the development of a cultural tradition and of means of communication. These help the below-average individual to reproduce as successfully as the above-average one. We are not interested here in discussing whether this development is good or bad and whether anything should be done about it. This is the domain of the politician and the social scientist. All we want to do is to point out the almost abrupt flattening out of an exceedingly steep evolutionary advance, a phenomenon that must have a biological interpretation. The fact is that man's biological evolution seems to have come to a standstill. As evolutionists and historical geologists we must attempt to decide what factors were responsible for the drastic reduction of selection pressure after the evolution of *Homo sapiens* was completed.

15-8 EPILOGUE

Such is our understanding of the geological and biological history of the earth. Progress in knowledge since the time of William Smith has been significant, and for most of the major features of the earth we have a perspective of temporal and spatial relations. At best the available data allow only the creation of a framework of thought. However, geology as a science is still in its infancy—rich in controversy and constantly deriving new impetus from a varied assortment of allied sciences. Advancement of our knowledge of the earth's history, be-

cause of the nature of the record, can never be equal in all parts of the time scale. This restriction has been admirably stated by the late G. K. Gilbert, of the U.S. Geological Survey, as follows:

"When the work of the geologist is finished and his final comprehensive report written, the longest and most important chapter will be upon the latest and shortest of the geologic periods. The chapter will be longest because the exceptional fullness of the record of the latest period will enable him to set forth most completely its complex history. The changes of each period—its erosion, its sedimentation, and its metamorphism—obliterate part of the records of its predecessor and of all the earlier periods, so that the order of our knowledge of them must continue to be, as it now is, the inverse order of their antiquity."

SUGGESTED READINGS

Brooks, C. E. P., *Climate Through the Ages: A Study of the Climatic Factors and Their Variation* (Yale Univ. Press, New Haven, 1928).

Charlesworth, J. K. *The Quaternary Era*, 2 vols. (Arnold, London, 1957).

Clark, W. E. Le Gros, *History of the Primates: An Introduction to the Study of Fossil Man* (Univ. of Chicago Press, Chicago, 1957).

———, *The Fossil Evidence for Human Evolution*, 2nd ed. (Univ. of Chicago Press, Chicago, 1964).

Coleman, A. P., *Ice Ages, Recent and Ancient* (Macmillan, New York, 1926).

Coon, C. S., *The Story of Man* (Knopf, New York, 1954).

Daly, R. A., *The Changing World of the Ice Age* (Yale Univ. Press, New Haven, 1934).

Dobzhansky, Theodosius, *Evolution, Genetics, and Man* (Wiley, New York, 1957).

Flint, R. F., *Glacial and Pleistocene Geology* (Wiley, New York, 1957).

Howell, F. C., and F. Bourlière, eds., *African Ecology and Human Evolution* (Aldine, Chicago, 1963).

Martin, P. S., and H. E. Wright, eds., *Pleistocene Extinctions, A Search for a Cause* (Yale Univ. Press, New Haven, 1967).

Oakley, K. P., *Man the Toolmaker* (British Museum, Natural History, London, 1952).

———, *Frameworks for Dating Fossil Man* (Aldine, Chicago, 1964).

Rankama, K., ed., *The Quaternary: 1, Introduction Quaternary of Sweden, Finland, and Norway* (1965); *2, France, British Isles, Netherland, and Germany* (1967) (Wiley, New York).

Sears, M., ed., *Progress in Oceanography: 4, The Quaternary History of Ocean Basins* (Pergamon Press, New York, 1967).

Schwarzbach, M., *Climates of the Past* (Van Nostrand, London, 1963).

Simpson, G. G., *The Meaning of Evolution: A Study of the History of Life and of Its Significance for Man* (Yale Univ. Press, New

Haven, 1949; Mentor Books, New York, 1951).

Symposium, Time and Stratigraphy in the Evolution of Man (Nat. Acad. Sci., Division of Earth Sciences, Publ. 1469, 1967).

Tax, Sol, ed., *The Evolution of Man: 2, Evolution After Darwin* (Univ. of Chicago Press, Chicago, 1960).

Tobias, P. V., "Early Man in Africa," *Science*, **149** (1965).

Washburn, S. L., ed., *Classification and Human Evolution* (Aldine, Chicago, 1963).

Wendt, Herbert, *In Search of Adam: The Story of Man's Quest for the Truth About His Earliest Ancestors* (Houghton Mifflin, Boston, 1956).

West, R. G., *Pleistocene Geology and Biology* (Wiley, New York, 1968).

Zeuner, F. E., *Dating the Past, An Introduction to Geochronology*, 4th ed. (Methuen, London, 1958).

————, *The Pleistocene Period*, 2nd ed. (Hutchinson Scientific and Technical, London, 1964).

Scientific American Offprints

140 John Napier, *The Evolution of the Hand* (December, 1962).

601 Sherwood L. Washburn, *Tools and Human Evolution* (September, 1960).

602 Marshall D. Sahlins, *The Origin of Society* (September, 1960).

604 William W. Howells, *The Distribution of Man* (September, 1960).

609 Theodosius Dobzhansky, *The Present Evolution of Man* (September, 1960).

630 William W. Howells, *Homo Erectus* (November, 1966).

632 Wilton M. Krogman, *The Scars of Human Evolution* (December, 1951).

820 J. Desmond Clark, *Early Man in Africa* (July, 1958).

832 Robert Broom, *The Ape-Men* (November, 1949).

844 J. E. Weckler, *Neanderthal Man* (December, 1957).

1070 John Napier, *The Antiquity of Human Walking* (April, 1967).

AN INTRODUCTION TO
ANIMALS AND PLANTS

The nature and general significance of the fossil record form the subject matter of Chapter 2, and the character and evolution of the animals and plants of each major division of the geologic column are the subjects of Chapters 7, 10, 14, and 15. These chapters presume at least an elementary knowledge of the morphology and relationships of the major groups of animals and plants. This appendix is included for those who have a limited or no familiarity with this aspect of the subject. The emphasis here, especially with the invertebrate phyla, is on morphology, but it includes brief comments on habitat. The diversity of life and methods of classification are discussed in § 2-1. The relationships of the animal phyla are discussed in § 2-4, and the nature of the Phanerozoic fossil record of each major group of animals and plants is discussed in § 2-6.

For reference purposes a simplified classification of animals and plants is given here. All the genera of animals and plants mentioned in the text or in the figure legends are included in their proper places in the scheme of classification. The generic names are italicized.

KINGDOM PROTISTA

Phylum Schizomycetes — bacteria
Phylum Mastigophora — flagellates: *Euglena, Trypanosoma*
Phylum Sarcodina — foraminifers, radiolarians: *Actinomma, Amoeba, Globigerina, Nummulites*
Phylum Sporozoa — parasitic "protozoans"
Phylum Ciliophora — ciliated "protozoans": *Paramecium, Vorticella*
Phylum Myxomycetes — slime molds

KINGDOM ANIMALIA

Phylum Archaeocyatha—primitive, sponge-like organisms: *Ajacicyathus, Archaeocyathus* (L.–M. Cam.)

Phylum Porifera—sponges (Cam.–Rec.)
  Class Demospongea—with siliceous spicules: *Defordia, Heliospongia*
  Class Hyalospongea—with siliceous spicules, all triaxons
  Class Calcispongea—with calcareous spicules: *Girtyocoelia*

Phylum Coelenterata—corals, hydras, jellyfish, etc.
  Subphylum Cnidaria (Precam.–Rec.)
    Class Protomedusae (Precam.–Ord.)
    Class Dipleurozoa (L. Cam.)
    Class Scyphozoa—jellyfish, etc. (Cam.–Rec.)
    Class Hydrozoa—hydras, etc. (L. Cam.–Rec.)
      Order Stromatoporoidea—extinct group of hydrozoans: *Stromatoporoides* (Cam.–Cret.)
    Class Anthozoa—corals (Ord.–Rec.)
     Subclass Ceriantipatharia—"black corals" (Mio.–Rec.)
     Subclass Octocorallia—with 8 tentacles and mesenteries (?Sil.–Rec.)
     Subclass Zoantharia—mesenteries in cycles of 4, 6, or 8 (Ord.–Rec.)
      Order Rugosa—septa and mesenteries in cycles of 4: *Araeopoma, Goniophyllum, Omphyma, Palaeophyllum, Xylodes* (Ord.–Perm.)
      Order Schleractinia—septa and mesenteries in cycles of 6: *Favia, Latomeandra, Margosmilia, Rhipidogyra, Stylosmilia, Thecosmilia* (M. Trias.–Rec.)
      Order Tabulata—extinct group, septa absent or rudimentary: *Favosites, Syringopora, Thamnopora* (Ord.–Perm., Trias?, Eoc.)
     Subclass Ctenophora (Rec.)

Phylum Platyhelminthes—flatworms (no fossils)
Phylum Nemertinea—ribbonworms (no fossils)
Phylum Nemathelminthes—roundworms (no fossils)
Phylum Gordiacea—horsehair worms (no fossils)
Phylum Acanthocephala—spiny-headed worms (no fossils)
Phylum Trochelminthes—rotifers (no fossils)
Phylum Chaetognatha—arrow-worms (one fossil from Middle Cambrian)

Phylum Bryozoa—moss animals
  Class Phylactolaemata—fresh-water forms (no fossils)
  Class Gymnolaemata—almost exclusively marine forms having row of tentacles around the mouth (Ord.–Rec.)
    Order Ctenostomata: *Vinella* (Ord.–Rec.)
    Order Cyclostomata: *Petalopora* (Ord.–Rec.)
    Order Trepostomata: *Monticulipora* (Ord.–Perm.)
    Order Cryptostomata: *Polypora* (Ord.–Perm.)
    Order Cheilostomata: *Sertella, Bugula* (?M. Jur., Cret.–Rec.)

Phylum Brachiopoda—brachiopods (L. Cam.–Rec.)
  Class Inarticulata—without hinge line, teeth, or sockets: *Discina, Lingula*
  Class Articulata—with hinge line, teeth, and sockets: *Atrypa, Aulosteges, Conchidium, Dialasma, Glasia, Leptaena, Leptodus, Neospirifer, Pentamerus, Prorichtofenia, Productus, Richtofenia, Strophomena, Spirifer*

Phylum Mollusca
  Class Monoplacophora (L. Cam.–Rec.)
  Class Amphineura—chitons (L. Cam.–Rec.)
  Class Scaphopoda—tusk-shells (?Ord., Dev.–Rec.)
  Class Pelecypoda—oysters, mussels, scallops, etc.: *Barbatia, Caprinula, Cubiostrea, Exogyra, Gryphaea, Hippurites, Inoceramus, Monotis, Mytilus, Ostrea, Pecten, Trigonia* (Ord.–Rec.)

Class Gastropoda—snails, slugs, limpets, etc.: *Basilissa, Buccinum, Cirsotrema, Cocculina, Neptunea, Plicifusus, Volvula* (L. Cam.–Rec.)

Class Cephalopoda (U. Cam.–Rec.)

Subclass Orthoceratoidea—*Orthoceras* (U. Cam.–U. Ord.)

Subclass Actinoceratoidea (M. Ord.–U. Carb.)

Subclass Endoceratoidea (Ord.–?M. Sil.)

Subclass Nautiloidea—*Cooperoceras, Cyrtoceras, Metacoceras, Nautilus, Phragmoceras, Stenopoceras* (L. Ord.–Rec.)

Subclass Bactritoidea (Ord.–Trias.)

Subclass Ammonoidea—complex septa, ventral siphuncle: *Adrianites, Anarcestes, Androgynoceras, Artinskia, Baculites, Douvilleiceras, Epiwocklumaria, Eumorphoceras, Liparoceras, Lytoceras, Manticoceras, Oistoceras, Ophiceras, Placenticeras, Phylloceras, Protrachyceras, Schistoceras, Tetraspidoceras, Timorites, Waagenoceras* (Dev.–Cret.)

Subclass Coleoidea—squids, octopods, etc.: *Belemnites* (Miss.–Rec.)

Phylum Annelida—segmented worms: *Canadia* (?Precam.–Rec.)

Phylum Arthropoda

Superclass Trilobitomorpha (Cam.–Perm.)

Class Trilobitoidea (Cam.–Dev.)

Class Trilobita—extinct trilobites: *Actinurus, Asaphus, Aulacopleura, Ceratarges, Cheirurus, Condylopyge, Cyclopyde, Deiphon, Ellipsocephalus, Harpes, Illaenus, Isotelus, Lonchodomas, Megalaspis, Neolenus, Ogygopsis, Olenellus, Paradoxides, Pemphigaspis, Proetus, Radiaspis, Staurocephalus, Triarthrus, Trimerus* (Cam.–Perm.)

Superclass Chelicerata (Cam.–Rec.)

Class Merostomata—eurypterids, horseshoe crabs, etc. (Cam.–Rec.)

Class Arachnida—spiders (Sil.–Rec.)

Superclass Pycnogonida (Dev.–Rec.)

Superclass Crustacea (Cam.–Rec.)

Class Cephalocarida (Rec.)

Class Branchiopoda (L. Dev.–Rec.)

Class Mystacocarida (Rec.)

Class Ostracoda (L. Cam.–Rec.)

Class Euthycarcinoidea (Trias.)

Class Copepoda (Mioc.–Rec.)

Class Branchiura (Rec.)

Class Cirripedia—barnacles (U. Sil.–Rec.)

Class Malacostraca (L. Cam.–Rec.)

Superclass Onychophora—worm-like animals: *Aysheaia* (?Precam., Cam.–Rec.)

Superclass Myriapoda (U. Sil.–Rec.)

Class Archipolypoda (U. Sil.–U. Carb.)

Class Diplopoda—millipedes (U. Carb.–Rec.)

Class Pauropoda (Rec.)

Class Chilopoda—centipedes (Cret.–Rec.)

Class Symphyla—symphlans (Olig.–Rec.)

Superclass Hexapoda (U. Carb.–Rec.)

Class Protura (Rec.)

Class Collembola (Rec.)

Class Diplura (Rec.)

Class Thysanura (Olig.–Rec.)

Class Insecta—insects (U. Carb.–Rec.)

Superclass Pentastomida (Rec.)

Superclass Tardigrada (Rec.)

Phylum Echinodermata

Subphylum Homalozoa (M. Cam.–M. Dev.)

Class Homostelea (M. Cam.)
Class Stylophora: *Ceratocystis* (Cam.–M. Dev.)
Class Homoiostelea (U. Cam.–L. Dev.)
Class Ctenocystoidea: *Ctenocystis* (M. Cam.)
Subphylum Crinozoa (L. Cam.–Rec.)
Class Eocrinoidea: *Gogia* (L. Cam.–M. Ord.)
Class Cystoidea: *Pleurocystites* (L. Ord.–U. Dev.)
Class Parablastoidea (M. Ord.)
Class Blastoidea—blastoids: *Pyramiblastus* (M. Sil.–U. Perm.)
Class Paracrinoidea (L. Ord.–M. Ord.)
Class Crinoidea—crinoids: *Botryocrinus, Calceolispongia, Scyphocrinus* (L. Ord.–Rec.)
Subphylum Asterozoa (L. Ord.–Rec.)
Class Stelleroidea—starfish, brittle stars: *Villebrunaster* (L. Ord.–Rec.)
Subphylum Echinozoa (L. Cam.–Rec.)
Class Helicoplacoidea: *Helicoplacus* (L. Cam.)
Class Edrioasteroidea: *Carneyella* (L. Cam.–Carb.)
Class Edrioblastoidea (M. Ord.)
Class Ophiocistoidea: *Volchovia* (L. Ord.–U. Sil.)
Class Cyclocystoidea (M. Ord.–M. Dev.)
Class Camptostromoidea (L. Cam.)
Class Echinoidea—sea urchins: *Bothriocidaris* (L. Ord.–Rec.)
Class Holothuroidea—sea cucumbers (Ord.–Rec.)
Phylum Chordata
Subphylum Hemichordata—acorn worms
Class Enteropneusta (Rec.)
Class Pterobranchia (L. Ord.–Rec.)
Class Graptolithina—graptolites (Cam.–L. Carb.)
Order Dendroidea—dendroids: *Dictyonema*
Order Graptoloidea—graptoloids: *Dichograptus, Didymograptus, Diplograptus, Goniograptus, Hallograptus, Leptograptus, Monograptus, Paraplectograptus*
Subphylum Cephalochordata: *Amphioxus*
Subphylum Urochordata—sea squirts
Subphylum Vertebrata—vertebrates
Superclass Pisces
Class Agnatha—jawless fish (L. Ord.–Rec.)
Subclass Monorhina (M. Sil.–Rec.)
Order Osteostraci: *Hemicyclaspis* (U. Sil.–L. Dev., ?M. Dev.)
Order Anaspida: *Birkenia* (M. Sil.–U. Dev.)
Order Cyclostomata—modern lampreys and hagfishes (Rec.)
Subclass Diplorhina (L. Ord.–U. Dev.)
Order Heterostraci: *Anglaspis* (L. Ord.–U. Dev.)
Order Coelolepida: *Lanarkia* (U. Sil.–L. Dev.)
Class Placodermi—early jawed fish (Dev.)
Order Petalichthyida—armored fishes related to arthrodires: *Lunaspis* (Dev.)
Order Rhenanida—shark-like placoderms: *Gemuendina* (Dev.)
Order Arthrodira—armored fishes with jointed necks: *Dunkleosteus* (Dinichthys) (Dev.)
Order Phyllolepida—degenerate forms (M.–U. Dev.)
Order Ptyctodontida—with reduced bony armor (M.–U. Dev.)
Order Antiarchi—with jointed, movable pectoral spines: *Pterichthyodes* (Dev.)
Class Chondrichthyes—sharks and related fishes (M. Dev.–Rec.)
Subclass Elasmobranchii—sharks, skates, rays: *Cladoselache* (M. Dev.–Rec.)

Subclass Holocephali—chimaeras or "ratfishes" (U. Dev.–U. Perm.)
Class Osteichthyes—bony fishes (U. Sil.–Rec.)
　?Subclass Acanthodii—primitive forms: *Climatius* (U. Sil.–L. Perm.)
　Subclass Actinopterygii—ray-finned fishes (U. Dev.–Rec.)
　　Infraclass Chondrostei—less primitive forms (U. Dev.–Rec.)
　　Infraclass Holostei—primarily a Mesozoic radiation (U. Perm.–Rec.)
　　Infraclass Teleostei—successful group of modern fishes: *Amphistium* (U. Trias.–Rec.)
　Subclass Sarcopterygii—lobe-finned or air-breathing fishes (L. Dev.–Rec.)
　　　Order Crossopterygii—progressive air-breathing fishes: *Eusthenopteron, Latimeria, Osteolepis* (M. Dev.–Rec.)
　　　Order Dipnoi—lungfishes: *Epiceratodus, Lepidosiren, Protopterus* (L. Dev.–Rec.)
Class Amphibia (U. Dev.–Rec.)
　Subclass Labyrinthodontia (U. Dev.–U. Trias.)
　　　Order Icthyostegalia—earliest amphibians: *Ichthyostega* (U. Dev.–L. Miss.)
　　　Order Temnospondyli—Paleozoic amphibians, now extinct: *Benthosuchus, Buett-neria, Eryops* (U. Miss.–U. Trias.)
　　　Order Anthracosauria—ancestral group of reptiles: *Archeria, Seymouria* (L. Penn.–U. Perm.)
　Subclass Lepospondyli (L. Miss.–L. Perm.)
　　　Order Nectridea—eel-like types (L. Penn.–L. Perm.)
　　　Order Aistopoda—limbs reduced or lost (L. Miss.–L. Perm.)
　　　Order Microsauria—lizard-like amphibians (L. Miss.–L. Perm.)
　Subclass Lissamphibia—modern amphibians (M. Jur.–Rec.)
　　Superorder Salientia (M. Jur.–Rec.)
　　　Order Anura—frogs and toads (M. Jur.–Rec.)
　　Superorder Caudata (U. Jur.–Rec.)
　　　Order Urodela—salamanders (U. Jur.–Rec.)
　　　Order Apoda—tropical limbless forms (Rec.)
Class Reptilia (L. Penn.–Rec.)
　Subclass Anapsida—reptiles with a solid skull roof (L. Penn.–Rec.)
　　　Order Cotylosauria—primitive reptiles: *Hylonomus* (L. Penn.–U. Trias.)
　　　Order Mesosauria—ancient aquatic reptiles: *Mesosaurus* (L. Perm.)
　　　Order Chelonia—turtles (M. Perm.–Rec.)
　Subclass Lepidosauria—primitive diapsids, with two temporal openings in skull (M. Perm.–Rec.)
　　　Order Eosuchia—early diapsids (M. Perm.–Eoc.)
　　　Order Squamata—lizards and snakes: *Mosasaurus, Tylosaurus* (U. Trias.–Rec.)
　　　Order Rhynchocephalia—rhynchocephalians: *Sphenodon* (L. Trias.–Rec.)
　Subclass Archosauria—advanced diapsids (U. Perm.–?Paleoc.)
　　　Order Thecodontia—ancestral types: *Euparkeria* (U. Perm.–U. Trias.)
　　　Order Crocodilia—crocodiles (M. Trias.–Rec.)
　　　Order Pterosauria—flying reptiles: *Pteranodon, Pterodactylus, Rhamphorhynchus* (L. Jur.–U. Cret.)
　　　Order Saurischia—saurischian dinosaurs: *Brachiosaurus, Brontosaurus, Cerato-saurus, Coelophysis, Compsognathus, Coelurus* (Ornitholestes), *Ornithosuchus, Ornithomimus* (Struthiomimus), *Tyrannosaurus* (M. Trias.–U. Cret.)
　　　Order Ornithischia—ornithischian dinosaurs: *Ankylosaurus, Camptosaurus, Corythosaurus, Iguanodon, Pachycephalosaurus, Parasaurolophus, Protoceratops, Stegosaurus, Triceratops, Anatosaurus* (Trachodon) (U. Trias.–U. Cret., ?Paleoc.)
　Subclass Euryapsida—mainly marine reptiles (L. Perm.–U. Cret.)
　　　Order Araeoscelidia—ancestral group (L. Perm.–U. Cret.)

Order Sauropterygia—shore dwelling and marine forms: *Elasmosaurus,*
    Nothosaurus (L. Trias.–U. Cret.)

Order Placodontia—mollusk-eating types: *Placodus* (Trias.)

Subclass Ichthyopterygia—marine fish-like reptiles (M. Trias.–U. Cret.)

Order Ichthyosauria—ichthyosaurs: *Ichthyosaurus* (M. Trias.–U. Cret.)

Subclass Synapsida—mammal-like reptiles (L. Penn.–M. Jur.)

Order Pelycosauria—early mammal-like reptiles: *Dimetrodon, Edaphosaurus,*
    Ophiacodon (L. Penn.–U. Trias.)

Order Therapsida—varied mammal-like reptiles: *Cynognathus, Kannemeyeria,*
    Moschops (M. Perm.–M. Jur.)

Class Aves—birds (U. Jur.–Rec.)

Subclass Archaeornithes—primitive, toothed birds of the Jurassic: *Archaeopteryx* (U. Jur.)

Subclass Neornithes—all other birds (L. Cret.–Rec.)

Class Mammalia—mammals (U. Trias.–Rec.)

Subclass Prototheria—egg-laying mammals (Pleist.–Rec.)

Order Monotremata—*Echidna, Ornithorhynchus* (Pleist.–Rec.)

Subclass uncertain

Order Docodonta (U. Trias.–U. Jur.)

Order Triconodonta (M. Jur.–L. Cret.)

Subclass Allotheria (U. Jur.–U. Eoc.)

Order Multituberculata—multituberculates (U. Jur.–U. Eoc.)

Subclass Theria—most mammals (M. Jur.–Rec.)

Infraclass Trituberculata—the first therians (M. Jur.–L. Cret.)

Order Symmetrodonta—symmetrodonts (U. Jur.–L. Cret.)

Order Pantotheria—ancestors of marsupials and placentals (M. Jur.–L. Cret.)

Infraclass Metatheria—marsupials (U. Cret.–Rec.)

Order Marsupialia (U. Cret.–Rec.)

Infraclass Eutheria—placental mammals

Order Insectivora—insectivores, the most primitive placentals: *Zalambdalestes*
    (L. Cret.–Rec.)

Order Tillodontia—tillodonts (U. Paleoc.–U. Eoc.)

Order Taeniodontia—taeniodonts (L. Paleoc.–U. Eoc.)

Order Chiroptera—bats (U. Eoc.–Rec.)

Order Primates—tupaioids, lemurs, tarsiers, apes, monkeys, man: *Australopithecus,*
    Galago, Homo, Hylobotes, Macaca, Notharctus, Pan, Pliopithecus, Proconsul,
    Propliopithecus, Tarsius, Tupaia (M. Paleoc.–Rec.)

Order Creodonta—primitive flesh eaters: *Hyaenodon, Oxyaena* (U. Cret.–E. Plioc.)

Order Carnivora—carnivores: *Dinictis, Felis, Hoplophoneus, Smilodon*
    (M. Paleoc.–Rec.)

Order Condylarthra—ancestral ungulates: *Phenacodus* (U. Cret.–U. Mioc.)

Order Amblypoda—large, primitive ungulates: *Barylambda, Coryphodon,*
    Uintatherium (M. Paleoc.–L. Olig.)

Order Proboscidea—elephants and related forms: *Amebelodon, Moeritherium,*
    Mastodon, Trilophodon (U. Eoc.–Rec.)

Order Sirenia—sea-cows (L. Mioc.–Rec.)

Order Desmostylia—desmostylids (L. Mioc.–Pleist.)

Order Hyracoidea—hyraxes (L. Mioc.–Rec.)

Order Embrithopoda—arsinoitheres of Egypt (L. Olig.)

Order Notoungulata—South American hoofed mammals: *Toxodon* (U. Paleoc.–Pleist.)

Order Astrapotheria—South American hoofed mammals: *Astrapotherium*
    (U. Paleoc.–U. Mioc.)

Order Litopterna—South American hoofed mammals: *Macrauchenia* (U. Paleoc.–
    Pleist.)

Order Perissodactyla—odd-toed hoofed mammals: *Brontotherium, Equus, Hipparion, Hippidium, Hyracotherium* (Eohippus), *Hypohippus, Hyracodon, Indricotherium, Merychippus, Mesohippus, Metamynodon, Miohippus, Moropus, Pliohippus, Rhinoceros, Tapir, Teleoceras* (L. Eoc.–Rec.)

Order Artiodactyla—even-toed hoofed mammals: *Alticamelus, Archaeotherium, Bothriodon, Dinohyus, Diceratherium, Hippopotamus, Merycoidodon, Palaeochoerus, Procamelus, Stenomylus, Sus, Syndyoceras, Synthetoceras, Tragulus* (L. Eoc.–Rec.).

Order Edentata—New World edentates (U. Paleoc.–Rec.)

Order Pholidota—pangolins (Olig.–Rec.)

Order Tubulidentata—aardvarks (L. Eoc.–Rec.)

Order Cetacea—whales, dolphins, porpoises (L. Eoc.–Rec.)

Order Rodentia—rodents (U. Paleoc.–Rec.)

Order Lagomorpha—hares, rabbits (U. Paleoc.–Rec.)

KINGDOM PLANTAE

Phylum Thallophyta—algae, fungi, lichens

Phylum Bryophyta—liverworts and mosses

Phylum Tracheophyta—vascular plants

Subphylum Psilopsida—most primitive tracheophytids: *Aneurophyton, Asteroxylon, Pseudosporochnus, Psilophyton, Rhynia, Sciadophyton, Taeniocrada, Zosterophyllum*

Subphylum Lycopsida—lycopsids: *Archaeosigillaria, Duisbergia, Lepidodendron, Pleuromeia, Protolepidodendron, Sigillaria*

Subphylum Sphenopsida: *Calamites, Calamophyton, Equisetum, Hyenia, Schizoneura, Sphenophyllum*

Subphylum Pteropsida

Class Filicineae—ferns: *Cladoxylon, Protopteridium*

Class Gymnospermae—plants with naked seeds

Order Gnetales—close resemblance to angiosperms, poor fossil record

Order Coniferales—conifers: *Voltzia*

Order Ginkgoales—ginkgos: *Ginkgo*

Order Cordaitales—cordaites: *Cordaites*

Order Cycadales—living cycads

Order Cycadeoidales—extinct cycad-like plants: *Williamsonia, Williamsoniella*

Order Cycadofilicales—primitive gymnosperms: *Barrandeina, Gangamopteris, Glossopteris, Neuropteridium, Sphenopteris*

Class Angiospermae—flowering plants

Subclass Monocotyledonae—seed with single cotyledon

Subclass Dicotyledonae—seed with two cotyledons

KINGDOM PROTISTA

Several phyla of Protista—the Mastigophora, Sarcodina, Sporozoa, and Ciliophora—are much like animals and are often classified as animals in the phylum Protozoa. These forms usually consist of a lump of protoplasm with a nucleus and are generally microscopic in size, ranging from less than one micron to several centimeters.

The Mastigophora have a cell wall of fixed shape and one or more whip-like flexible projections (flagellae) used for locomotion (Fig. A-1GH). The Sarcodina are distinguished by a changeable cell form, which is associated with mobile extensions of the body (Fig. A-1ABCD). The Sporozoa are parasitic forms having a fixed cell wall but no hard parts. The Ciliophora have abundant short thread-like processes, called cilia,

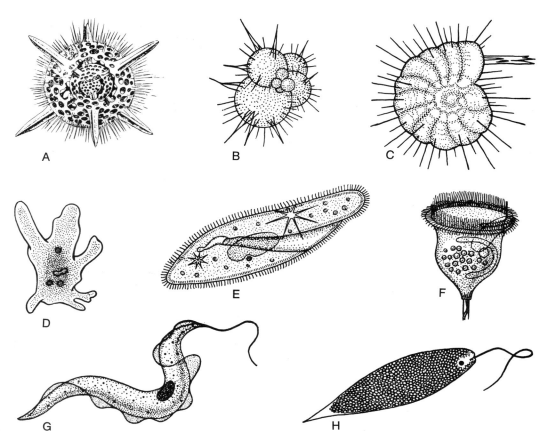

FIGURE A-1

Representative members of three phyla of the Protista.

A. *Actinomma*, a typical radiolarian (phylum
   Sarcodina); parts of the lattice shells are broken
   away to reveal the interior construction.
B. Typical foraminifer (phylum Sarcodina).
C. Typical foraminifer (phylum Sarcodina).

D. *Amoeba*, a shell-less type of the phylum Sarcodina.
E. *Paramecium* (phylum Ciliophora).
F. *Vorticella* (phylum Ciliophora).
G. *Trypanosoma* (phylum Mastigophora).
H. *Euglena* (phylum Mastigophora).

(A from Vol. D-3 of *Treatise on Invertebrate Paleontology*, Geol. Soc. Am. and Univ.
of Kansas Press, 1954; B–H from G. Hardin, *Biology: Its Human Implications* (2nd ed.)
by Garrett Hardin. W. H. Freeman and Company, San Francisco, 1952.)

which cover the exterior of the definitely shaped
cell wall (Fig. A-1EF).

The Mastigophora and Sarcodina include organisms that possess hard parts capable of preservation. The most important by far belong to
the order Foraminifera, of the phylum Sarcodina;
next in importance is the order Radiolaria, of the
same phylum.

The foraminifers usually secrete a shell, which
may be calcareous, chitinous, or agglutinated —

that is, composed of small particles picked up
from the bottom and cemented together. The first
two materials are secreted directly by the organism. Foraminifers are entirely aquatic, most of
them living in marine waters, but a few live in
fresh waters.

The foraminifers are among the most useful
fossils in straitigraphic work because they are
widely distributed in marine sediments of the
geologic column, especially from the Carbonifer-

ous to the Cenozoic. They are of special importance in the straitigraphic zoning of rocks penetrated by deep borings in the course of petroleum exploration.

Radiolarians are sarcodinids with thread-like, radically directed pseudopodia. The nucleus is covered by a chitinous material. The internal, delicately complex skeleton is composed of silica and consists of an intricate lattice-work having a globular, discoid, stellate, or conical shape, bearing spines or lacking them (Fig. A-1A). The shape and architecture of the skeleton vary greatly. Radiolarians are entirely marine and are known to have lived from the Cambrian to the present time. They are abundant in existing seas.

KINGDOM ANIMALIA

Phylum Archaeocyatha

Archaeocyathids are exclusively Cambrian marine organisms of world-wide distribution. Their position within the animal kindgom is in doubt, and in the past they have been assigned to the calcareous algae, the foraminifers, the calcareous sponges, the siliceous sponges, and the corals; but they are now generally classed as an independent phylum.

The skeletons of archaeocyathids are commonly calcareous and cone-, goblet-, or vase-shaped, but irregular, crenulate, saucer-like, and conical forms are found. Simpler types consist of outer and inner conical cups with varied types of structural elements between them (Fig. A-2). Both the outer and the inner walls are perforated by numerous pores. The space between the walls (the intervallum) contains vertical partitions (parieties) and transverse partitions (tabulae). All skeletal elements of the intervallum and the side of the inner wall facing the central cavity are judged to have been covered by living tissue while the animals were growing. Evidence of a coelenteron is lacking, and it seems likely that assimilation of food was confined mainly to the intervallum. The anatomy of the Archaeocyatha is obviously quite unlike that of the Coelenterata, and it also includes structures that are not characteristic of the Porifera, for sponges lack concentric walls with intervening partitions and have skeletons composed of distinct spicules.

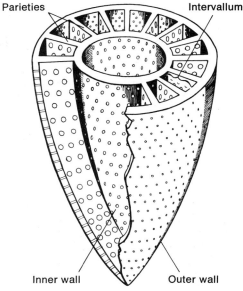

Parieties   Intervallum

Inner wall   Outer wall

FIGURE A-2

Diagrammatic sketch of a typical archaeocyathid (*Ajacicyathus*). (Redrawn from V. J. Okulitch, 1943.)

FIGURE A-3

Diagrammatic sketch of a typical simple sponge, showing the main morphological features and the directions of water currents. (From G. Hardin, *Biology: Its Human Implications*, 2nd ed., W. H. Freeman and Company, San Francisco, 1952.)

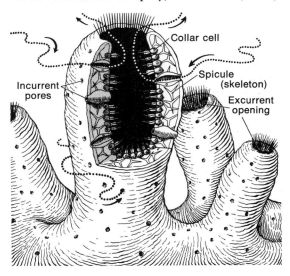

Collar cell

Incurrent pores

Spicule (skeleton)

Excurrent opening

Phylum Porifera

The sponges are aquatic, dominantly marine invertebrates, which are ranked next above the Protista in classification. They are the simplest multicellular animals. Grouped with the sponges are several sponge-like fossils of doubtful classificatory status but of geological importance.

The body of a simple sponge is somewhat like a vase, open at the top, attached at the bottom, the sides perforated with numerous pores (Fig. A-3). There are no internal organs, no nervous tissue, no circulatory or digestive system, such as occur in higher invertebrates. The vase-shaped sac is covered and protected on the outside by flattened covering cells (ectoderm). The inside is lined with specialized collar cells (called choanocytes), each of which has a long flagellum. The movement of the flagella creates the water current that passes through the sponge, entering by the minute pores, passing through the main cavity, and leaving by way of the large hole at the top of the sac-like body. As the water passes through, the collar cells capture and ingest food organisms.

Between the outer covering cells and the inner collar cells is a non-living jelly-like material containing various specialized freely moving cells (called mesenchyme cells), some of which receive partly digested food particles from the collar cells, complete the digestion, and carry the digested food from one place to another. Other of these cells secrete needles, called spicules, of calcium carbonate or silica.

Three general types of body structure are recognized (Fig. A-4), the simplest being the sac-like form discussed above. This body type is called the ascon type. To increase the area of collar cells, some sponges have crenulated body walls giving rise to chambers lined by collar cells. This is the sycon type. In the most advanced types the body wall is perforated with a complex system of canals and chambers with collar cells; this is the rhagon type.

The body of a sponge is supported by an internal structure composed of organic fibers or crystalline spicules or a combination of the two. The familiar bath sponge has a skeleton of organic material called spongin. In other sponges the skeleton is made of spicules of calcium carbonate or silica secreted by the special mesenchyme cells. The spicules may be either discon-

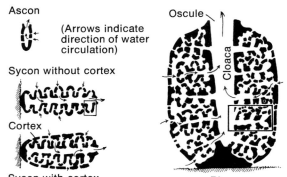

FIGURE A-4

Diagrammatic sections showing types of sponge architecture. The ascon type comprises a single chamber lined by collar cells. The sycon type is composed of several chambers opening into a single excurrent passageway; the area outlined in the form without a cortex (a thick, leathery, external cover) shows the correspondence to the ascon structure. The rhagon type consists of many sycon-like elements (one of them outlined) arranged round a central excurrent passage (cloaca). (After M. W. de Laubenfels in Vol. E of *Treatise on Invertebrate Paleontology*, Geol. Soc. Am. and Univ. of Kansas Press, 1955.)

FIGURE A-5

Fundamental structure of the coelenterate polyp (A), and medusa (B), and cross-section (C) of the wall of a coelenterate, showing the stinging capsules that are produced only in this phylum. (After G. Hardin, *Biology: Its Human Implications*, 2nd ed., W. H. Freeman and Company, San Francisco, 1952.)

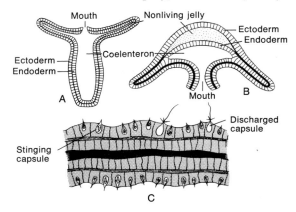

nected or firmly bound together, and their shape is highly variable. The classification of the Porifera is based mainly on the nature and composition of the skeleton.

Phylum Coelenterata

The phylum Coelenterata includes such common forms as the hydra, corals, and jellyfishes and, in addition, a great host of fossil forms that have left no living representatives. The coelenterates exemplify a grade of animal structure that is characterized by two peculiarities: (1) it is composed of tissues not constituted into organs; and (2) its parts are arranged on a radially symmetrical plan. Throughout their history the Coelenterata have been aquatic; though they are predominantly marine, there are numerous fresh-water forms. This phylum has left one of the most nearly complete of fossil records; it has been studied widely and rather thoroughly; and it is one of the important groups to the historical geologist both for stratigraphy and in the interpretation of past environments.

The individual coelenterate is one of two definite morphological types (Fig. A-5): (1) a hollow, sac-like polyp; and (2) an umbrella-shaped medusa. Polyps are invariably attached and frequently secrete a skeleton. Medusas are usually free-swimming and without skeletal hard parts.

The polyp usually has a hollow, sac-like body with an opening at the top. Surrounding this opening, called the mouth, may be one or more circlets of tentacles, which are either hollow outgrowths of the body wall or solid. The body may be spherical, cylindrical, or disk-shaped, almost always exhibits well-developed radial symmetry — though that may be modified sometimes by bilateral symmetry — and has a single hollow internal cavity called the coelenteron. The water enters the coelenteron through the mouth and is expelled through the same opening after the cells lining the coelenteron have extracted the food.

The tentacles and body wall are composed of three layers, the ectoderm, mesoglea, and endoderm. The Coelenterata differ from the Porifera, however, in having a solid instead of a perforated wall and in having few or no cells developed in the middle layer (mesoglea).

The ectoderm secretes the external, supporting and protective, stony or chitinous skeleton. Muscles and nerves, always of very simple organization, may also be found in the ectodermal cells. Most coelenterates have their surfaces covered with minute stinging capsules, called cnidoblasts (Fig. A-5C). These capsules are also found on the endoderm of certain forms. They are filled with an irritating or poisonous liquid, and in addition contain a spirally coiled filament (nematocyst), which, when released, shoots out like a watch spring. The capsule has a sensory projection on the exterior, which, when disturbed, releases the filament and the poisonous liquid. This action paralyzes small organisms, which then serve as food. The cnidoblasts also perform a protective function.

The inside of the coelenteron and of the hollow tentacles is lined with endodermal cells. These cells, sometimes equipped with cnidoblasts, acquire and assimilate the food; excrete the waste products, which are discharged ultimately through the mouth; and produce the sexual organs. Sexual products also leave the organism through the mouth after having been discharged into the coelenteron from the endoderm.

The mesoglea is a mass of unorganized protoplasmic material of jelly-like consistency. It is usually not well developed and in some forms is absent. It does not, as in the Porifera, have important functions.

The body wall may be smooth, or it may be folded and minutely crenulated, especially in the lower half. Crenulation is reflected in the external skeleton where it gives rise to certain special structures.

The tentacles are extended when the animal is feeding. They are covered with cnidoblasts and aid in seizing and killing food. When the animal is disturbed, the tentacles are retracted, covering and protecting the mouth.

The typical medusa resembles an umbrella with tentacles hanging down along the rim and with the mouth at the end of a short central tube on the under side. A prominent, inwardly projecting shelf, the velum, is found in some forms. Four (or some multiple of four) radial canals lead from the coelenteron to a circular canal lying in the bell margin. All medusas are more than 90 percent water by weight.

Coelenterates having only medusoid forms reproduce sexually. Those having only polyp forms reproduce either sexually or both sexually and asexually. Those having alternating polyp and medusoid stages have asexual reproduction in the polyp stage and sexual reproduction in the medusoid stage (Fig. A-6).

In asexual reproduction buds are formed in the body wall of the mother polyp, and sometimes there may be as many as three generations in one organism. In sexual reproduction the sexual elements join to form an embryo (planula), which develops directly into a medusa without an intermediate polyp stage. In this type of reproduction there is no sessile stage; all individuals are free-moving.

In reproduction by alternation of generations some of the buds of the colony produce medusas, or little jellyfish, which are released from the bud and become free-swimming individuals. After a developmental period they give off sexual products that come in contact with similar products from other medusas. The union of male and female elements produces an embryo (planula), which, after development as an active, free-swimming individual, becomes attached and develops into a sessile polyp, which will develop a new colony.

The colonial habit is very well developed among the Coelenterata, especially in the reef-building corals, which during the geological past have built and are still building great calcareous reefs partly composed of their stony skeletons (Fig. 7-5, p. 227). There are and have been many coelenterates without the colonial habit—for example, the cup corals of the Paleozoic and the fresh-water hydra of the present.

Most sessile coelenterates develop a protective covering commonly called a skeleton. This so-called skeleton is not to be confused with the internal skeleton of the Porifera or of such highly developed animals as the vertebrates. Usually the coelenterate skeleton is secreted by the ectodermal cells, but in a few species skeletal material in the form of calcareous particles may be secreted by the mesoglea. Coelenterates build their skeletons of chitin or calcareous material.

The embryonic skeleton of a coelenterate has the shape and appearance of a little hollow cup or cone and is called the prototheca. From this simple beginning may arise a great variety of different skeletal types, depending upon the subsequent development of the polyp. In maturity the skeleton may still be a single, isolated cone, now partly filled, however, with internal struc-

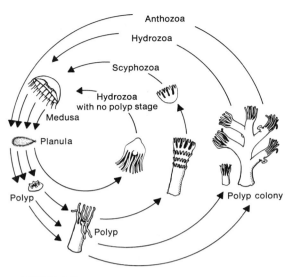

FIGURE A-6

Life cycles of the main groups of coelenterates. (Redrawn from D. Hill and J. W. Wells in Vol. F of *Treatise on Invertebrate Paleontology,* Geol. Soc. Am. and Univ. of Kansas Press, 1956.)

tures; or it may have the form of a bush or tree; it may be shaped like a biscuit or cabbage head; or it may be composed of a large number of loosely or closely packed cylindrical or prismatic tubes, in which case the entire structure will exhibit a variety of shapes—"honeycomb coral," "chain coral," "organ-pipe coral," etc. (Fig. 7-6, p. 228).

As already stated, the coelenteron, or digestive cavity, has either a smooth, unruffled surface or is crenulated or folded, and it has partition-like extensions from the body wall inward toward the central portion of the cavity (Fig. A-7). These extensions of the body are called mesenteries. Their function is to increase the digestive surface of the coelenteron and thus to make it possible for more of the sea water with its contained nutrients to come in contact with that surface.

In some kinds there is an infolding of the body walls between the mesenteries, which secrete on their outer surface thin blades of calcite. These blades are usually arranged in cycles of four or six, or multiples of these numbers, and are called septa. Septa divide the conical or cylindrical skeletons into more or less equal parts. They are best developed in the lower and older part of the skeleton.

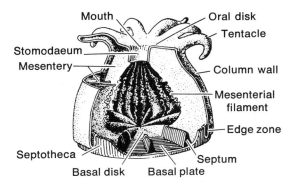

Relations of polyp and skeleton in a typical coral. The mesenteries are not shown between the septa in the foreground. (After *Treatise on Invertebrate Paleontology*, Geol. Soc. Am. and Univ. of Kansas Press, 1956.)

As the polyp grows and buds, it moves upward in its tubular or conical skeleton (corallite), at the same time building a transverse floor beneath itself. If this floor extends entirely across the corallite, it is called, whether it is plane, convex, or concave, a tabula. If it extends only part of the way across the corallite, it is called a dissepiment. The endotheca comprises dissepiments inside the wall of a corallite, the exotheca those outside the wall. Dissepiments give corallites a vesicular structure. The upper part of the corallite, which is occupied by the polyp, is called the calice. A central axial structure, originating from various modifications of the inner ends of the septa, is called the columella (Fig. A-8). The adult skeleton, whether consisting of one or many individual corallites, is called a corallum. The skeleton may remain conical to maturity; it may be joined to other small cups or cones by stony connective material, which in life was covered by living matter called coenosarc, in which cases the entire colony will have a bushy or massive, hemispherical appearance; or the successive cones may grow upward into cylinders, and these by packing may form a compact colony of hexagonal prismatic corallites resembling a honeycomb. In those forms in which the corallites are not tightly packed, the space between adjacent corallites may be filled with extraneous calcareous deposits or it may be open.

The Coelenterata are divided into three major classes, the Hydrozoa, the Scyphozoa, and the Anthozoa. Each of these classes contains many orders, but only those of prime importance in the fossil record will be mentioned here. The most important class is the Anthozoa (corals); the Hydrozoa have a fair geologic record, but the Scyphozoa (jellyfish) have a very sparse record.

CLASS HYDROZOA

The Hydrozoa are animals with the general characteristics of the coelenterates. The coelenteron is undivided and has no mesenteries and hence has no septa in the skeleton. Some kinds have alternation of generations, in which the polyp stage alternates with a medusoid stage. The medusas produce ova and sperms from the union of which the polyp develops, and the polyps in their turn bud off medusas. The medusas are developed on a portion of a polyp colony where they may be surrounded with a transparent vaselike structure known as the gonotheca. The other polyps of the colony, living in transparent vaselike cups called hydrotheca, devote themselves to animal activities other than reproduction. In forms whose polyp stage is wanting, budding does not occur, and the sexually produced

General relations of polyp and corallum. (After *Treatise on Invertebrate Paleontology*, Geol. Soc. Am. and Univ. of Kansas Press, 1956.)

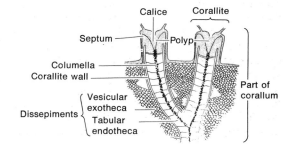

individual is a little jellyfish, or medusa. In forms that have no medusoid stage, the polyps develop the new generation both by budding and by sexual reproduction.

In hydrozoans that have asexual reproduction without detachment from the parent, a colony results; and when budding without any restriction, the colony may grow indefinitely. The great development of coenosarc—the connective tissue between adjacent polyps—in the Hydrozoa makes it improbable that individual corallites will come in contact with one another; hence the prismatic forms will not be developed. The corallites are usually cylindrical and are embedded in stony matter secreted by the under surface of the coenosarc.

Hydrozoa are extremely common in modern marine waters, and these forms secrete stony skeletons. Some varieties live in fresh water, both polyp and medusoid stages being found, but none secretes stony matter. Hydrozoans are important constituents of modern reefs.

A very important order of Hydrozoa was the extinct Stromatoporoidea, which were rock-builders in the Paleozoic and Mesozoic. The stromatoporoids built a nodular or hemispherical colony composed of a number of wavy, concentric layers of calcium carbonate. In life the entire surface of the colony was covered with minute pores, through which the different kinds of polyps protruded. All the polyps were connected at the base by a delicate system of minute tubes, which ramified the laminations. The polyps built around them small tubular or solid pillars that were more or less radially disposed with respect to the laminations. Hence a section cut across a stromatoporoid colony will show a number of concentric laminae, the space between them appearing to be divided into small cells or rooms by the radial pillars. The pillars and concentric layers were probably built during definite growth stages. The stromatoporoids are now extinct but were important reef-builders in the Silurian and Devonian seas. Their total range was from the Cambrian to the Cretaceous. Certain living hydrozoans are very much like their ancient relatives.

CLASS SCYPHOZOA

The Scyphozoa include the large jellyfishes of the modern oceans as well as certain fossil forms. There is no polyp or fixed stage in this class. The animal secretes no hard parts and hence builds no skeleton. Even though its body is often as much as 99 percent water, impressions of the animal may be left on the mud and, under unusual conditions, preserved. The hollow interior may be filled with mud or sand and later form a mold. Such rare fossils have been found in rocks as old as Late Precambrian (Fig. 2-27, p. 52).

CLASS ANTHOZOA

This class of coelenterates includes the very familiar stony corals and fleshy sea anemones—common in the warmer parts of the present oceans—and several large groups of corals long since extinct.

The organism is built on much the same general plan as a simple hydrozoan but differs from it in several respects. It is a columnar body bearing a crown of numerous circlets of tentacles at one end and attached to the bottom by the other. Lying within the ring of tentacles is a small muscular area, in the middle of which is the slit-like mouth. The mouth is sometimes invaginated to form a tubular gullet, which hangs down into the coelenteron (Fig. A-7).

The sac-like body of the Anthozoa is folded radially into mesenteries; hence most of the skeletons have or had septa. Most forms also show some tabulae and dissepiments.

Reproduction takes place both asexually and sexually. New individuals may be formed directly by budding and ordinarily remain attached to the parent, thus helping to build a colony. In sexual reproduction a free-swimming larva is formed. This larva ultimately attaches itself to some foreign object and develops into a new individual. When the larva settles down, it assumes the shape of a thin biscuit, modifying that general shape to conform to any irregularities of the object to which it has become attached. It first secretes a basal plate, and soon afterward the basal portion of the animal develops four or six folds, between which are soon formed a corresponding number of blade-like radial upgrowths of the basal plate. These are the primary septa of the skeleton, and ultimately they are united at their outer edges by a wall, or theca. This complete structure, now roughly conical or cup-shaped, constitutes the embryonic skeleton, or prototheca. As the animal grows, several cycles of folding may be developed and a corresponding number of septa brought into existence. The number of mesenteries and septa will be a multiple of the number of the primary ones. Small dissepiments and larger tabulae may be built under part or all of the polyp as it grows upward.

The colonial habit is intensely developed in many corals. Only a few forms are solitary. The coralla show a wide variation in size, shape, and character. There are the simple, solitary, conical type, the bushy form with loosely connected tubes or branches, the compact, massive form, etc. The individual corallites may be simple cups or cones, simple or tabulated tubes, or prisms of various kinds. The skeletal material is either chitinous or calcareous; the latter kind is often replaced by silica during fossilization.

There are three subclasses of Anthozoa, each of which has a large number of subdivisions. The groups we shall discuss are the subclass Octocorallia and the subclass Zoantharia and three of its orders: the Rugosa, the Scleractinia, and the Tabulata.

SUBCLASS OCTOCORALLIA

The Octocorallia are almost invariably colonial. Living forms are provided with eight mesenteries and eight plumose tentacles; the tentacles are arranged in a single circle round the mouth. It is possible, therefore, to distinguish living octocorallians from other living corals by the number, arrangement, and character of mesenteries and tentacles. The skeleton shows considerable variation in structure. It may be composed of detached calcareous elements loosely scattered throughout the body wall; it may consist of a meshwork of horny or calcareous axes about which polyps are deployed; or it may consist of tabulated tubular corallites grown together in various ways.

The first Octocorallia are possibly from the Silurian, but the definitely established range is from the Permian to the Recent. At no time in the geologic past were they very abundant or important.

SUBCLASS ZOANTHARIA

This subclass, which includes both corals and sea anemones, contains seven orders, but we shall discuss only three of them, the Rugosa, the Scleractinia, and the Tabulata. The Rugosa and the Tabulata are exclusively Paleozoic forms; the Scleractinia range from the Mesozoic to the Recent and make up the bulk of modern corals. These groups differ mainly in the mode of septal appearance.

Order Rugosa. The Rugosa are exclusively Paleozoic, extinct corals with solitary or composite coralla (Fig. 7-6BEF, p. 228). The solitary corals are called cup corals. The colonial forms, composed of composite coralla, may be either branching or compact and massive. The septa are arranged in cycles of four, with radial symmetry modified by bilateral symmetry.

In the cup corals the four primary septa divide the circular corallum into four quadrants, which are not all of the same size. The main septum, which is usually longer than the other three, is called the cardinal septum; the short one diametrically opposite is called the counter; and the two other septa are called alar septa. The place of the cardinal or counter septum may be marked by a shallow depression called a fossula.

In colonial types of Rugosa, in which the corallites are more or less tightly packed, the septa are all of the same size, and the radial symmetry is not modified. Cardinal, counter, and alar septa usually cannot be distinguished.

The Rugosa are the most important Coelenterata of the Paleozoic. They made their first appearance in the Middle Ordovician and died out in the Permian.

Order Scleractinia. This order is by far the most numerous and varied of all the living Anthozoa. It is distinctive in that the septa are arranged in cycles of six. The mesenteries and septa exhibit radial symmetry modified by bilateral symmetry. There are numerous tentacles, arranged in circles round the mouth. Both solitary and colonial forms are included in this order (Fig. 10-7, p. 350).

The scleractinians range from the Middle Triassic to the Recent and have been important reef-builders during their entire history.

Order Tabulata. The Tabulata constitute an extinct order of Paleozoic and very early Mesozoic corals, which are named for the excellent development of their tabulae. They were invariably colonial, and they built skeletons composed of long tubular or prismatic corallites, always with well-developed tabulae and frequently with septa (Fig. 7-6, p. 228). The individual polyp lived only in the upper story of the stony, calcareous corallite.

In one group of Tabulata, the Favositidae, the walls between the six-sided prismatic corallites are perforated with small holes called mural pores (Fig. 7-6C). There may be none, or more than one, between adjacent tabulae. They are thought to have been formed by the building of the corallite wall around buds that later died, leaving the pore. In a second group, the Syringoporidae, the tabulae are funnel-shaped instead of transverse, and the colony has the appearance of organ pipes held together by cross-bars; hence the name "organ-pipe coral" (Fig. 7-6D). Still a third group, the Halysitidae, have tubes of elliptical or circular cross-section so joined that a cross-section resembles a chain; hence the name "chain coral" (Fig. 7-6G).

Worms

The term "worms" is here used to include a large group of worm-like animals referable to several phyla. Because of their lack of hard parts for preservation and, consequently, because of their very fragmentary and incomplete geologic record (being the least known of all fossil animals), they will be considered together.

A worm may be defined as a long, soft creature, usually without hard parts, which usually moves by wiggling. Some worms are segmented; others are not. Some have appendages; others are without them.

Owing to the soft-bodied character of worms in general, it is to be expected that fossil remains of them are uncommon, or even rare. Under exceptional conditions, however, the most delicate structures may be preserved. Generally the fossil records consist of: (1) impressions of the body (Fig. 2-5, p. 28); (2) tracks and trails made by the animals as they crawled over muddy bottoms (Fig. 2-15, p. 33); (3) tubes and burrows in which they lived or through which they moved (Fig. 2-15); (4) coprolites, or fossil excreta, which are composed of mud passed through the intestinal tract of the worm (Fig. 2-15); or (5) tiny chitinous jaws. A few parasitic forms have been found preserved (in their hosts) in amber.

This great group of organisms is of considerable interest in several ways in the study of fossils. Though there is no accurate fossil record of them, there is every reason to suppose that they have been of considerable importance throughout most of geologic time. Their part in chewing up materials on the bottoms of ancient seas, for instance, has been a very important one.

Worms were present in the Precambrian, for trails similar to modern worm trails have been found, and other lines of evidence strongly suggest that worms were living at that time. The affinities of such worms are unknown, however. From the Cambrian to the present time worms have been represented by many large groups, some of which are still living and others of which have long since become extinct.

The worms include several independent, but somewhat related, phyla: (1) Platyhelminthes; (2) Nemertinea; (3) Nemathelminthes; (4) Gordiacea; (5) Acanthocephala; (6) Trochelminthes; (7) Chaetognatha; (8) Annelida; and (9) Onychophora.

Platyhelminthes (flatworms). This group includes bilaterally symmetrical, flattened worms that are unsegmented, have no hard supporting skeleton, and are mainly parasitic. The liver-flukes and tapeworms are representatives of this phylum. No fossils of the flatworms have been found.

Nemertinea (ribbonworms). This group is often included in the Platyhelminthes, but various important differences, especially in the nervous system, warrant placing them in a separate phylum. They have soft, flat, unsegmented bodies, and the many known species are mainly marine in habitat. No fossil ribbonworms are known.

Nemathelminthes (roundworms). To this phylum belong the very small hair-like worms, which are elongated, are pointed at both ends, have no supporting skeleton, and are mainly internal parasites.

Gordiacea (horsehair worms). These are free-living and parasitic worms with a long, slender, thread-like body like that of the roundworms (Nemathelminthes). No fossil horsehair worms are known.

Acanthocephala (spiny-headed worms). These parasitic worms have a proboscis bearing rows of recurved spines. No fossils of them have been found.

Trochelminthes (wheelworms). These interesting worms have bands of small cilia on the anterior apex or on their body surface, which, by

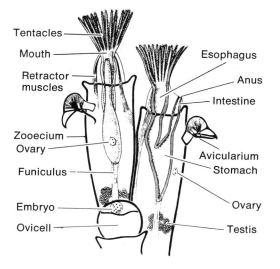

Tentacles
Mouth
Retractor muscles
Zooecium
Ovary
Funiculus
Embryo
Ovicell

Esophagus
Anus
Intestine
Avicularium
Stomach
Ovary
Testis

FIGURE A-9

Anatomy of two Holocene individual bryozoans (*Bugula*), highly magnified. (After *Treatise on Invertebrate Paleontology*, Geol. Soc. Am. and Univ. of Kansas Press, 1953.)

their vibration, give the animal the appearance of a vibrating wheel. They have no hard skeleton and hence have left no fossils.

Chaetognatha (arrow-worms). Torpedo-shaped, planktonic, marine worms with bristle-like jaws about the mouth. They are unsegmented and lack cilia. One of the fossils in the famous Middle Cambrian Burgess shale fauna is thought to belong to this group. No other fossils of this group have been recorded.

Annelida. The Annelida comprise the more highly developed, or segmented, worms. They may be either active or sedentary and may swim about or live in burrows in the soil or in the mud and sand at the bottom of water. The familiar earthworms and leeches and most fossil worms belong to this phylum.

Phylum Bryozoa

These tiny invertebrates, averaging less than one millimeter in length, are aquatic colonial animals of which the great majority live in the sea and only a few kinds inhabit fresh waters. Most colonies are an inch or so in diameter but in some forms may reach a foot. Bryozoans are found in present seas in all latitudes and at depths down to at least 18,000 feet. They are most abundant in shallow seas of temperate and tropical zones. The colonies grow on the sea

bottom attached to rocks or the shells of other invertebrates, and they prefer clear rather than muddy water. Their food consists almost exclusively of microscopic floating organisms, such as diatoms and radiolarians.

The individual bryozoan (zooid) is considerably more advanced in body organization than the coelenterates, which it resembles in having tentacles and a tubular or sac-like body (Fig. A-9). The tentacles are joined at their base to form a ring round the mouth and at times also round the anus. Suspended in the hollow body is a U-shaped digestive tube opening at the surface in the mouth and anus. Part of this tube forms the esophagus, part of it forms the stomach, and part of it forms the intestines. Surrounding the tube is a liquid-filled body cavity, which contains reproductive elements. Several sets of muscles serve for protrusion and retraction of the animal from the membranous or calcareous skeleton. In some of the more highly developed bryozoans, the zooecium—the tube in which the animal lives—bears peculiar processes called avicularia, which look like birds' heads. Each avicularium is attached to the wall of the zo-oecium by a flexible neck and has movable jaws that are used for seizing passing organisms.

The skeleton, an external one, is similar to that built by the majority of corals, and is secreted intracellularly, or within the cells, by the fleshy wall that lines the zooecium. Many zooecia grow together to form a colony called a zoarium. The skeletal material may be chitinous, calcareous, membranous, or gelatinous. The zooecia may be box-shaped, cylindrical, or prismatic, and they may be closely packed or separated by extraneous material.

The zoarium may be encrusting, branching, tree-like, fan-like, or just an irregular, nodular mass. An entire geological formation may consist of little else than the fragments of zoaria.

The zooecia, which house the zooids, are often

divided into successive chambers, either irregularly or regularly shaped, by partial or complete, curved or straight, transverse partitions. The partial, curved partitions are called cystiphragms; the straight, complete partitions are called diaphragms. Very minute, tabulated tubes between zooecia, called mesopores, are mere interspaces between the adjacent zooecia, and their chief functions are to fill up the space and by means of their cross partitions to support the walls of the surrounding zooecia.

The main divisions of the Bryozoa are as follows:

CLASS PHYLACTOLAEMATA

These are fresh-water bryozoans having a horseshoe-shaped loop of tentacles round the mouth, which is protected by an overhanging lip. They are Recent and are not known as fossils.

CLASS GYMNOLAEMATA

These almost exclusively marine bryozoans have a circular row of tentacles round the mouth; they range from the Ordovician to the Recent.

Order Ctenostomata. This order includes the simplest and earliest known forms. The skeleton has the form of a simple tube with small holes, in which the zooids lived, being connected by living tissue that filled the tube (Fig. A-10A). In some skeletons the tubular forms unite laterally to form sheets. The walls of the tubes are usually quite soft and uncalcified. The fossil record is rather poor. The name, meaning comb-mouth, comes from processes resembling the teeth of a comb that close the aperture when the tentacles are retracted. The order ranges from the Ordovician to the Recent.

Order Cyclostomata. The cyclostomatous bryozoans have simple zooecia with lidless circular apertures (Fig. A-10B). The name means circular mouth. Since the skeletal material is calcareous, they have left a fossil record. They made their appearance in the Middle Ordovician and are still living, but never at any time did they become an important group.

Order Trepostomata. This is a totally extinct Paleozoic group in which the zoaria are composed of long, tabulated, prismatic or cylindrical zooecia, which are usually packed together more or less tightly (Fig. A-10C). Cystiphragms and diaphragms are present and often numerous. The order first appeared in the Ordovician and had become extinct by the end of the Paleozoic.

Order Cryptostomata. In this order the zoaria are branching, bush-like, fan-shaped, dagger-shaped, or cone-shaped (Fig. A-10D). The true aperture to the zooecium is hidden, being located at the bottom of a vestibular shaft that is surrounded by either vesicular tissue or solid calcareous material. This order made its appearance in the Early Ordovician, reached its apex in the Devonian and Mississippian, when the lace-like zoaria are present in almost every kind of rock, and had become extinct by the end of the Paleozoic. The name means hidden mouth.

Order Cheilostomata. The cheilostomatous bryozoans are the modern and most highly specialized forms. The zooecia are shaped like urns or irregular boxes and are usually placed side by side to form a compact zoarium (Fig. A-10E). The aperture is smaller than the zooecium and has a lid, or operculum, which is opened and closed by a compensating sac located in the interior of the zooecium. The order first appeared in the Jurassic and is the predominant form today. The name means rimmed mouth.

Phylum Brachiopoda

One group of invertebrates that is of special interest to the historical geologist is the brachiopods, which are exclusively marine animals. They made their first appearance in the Cambrian and are still to be found, though several of the larger divisions have long been extinct. The phylum developed rapidly during the Cambrian, and by the Ordovician all the important subdivisions were represented. The brachiopods were very important members of the Paleozoic invertebrate faunas, but since that time they have gradually been declining.

The brachiopod lives inside a bivalved shell composed of chitin, phosphatic material, or calcareous material. The valves open along a hinge line, similar to that of the clams, and when

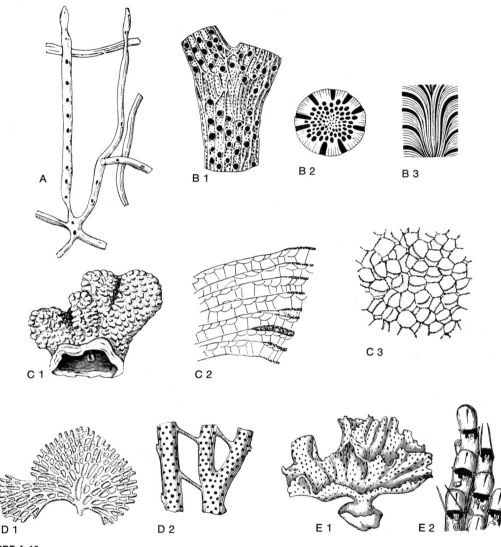

FIGURE A-10

Representative bryozoans.

A. *Vinella*, class Ctenostomata.
B. *Petalopora*, class Cyclostomata (1, surface; 2, transverse cross-section; 3, longitudinal section).
C. *Monticulipora*, class Trepostomata (1, massive zoarium; 2, longitudinal section; 3, transverse section).
D. *Polypora*, class Cryptostomata.
E. *Sertella*, class Cheilostomata (1, massive zoarium; 2, enlargement of individual zooecium).

(From Vol. G of *Treatise on Invertebrate Paleontology*, Geol. Soc. Am. and Univ. of Kansas Press.)

closed effectively protect the animal on the inside (Fig. A-11). The valves are designated as ventral (pedicle) and dorsal (brachial), depending on the relation they bear to the ventral and dorsal parts of the animal's body and to the various internal organs. Certain important animal structures are fastened to the inside of the valves. Among these are the various muscles, which are used in the opening and closing of the shell, and the pedicle, which is a fleshy stalk protruding from between the two valves and serving to attach the animal to the bottom.

The inside of the shell is lined with a thin membranous mantle, which secretes the shell material and whose surface features may leave permanent structures on the inside surface of or in the valves. The cavity between the two valves is divided into a small posterior and a large anterior region by a transverse membrane. Lying in the posterior cavity (the end by which the animal is attached) is the actual body of the animal. This body consists of heart, kidney, alimentary tract, and digestive glands. The alimentary tract opens into the anterior (mantle) cavity by the mouth.

In the mantle cavity are two ciliated, horseshoe-shaped processes called lophophores, which arise from the sides of the mouth and extend toward the anterior margin of the shell. In some forms the lophophores are coiled into a spiral, which can be unwound when the animal feeds. In the most complex group of brachiopods the lophophores are supported by slender axes of calcium carbonate called brachidia, which range from two simple rods, extending marginally from the beak of the dorsal valve, to complicated, spirally coiled supports.

The valves of brachiopods are opened and closed by muscles. In the very simple forms, which have no definite hinge line, the valves are free to move in almost any direction in relation to each other, and a very complicated set of muscles is present to shift the valves. In the more advanced types, however, there are a hinge line and hinge teeth that lock the valves; the valves can be opened and closed, but they cannot be slipped about as in the simple forms, and hence the muscular system is simpler.

In the forms that have no hinge line or hinge teeth (class Inarticulata), the two valves may slide lengthwise or sidewise or may by the swell-

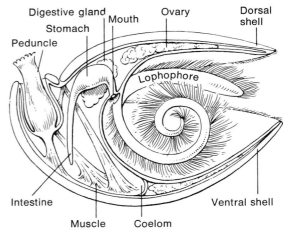

FIGURE A-11

Anatomy of the brachiopod. (Redrawn from Ralph Buchsbaum, *Animals Without Backbones*, Univ. of Chicago Press, Chicago, 1938.)

FIGURE A-12

Diagram of the muscular system of an articulate brachiopod. (Redrawn from W. H. Twenhofel and R. R. Shrock, *Principles of Invertebrate Paleontology*, McGraw-Hill, New York, 1935.)

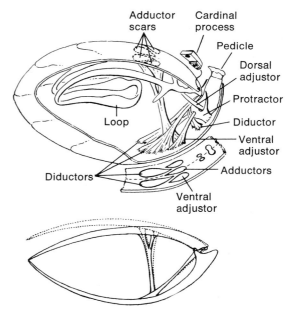

ing of certain muscles be pushed directly apart. In the forms that have a definite hinge line with teeth along it (class Articulata), the valves are pulled apart and together by the contraction of two different sets of muscles (Fig. A-12). The diductors open the shell; the adductors close it. The muscles are either stretched or contracted when the valves are not in contact and are relaxed when the shell is closed. This situation is the reverse of that found in the pelecypods. For this reason brachiopod shells are usually found complete: the relaxation coming from death causes the shell to close and remain so.

The place of attachment of the muscles to the inside of the valves is marked by a slightly depressed or elevated area called a muscle scar. In the articulate forms there are muscle scars on each valve. The diductors are attached to the inside of the ventral valve and to a process, called the cardinal process, at the beak of the dorsal valve. The adductors are two single muscles where they are attached to the ventral valve, but each of these splits before becoming attached to the dorsal valve. Hence there are four adductor scars on the dorsal valve and two adductor and two diductor scars on the ventral valve. In addition there is a series of muscles that provide for movement on the pedicle (the adjustors and protractor of Fig. A-12).

The shell of a brachiopod consists of three parts: (1) a convex ventral valve fastened to the ventral, or abdominal, part of the animal; (2) a convex dorsal valve fastened to the dorsal side of the animal; and (3) a plate, called the deltidium, that surrounds the pedicle and partly or completely closes the pedicle opening. The deltidium has been lost from most shells. The two valves of a brachiopod are equilateral but are of different sizes and shapes. The ventral valve is almost always larger and is usually more convex, and its pointed part, or beak, projects over the rounded beak of the dorsal valve. The brachiopod shell is bilaterally symmetrical, the plane of symmetry passing through the beaks of the valves and the anterior margin.

Most inarticulate brachiopods have chitinophosphatic hard parts — composed mainly of calcium phosphate and chitinous organic matter. The hard parts of all articulate brachiopods are composed almost exclusively of calcium carbonate. Calcareous shells consist of an outer, finely laminated layer and an inner, thicker layer of obliquely inclined prisms or fibers. Three major groups of calcareous shells are recognized: the punctate, the pseudo-punctate, and the impunctate. The punctate shells have fine perforating tubes from the interior of the shell almost to the exterior. The pseudo-punctate shells lack pores, but the inner fibrous layer has rod-like bodies of calcite that commonly form projections on the shell interior. In the impunctate shell the layers are solid.

Primitive brachiopods ordinarily have smooth shells, but the more advanced types have the exterior surfaces wrinkled or corrugated in several different ways. Representative Paleozoic brachiopods are illustrated in Figure 7-7 (p. 229). Concentric growth lines formed by successive additions of shell matter along the shell margin are usually present. Radial striations or plications are small, but large ridges extend radially outward from the older portion (the beak) of the shell. The surface of the shell may present a reticulated appearance if both growth lines and striations or plications are present. Some brachiopod shells, in addition to having radial plications or striations, have along the plane of symmetry a very large fold on one valve and a corresponding depression, or sinus, on the other valve.

The pointed part of a valve is called the beak (Fig. A-13). Between the beak and the hinge line on each valve there may be a curved or flat sur-

FIGURE A-13

External features of a brachiopod shell. (After H. W. Shimer, 1914.)

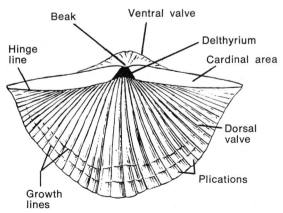
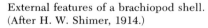

face, called the cardinal area, which is usually more prominent on the ventral valve. The pedicle opening, through which the pedicle emerges, may be shared by both valves or may be limited to the ventral valve; it is never limited to the dorsal valve alone. If this pedicle opening is triangular, it is called a delthyrium. In forms that cement the ventral valve to some object on the bottom the pedicle opening will be grown shut and lost. In still other forms, especially the more advanced brachiopods, the pedicle opening may be partly or completely closed by one or more plates.

The main divisions of the Brachiopoda are:

CLASS INARTICULATA

The valves are unhinged, lack teeth and sockets, and are generally chitinophosphatic. The range is from the Early Cambrian to the Recent.

CLASS ARTICULATA

The valves are hinged and calcareous and generally bear well-defined teeth and sockets. The range is from the Early Cambrian to the Recent.

Phylum Mollusca

The phylum Mollusca includes a great host of animals that are fundamentally alike in their internal organs but radically different in skeletal structure. Such familiar forms as snails, slugs, clams, oysters, squids, the chambered nautilus, and the octopus belong to the phylum. The shells built by the mollusks range in size from tiny forms not more than one millimeter across to the giant shells, built by the cephalopods of the Ordovician, that reached lengths of fifteen feet and more. The squids differ from most of the mollusks in that the shell is internal. In practically all mollusks the shell is calcareous. The octopus has no shell.

A mollusk is usually elongated in form, having an originally symmetrical organization. The visceral part of the animal consists of the vital organs, such as the heart, the nervous system, and the alimentary tract. The alimentary tract may be a more or less straight tube, locally enlarged to form a stomach, with the mouth at one end and the anus at the other; or, in many gastropods, it

may be twisted into a shape like the number eight (Fig. A-14).

Certain mollusks have a ribbon-like structure in the mouth. This organ, roughly comparable to a tongue, is called a radula and is beset with tiny chitinous teeth. It is used as a rasp or file for burrowing, etc.

The covering of the upper, or dorsal, part of the animal is extended on either side as a free fold round the body, enclosing a space called the mantle cavity. The fold itself is called the mantle and is largely responsible for the secretion of the shell.

The under surface of the mantle within the mantle cavity contains the organs of respiration, called gills, or, in the land-inhabiting, air-breathing forms, is modified into lung-like structures.

In all the mollusks except the pelecypods the front (anterior) part of the body is modified into a head, equipped usually with sensory appendages or organs, and containing the radula. The under (ventral) part of the body is produced into a variously modified organ called the foot, which may function in crawling, creeping, jumping, or swimming.

Amphineura, Scaphopoda, and Cephalopoda are exclusively marine in their distribution. The pelecypods, however, have adapted themselves to brackish and fresh-water environments, in addition to the marine environment, and many gastropods have gone even further, adapting themselves to a terrestrial habitat. The mollusks began to populate fresh waters in the Devonian, and they invaded the land during the Carboniferous. Since the Carboniferous the land population has increased more rapidly and with more diversification than the fresh-water forms, which have evolved more slowly.

If variety of shell form, diversification of structure, and success of adaptation are considered, the Mollusca stand foremost among the invertebrates.

The phylum is divided into five classes, primarily by the nature of the foot and certain other soft parts. The recognized classes are as follows:

> Monoplacophora—monoplacoids
> Amphineura—chitons
> Pelecypoda—Mussels, oysters, etc.
> Scaphopoda—tusk-shells
> Gastropoda—snails, slugs, etc.
> Cephalopoda—squids, octopus, etc.

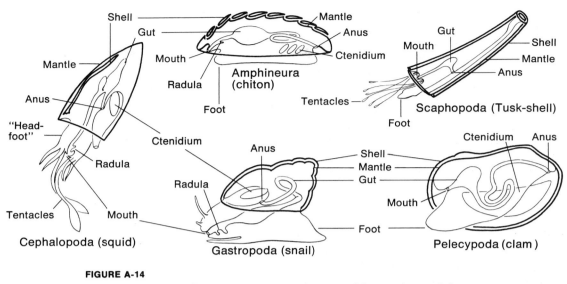

FIGURE A-14

The classes of mollusks, showing the main features of the anatomy and the relation of the shell. (After G. G. Simpson, C. S. Pittendrigh, and L. H. Tiffany, *Life: An Introduction to Biology*, Harcourt, Brace, New York, 1957.)

CLASS MONOPLACOPHORA

This class includes bilaterally symmetrical mollusks with paired gills, muscles, and other anatomical structures and a single shell (Fig. A-15). Two living species of this group are known, both from extremely deep waters. The fossil record is confined to extinct Paleozoic groups which lived in shallow epicontinental seas.

CLASS AMPHINEURA

This class includes the chitons, or "sea mice," and a second division of worm-like creatures that possess no shells. The Amphineura are usually considered the most primitive of the Mollusca. There are several hundred living species, widely distributed and living at all depths. They are slow, sluggish, and small.

The body is quite simple in plan, and the calcareous shell consists of eight plates fastened together along the back of the animal (Fig. A-15). The geologic range is from the Ordovician to the Recent.

CLASS PELECYPODA

Pelecypods possess a calcareous shell composed of two generally equal, more or less convex valves and a ventral, hatchet-shaped foot used for crawling. They lack the head and radula of other molluscan groups (Fig. A-14). Figures A-14 and A-16 show the general characteristics of the body and the various structures of the interior and exterior of one of the valves.

Lying next to the right and left valves are the lobes of the mantle, which, below the line of attachment to the valves, hang downward, curtain-like, beyond the ventral margin of the shell. The ends of these mantle lobes secrete the outside and middle layers of the shell along the margin, and the mantle lobes themselves secrete the inner mother-or-pearl layer.

The pallial line marks the junction of the mantle with the shell. In some forms the posterior parts of the mantle lobes are fused into a single or double tube called the siphon, through which water is taken in and expelled (Fig. A-14). The mantle, when this structure is present, is pulled away from the shell slightly, and an indentation, called the pallial sinus, is formed in the pallial line.

The valves are held together partly by teeth along the hinge line but mainly by several muscular structures. Above the hinge line the two valves are fastened together by a muscular ligament, below the hinge line by one or two internal transverse muscles, the adductor muscles. The valves are pulled together by the contraction of the adductors, and at the same time the ligament is stretched. Hence, when the shell of a pelecypod is closed, both muscles are in a strained condition. Upon the relaxation of the adductors and the ligament, the valves open. This system of muscles is almost exactly the opposite of that found in the brachiopods. Whereas fossil brachiopods are usually found with the shell closed, fossil pelecypods are usually found with the shell open or with the valves completely detached from each other.

The valves of the pelecypod shell are two equal, asymmetrical, convex plates, attached to the right and left sides of the organism, and opening along a more or less straight hinge line. The pointed part of the valve is called the umbo. The numerous other structures that should be noted in the study of pelecypod shells are shown in Figure A-16.

The hinge-line structure of a pelecypod is of paramount importance in classification. The valves of almost all pelecypods are locked along the hinge line by teeth and corresponding sockets. The arrangement and character of these teeth are called the dentition. There are nine or ten different types of dentition, all representing modifications of a simple type developed early in the history of the class.

The shell of the pelecypod is composed of three layers, as shown in Figure A-16: (1) an

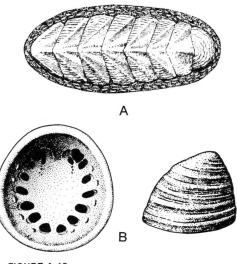

FIGURE A-15

A. Dorsal view of one of the Amphineura
B. Side and ventral view of *Archaeophiala*, an Ordovician monoplacoid.

outer chitinous layer called the periostracum; (2) a middle layer of calcite prisms disposed at right angles to the surfaces of the shell, and called the prismatic layer; and (3) an inner layer of calcite laminae called the laminated layer. The laminated layer is also spoken of as the mother-of-pearl layer.

FIGURE A-16

Interior view (left) and exterior view (center) of a pelecypod shell, showing the principal structures, and transverse section (right) showing the relations of the animal to the shell. (Redrawn from Twenhofel and Shrock, 1935.)

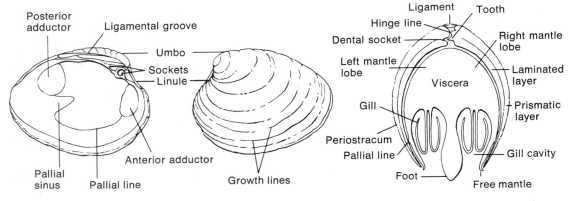

CLASS SCAPHOPODA

Scaphopods are exclusively marine, bilaterally symmetrical mollusks that generally inhabit deep water. Their calcareous shell is a curved, tubular structure, open at both ends and tapering. In life the animal is embedded in the bottom sediment, head down, with the narrow end of the shell projecting into the water (Fig. A-14). Fossil forms are not common.

CLASS GASTROPODA

Gastropods are predominantly aquatic mollusks, but many forms have adapted themselves to a land habitat. The animal's worm-like body lacks the bilateral symmetry of most mollusks because of the twisted shape of certain organs. There is a well-developed head with stalked eyes, and a radula is present in the mouth cavity (Fig. A-14). The ventral part of the animal is modified into a foot, which is used for creeping, jumping, and burrowing. This foot may also take the form of a fin or wing-like structure, either of which is used in swimming. The mantle is never divided into two lobes; it secretes a hollow, conical, calcareous, unchambered shell, which is usually but not always coiled.

The gastropod shell begins with the protoconch (embryonic shell), which consists of several minute whorls or coils. This protoconch is rarely preserved and is quite different from the remainder of the shell.

In the simplest gastropod shell the fundamental conical shape is apparent. When coiling takes place, a shell may be loosely coiled about an imaginary axis, in which case the successive whorls will not be in actual contact, or it may be tightly coiled, with the whorls on the exterior visibly in contact along a definite line. If this tight coiling takes place round an imaginary axis, a solid spiral structure, called the columella, is formed along the axis by the fusing of the inside parts of the whorls, and the shell is an imperforate one. If the coiling takes place about an imaginary hollow cone, however, the shell is said to be perforate, and the opening at its base or at the larger opening of the imaginary cone is called the umbilicus. If the coiling is so tight that the last whorl completely conceals all the preceding whorls, the shell is said to be convolute. Coiling may take place in either a right-handed or a left-handed manner. In determining whether the shell is right-handed or left-handed, point the apex of the shell up and rotate the shell on the axis of coiling until the aperture is visible. If it lies on the left side, the shell is left-handed; if it lies on the right side, the shell is right-handed. Representative gastropod shells are shown in Figure A-17.

The shell of the gastropod is secreted by the mantle, which surrounds the viscera. It is composed largely of calcium carbonate, arranged in two layers of inclined prisms of calcite, with an external periostracum of chitinous horny material.

CLASS CEPHALOPODA

The cephalopods are the most highly developed of all mollusks and are exclusively marine. They constitute a very old group, which appeared first in the Late Cambrian. The animal may have a shell, or it may be naked. The famous chambered nautilus represents the first type, the octopus the second type.

The cephalopod is a bilaterally symmetrical animal and usually has a bilaterally symmetrical shell. There is a well-differentiated head with prominent eyes (Fig. A-14). The mouth is located at the base of a circlet of tentacles, has horny, beak-like jaws with calcified tips, and possesses a radula.

The anterior part of the foot is modified into arms, or tentacles, which surround the mouth and perform a prehensile function. The posterior part of the foot is modified into a leaf-like or tube-shaped process called the hyponome. Through the hyponome water is brought into and expelled from the mantle cavity. A violent expulsion of the water through the hyponome propels the animal forward. The body is fastened to the shell by means of muscles, which may leave scars, and by means of a fleshy tube, called the siphuncle, which extends from the body back through all of the abandoned hollow chambers of the shell to the embryonic shell. Hence, in life, all the chambers of the shell are in communication.

The shell of a cephalopod is a straight or coiled, tapering cone, closed at one end (which is the older and smaller part of the shell), divided by transverse partitions into chambers that are in communication with one another and with

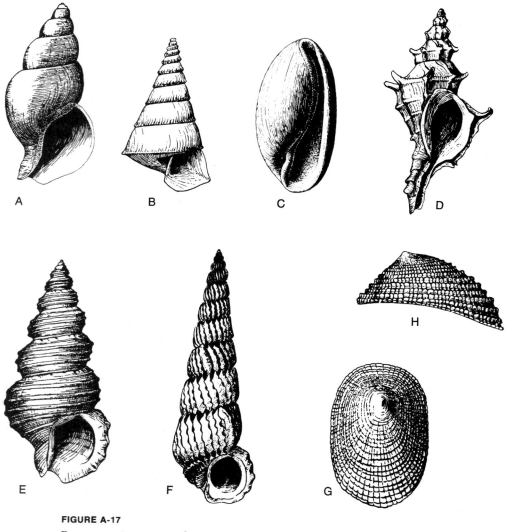

FIGURE A-17

Representative gastropods.

A. *Plicifuses.*	C. *Volvula.*	E. *Buccinum.*	G. *Cocculina.*
B. *Basilissa.*	D. *Neptunea.*	F. *Cirsotrema.*	

the larger living chamber at the large end of the shell, and composed of calcareous material. The shell is composed of an outer layer of imbricated laminae of calcite and an inner layer of mother-of-pearl arranged in thin parallel laminae. The interior of the shell is often iridescent.

The chambered portion of the cephalopod shell is called the phragmocone; the individual chambers are called camerae; the dividing transverse partitions are called septa. The junction of a septum with the internal surface of the shell is marked by a smooth or wrinkled line called the suture. The embryonic shell is called the protoconch, and the last, large chamber, occupied by the animal in life, is called the living chamber (Fig. A-18).

The primitive and simple shells have simple sutures, which, when projected on a plane, are either straight or slightly curved lines; in the complex forms, however, the suture becomes very much wrinkled. The folds of a suture that are convex toward the aperture of the shell are called saddles; the folds convex toward the protoconch are called lobes.

The main subdivisions of the class Cephalopoda are as follows.

SUBCLASS ORTHOCERATOIDEA

This group is regarded as the central cephalopod stock from which all other cephalopods were derived. They are forms with predominantly straight shells and simple septa. Some groups have internal calcareous deposits. The group was especially diverse in the early Paleozoic, then declined and survived into the Triassic.

SUBCLASS ACTINOCERATOIDEA

These are primarily straight-shelled forms, some of which attained a length of twenty feet. They are characterized by voluminous and complex cameral and siphuncular deposits. The group first appears in the Early Ordovician and became extinct in the Pennsylvanian. The greatest evolutionary radiation was in the Middle and Upper Ordovician.

SUBCLASS ENDOCERATOIDEA

This group is primarily a straight form with large siphuncles that contain specialized cone-like deposits. Individuals are known with a length of thirty feet. Greatest radiation of the group was in the Early Ordovician, but the group became extinct during the Silurian.

FIGURE A-18

Diagrammatic ventral (A), cross-sectional (B), and lateral (C) views of a typical ammonoid, *Manticoceras*, and (D) enlarged representation of a suture of the same. The upper parts of A and C portray the exterior of the test and show the growth lines; the lower parts represent the internal mold with the sutures. (Courtesy of A. K. Miller and W. M. Furnish.)

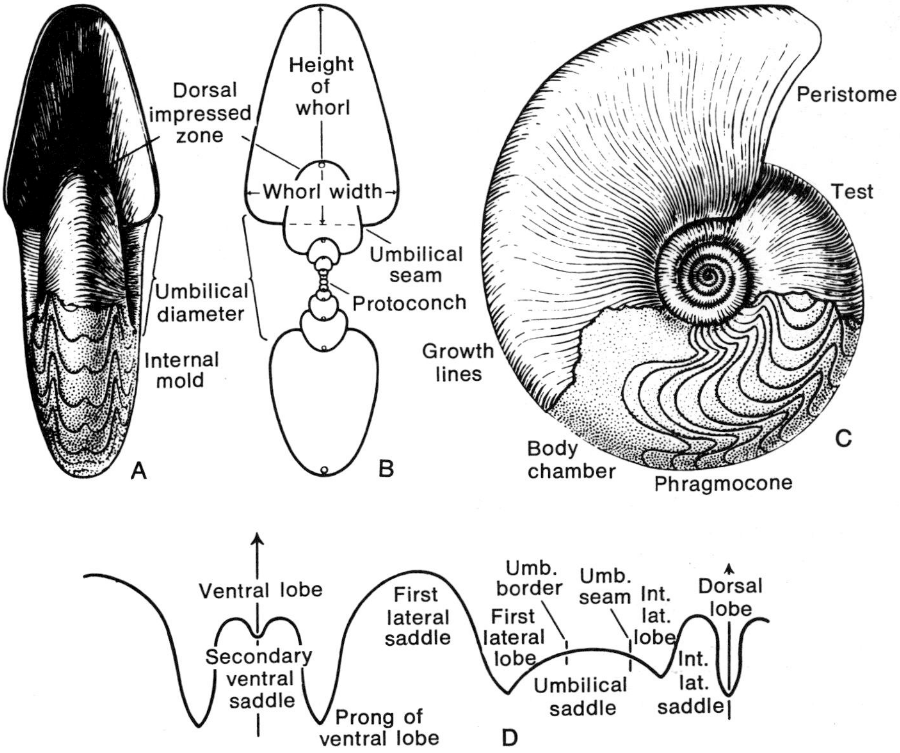

SUBCLASS NAUTILOIDA

These cephalopods have simple or broadly undulating smooth sutures and, generally, centrally located siphuncles. The shells are straight to tightly coiled, and some groups have internal calcareous deposits in addition to septa. The nautiloids were particularly abundant and widespread during the early and middle Paleozoic, after which they continued to decline steadily; only the modern chambered nautilus survives. Representative Paleozoic nautiloids are illustrated in Figures 7-4 (p. 226) and 7-39 (p. 257).

SUBCLASS BACTRITOIDEA

These are generally small, straight-shelled forms with a marginal siphuncle and simple septa. The group is of special importance since they are regarded as the ancestors of the ammonoids as well as the Coleoidea, two large and important groups of cephalopods.

SUBCLASS AMMONOIDEA

These cephalopods generally have complex sutures and small siphuncles in an extreme ventral position. The morphology of a typical early ammonoid is shown in Figure A-18. The ammonoids of the Paleozoic, with some exceptions, have sutures with smooth lobes and saddles, a type called goniatitic (Fig. 7-36, p. 354). Most of the Triassic ammonoids have sutures with smooth saddles and denticulated lobes, a type called ceratitic (Fig. 10-2A, p. 343). Other Triassic ammonoids and most of those of the Jurassic and Cretaceous have sutures in which both the lobes and the saddles are denticulated, a type called ammonitic (Fig. 10-2AB, p. 343).

SUBCLASS COLEOIDEA

All living cephalopods with the exception of the genus *Nautilus* belong to the subclass Coleoidea. These cephalopods have two gills and from eight to ten arms around the mouth. The shell is internal and may or may not be chambered. In some forms the shell is absent altogether. The group ranges from the Mississippian to the Recent. The subclass is divided into six orders, three of which are extinct.

Order Aulacocerida. These are primitive forms that are derived from bactritids. They have a shell with chambered portion, body chamber, and massive calcareous deposits around the chambered portion. This group became extinct in the Jurassic.

Order Belemnitida. The shell here is internal with a chambered portion surrounded by a counterweight. The group originated from the bactritids and was particularly abundant in the Jurassic and Cretaceous.

Order Phragmoteuthida. This order has an internal shell with a short, stumpy chambered portion with a fan-shaped projection. The group originated from the Belemnitida and are known from Triassic and Jurassic formations. It died out during Jurassic time.

Order Teuthida. This group includes the well-known squids. The group is especially well represented in Jurassic and Cretaceous formations.

Order Sepiida. The internal shell consists of a flat shield containing remnants of chambers.

Order Octopoda. Here belong the octopus and related forms that have no shell and eight arms.

Phylum Arthropoda

The phylum Arthropoda (meaning joint-footed) includes the most highly developed of all invertebrate animals. Many familiar animals, both aquatic and terrestrial, belong in this group — pillbugs, centipedes, millipedes, scorpions, spiders, and the great host of insects. An extinct group, the Trilobita, includes a great variety and number of crayfish-like animals that were very important during the early Paleozoic.

Arthropods show a great range in size, from microscopic insects to eurypterids more than five feet long and the giant Japanese crab that can span eleven feet with its claws. They have developed great numbers of species and still greater numbers of individuals, have invaded and conquered almost every conceivable environment, and have had a very long geologic history, beginning almost certainly in the Precambrian and persisting in great numbers to the present time. Throughout most of their history

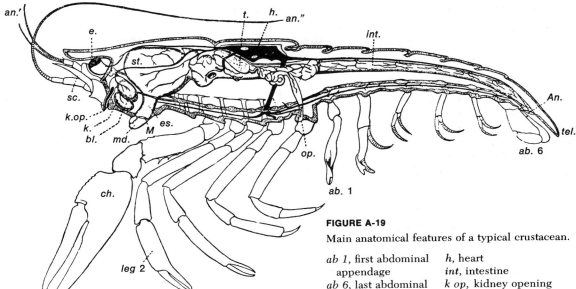

FIGURE A-19

Main anatomical features of a typical crustacean.

ab 1, first abdominal appendage	*h*, heart
ab 6, last abdominal appendage	*int*, intestine
an ', antennule	*k op*, kidney opening
an", antenna	*leg 2*, first walking leg
An, anus	*m*, mouth
bl, bladder	*md*, mandible
ch, chela	*op*, genital opening
e, eye	*sc*, scale of antenna
es, esophagus	*st*, stomach
	t, testis
	tel, telson

(From L. A. Borradaile and F. A. Potts, *The Invertebrates*, Macmillan, New York, 1936.)

the arthropods have been equipped with an external covering of various types and forms, the material of which is and usually has been chitin.

The arthropod is a very complex invertebrate animal, with an elongated, bilaterally symmetrical, transversely segmented body. The body segments may be alike or different, and in some forms several segments may be fused together. The mouth and the anus are at opposite ends of the elongated body (Fig. A-19).

Each typical body segment bears a pair of appendages. Each appendage consists of two or more limb segments separated from one another by movable joints and acted upon by special muscles. These appendages, variously modified, are used for locomotion, respiration, mastication, etc.

The animal possesses two types of breathing organs, gills and tracheae (minute tubes perforating the body wall); it has a well-developed nervous system and well-developed eyes, which are sometimes very complex.

The body is covered by a chitinous dorsal shield called the carapace. This is usually composed of a number of plates, but in some unusual forms it may take the form of a bivalved shell; and some arthropods are naked. Growth takes place only immediately after periods of "molting" or "shedding," for the animal sheds its skeleton periodically, then grows rapidly before building another, larger skeleton. The more

rapid growth is in the youthful stages. In maturity growth practically ceases.

In many of the more advanced arthropods partial fusion of some of the skeletal segments may take place, giving rise to a head, a thorax, and an abdomen. Fusion may even go so far as to produce a cephalothoracic plate and an abdominal plate, both showing but obscurely the original segmentation.

A complete classification of the Arthropoda is given on page 589. Most of the classes are not known or are very poorly known in the fossil record. We shall summarize briefly the characteristics of only the more important groups represented in the fossil record.

CLASS TRILOBITOIDEA

Trilobite-like arthropods are known primarily from the Middle Cambrian Burgess Shale. Representative forms are illustrated in Figure 2-5EF, p. 28.

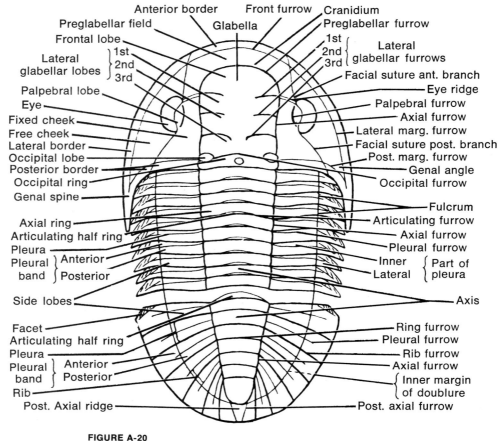

FIGURE A-20

Diagram showing the main morphological features of a trilobite.
(After E. Warburg.)

CLASS TRILOBITA

These extinct marine crustaceans had a variable number of body segments. The body was covered by a hard dorsal shield, composed of chitin and divisible longitudinally into three lobes (hence the name), a central ridge and two side lobes. The shield is further divided into a front portion, called the head or cephalon; a middle, thoracic portion; and a hind portion, called the pygidium. The animal also possessed a large number of paired appendages, which are rarely preserved. Eyes were well developed in many forms (Fig. A-20).

The cephalon is covered ordinarily with three plates, a central plate called the cranidium and two lateral plates called free cheeks. The back

part of a free cheek may be drawn out into a genal spine. The free cheeks carry the eyes and are separated from the cranidium by the facial sutures. The lateral parts of the cranidium are called fixed cheeks; the central ridge, which may or may not be lobed, is called the glabella.

Specimens preserving appendages are very rare. The cephalon has five pairs of appendages, and the thorax and pygidium have one pair of biramous appendages on each segment.

The trilobites dominated the organic world during the Cambrian, were abundant but subordinate during the Ordovician and Silurian, and began a rapid decline in the Devonian. They were rare during the Carboniferous and became extinct during the Permian. Representative trilobites are illustrated in Figure 7-2, p. 224;

habitat reconstructions are shown in Figures 7-1, 7-3, 7-4, (pp. 222, 225, 226), and 7-35 (p. 253).

CLASS MEROSTOMATA

This is an ancient race of marine arachnoids, which made its appearance in the Cambrian, flourished during the middle Paleozoic, and declined through the Mesozoic to its present single surviving genus, the familiar horseshoe crab, *Limulus* (Fig. 2-14, p. 32). An extinct Paleozoic group, the eurypterids, included some of the largest arthropods known, up to ten feet in length (Fig. 7-12, p. 233). The eurypterids were especially common in the Silurian and Devonian.

CLASS ARACHNIDA

This group includes the air-breathing forms, such as the spiders, scorpions, mites, and ticks. The fossil record is very sparse.

CLASS OSTRACODA

Ostracods are small, lentil-shaped crustaceans having a bivalved carapace that encloses an indistinctly segmented body. They are aquatic, mostly marine animals. The carapace is chitinous or calcareous and consists of two valves, a right and a left, which are articulated along the dorsal edge. Ostracods are widely used for stratigraphic correlation.

CLASS CIRRIPEDIA

These, the barnacles, are sessile, mostly hermaphroditic animals, enclosed in a membranous mantle, which is often covered with calcareous plates.

CLASS MALACOSTRACA

This group includes the well-known crabs, crayfishes, lobsters, and shrimps. The fossil record, however, is not extensive.

CLASS INSECTA

This group includes a larger number of species than any other class of animal. The fossil record is rather good but is of little use in stratigraphic studies.

Phylum Echinodermata

The Echinodermata are exclusively marine animals, which, because of their excellent preservation and fairly abundant occurrence in the geologic record, have been studied extensively and comprise one of the more important and useful phyla of fossil organisms. Starfish, sea cucumbers, sea urchins, and sea lilies are representative of extant groups of Echinodermata. During the geologic past, however, there were other groups, now long extinct, whose existence and characteristics are known only from the fossil record.

The body may be spherical, pear-shaped, or star-shaped, and it has two openings: a mouth and an anus. It shows a considerable advance over the Coelenterata in the introduction of a true body cavity with a stomach, an intestinal tract separating the mouth from the anus, and a distinctly localized nervous system.

The earlier echinoderms almost invariably exhibit radial (pentamerous) symmetry, but later forms usually show bilateral symmetry. The arrangement and position of the mouth and anus, as well as the various structures on the test, change with alteration of the symmetry from radial to bilateral. Sessile forms usually have radial symmetry whereas forms that move about actively gradually change to bilateral symmetry, the mouth migrating toward the front of the animal and the anus migrating toward the rear. In the bilaterally symmetrical animals the plane of symmetry passes through the mouth and anus.

Peculiar to the Echinodermata is the so-called water-vascular system, which performs the function of locomotion. In the simplest type of water-vascular system, illustrated by that of a starfish (Fig. A-21), water is taken in through a sieve plate, called the madreporite, at the top of a long stony canal that leads down from the top of the animal and opens directly into a circular hollow that communicates with the five radial tubes that lie in the five arms. These arms are called ambulacra, and in forms that do not have free arms the areas corresponding to them are called ambulacral areas. The areas between the ambulacra are called interambulacral areas. The water is now injected into one or more of the ambulacra. In many forms the surface of the ambulacra is covered with minute bag-like structures, called tube feet, into which the water may be pumped. These minute feet stick fast to the bottom; as they are contracted and expanded by the water and as their position is changed by muscles, the animal slowly moves in the desired direction.

In many attached forms of Echinodermata a current of water is kept in motion along the ambulacral areas by various structures, and in this way food is brought to the mouth.

The test consists of calcareous plates covered by a leathery or fleshy skin and is usually a hollow, pear-shaped, spherical, or star-shaped structure. It is composed, in most forms, of a series of calcareous plates arranged in either a symmetrical or an asymmetrical manner. In some forms the plates are held together by muscles and fall apart when the animal dies; in other forms, however, the plates are firmly cemented together, and the test has a rigid, box-like structure that is capable of retaining its form more or less perfectly upon burial. The plates, upon fossilization, behave mineralogically and optically as calcite crystals and always exhibit typical calcite cleavage. In this important respect skeletal elements of the Echinodermata differ from those of all other phyla.

In addition to the main, box-like test there may be segmented, movable appendages composed of calcareous plates; segmented stems and roots, of similar material, for attachment; calcareous spines ornamenting the surface; hard, resistant teeth; and, in a few forms without a definite hollow test, spicules that become detached when the animal dies.

Without exception the Echinodermata are marine animals, living in depths varying from the tide mark to more than three miles, and in all latitudes. Consequently an abundance of echinoderm fragments in a geological formation stamps it as marine in origin. Most echinoderms are gregarious, and their remains may therefore be found in great abundance in small areas and may be missing everywhere between. Fragments of echinoderms are always found associated with the corals in the ancient coral reefs of the Silurian and Devonian, and many echinoderms may be seen on present reefs.

The Echinodermata are of considerable importance in the alteration of sediments, for some of them can chew up hard rock and even affect it chemically when it passes through their bodies. They can cut the hardest rock with their jaws or teeth, and they are sometimes found in holes that

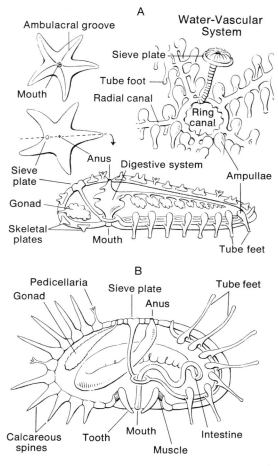

FIGURE A-21

Diagrams showing the main anatomical features of a starfish (A) and a sea urchin (B). (After G. G. Simpson, C. S. Pittendrigh, and L. H. Tiffany, *Life: An Introduction to Biology*, Harcourt, Brace, New York, 1957.)

they have cut into rock. They are also important in present seas as scavengers, and they undoubtedly carried on similar activities in ancient seas.

Subphylum Homalozoa

This group has primitive echinoderms with an asymmetrical plated, flattened skeleton and generally with a tail-like appendage. The margin of the skeleton is made up of large calcareous plates enclosing an inner area with smaller plates. The group first appeared in the Middle Cambrian and became extinct in the Middle Devonian. This group is generally known as "carpoids."

CLASS HOMOSTELEA

This group has an anterior margin with a concave food groove leading to the mouth. The tail-like appendage is gradational with the skeleton. The few known representatives of this class are of Middle Cambrian age.

CLASS STYLOPHORA

These are asymmetrical flattened echinoderms with a distinctive jointed, food-gathering appendage. A typical form is illustrated by *Ceratocystis* in Figure 7-9K, p. 000. The group ranges from the Middle Cambrian to the Middle Devonian.

CLASS HOMOIOSTELEA

The skeleton is made up of numerous small plates, and there is an arm-like appendage at one end of the body and a tail-like appendage at the other end. The class ranges from the Upper Cambrian to the Lower Devonian.

CLASS CTENTOCYSTOIDEA

This group differs from the above classes in having no appendages, near bilateral symmetry, and a very distinctive feeding apparatus consisting of a grill-like array of movable, blade-like plates. This group is known only from the Middle Cambrian strata of northern Utah. A typical form is illustrated by *Ctenocystis* in Figure 7-9G, p. 231.

Subphylum Crinozoa

This group has a globular plated skeleton with an attachment stem or appendage; their feeding apparatus consists of movable pinnule-bearing arms or brachioles. The group ranges from the Early Cambrian to the Recent.

CLASS EOCRINOIDEA

These are primitive, extinct echinoderms consisting of a globular body made up of numerous pore-bearing, calcareous plates. The body is attached to the sea bottom by a hollow column or stem that served for temporary or permanent fixation of the organism. The oral side of the globular body has a ring of slender food-gathering brachioles. A typical representative of this group is *Gogia* illustrated in Figure 7-9C, p. 231.

CLASS CYSTOIDEA

The cystoids were stemmed or stemless echinoderms with a globular, cylindrical, or flattened calyx consisting of calcareous plates more or less irregularly arranged. The pentamerous symmetry that characterizes other classes is often not well developed here. The number of calcareous plates is highly variable and may be as high as 200. Many calyx plates are perforated by tiny tubes called thecal pores or rhombs, which were probably used for respiration. These distinct types of respiratory structures are used as a basis of classification into subclasses.

The mouth is located near the center of the upper surface of the calyx, and the anus is usually lateral. The ambulacral areas, or food grooves, usually radiate from the mouth and bear finger-like appendages. A typical cystoid, *Pleurocystites*, is illustrated in Figure 7-9B, p. 231. The cystoids range from the Early Ordovician to the Late Devonian and are especially abundant in the Ordovician and Silurian.

CLASS PARABLASTOIDEA

These are blastoid-like, stemmed echinoderms but with more numerous plates. The deltoids are large triangular plates bearing respiratory structures. This group is confined to the Middle Ordovician.

CLASS BLASTOIDEA

The test of the blastoid very closely resembles a flower bud. The rigid, radially-symmetrical calyx was composed of thirteen plates arranged in three circles (see *Pyramiblastus*, Fig. 7-9D, p. 231). The three basal plates were attached to the stem. The next circle of plates consisted of large forked radials that enclosed the ambulacral areas. Above these were five small diamond-shaped plates, called deltoids. Beneath the ambulacral areas was a series of complex, pleated folds of thin calcite, called hydrospires, which served to aerate the interior body liquids: water entered the hydrospires through small pores along the edge of the ambulacral areas, flowed upward, and left the body at small openings around the summit, called spiracles.

Each ambulacral area was lined on each side with numerous, thin, plated, food-gathering appendages called brachioles, which connected with the food grooves on the ambulacral area. The Blastoidea ranged from the Middle Silurian to the Late Permian. The acme of their development was in the Mississippian.

CLASS PARACRINOIDEA

This is a poorly known, extinct group of echinoderms known only from the Middle Ordovician. They are similar to some stalked groups of echinoderms but differ in several significant features. They differ from the blastoids, cystoids, and eocrinoids by the uniserial nature of their ambulacral areas, as well as by the presence of pinnules. They differ from the crinoids in having a different type of skeletal organization with irregularly arranged plates.

CLASS CRINOIDEA

This group includes the only surviving members of the stalked echinoderms. The entire skeleton is composed of three important structural elements (Fig. 7-9A, p. 231): (1) the stem and root system, which attaches the animal to the sea bottom: (2) the cup-shaped calyx, made up of regularly arranged plates and housing the animal body; and (3) the five (or multiples of five) movable arms, which can be regarded as extensions of the body and serve the important function of food gathering. The arms have thin side branches called pinnules.

The regular arrangement of the plates in crinoids is characteristic. The fact that each basic type of plate has a special name leads to a very complex nomenclature. The class is divided into four important orders according to the relations and arrangements of the plates that comprise the calyx.

Crinoids first appear in rock formations of the Early Ordovician age and range to the Recent. The acme of their development was in the Mississippian. A reconstruction of a Mississippian sea bottom densely populated by crinoids is shown in Figure 7-38, p. 256.

Subphylum Asterozoa

These are echinoderms that are characterized by a generally depressed star-shaped body, composed of a central disc with a mouth on the underside and symmetrical radiating arms; the axial skeleton along the arms protects the radial water vessels and nerves; tube feet are normally confined to lower side of body. This group includes the starfish and brittle stars and ranges from the Early Ordovician to the Recent.

CLASS STELLEROIDEA

The Stelleroidea have a star-shaped test and body, consisting of a central disk and five or more arms (Fig. A-21). The mouth is in the center of the under side. The anus is a small hole in the middle of the upper side of the central disk. Starfishes, like other echinoderms, have a water-vascular system in which the radial tubes extend from near the mouth to the tips of the arms and bear the numerous branches known as tube feet. The food grooves lie on the under side of the arms, between the rows of tube feet, and radiate from the mouth. The skeleton is not a rigid box but is composed of many loosely joined plates in a leathery skin. For this reason well-preserved complete fossil starfish are very rare.

There are three recognized subclasses: (1) the Somasteroidia, a primitive group known only from the Lower Ordovician to the Late Devonian formations; (2) the Asteroidia, the true starfish, characterized by hollow arms, first appeared in the Ordovician and is an important component of modern marine faunas; and (3) the Ophiuroidia, which have slender, whip-like arms extending from a circular body chamber. The ophiuroids range from the Ordovician to the Recent.

Subphylum Echinozoa

These are echinoderms fundamentally globoid in shape, lacking arms or the outspread rays which characterize most crinozoans and asterozoans. The group is present in the lowest Cambrian strata and is represented today by the Echinoidea and Holothuroidea.

CLASS HELICOPLACOIDEA

These are free-living, generally fusiform echnoderms with a spirally pleated, expansible, and flexible test (Fig. 7-9F, p. 231). They are known from several genera in the Lower Cambrian strata of California and western Nevada.

CLASS EDRIOASTEROIDEA

The edrioasteroids range from the Cambrian to the Pennsylvanian. The calyx, consisting of an indefinite number of irregular plates, is discoid in shape and apparently was attached directly to the sea bottom, for there is no stem. Its upper surface bears straight or curved ambulacral areas, with no other plated appendages. A typical edrioasteroid is *Carneyella*, illustrated in Figure 7-9H, p. 231.

CLASS EDRIOBLASTOIDEA

These are globular, primitive echinoderms superficially resembling blastoids but more probably closely related to the edrioasteroids. There are five ambulacra bearing pores for soft food-gathering appendages. This group is known only from the Middle Ordovician.

CLASS OPHIOCISTOIDEA

This group consists of unattached pentaradiate echinoderms with a more or less depressed dome-shaped body entirely covered by a plated test or with a cover of plates on one side only and lacking arms or comparable projections (Fig. 7-9J, p. 231). The ventral side has a mouth apparatus consisting of five interradially disposed jaws. This is a small, relatively rare group known to range from the Lower Ordovician to the Upper Silurian and possibly into the Middle Devonian.

CLASS CYCLOCYSTOIDEA

This is a poorly known and rare group of Paleozoic echinoderms. The fossil record consists mainly of well-preserved rings of submarginal plates. These rings enclose disc-shaped areas in which the upper and lower covering plates were poorly calcified. The group ranges from the Middle Ordovician to the Middle Devonian.

CLASS CAMPTOSTROMOIDEA

These are free-living, apparently radially symmetrical echinoderms with a heavily plated, dome-shaped body with more than one layer and radial plated arms attached to the periphery of the body. It is an extremely rare group known only from the Early Cambrian of Pennsylvania.

CLASS ECHINOIDEA

The Echinoidea, a group of armless and stemless echinoderms, have a globular or discoid body encased in a calcareous test (corona), which is composed of a large number of plates that form a rigid structure (Fig. A-21). The corona consists of five ambulacral and five interambulacral areas, each made up of two or more rows of plates. The mouth is on the under side or, in bilaterally symmetrical forms, toward the front. The anus is on the upper side or, in bilaterally symmetrical forms, toward the rear. Tube feet line the ambulacral areas and are connected with a water-vascular system that has an outlet on the upper surface near the anus. Many echinoids have a unique masticatory apparatus, known as Aristotle's lantern, composed of forty calcareous pieces, of which five are powerful teeth arranged in a circlet round the mouth. The teeth are independently moved by means of powerful muscles and are capable of cutting many hard materials, including hard rocks like quartzite and granite.

The first echinoids appeared in Middle Ordovician formations, but it was not until the late Mesozoic and Cenozoic that they became abundant and important parts of the fossil record.

CLASS HOLOTHUROIDEA

The Holothuroidea differ greatly from the typical echinoderm. They have an elongated, cucumber-like body, with contractile tentacles surrounding the mouth. The surface of the animal is prickly and has the appearance of a cucumber, from which comes the common name, "sea cucumber." There is no test, but scattered throughout the animal's body are small, detached calcareous particles, usually microscopic in size. These rarely join to form a rigid body wall. The class appeared in the Cambrian and is living today. It has left no important fossil record.

Phylum Graptolithina

The graptolites are an extinct, marine, colonial group of animals with a chitinoid skeletal structure. There is much uncertainty as to the biological affinities of this group, and in recent years it has been argued that they belong to a subphylum of the Chordata. Since there is still much debate on this problem, the Graptolithina are here placed in their own phylum.

In fossilization the individuals are usually preserved as flattened films of carbon, resembling pencil marks; thus the name "graptolite" (*grapto*, write; *lite*, stone). Most graptolites were planktonic organisms, either attached to some floating object such as seaweed or with their own float apparatus (Fig. 7-10, p. 232). Many of the genera have world-wide distribution; this, allied to the fact that the phylum evolved very rapidly, makes the graptolites exceptionally important index fossils, especially for the Ordovician and Silurian.

There are two major classes of graptolites, the Dendroidea and the Graptoloidea. The dendroid graptolites derive their name from the many-branched tree-like mode of growth of the colony (which is called a rhabdosome). The individual branches of this tree-like structure are called stipes, and they diverge from an apex. Short cross structures connecting the stipes and strengthening the colony are called dissepiments. The stipes are made up of a series of overlapping tubes or cups, called thecae, in which the zooid (the animal) lived. At the apex of the stipes is a single cup (sicula) belonging to the initial zooid. The sicula is attached to a thread-like tube called the nema, which is the attachment structure for the whole colony.

The colony was produced asexually by budding. The first zooid secreted the first cup, the sicula. Successive budding built up the stipes and the whole rhabdosome.

The earliest known dendroids are Late Cambrian in age, and the latest are Mississippian. The dendroids do not compare with the graptoloids in geographic distribution or stratigraphic importance.

The graptoloids are essentially simplified dendroids. The earlier graptoloids had a fairly large number of stipes, but in the history of this group there was a marked tendency toward a reduction in the number of stipes (see p. 232 and Fig. 7-11). A change in the orientation of the stipes also took place. In the earliest forms the stipes simply hung below the nema and sicula, but gradually they became nearly perpendicular to the nema, and finally, coming all the way round, they became parallel to and attached to the nema. This change in the organization of the rhabdosome resulted in a great reduction in the number of thecae. Whereas in the dendroids the thecae are simple cups or tubes, among the graptoloids a great elaboration of thecal types is found.

The graptoloids are among the very best guide fossils for the Ordovician and Silurian. They are characteristically found in black shale facies and only occasionally in other rock types.

Phylum Chordata

FISHES[*]

Four classes of vertebrates are called fishes: (1) the class Agnatha includes the jawless fish represented today by the lampreys and hagfishes; (2) the class Placodermi includes the early jawed fish that became extinct at the end of the Paleozoic; (3) the class Chondrichthyes includes the sharks, rays, and skates; and (4) the class Osteichthyes includes the bony fishes. The Agnatha, Placodermi, Chondrichthyes, and certain groups of the Osteichthyes play an extremely important role in the evolution of Paleozoic life and are discussed in detail in §§ 7-2, 7-3, and 7-5.

The great majority of modern fishes belong to the group of bony fishes. The greater part of the body of a bony fish is made of muscles by which the animal swims. In most species the fins (with the exception of the caudal fin) serve only for steering. Ascent and descent in the water are aided by a compressible swim bladder, which is filled with air (Fig. A-22). Exchange of respiratory gases is effected by the gills, which are so constructed as to present a great amount of surface to the water that moves across them. The blood is pumped round the body by a simple two-chambered heart (one auricle, one ventricle). The gastrointestinal tract is simple and terminates in a chamber called the cloaca, into

---

[*] Parts of this section have been freely adapted from Garrett Hardin, *Biology: Its Human Implications*, 2nd ed. (W. H. Freeman and Company, San Francisco, 1952).

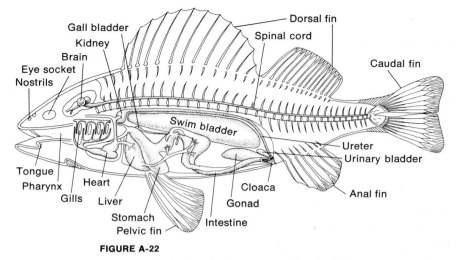

FIGURE A-22

Structure of a bony fish. (From G. Hardin, *Biology: Its Human Implications,* 2nd ed., W. H. Freeman and Company, San Francisco, 1952.)

which the products of both the reproductive system and the kidneys are poured.

CLASS AMPHIBIA[*]

The first class of vertebrates to achieve a fairly successful conquest of the land was the class Amphibia, which includes frogs, toads, and salamanders. Although many adult amphibians live well in rather dry environments, they must return to fresh water to breed, for their eggs are not equipped to withstand dryness. The name of the class is derived from the Greek *amphi,* meaning double, and *bios,* life. Amphibians live two lives: one in the water, one on land. The egg laid in the water develops into a sexually immature, or larval, form, with gills for exchange of respiratory gases. After a few weeks or months the larva undergoes metamorphosis and produces a sexually mature land-dwelling form with lungs (Fig. A-23). In salamanders the two life stages are usually noticeably different, though not so much so as in tailless amphibians (for example, frogs), which have a tailed, fish-like larval stage, the tadpole.

Larval and adult stages of amphibians differ in other important respects. A tadpole is a herbivore, feeding on microscopic algae; a frog is a strict carnivore, living on flies and other animals that attract its attention by moving.

Although adult amphibians hop or jump on land, their adaptation to terrestrial locomotion is only partial. The legs are produced at the side of the body rather than directly under it, where they could better bear the load; as a result, the amphibian belly is in contact with the substratum, a position that does not make for speedy locomotion. No doubt this failure to evolve a better type of limb arrangement is one of the reasons for lack of large species in the class. Also of probable importance is the fairly simple type of lung: though not so simple a sac as a swim bladder, it is much less complicated in structure than the mammalian lung, having relatively less surface available for gaseous exchange. Its deficiency is partly made up for by the moist, highly vascularized skin of frogs and salamanders, through which oxygen and carbon dioxide can be exchanged.

Amphibians, like fishes, have a cloaca, the common chamber for urinary, fecal, and reproductive products. The heart is more complicated than in fishes, being three-chambered (two auricles, one ventricle)—a difference that apparently makes for greater efficiency in a lung-possessing animal.

[*] Parts of this section have been freely adapted from Garrett Hardin, *op. cit.*

FIGURE A-23

Double life of a salamander: above, the larva, a water-dwelling animal
with external gills; below, the adult, a slimy-skinned, land-dwelling animal with
internal lungs. (Copyright by General Biological Supply House, Inc., Chicago.)

CLASS REPTILIA°

With the class Reptilia a truly land-dwelling type
of vertebrate was evolved. No longer is the skin
an organ for gaseous exchange. It is covered with
scales, and it prevents the loss of water (Fig.
A-24). Breathing is carried on entirely by lungs,
which are better developed than those of amphibians. In the circulation of blood there is a
close approach to the mammalian system of a
complete separation of pulmonary and systemic
circulations. There are two auricles and two
ventricles, and the latter two chambers are almost completely separated from each other.

Today the class is represented by turtles, croc-

° Parts of this section have been freely adapted from
Garrett Hardin, *op. cit.*

odiles, lizards, and snakes. During the Mesozoic,
however, the reptiles were the dominant vertebrate animals, and most of Chapter 10 is devoted
to their evolutionary radiation.

CLASS AVES

One of the two significant classes that evolved
from the reptiles was the birds, the other being
the mammals. Birds have feathers and are able
to fly. The oldest bird, *Archaeopteryx* (Fig. 10-35,
p. 376), of Jurassic age, was reptilian in many
respects except for the presence of feathers. With
the acquisition of feathers and the ability to fly
birds evolved remarkably keen sight and rapid
and varied reactions. Aside from the mammals,
they are the only organisms that maintain a con-

stant temperature. Their heart, like that of mammals, is four-chambered. Birds had evolved their basic structural plan by the Late Cretaceous, and since then their evolution has been in subtle features reflecting specific adaptations.

CLASS MAMMALIA

Mammals first appeared in the Late Triassic, and in the Cenozoic they underwent an adaptive radiation that is unparalleled in the animal kingdom. The Cenozoic can truly be called the age of mammals. The biology of mammals, and their similiarities to and differences from their ancestors, the reptiles, are discussed in § 14-3. The major groups of mammals are discussed in § § 14-4–14-17. The evolution of man is taken up in § § 15-5–15-7.

KINGDOM PLANTAE

Phylum Thallophyta

This phylum includes the simplest types of plants—the algae, fungi, and lichens, represented today by approximately 115,000 species. The phylum is not well represented in the fossil record, but certain algae secrete calcium carbonate and have been extremely important reef- and rock-formers throughout the phylum's history. Algal reefs are the oldest known fossils, being fairly common in rocks as old as Precambrian.

Phylum Bryophyta

The bryophytes, represented by liverworts and mosses, are more highly developed plants than the thallophytes. All bryophytes are small and confined to localities that are at least periodically quite wet. Though the phylum undoubtedly existed during the earlier periods of geological time, the plants have left fossils rarely and only in the younger rocks.

Phylum Tracheophyta

This phylum includes all the vascular plants. The conquest of land could not be complete until some means was developed whereby water could be rapidly raised from below the surface of the earth to a considerable height above it, thus making possible the evolution of large plants. The tissues through which liquids are transported upward and downward in plants constitute the

FIGURE A-24

Lizards, which are sometimes confused with salamanders. Like all reptiles (snakes, turtles, crocodiles), lizards have a dry, scaly skin. (Copyright by General Biological Supply House, Inc., Chicago.)

vascular system, and plants possessing a well-developed system of this sort are called vascular plants.

Subphylum Psilopsida

The earliest known vascular plants belonged to this group, which today is represented by only two genera. During the Devonian the psilopsids were a varied group, extremely widespread. They had no true roots; instead, a subterranean part of the stem served that need. The leaves were generally small, and many forms had no leaves. The psilopsids were the predominant land plants of the early and middle Paleozoic, after which they were replaced by more progressive plants. Typical Early Devonian psilopsids are illustrated in Figure 7-29 (p. 247, *Zosterophyllum, Psilophyton, Taeniocrada, Sciadophyton*), and Middle Devonian psilopsids are illustrated in Figure 7-30 (p. 248, *Aneurophyton, Pseudosporochnus, Asteroxylon, Rhynia*).

Subphylum Lycopsida

The lycopsids were extremely abundant and diverse in the Paleozoic, but today they are represented by only a few genera. They are spore-bearing plants with true roots and differentiated stems and leaves. Modern lycopsids, comprising four genera and approximately 660 species, are generally small plants. In the Carboniferous many lycopsids were enormous trees. Typical Paleozoic representatives are illustrated in Figures 7-29 (p. 247, *Protolepidodendron*), 7-30 (p. 248, *Archaeosigillaria, Duisbergia, Protolepidodendron*), 7-31 (p. 249, *Lepidodendron, Sigillaria*), and 7-33 (p. 251).

Subphylum Sphenopsida

All that survives of this once flourishing group is one genus, *Equisetum*, commonly known as "scouring rushes" because the plants contain silica and were once used for cleaning pots and pans. These are spore-bearing plants with vertically ribbed and jointed stems. The sphenopsids first appeared in the Middle Devonian. The modern *Equisetum* is only a few feet tall, but during the Paleozoic the sphenopsids grew to heights of several tens of feet. Typical ancient members of this group are shown in Figures 7-30 (p. 248, *Calamophyton, Hyenia*), 7-31 (p. 249, *Calamites*), 7-33 (p. 251), and 11-11 (p. 396, *Sphenophyllum, Schizoneura*).

Subphylum Pteropsida

This subphylum includes the ferns and the seed-bearing plants. It is such a large and diversified group that some of its major divisions will be taken up separately.

CLASS FILICINEAE

These are the true ferns, which reproduce by spores rather than by seeds. The ferns first appeared in the Middle Devonian and are today still abundant and diverse, especially in wet tropical forests. Their paleontological record is confusing because of the similarity of their leaves to those of the seed ferns (Cycadofilicales).

CLASS GYMNOSPERMAE

The predominant flora of the late Paleozoic and early Mesozoic consisted of the conifers, the cycadeoids, the cycads, the ginkgos, and the cordaites — all plants that had unenclosed, or naked, seeds. The name of the class is derived from Greek roots for "naked seeds." These were the first seed plants, and by the Carboniferous they had undergone a very extensive adaptive radiation, evolving many different adaptive types. Many of the gymnosperm groups declined greatly in numbers and diversity or became extinct when the angiosperms (plants with enclosed seeds) came on the scene; the most widespread gymnosperms today are the conifers.

The major orders of gymnosperms are as follows:

Order Cycadofilicales. These very primitive gymnosperms had fern-like leaves. The seeds grew singly attached along a leaf in various ways.

The order was very widespread and abundant in the late Paleozoic and early Mesozoic but is now wholly extinct. Typical representatives are illustrated in Figure 11-11 (p. 396, *Glossopteris, Sphenopteris, Gangamopteris*).

Order Cordaitales. This was one of the predominant orders of the Carboniferous forests, but it has been extinct since the Triassic. Typical representatives are illustrated in Figures 7-31 (p. 249) and 7-33 (p. 251) and described in § 7-10.

Order Gnetales. This is a small, peculiar group of plants with a very poor fossil record. Presumably the order is a very old one, but the oldest known fossils are from the Pliocene.

Order Coniferales. The commonest gymnosperms living today belong to this order, which includes such familiar trees as the pines, firs, cedars, cypresses, redwoods, and junipers. Approximately 450 species of Recent Coniferales are known. The order first appeared in the Late Carboniferous and reached a climax in the mid-Mesozoic. It has since declined in variety and numbers. The largest trees known, the *Sequoia*, belong to this order. A few conifers are illustrated in Figure 10-9 (p. 353).

Order Ginkgoales. This order survives in a famous "living fossil," *Ginkgo biloba*, commonly known as the maidenhair tree. The ginkgos are represented mainly in the form of cultivated plants in China but are now fairly common in American parks and streets. They first appeared in the Permian and were particularly widespread during the Mesozoic.

Order Cycadales. This relic order and the extinct cycadeoids are closely similar in general appearance and structure but differ in the organization of the reproductive organs. About ninety species of cycads are known, confined to warm places and nowhere abundant. The order first appeared in the Triassic.

Order Cycadeoidales. During the Mesozoic these cycad-like plants were extremely varied and widespread but then became extinct. They differed from the living cycads mainly in the presence of a flower-like organ that bore the seeds. Typical Mesozoic representatives of this group are illustrated in Figure 10-9 (p. 353, *Williamsonia* and *Williamsoniella*) and are described in § 10-2.

CLASS ANGIOSPERMAE

This class is the predominant element of the modern world floras. The angiosperms underwent a most remarkable adaptive radiation in the Late Cretaceous, and most of the modern orders of plants were already well established by the early Cenozoic. With the rise of the angiosperms the gymnosperms declined drastically: there are only about 600 species of Recent gymnosperms but more than 250,000 species of Recent angiosperms. Most of the Recent plants known to the average person, except the conifers, are angiosperms.

The angiosperms have true flowers and enclosed seeds, in contrast to the naked seeds of gymnosperms. The class represents the highest level of plant evolution attained up to now.

Correlation Charts

Rock formations are the means by which the historical geologist interprets the earth's history. Tens of thousands of formations have been named, described, and mapped. Throughout the text formational names have been avoided, except in special cases, so as not to encumber the presentation of the principal theme — continental evolution. For those who desire to carry their studies one step further a series of twenty-three correlation charts is presented here. For each geologic system there is a chart of fifteen stratigraphic sections for North America and a chart of fifteen stratigraphic sections for the continents outside North America. The series and stages are indicated. The ordinate of the charts is time, not thickness of formations. If the information is available, the average thickness of a formation is included. The following abbreviations are used, and the major kinds of rock are indicated by the following patterns.

	Abbreviation	Lithologic Symbol		Abbreviation
Redbeds			Coal	c.
Sandstone	Ss.		Quartzite	Qtzite
Shale	Sh.		Formation	Fm.
Limestone Dolomite	Ls. Dol.		Group	Gr.
Conglomerate	Congl.		Stage	St.
Chert	Ch.		Horizon	H.
Volcanic rocks	Volc.			
No record				

The Cambrian in North America

Series	Stages	S.E. New-foundland	W. New-foundland	S.E. New Brunswick	Western Vermont	Clinton County, New York	S. Pennsylvania	Tennessee
Upper (Croixan)	Trempeal-eauan	Elliott Cove Group	St. George Formation	Narrows Formation 500'	Clarendon Springs Formation 200'	Little Falls Formation 200'	Conococheague Limestone 2,000'	Copper Ridge Dolomite 2,000'
	Franc-onian					Theresa Formation		
	Dres-bachian		Petit Jardin Formation	Black shale Brook Fm. 25' / Agnostus Cove Fm. 200'	Danby Formation 700'	Potsdam Formation 200'		Nolichucky Shales 800'
Middle (Albertan)	Stages not yet defined	Manuels Brook Formation / Chamberlin's Brook Formation	March Point Sandstone ? / Hawkes Bay Formation	Hastings Cove Fm. 35' / Porter Road Fm. 75' / Fossil Brook Fm. 35'	Winooski Dolomite 600'		Elbrook Formation 2,000'	Maryville Limestone 600' / Rogersville Shale 250' / Rutledge Limestone 500' / Pumpkin Valley Formation 400'
Lower (Waucoban)		Brigus Formation / Smith Point Formation / Bonavista Formation	? / Forteau Formation / Bradore Formation	Hanford Brook Fm. 75' / Glen Falls Fm. 50 / Ratcliff Brook Fm. 2,000'	Monkton Quartzite 800' / Dunham Dolomite 2,000' / Cheshire Quartzite 1,000'		Waynesboro Formation 1,200' / Tomstown Dolomite 1,000' / Antietam Quartzite 700'	Rome Formation 1,000' / Shady Dolomite 1,100' / Chilowee Group

Wisconsin	Arbuckle and Wichita Mts., Okla.	Wind River Mts., Wyoming	Western Utah	Eureka, Nevada	Nopah Range, Calif.	Robson Peak, British Columbia	Eastern Greenland
Trempealeau Formation 200'	Butterfly Dol. 300						
	Signal Mt. Fm. 300		Notch Peak Formation 2,000'	Windfall Formation	Nopah Formation	Lynx Formation 5,000'	
Franconia Formation 100'	Royer Dol. 700						
	Fort Sill Ls. 224	Open Door Formation 300'		Dunderberg Sh.			
	Honey Creek Fm. 90'		Orr Fm. 2,000				
Dresbach Formation	Reagan Sandstone 460'	Du Noir Formation 500'	Weeks Formation 2,000'	Hamburg Dolomite	Dunderberg Shale		
			Marjum Formation 1,500'	Secret Canyon Shale	Bonanza King Formation	Titkana Formation 2,550'	Dolomite Point Formation 1,200' (exact horizon unknown)
		Depass Formation	Wheeler Formation 350'				
			Swasey Formation 400'			Tatei Fm. 1,000'	
			Dome Ls. 300	Eldorado Dolomite		Chetang Fm. 950'	
		Flathead Fm. 200	Howell Formation 800'	Pioche Sh.	Carrara Formation	Adolphus Fm. 400'	
			Tatow Ls. 150				Hyolithes Creek Formation 1,000'
			Pioche Formation 250'		Zabriskie Quartzite	Mahto Formation 1,200'	
			Prospect Mountain Quartzite 1,000'	Prospect Mountain Quartzite	Middle Wood Canyon Formation	Mural Formation 1,000'	Ella Island Formation 300'
							Bastion Formation 600'
					McNaughton Formation 400'		

631

Series	Estonia	Oslo region, Norway	Harlech, Northern Wales	Southern Wales	Bohemia	Central Massif, France	Anti-Atlas, Morocco
Upper	?	Olenid Series 120'	Lingula flags 4,500' — Dolgelly beds — Festiniog beds — Maentwrog beds	Lingula flags 2,000'	porphyries and porphyrites / Pavlosko Cgl.	S' – C⁴	unfossil-iferous sandstones
Middle		Paradoxides Series 100'	Menevian beds 300' — Clogan Slate — Cefn Goch Grit — Gamlan Flags and Grits — Barmouth Grits	Menevian Series 750' / Solva Series 1,500'	Ohrazenice Cgl. / Jince Fm / Tremošná Fm / Sadek Fm. / Hlubos Cgl.	C⁴ / C³ / C²	shales and archaeo-cyathid limestones / ?
Lower	Tiskri Ss. / Kakumagi beds and Mickwitzia Cgl. / Lükati beds / Blue clays / conglomerate and arkose	Holmia Series 150'	Harlech beds 7,000' — Manganese Shales — Rhinog Grits — Llanbedr Slate — Dolwen Grits	Caerfai Series 1,000'		C²⁻¹ / C¹	?

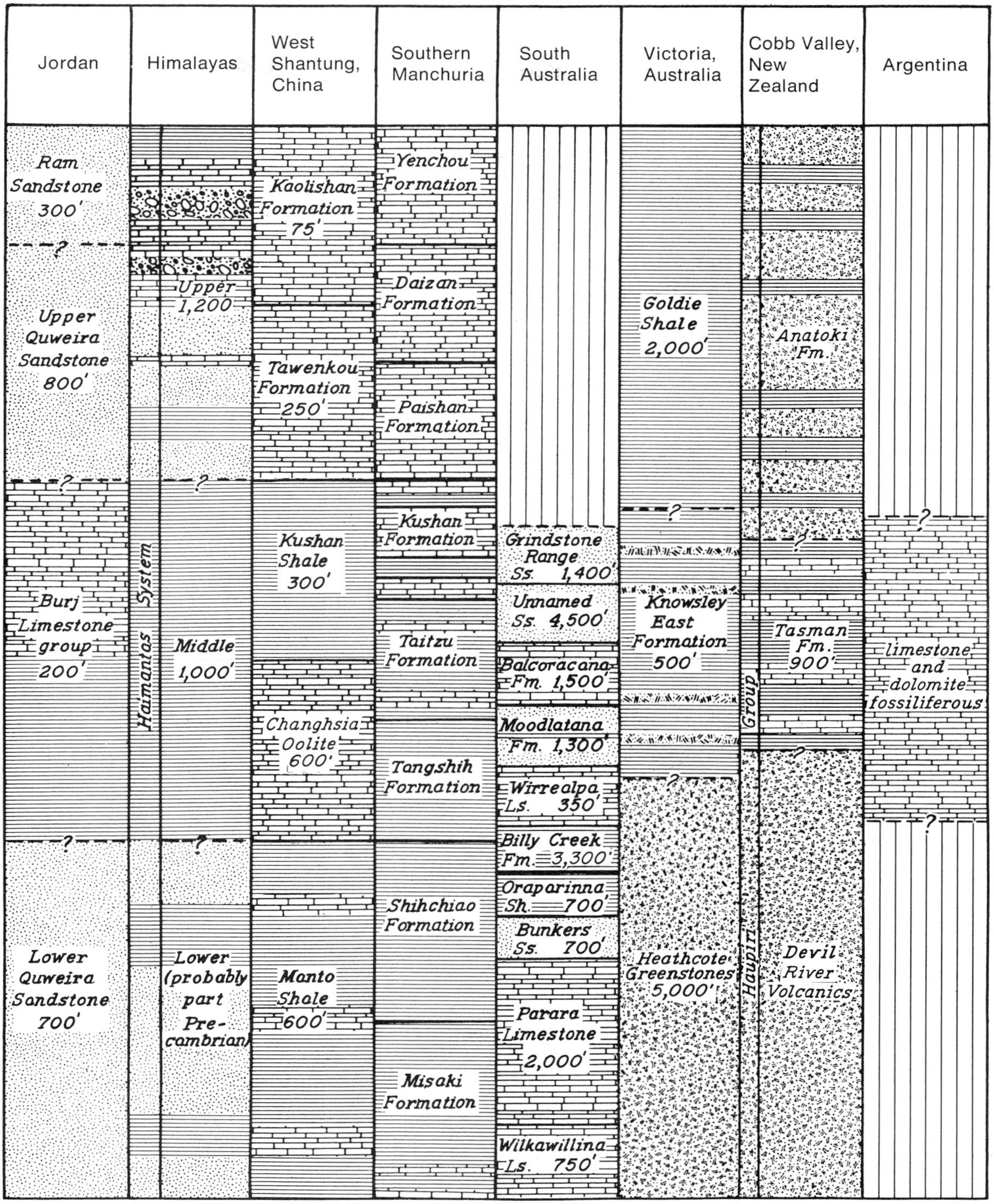

Jordan	Himalayas	West Shantung, China	Southern Manchuria	South Australia	Victoria, Australia	Cobb Valley, New Zealand	Argentina
Ram Sandstone 300'		Kaolishan Formation 75'	Yenchou Formation				
?			Daizan Formation		Goldie Shale 2,000'	Anatoki Fm.	
Upper Quweira Sandstone 800'	Upper 1,200	Tawenkou Formation 250'	Paishan Formation				
?	?	Kushan Shale 300'	Kushan Formation	Grindstone Range Ss. 1,400'	?	?	?
Burj Limestone group 200'	Middle 1,000'		Taitzu Formation	Unnamed Ss. 4,500'	Knowsley East Formation 500'	Tasman Fm. 900'	limestone and dolomite fossiliferous
				Balcoracana Fm. 1,500'			
		Changhsia Oolite 600'	Tangshih Formation	Moodlatana Fm. 1,300'			
				Wirrealpa Ls. 350'	?	?	
?	?			Billy Creek Fm. 3,300'			?
			Shihchiao Formation	Oraparinna Sh. 700'			
Lower Quweira Sandstone 700'	Lower (probably part Pre-cambrian)	Manto Shale 600'		Bunkers Ss. 700'	Heathcote Greenstones 5,000'	Devil River Volcanics	
				Parara Limestone 2,000'			
			Misaki Formation				
				Wilkawillina Ls. 750'			

Haimanas System

Haupiri Group

633

The Ordovician in North America

Series in N. America	Stages in N. America	Central Newfoundland	Gaspé, Quebec	Taconic area, W. Vermont	Northwestern New York	Central Pennsylvania	Virginia	Alabama
Cincinnatian	Gamachian	Present	? Whitehead Formation					
Cincinnatian	Richmondian	Present	Whitehead Formation		Queenston Red Shale 1,000'	Juniata Sandstone 1,000'	Juniata Formation	Fernvale equivalent
Cincinnatian	Maysvillian	Present			Oneida Sandstone 100'	Bald Eagle Ss.		Leipers equivalent
Cincinnatian	Edenian	Present			Lorraine Gr.: Pulaski Shale; Whetstone Gulf Shale	Reedsville Shale 1,000'	Martinsburg Formation 1,000'	
Champlainian (Mohawkian)	Trentonian	? Possibly absent ?			Holland Patent Sh.; Cobourg Ls.; Shoreham Ls.; Kirkfield Ls.; Rockland Ls. (Trenton Ls.)	Antes Shale; Coburn Ls.; Salona Ls.; Nealmont Limestone	Chickamauga Limestone	Cannon-Catheys equivalent; Curdsville Ls.; Tyrone Ls.; Lebanon Ls.
Champlainian (Mohawkian)	Black River	Bay of Exploits	Tetagouche Shale	Normanskill Sh. Ss.	Chaumont Ls.; Lowville Ls.; Pamelia Ls. (Black R. Ls.)	Bays Formation; Hunter Limestone; Hatter Limestone; Edinburg Formation	Bays Formation; Edinburg Formation	Ridley Ls.; Little Oak Ls.; Athens Sh.; Effna Ls.
Champlainian	Chazyan			?	Loysburg Limestone; New Market Limestone	Lincolnshire Ls.; Whistle Cr. Sh.; New Market Limestone	Lenoir Limestone 500'; Mosheim Limestone 100'
Canadian	(Stages not yet recognized)		Deepkill Shale; Macquereau Formation	Deepkill Shale 300'; Schaghticoke Shale	Bellefonte Dol.; Axeman Ls.; Nittany Dolomite; Stonehenge Ls.; Larke Dol.	Knox Dolomite (upper part)	Odenville Ls.; Newala Limestone 1,000'; Longview Ls. 500'; Chepultepec Dolomite 1,100'

634

Ordovician stratigraphic correlation chart

Upper Mississippi V.	Arbuckle Mountains	Ouachita Mountains	Utah	North-central Nevada	Canadian Rockies	Arctic islands, Canada	N.W. Greenland and Ellesmere Island	Stages in Great Britain
Maquoketa Formation 150'	Sylvan Sh.	Polk Creek Shale 100'	Fish Haven Dolomite	Valmy Formation 10,000'	Beaverfoot Limestone		Cape Calhoun Formation 750'	Ashgillian
					Wonah Quartzite	(lower) Cape Phillips Formation (graptolitic facies)		
Galena Ls. — Dubuque Dol. Ls.; Stewartville Dol. Ls.; Prosser Ls.; Decorah Sh. Ls.	Viola Limestone 700'	Big Fork Chert 700'				Cornwallis Formation 5,000'	Gonioceras Bay Formation	Caradocian
Platteville Fm.; Glenwood Fm.	*Simpson Group 1,500'* — Bromide Limestone; Tulip Cr. Ls.; McLish Ls.; Oil Cr. Ls.; Joins Ls.	Womble Shale 1,000'	Swan Peak Quartzite			Eleanor River Formation	Cape Webster Formation 800'	Llandeilian / Llanvirnian
St. Peter Ss. 150'					Glenogle Shale 2,000'		Nunatami Formation 400'	Arenigian
Prairie du Chien Gr. — Shakopee Dol.; New Richmond Ss.; Oneota Dolomite	*Arbuckle Group (upper)* — West Spring Creek Ls.; Kindblade Fm.; Cool Cr. Fm.; McKenzie Hill Fm.	Blakely Ss.; Mazarn Shale	Garden City Limestone 1,000'		Mons Ls.		Cape Weber Fm.; Nygaard Bay Ls.; Poulsen Cliff Sh.; Cape Clay Fm.; Cass Fjord Fm.	Tremadocian

The Ordovician outside North America

Stages in Europe	Estonia	Västergötland, Sweden	Oslo region, Norway	Scotland	Lake District, England	Wales	Bohemia
Ashgillian	Porkuni Stage / Lyckholm (Harju Series)	Dalmatina beds / Staurocephalus Sh. / Tretaspis beds	Tretaspis Series 350'	Upper Drummuck Group / Barren Flagstone Group	Ashgill Shale / Keisly Limestone	Ashgill Series 2,000	Kosov Quartzite / Králův Dvůr Sh & Ss.
Caradocian	Rakvere St. Vazalemma / Keila St. / Jõhvi St. / Idavere St. / Kukruse St. / Uhaku St. / Lasnamäe St. (Chasmops Series)	Slandrom Ls. / Macrourus Ls. / Ludibundus Ls. (Chasmops Limestone)	Chasmops Limestone and Shale / Ampyx Limestone 200' (Chasmops Series)	Whitehouse Group / Ardwell Group / Balclatchie Group / Benan Conglomerate etc. / Tappins Group	Applethwaite beds / Stockdale lavas / Stile End beds / Roman Fell beds	Caradoc Series 1,000' (Bala Series)	Bohdalec Shales / Loděnice Shales / Chrustenice Shales
Llandeilian	Azeri Stage	Orthoceras Limestone	Ogygiocaris Shale (Ogygiocaris Series)	? ? ?	Borrowdale Volcanics 10,000'	Llandeilo Series 2,500'	Dobrotiva Sh.
Llanvirnian	Kunda Stage		Upper Didymograptus Shale			Llanvirn Series 1,500'	Sarka Sh.
Arenigian	Megalaspis Limestone	Tetragraptus Sh.	Orthoceras Limestone / Lower Didymograptus Shale (Asaphus Series)	Ballantrae Volcanics	Skiddaw slates 3,000	Arenig Series 3,000'	Klabava Sh.
Tremadocian	Iru Fm. / Pakerort Sh. / Ss.	Ceratopyge Sh. / Ls. / Obolus Cgl. (Ceratopyge Ser.)	Ceratopyge Limestone / Dictyonema Shale	? ?	? ?		Milina Sh. Ss. / Trenice Cgl.

636

Stratigraphic correlation chart:

E. Urals, U.S.S.R.	Kazakh S.S.R.	Spiti, Himalayas	Southern China	South Victoria, Australia	Western Australia	Western Argentina	Eastern Peru	Series in N.A.
	Chokparsky Formation	Age span of these units very uncertain		Bolindian				Cincinnatian
	Kizilsysky Formation 3,000'	gray limestone 100'	Wufeng Shale 20'					
		?						
	Dulankarinsky Formation 1,400'		Yentsin Formation 50'			Trapiche Formation 2,000'		
	Otarsky Formation 1,000'	black fossiliferous limestone 200'	Pagoda Limestone 50'	Eastonian Slate				Champlainian
						Las Plantas Shale 900'		
	Anderkensky Formation 1,300'					Las Vacas Conglomerate 1,000'		
			Miaopo Shale	Gisbornian Slate				
	Beck Series 1,800'	?	Shihtzupu Fm. 100					
	Karakansky 550		Kuniutang Ls.			San Juan Limestone 4,500'		
	Kopalinsky Formation 500'		Yangtzeella beds 100'	Darriwilian Slate	Gap Creek Fm. 650'		Contaya Formation 1,000' ±	Canadian
		flaggy quartzite, siliceous shale, and basal conglomerate 1,500'	Meitan Shale 40'	Yapeenian Slate		Gualcamayo Shale 1,500'		
Mayachnaya Formation 1,500'	Kurdaisky Formation 1,500'		Hunghuayuan Limestone 100	Castlemanian Slate	Emanuel Fm. 2,000'			
				Bendigonian Slate				
Rimnitsky Series 1,500'	Agalatasky Kandiktasky Fm. 600'		Panho Formation 600'	Lancefieldian Slate				

637

The Silurian in North-America

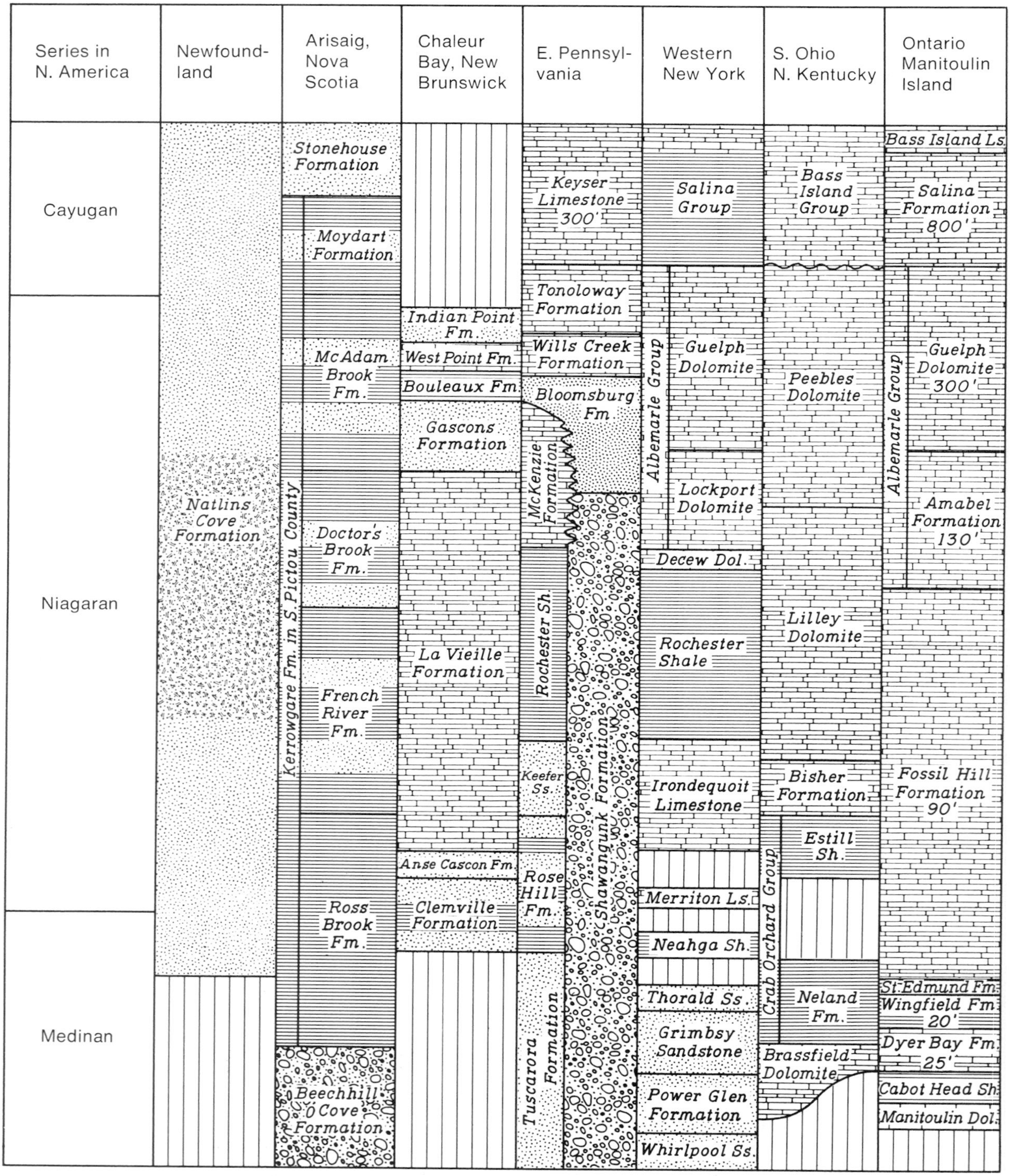

Series in N. America	Newfound-land	Arisaig, Nova Scotia	Chaleur Bay, New Brunswick	E. Pennsyl-vania	Western New York	S. Ohio N. Kentucky	Ontario Manitoulin Island
Cayugan	Natlins Cove Formation	Stonehouse Formation / Moydart Formation	*(vertical ruling)*	Keyser Limestone 300'	Salina Group	Bass Island Group	Bass Island Ls. / Salina Formation 800'
Niagaran	Natlins Cove Formation	McAdam Brook Fm. / Doctor's Brook Fm. / French River Fm.	Indian Point Fm. / West Point Fm. / Bouleaux Fm. / Gascons Formation / La Vieille Formation	Tonoloway Formation / Wills Creek Formation / Bloomsburg Fm. / McKenzie Formation / Rochester Sh. / Keefer Ss.	Guelph Dolomite / Lockport Dolomite / Decew Dol. / Rochester Shale / Irondequoit Limestone	Peebles Dolomite / Lilley Dolomite / Bisher Formation / Estill Sh.	Guelph Dolomite 300' / Amabel Formation 130' / Fossil Hill Formation 90'
Medinan	*(vertical ruling)*	Ross Brook Fm. / Beechhill Cove Formation	Anse Cascon Fm. / Clemville Formation	Rose Hill Fm. / Shawangunk Formation / Tuscarora Formation	Merriton Ls. / Neahga Sh. / Thorald Ss. / Grimbsy Sandstone / Power Glen Formation / Whirlpool Ss.	Neland Fm. / Brassfield Dolomite	St. Edmund Fm. / Wingfield Fm. 20' / Dyer Bay Fm. 25' / Cabot Head Sh. / Manitoulin Dol.

Kerrowgare Fm. in S. Pictou County

Albemarle Group

Crab Orchard Group

638

Hudson Bay	Western Tennessee	Northern Illinois	New Mexico, West Texas	West Utah, East Nevada	Roberts Mountains, Nevada	Southeast Alaska	N. Canada Innuitian region	Stages in Europe
	Rockhouse Fm.							Pridolian
	Decatur Limestone			Sevy Dolomite		unnamed slate and graywacke		Ludlovian
	Brownsport Formation				Lone Mountain Dolomite			
	Dixon Ls.	Racine Dolomite	─ ? ─	─ ? ─				
	Lego Limestone					─ ? ─		
	Waldron Sh.					unnamed limestone	Cape Phillips Formation	Wenlockian
	Laurel Limestone	Waukesha Dol.		Laketown Dolomite	Roberts Mountains Formation	─ ─ ? ─	Graptolitic shale + sandstone (detailed zonation within Silurian unknown)	
Attawapis-kat Coral reef		Joliet Dolomite	Fusselman Dolomite					
Ekwan R. Ls.	Osgood Formation				?	unnamed black slate and graywacke		Llandoverian
Severn River Ls.								
Port Nelson Ls.	Brassfield Dolomite	Kankakee Dolomite		─ ? ─				
		Edgewood Dolomite						

Stages in Europe	Oslo region, Norway	Moffat, Scotland	Lake district, England	Southern Wales	Thuringia, Germany	Central Bohemia	Alps
Pridolian	Ringerike Group			Red Marls / Temeside beds / Downton Castle Ss.	Ocker Kalke	Přídolí beds	
Ludlovian	? / favositids at top		Kirby Moor Flags / Bannisdale Slates / Coniston Grits	Ludlow Shales including Aymestry Limestone	Obere Graptolithen-schiefer	Kopanina Limestone	
Wenlockian	Spiriferid Group		Coldwell beds / Brathay Flags 1,000'	Wenlock Limestone / Wenlock Shales		Motol beds 500'	lime-stone facies / shale facies
Llandoverian	? / Pentamerus Group / ? / Stricklandin Group	Gala Group 4,000' / Birkhill Group 75'	Browgill beds / Stockdale Shales / Skelgill beds	Llandovery Series	Untere Graptolithen-schiefer	Zelkovice beds 200'	

Note: "Budñany Limestone" and "500'" labeled in Central Bohemia column (Pridolian–Ludlovian); "Liten beds" labeled in Central Bohemia column (Wenlockian–Llandoverian).

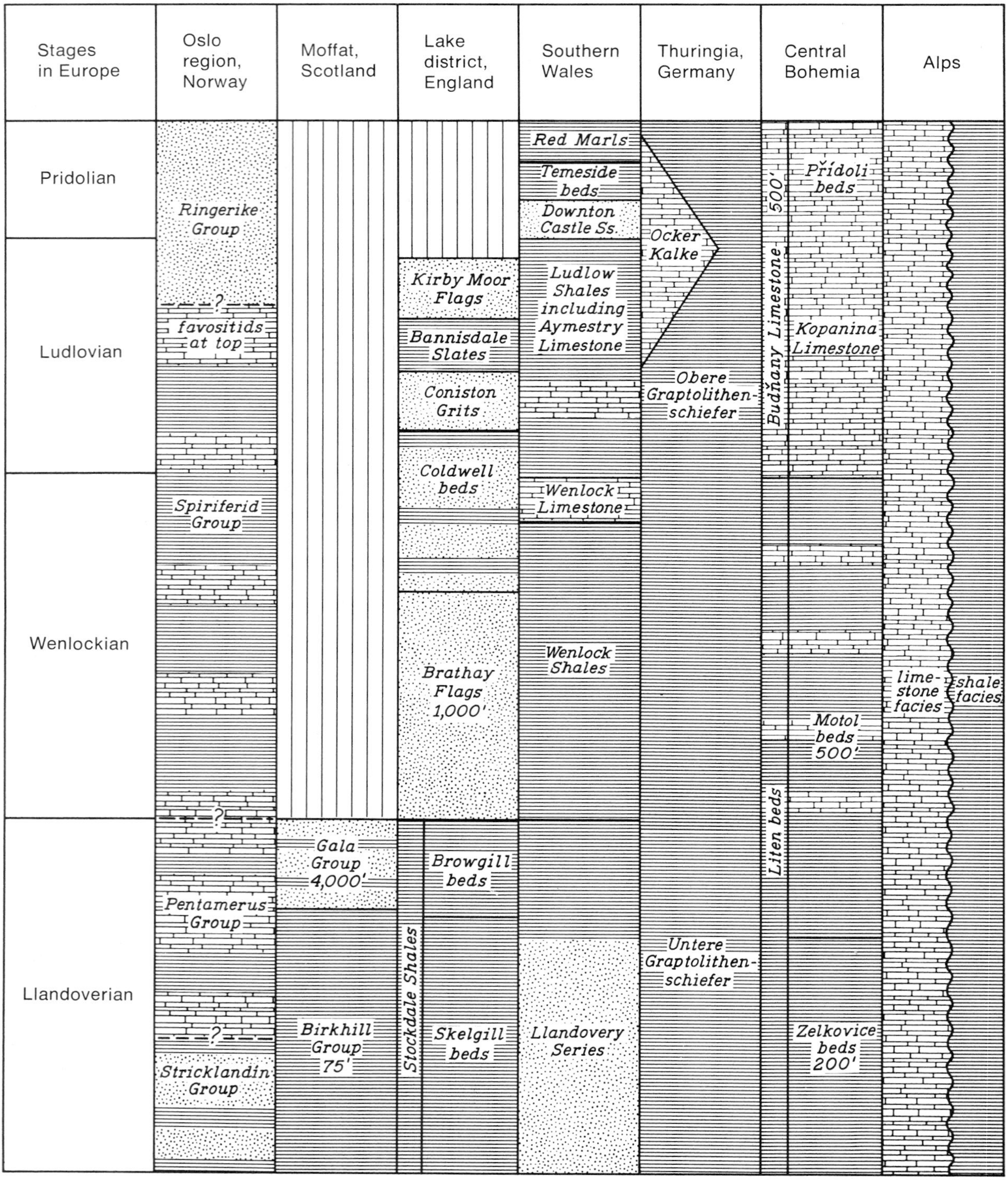

Ural Mts., Russia	Saudi Arabia	Burma	Yangtze River Gorge, China	N.E. Siberia	N.E. Japan	Melbourne area, Victoria, Australia	Amazon R., Brazil	Series in N. America
5,000' ±	Sharawa Fm.	Zebingyi Series; Konghsa Marls; Namhsim Ss (Namhsim Series); Graptolite Stage (Panghsa-pye Series); Trilobite Stage	Shamao Group; Lo-jo-ping Group; Lungmachi Shale	Takainari Series; limestone with corals 3,000'; ? ; limestone with corals 1,500'; ? ; calcareous + graptolitic shales 5,000'	Kawauchi Series; ?	Yering Group (Dev. in part) 10,000'; Melbourne Group	Trombetas Series	Cayugan; Niagaran; Medinan
	Tabuk Formation 2,000' (upper part) ?							

(Stratigraphic correlation chart)

The Devonian in North America

Series in N. America	Stages in North America	New Brunswick	Catskills, New York	Buffalo, New York	Central Ohio	Western Maryland	Western Tennessee	Southern Indiana
Upper	Bradfordian	Perry Formation		Oswayo Fm.	Ohio Black Shale	Hampshire Formation	Chattanooga Shale	Gassaway Member
Upper	Bradfordian	Perry Formation		Cattaraugus Formation	Ohio Black Shale	Hampshire Formation	Chattanooga Shale	Gassaway Member
Upper	Cassadagan	Perry Formation		Chadakoin Formation	Ohio Black Shale	Hampshire Formation	Chattanooga Shale	New Albany Shale
Upper	Cassadagan	Perry Formation		Northeast & Westfield Sh.			Chattanooga Shale	New Albany Shale
Upper	Cassadagan	Perry Formation		Perrysbury Formation	Olentangy Shale	?	Dowelltown Member	New Albany Shale
Upper	Cohoctonian	Perry Formation	Wittenburg Cgl.	Java Fm. / West Falls Formation	Olentangy Shale	Chemung Formation		
Upper	Cohoctonian	Perry Formation	Walton Sh.	West Falls Formation		Parkhead Ss.		
Upper	Fingerlakesian	Perry Formation	Oneonta Fm.	Sonyea Fm.		Brallier Fm.		
Upper	Fingerlakesian	Perry Formation	Oneonta Fm.	Genessee Fm.		Brallier Fm.		
Middle	Taghanican		Mahantango Fm. (Hamilton Group)	Hamilton Group	Olentangy Sh.	Harrell / Mahantango Fm. (Hamilton Group)		North Vernon Fm.
Middle	Tioughniogan		Mahantango Fm. (Hamilton Group)	Hamilton Group	?	Mahantango Fm. (Hamilton Group)		North Vernon Fm.
Middle	Cazenovian		Marcellus Fm. (Hamilton Group)	Hamilton Group	Delaware Ls.	Marcellus Sh. (Hamilton Group)		North Vernon Fm.
Middle	Onesquethawan		Onondaga Limestone	Onondaga Limestone	Columbus Limestone			Tioga Bentonite / Jeffersonville Limestone
Middle	Onesquethawan		Onondaga Limestone	Onondaga Limestone	?			Geneva Dol. / Dutch Crk.
Lower	Espusian	Campbellton Fm.	Bois Blanc Fm.? / Schohari Fm.		Needmore Shale	Camden Chert	Clear Creek Formation	Dutch Crk.
Lower	Espusian	Campbellton Fm.	Carlisle Center Fm.			Camden Chert	Clear Creek Formation	
Lower	Espusian	Campbellton Fm.	Esopus Sh.			Camden Chert	Clear Creek Formation	
Lower	Deerparkian	Campbellton Fm.	Oriskany Ss.		Oriskany Ss.	Harriman Fm.	Backbone Ls.	
Lower	Deerparkian	Campbellton Fm.			Shriver Chert	Flat Gap Ls.	Grassy Knob Chert	
Lower	Deerparkian	Campbellton Fm.			Licking Creek Ls.		Grassy Knob Chert	
Lower	Helderbergian	Dalhousie Fm.?	Helderberg Group		New Scotland Ls.	Ross Ls. (Ross Fm.)	Bailey Formation	
Lower	Helderbergian	Dalhousie Fm.?	Helderberg Group		Elbo Ridge Ss.	Birdsong Sh. (Ross Fm.)	Bailey Formation	
Lower	Helderbergian	Dalhousie Fm.?	Rondout Ls.		Keyser Ls.		Bailey Formation	

North-central Iowa	Idaho	Eureka, Nevada	Confusion Range, Utah	Shasta district, California	MacKenzie River region, Canada	W. Brooks Range, Alaska	Ellesmere Island, Canada	Stages in Europe
English River Fm.	Three Forks Fm.	Pilot Shade	siltstone unit (Pilot Shale)			Noatak Sandstone		Famennian
Maple Mill Formation					Imperial Formation		Okse Bay Formation 6,150'	
Aplington Formation								
Sheffield Formation	Trident Member		dolomitic siltstone unit			Hunt Fork Sh.		Frasnian
Lime Creek Formation	Birdbear Formation	Devils Gate Limestone			Canol Formation			
Shell Rock Formation	Jefferson Formation / Lower Member		Guilmette Formation			Skajit Limestone		Givetian
Cedar Valley Formation				Kennett Formation	Hare Indian Fm.			
Wapsipinicon Formation	Channel-fill		Simonson Dolomite	Balaklala Rhyolite	Ogilvie Formation	?	Blue Fiord Fm.	Eifelian
La Porte City Formation	Beartooth Butte Formation	Nevada Formation	?	Copley Greenstone			Eids Formation	Emsian
			Sevy Dolomite		Gossage Formation			Siegenian
				?			Cape Phillips Formation	Gedinnian

643

The Devonian outside North America

Series in Europe	Stages in Europe	Northern Devon, England	Central-South Scotland	Dinant basin, Belgium	Bohemia	Alps	Russian Platform (margin)	Tien Shan Ural Mts., Russia
Upper	Famennian	Pilton beds 1,600'; Baggy beds 1,400'; Upcott beds 800'; Pickwell Down Sst. 4,000'; Morte Slates	Upper Old Red Sandstone	biostrome; biostrome; biostrome			Vladimirvolyn Formation; Torchyn Fm.; Litovezh Fm.; Zadonsko-yelets H.	gray fossiliferous limestone 1,000'
Upper	Frasnian	Ilfracombe beds					Evlano H.; Voronezh H.; Zolochev Fm; Remezov H.; Kynov H	Kurchavai Suite 1,000'
Middle	Givetian				Srbsko Formation	Complete Devonian represented in a variety of carbonate facies	Estherian H.; Strutin H.; Pelchin H.	Yauruntuz Suite 1,200'
Middle	Eifelian	Hangman Grits 4,000'; Lynton beds			Chotec Limestone; Trebotov Ls.; Daleje Sh.		Lapushan Formation 300'	Katran Suite 1,500'
Lower	Emsian	?			Zlichov Limestone		Dnestrov Series 3,000'	Tai Bulak Suite 1,200'
Lower	Siegenian		Lower Old Red Sandstone 18,000'		Upper Koneprusy Limestone			
Lower	Gedinnian				Lower Koneprusy Limestone			

Stratigraphic correlation chart

Lena River Delta	Himalayas	Yunnan, China	Kitakami Mts., Japan	Victoria, Australia	New South Wales, Australia	Southern Peru	Maranhao basin, Brazil	Stages in N. America
Abalaekh Suite 600' ?	Muth Quartzite up to 3,000' (lower part may be upper Silurian, upper part may be Early Carboniferous)	Hsikuangshan Group	Tobigamori Series	*Avon River Group 13,800* Snowy Plains Fm. — Mt. Kent Conglomerate — Wellington Rhyolites — Moroka Glen Fm.	? — Cunningham Formation — Merrions Tuff — Crudine Group	Cabanillas Group 10,000'	*Pimenteiras Formation* Longá Shale — Cabeças Ss. — Picos Shale — Itaim Ss.	Bradfordian
gray calcareous ss. 90'								Cassadagen
240'		Shetienchiao Group						Cohoct.
450'								Finger.
		Tungkang-ling Stage						Tag.
600'								Tioug.
		Yükiang / Tiaomachien / Yükiang	Nakazato Series 2,300'					Caz.
?								Onesque.
		Szupai Stage						Epusian
grey dolomites			Ono Series					Deerp.
		Lunghuashan Formation	Takainari Series					Held.

645

The Mississippian in North America

Stages in N. America	Nova Scotia	Pennsylvania (Harrisburg)	Western Maryland	Eastern Tennessee	North-Central Alabama	West-central Kentucky	Illinois
Chesterian	Canso Group / Mabou Formation		Mauch Chunk Shale 1,100'	Pennington Shale 1,000'	Pennington Shale 200' / Bangor Ls. 400' / Gasper Formation 100'	Leitchfield Formation 500' / Glen Dean Ls. / Hardinsburg Ss. / Golconda Fm. 200' / Cypress Ss. / Girkin Formation	Kinkaid Ls. / Degonia Ss. / Clore Ls. 30' / Palestine Ss. / Menard Ls. 80' / Baldwin Fm. 75' / Okaw Limestone 200' / Ruma Fm. 50' / Paint Creek Fm. / Yankeetown Ch. / Renault Fm. 100' / Aux Vases Ss.
Meramecian	Windsor Group 1,500'	Mauch Chunk Shale 1,600'	Greenbrier Limestone 800' / Loyalhanna Ls. 40'	Newman Limestone 4,000'	Ste. Genevieve Limestone / St. Louis Limestone 250'	Ste. Genevieve Limestone / St. Louis Limestone 250' / Salem Ls.	Ste. Genevieve Limestone / St. Louis Limestone 250' / Salem Ls.
Osagean	Ainslie Ss.	Cave Mt. Ss. / Peters Mountain Sandstone	Maccrady Shale 700' / Pinkerton Ss. 125 / Meyers Sh. 800'	Fort Payne chert	Warsaw Limestone 80' / Fort Payne chert 200'	Warsaw Ls. / Muldraugh Fm. 100' / Floyds Knob Fm. 4' / Brodhead Fm.	Warsaw Ls. / Keokuk Limestone / Burlington Limestone
Kinder-hookian	Horton Group 5,000'	Pocono Group 2,000' / Second Mt. Ss.	Pocono Group / Hedges Sh. 170' / Purslane Ss. 250' / Rockwell Fm. 500'	Grainger Shale 800' / Big Stone Gap Sh. / Olinger Sh. / Cumberland Gap Sh. / Chattanooga Sh.	Chattanooga Shale 15'	New Providence Shale 120' / New Albany Shale (upper + middle members)	Chouteau Ls. 50' / Maple Mill Shale / Louisiana Ls.

Velma, Southern Oklahoma	Marathon region, Texas	South-western Montana	Central Montana	Oquirrh Mts., Utah	North-eastern Nevada	Klamath Mts., Calif.	Northern Alaska	Stages in Europe

Velma, Southern Oklahoma: Goddard Shale; Caney Shale 200'; Sycamore Limestone 300'; Woodford Shale 600'

Marathon region, Texas: Tesnus Formation; ?; Caballos chert 100'

South-western Montana: Amsden Formation; ?; Madison Group — Mission Canyon Limestone 700'; ?; Lodgepole Limestone 800'

Central Montana: Lower Amsden Formation 100'; ?; Big Snowy Group — Heath Formation 450'; Otter Formation 500'; Kibbey Sandstone 125'; ?; St. Charles Limestone 950'; ?; Madison Group — Mission Canyon Limestone 700'; ?; Woodhurst Limestone; ?; Paine Shale

Oquirrh Mts., Utah: Manning Canyon Shale 1,000'; Great Blue Limestone 4,000'; Humbug Formation 700'; ?; Deseret Limestone 600'; Madison Limestone 500'

North-eastern Nevada: Tonka Formation 2,500'; White Pine Shale

Klamath Mts., Calif.: Baird Shale 600'; ?; Bragdon Formation 6,000'

Northern Alaska: Lisburne Group — Alapah Limestone 1,000'; ?; Wachsmuth Limestone 1,200'; ?; Kayak Shale 1,000'

Stages in Europe: Viséan; Tournaisian

647

Stages in N. America	Nova Scotia	Pennsylvania	W. Virginia	Kentucky	Alabama	S. Oklahoma	Marathon region, W. Texas
Virgilian			*Monongahela:* Waynesburg c., Uniontown c., Sewickley Ss., Sewickley c., Redstone c., Pittsburgh c., L. Pittsburgh Ss., L. Pittsburgh Ls., Connellsville Ss.				
Missourian		*Conemaugh Series:* Ames Ls., Bakerstown c., Mahoning Ss.	*Conemaugh Series 800':* Clarksburg Ls., Morgantown Ss., Two-mile Ls., Bakerstown c., Brush Cr. Ls., Mahoning Ss.				Gaptank Formation 2,000'
Desmoinesian	Pictou Group 8,000'	*Allegheny Series:* U. Freeport c., U. Kittann'g c., M. Kittann'g c., L. Kittann'g c., Kittanning Ss., Vanport Ls., Clarion coal, Clarion Ss., Brookville c., Homewood Ss.	*Allegheny Series 350':* U. Freeport c., E. Lynn Ss., No. 5 Block c.	Bryson F'm., Hignite Formation 450', Catron Fm. 400'			
Atokan	Shulie Formation 2,000'; Joggins Formation 7,000'	*Pottsville Series:* Mercer Shale, Connoansa Ss.	*Pottsville Series:* Kanawha Group 2,000', Hance Fm. 600'	Mingo Formation 950'		Atoka Formation 7,000'	Haymond Formation 2,000'
Morrowan	Boss Pt. Formation; Sharon Sh.; Clairmont Formation; Shepody Formation 3,000'	Okan Congl.	New River Group 1,000'; Pocahontas Group 700'	Lee Formation 1,400'	Pottsville Formation 9,000'; Parkwood Formation	Johns Valley Shale 2,500'; Jackfork Sandstone 8,000'; Stanley Shale 7,000'; Hot Spgs. Ss.	Dimple Limestone 1,000'; Tesnus Formation 3,000'

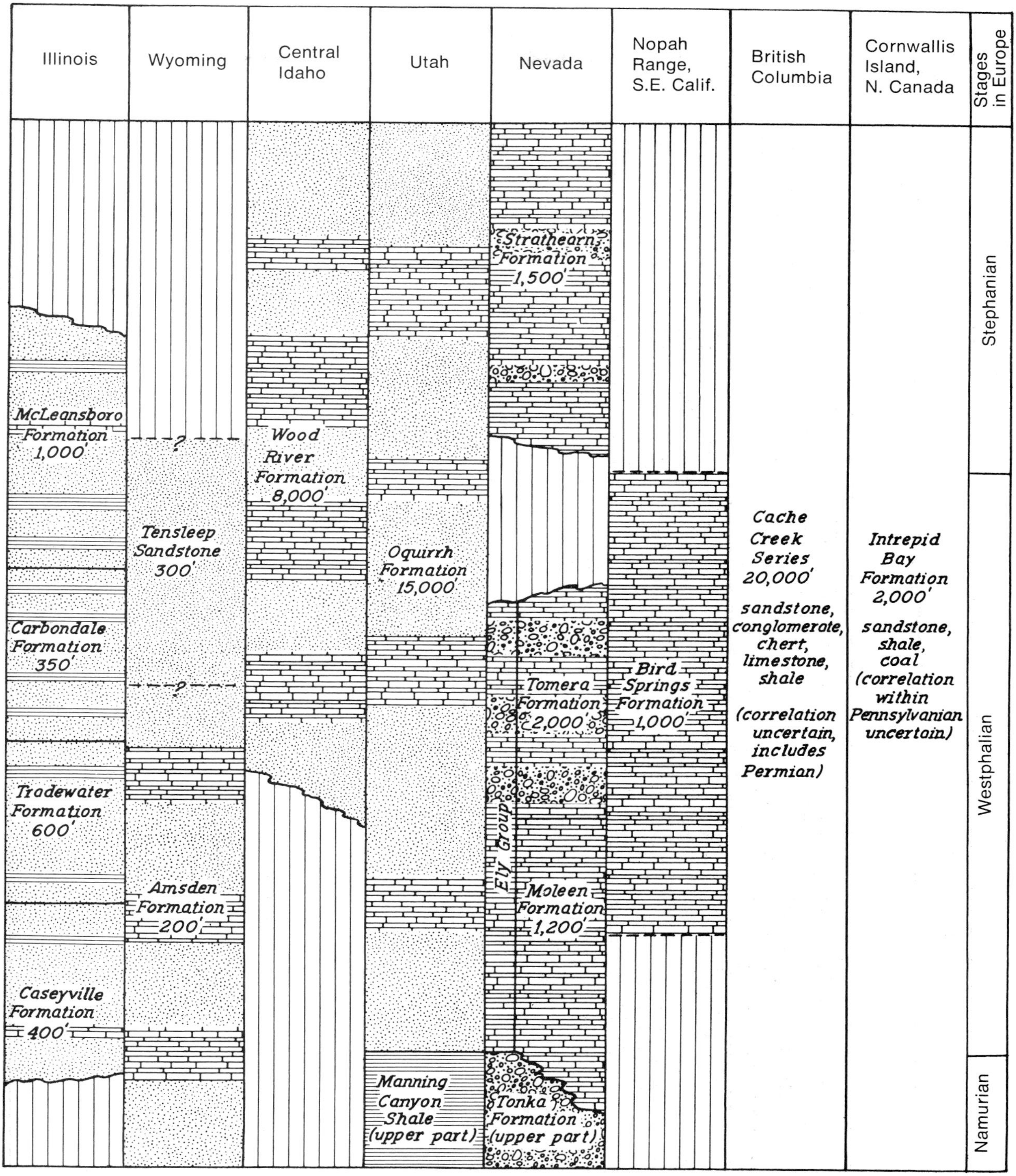

Illinois	Wyoming	Central Idaho	Utah	Nevada	Nopah Range, S.E. Calif.	British Columbia	Cornwallis Island, N. Canada	Stages in Europe

Column headers and labels within the chart:

Illinois:
- McLeansboro Formation 1,000'
- Carbondale Formation 350'
- Tradewater Formation 600'
- Caseyville Formation 400'

Wyoming:
- ?
- Tensleep Sandstone 300'
- ?
- Amsden Formation 200'

Central Idaho:
- Wood River Formation 8,000'

Utah:
- Oquirrh Formation 15,000'
- Manning Canyon Shale (upper part)

Nevada:
- Strathearn Formation 1,500'
- Tomera Formation 2,000'
- Ely Group
- Moleen Formation 1,200'
- Tonka Formation (upper part)

Nopah Range, S.E. Calif.:
- Bird Springs Formation 4,000'

British Columbia:
- Cache Creek Series 20,000'
- sandstone, conglomerate, chert, limestone, shale
- (correlation uncertain, includes Permian)

Cornwallis Island, N. Canada:
- Intrepid Bay Formation 2,000'
- sandstone, shale, coal
- (correlation within Pennsylvanian uncertain)

Stages in Europe:
- Stephanian
- Westphalian
- Namurian

The Lower Carboniferous outside North America

Stages in Europe	Bristol, England	Scotland	Belgium	Massif Central, France	Thuringia, Germany	South Libya	coal basin, U.S.S.R.
Viséan	D₂ Upper Cromhall Ss. and Hotwells Ls. D₁ S₂ Clifton Down Ls. C₂S₁ Goblin Combe Oolite Clifton Down Muds. Gully Oolite C₁	Lower Limestone Group Oil-shale Group 1,000' Calciferous sandstone Series / Clyde Plateau lavas	black limestone + shale Great Breccia cherty limestone oolitic limestone Black Marble of Dinant	All of Culm facies with volcanics	Kulm-grauwacke 3,000' Wechsel beds 2,000'	? fossiliferous sandstone + limestone 200'	Serpukhovian Formation Okian Formation Tula Horizon 150' Coal-bearing beds
Tournaisian	Z₂ Black Rock Ls. and Dol. Z₁ K Lower Limestone Shale Shire-hampton beds	Lower Cement-stone Group 700' ?	black limestone Crinoid limestone calc. shale Crinoid limestone shale Limestone of Etroeungt		Dach Shale 400' Russ Shale 30' Gattendorfia Limestone 10'	marine sandstone + shale 100' non-marine sandstone + shale 120' marine sandstone + shale 100'	Cherepetsky Horizon 90' Upa Horizon 15' Malerka Horizon 30' Ozersko-chovansky beds 60'

E. Slope, Ural Mts.	Himalayas	Kitakami Mountains, Japan	Western Australia	Queensland, Australia	New South Wales, Australia	Spitsbergen	Peru	Stages in N. America
Egorshinsk coal-bearing Fm.		Onimaru Formation 200'	Yindagindy Formation	Baywulla Formation	Gilmore volcanics 3,000'			Chesterian
					Wallaringa Formation			
		Ôdaira Formation 2,250'	Williambury Formation		Wiragulla beds		?	
	?				Ararat Fm.	Billefjorden Sanstones 2,000'	Ambo Group 3,000' (age span within system not known)	Meramecian
		Arisu Formation 3,000'		Tellebang Formation	Bingleburra Formation			Osagean
Bobrovka coal-bearing Fm.	Ice Lake Formation 1,000'						?	
	?	Hikoroichi Formation 2,350'	Moogooree Limestone					Kinderhookian

651

The Upper Carboniferous outside North America

Stages in Europe	Scotland	North-East England	Belgium	Saar region, Germany	South-eastern France	West-central Algeria	Moscow coal basin, U.S.S.R.
Stephanian				Ottweiler beds: Breiten-bach beds 600'; Potsberg beds 3,000'; Sarrelouis beds 1,600'; Holz Congl. 10'-600'	St. Etienne Stage; Rive-de-Gier Stage	? Stéphanien inférieur ? 3,000'	Pseudo-fusulina Horizon 150'; Gzel Horizon; Kasimov Horizon 200'
Westphalian	Barren Red coal measures 300'; Productive coal measures 3,000'	Coal measures: Upper 2,500'; Middle 3,000'; Lower 1,000'	Flenu; Charleroi; Châtelet	Saarbruck beds: LaHouve beds 6,000'; Sulzbach beds 2,000'; St. Ingbert beds 1,500'	La Houve Stage	Westphalien D 1,000'; Westphalien C 2,500'; Kénadzien 3,000'	Moscovian Stage: Myachkovian Horizon 75'; Podolskian Horizon 150'; Kashirian Horizon 150'; Vereyan Horizon 90'; Upper Bashkirian
Namurian	Millstone Grit; Upper Limestone Gr.; Limestone coal Gr.	Millstone Grit 2,000'	Andenne Ss.; Chokier beds			Namurian 2,000'	Vysokovskaya Fm.; Protva Fm. 30'; Serpukhovian Fm. 120'

E. Slope, Ural Mts.	Himalayas	Yangtze Valley, China	Kitakami Mountains, Japan	Western Australia	New South Wales, Australia	Spitsbergen	Peru	Stages in N. America

The stratigraphic columns contain the following labeled units:

- E. Slope, Ural Mts.: *Kostromikhino Fm.*, *Bursun Fm.*
- Himalayas: *Po Series*, *Fenestella shale + quartzite*
- Yangtze Valley, China: *Chuanshan Limestone 100'*, *Huanglung Limestone 300'*
- Kitakami Mountains, Japan: *Nagaiwa Formation 1,400'*
- Western Australia: *Anderson Formation*
- New South Wales, Australia: *Seaham Formation*, *Paterson volcanics*, *Mount Johnstone Formation*, *Gilmore Volcanics 3,000'*
- Spitsbergen: *Wordie-kammen Ls.*, *Black Craig Ls.*, *Passage beds 600'*, *Lower Gypsiferous Series 500'*, *Billefjorden Ss.*
- Peru: *Tarma Group*
- Stages in N. America: Virgilian, Missourian, Desmoinesian, Atokan, Morrowan

653

The Permian in North America

Stages in North America	Ohio and W. Virginia	Nebraska-Kansas	Oklahoma	North-central Texas	Western Texas — Shelf	Western Texas — Shelf-basin Margin	Western Texas — Basin
Ochoan							Dewey Lake Fm.; Rustler Fm. 400'; Salado Fm. 2,000'; Castile Fm. 1,500' (Ochoa Series)
Guadalupian		Quartermaster	Quartermaster Formation		Tansill Fm.; Yates Fm.; Seven Rivers Formation	Capitan Limestone 2,000'	Bell Canyon Formation 700'
Guadalupian		Whitehorse Sandstone 270'	Rush Sp'gs. Sandstone; Marlow Formation (Whitehorse Gr.)	Whitehorse Group	Queen Sandstone; Grayburg Formation	Goat Seep Limestone 1,000'; Getaway Limestone 200'	Cherry Canyon Formation 1,000'
Guadalupian		Dog Cr. Shale	Dog Cr. Shale	Dog Cr. Shale	San Andres Ls. 1,500'	Cherry Canyon; Brushy Canyon	Brushy Canyon Fm. 1,000
Leonardian		Blaine Fm.; Flowerpot Sh.; Cedar Hill Sandstone; Salt Plain Fm.; Harper Ss.; Stone Corral Dol.; Ninnescah Sh.	Blaine Gyps.; Flowerpot Sh.; Cedar Hills Sandstone; Hennessey Shale 500'; Garber Ss. (El Reno Group)	Blaine Fm.; Flowerpot Sh.; San Angelo Sandstone; Clear Fork Group 1,000'; Lueders Ls.	Yeso Formation 1,000'	Victorio Peak Limestone 1,000'	Bone Spring Limestone 2,000'
Leonardian				Clyde Fm.; Belle Plains; Admiral Fm.; Putnam Fm.			
Wolfcampian	Dunkard Group	Chase Group 300'; Council Grove Group 300'; Admire Gr. 200'	Chase Group; Council Grove Group; Admire Gr.	Moran Formation; Pueblo Formation 200' (Wichita Group)	Hueco Limestone 650'+	Hueco Limestone 250'	Hueco Limestone 500'

654

Arizona	Wyoming	Salt Lake City, Utah	Western Utah	Nevada	California	British Columbia	Eastern Greenland	Stages in Europe
				Koipato Formation, 5,000			Martinia Limestone	Tartarian (part)
							Posidonia Shale	Kazanian
							Pebbled Dolomite	
					Dekkas Andesite 1,000'		conglomerate	
	?	?	? Gerster Formation 1,000'		Nosoni Formation 900'	Cache Creek Group (volcanic, clastic sed. rocks, ls., chert; includes Miss., Penn., and Perm.) details of succession and correlation not known		Kungurian
	Phosphoria Formation 800'		Plympton Formation 700'	?				
Kaibab Limestone 600'		Park City Formation 1,500'	Kaibab Limestone 500'					Artinskian
Toroweap Formation 300'	?	Diamond Cr. Sandstone 800'	Arcturus Formation 2,700'		McCloud Limestone 2,000'			
Coconino Sandstone 350				?				
Hermit Sh 225'		Kirkman Limestone 1,600						Sakmarian
Supai Formation 800'		Ely Limestone (upper part)	Havallah Formation 10,000'					
		Oquirrh Formation 18,000'						Asselian

Stages in Europe	England	Germany	Southern Europe	Russian Platform	Ural Mts., Russia	Salt Range, Pakistan	Himalayas, India
Tartarian (part)	marls, dolomite, salt, gypsum	Zechstein	Bellerophon Limestone (E. Alps)	Tartarian 450'	Tartarian 300'		
Kazanian	Magnesian Limestone		Grödener Formation (E. Alps)	Conchifera beds 30' / Spirifer beds 100' (Kazanian)	Kazanian 600'	Chhidru Fm. 200' (Zaluch Group / Kuling System)	Productus shales 200'
Kungurian	Rotliegend	Kreuznach beds / Wadern beds	Sosio Limestone (Sicily)		Kungurian 2,000'	Wargal Ls. 500'	Calcareous sandstone 100'
Artinskian		Tholey beds	Trogkofel Limestone (Carnic Alps)	Shustovo-Deniatino beds 50'	Artinskian 3,000'	Amb Fm. 200'	
Sakmarian		Lebach beds	Schwagerinen Limestone (Carnic Alps)	Schwagerina beds 30'	Sakmarian 1,500'	Sardi Fm. / Warchha Ss. (Nilawan Group)	
Asselian		Cusel beds				Tobra Fm.	

Southern China	Timor	Kitakami Mountains, Japan	Western Australia	Eastern Australia	New Zealand	Central Spitsbergen	Peru	Stages in N.A.
Changhsing Limestone	Upper Shihotse Series 2,000'	Toyoma Series	Liveringa Formation 1,400'	Newcastle Coal Measures	Stephens Fm. 3,000'			Ochoan
Wuchiaping Limestone	Amarassi	Kanokura Series		Tomago Coal Measures 2,000'	Waiua Fm. 1,800'		Mitu Group 12,000'	
Maokou Limestone 300'	Basleo			Maitland Group	Greville Fm.			Guadalupian
	Tae Wei				Little Ben Ss.	Brachiopod Cherts 1,000'		
					Tramway Ss.			
					Wooded Peak Ls.			
					Patuki Volcanics			
					Dun Mt. Ultramafics			
Chihsia Limestone	Bitauni		Noonkanbah Formation 1,400'	Greta Coal Measures 250'	Patuki Volcanics 3,000'			Leonardian
			Poole Sandstone 1,000'	Gyarran Volcanics	Rai Sandstone 7,000'	Upper Gypsiferous Series 900'	Copacabana Group (upper part) 6,000'	
Maping Limestone 400'	Somohole	Sakamotozawa Series		Dalwood Group 5,000'		Upper Wordie-kammen Limestone 400'		Wolfcampian
Chuanshan Limestone 100'			Grant Formation 3,500'		Croisilles Volcanics 2,500'			

Tuffaceous marls with volcanics-in regions of great structural complexity

657

The Triassic in North America

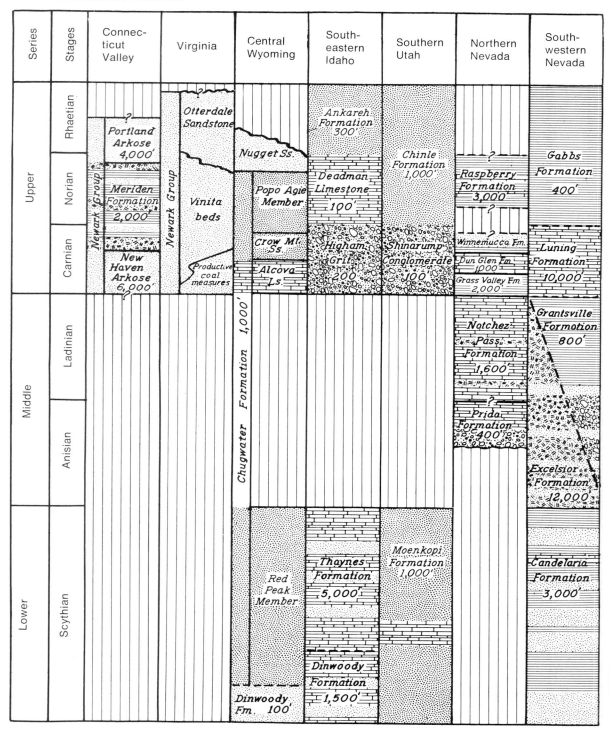

Series	Stages	Connec-ticut Valley	Virginia	Central Wyoming	South-eastern Idaho	Southern Utah	Northern Nevada	South-western Nevada
Upper	Rhaetian	Portland Arkose 4,000'	Otterdale Sandstone	Nugget Ss.	Ankareh Formation 300'	Chinle Formation 1,000'	Raspberry Formation 3,000'	Gabbs Formation 400'
Upper	Norian	Meriden Formation 2,000'	Vinita beds	Popo Agie Member	Deadman Limestone 100'		Winnemucca Fm.	Luning Formation 10,000'
Upper	Carnian	New Haven Arkose 6,000'	Productive coal measures	Crow Mt. Ss. / Alcova Ls.	Higham Grit 200'	Shinarump Conglomerate 100'	Dun Glen Fm. 1000' / Grass Valley Fm. 2,000'	
Middle	Ladinian			Chugwater Formation 1,000'			Notchez Pass Formation 1,600'	Grantsville Formation 800'
Middle	Anisian						Prida Formation 400'	Excelsior Formation 12,000'
Lower	Scythian			Red Peak Member / Dinwoody Fm. 100'	Thaynes Formation 5,000' / Dinwoody Formation 1,500'	Moenkopi Formation 1,000'		Candelaria Formation 3,000'

Inyo Range, S.E. Calif.	Shasta County, Calif.	Vancouver Island	North-western British Columbia	Southern Alaska	Northern Alaska	Eastern British Columbia	Eastern Greenland
	? Modin Formation 5,000'						Cape Stewart Formation 600'
	? Brock Shale 400'	Vancouver Volcanics	sedimentary group	McCarthy Fm ?	?	Pardonet beds 2,000'	?
? 300	Hosselkus Limestone 250'	Quatsino Ls. Karmutsen Volcanics	volcanic group	Nizina Ls. Chitistone Limestone ?	Shublik Formation	"Gray beds" 2,500' Liard Formation 1,000'	
	Pit Shale					Toad Formation 1,000'	
	Bully Hill Rhyolite ?			Nikolai Greenstone (Permian or Triassic)			
Inyo Formation 500 ?						Grayling Formation 1,000'	Anodontophora beds Proptychites beds Ophiceras beds Glyptophiceras beds
				Sadlerochit Formation			

659

Series	Stages	Central England	Germany	France	North-eastern Alps	Dolomite Alps	Saudi Arabia	Spiti, Himalayas
Upper	Rhaetian	Langport beds 15' / Westbury beds 20' / Sully beds 6'	Rät 100'	Rhétien	Kossen Formation 300'	?	Minjur Sandstone 1,000' (may be Lower Jurassic)	Megalodon Limestone
Upper	Norian	Keuper Marls 600'	Keuper / Gips Keuper 500'	Keuper 600'	Haupt Dolomite 900'	Dachstein Dolomite 1,800'		Quartzite series / Monotis beds / Juvavites beds 2,000'
Upper	Carnian				Raibl Formation 600'	Raibl Formation 150'	?	Dolomite Ls. / Tropites Sh. / Gray beds / Halobia Ls. 1,400'
Middle	Ladinian	Keuper Ss 150'	Letten-kohle / Haupt-muschelkalk 250'	Lettenkohle 75' / Muschelkalk supérieur 250'	Wetterstein Dolomite 900'	Cassian Formation 600' / Wengen Formation 600' / Buchenstein Formation 100'	Jilh Formation 1,000'	Daonella Ls. / Daonella Shales 300'
Middle	Anisian		Muschelkalk / Anhydrit gruppe 100' / Wellenkalk 150'	Muschelkalk moyen 350' / Muschelkalk inférieur 200'	Dolomites 150'	Mendel Dolomite 200' / Lower Muschelkalk 75'		Niti Limestone 100'
Lower	Scythian	Upper Mottled Sandstone 300' / Pebble beds 450' / Lower Mottled Sandstone 300'	Bunter Sandstone 1,200'	Grès bigarré 300' / Grès vosgien 1,200'	Werfen Formation 200'	Campil Formation 300' / Seis Formation 300'	?	Hedenstroemia beds 40' / Meekoceras beds 3' / Otoceras beds 3'

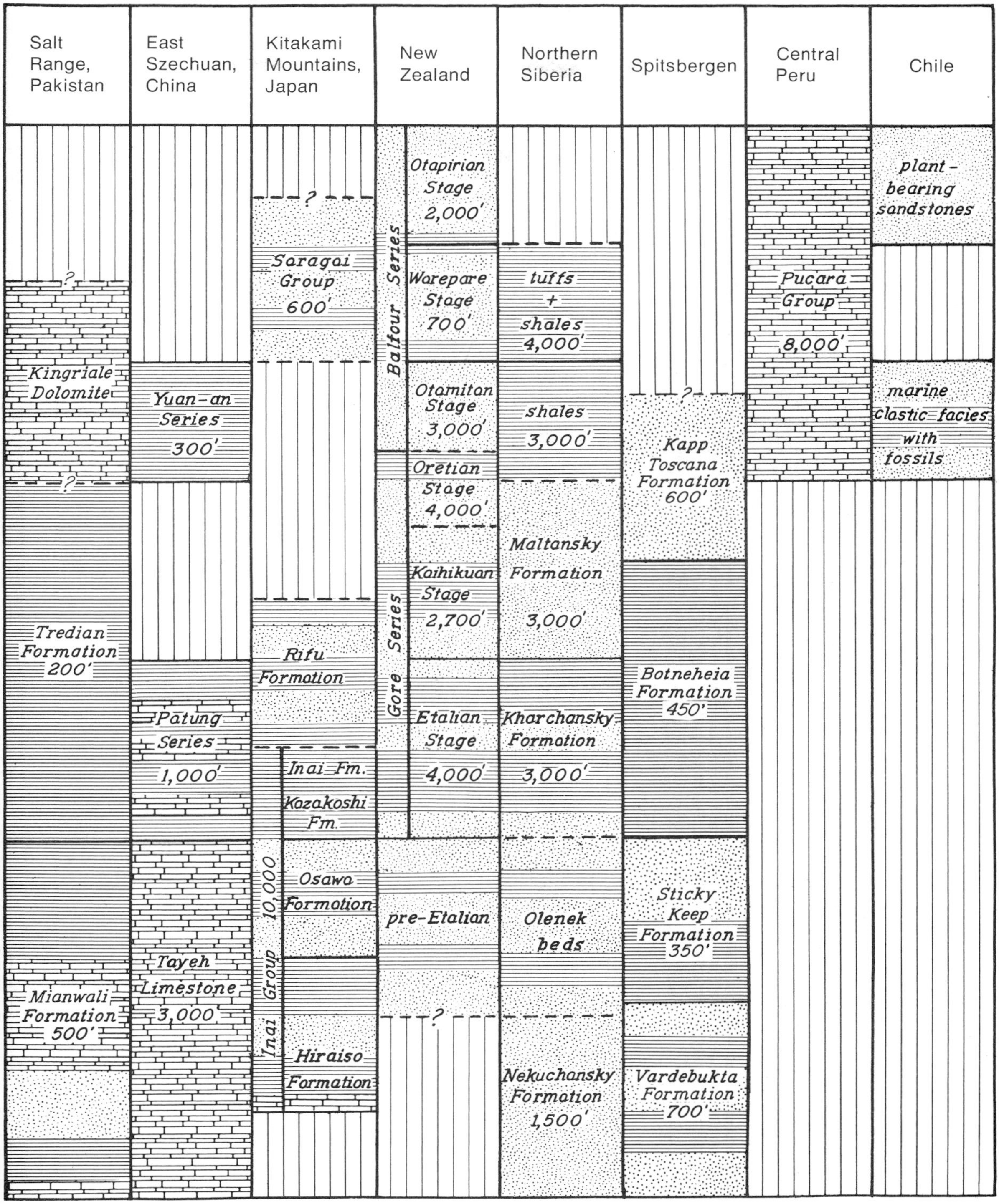

Salt Range, Pakistan	East Szechuan, China	Kitakami Mountains, Japan	New Zealand	Northern Siberia	Spitsbergen	Central Peru	Chile
		?	Otapirian Stage 2,000'				plant-bearing sandstones
		Saragai Group 600'	Warepare Stage 700'	tuffs + shales 4,000'		Pucara Group 8,000'	
Kingriale Dolomite ?	Yuan-an Series 300'		Otamitan Stage 3,000'	shales 3,000'	? Kapp Toscana Formation 600'		marine clastic facies with fossils
			Oretian Stage 4,000'				
				Maltansky Formation 3,000'			
Tredian Formation 200'		Rifu Formation	Kaihikuan Stage 2,700'				
	Patung Series 1,000'			Botneheia Formation 450'			
		Inai Fm. Kozakoshi Fm.	Etalian Stage 4,000'	Kharchansky Formation 3,000'			
		Osawa Formation	pre-Etalian	Olenek beds	Sticky Keep Formation 350'		
Mianwali Formation 500'	Tayeh Limestone 3,000'	Hiraiso Formation	?	Nekuchansky Formation 1,500'	Vardebukta Formation 700'		

New Zealand groupings: Balfour Series, Gore Series

Japan: Inai Group 10,000'

661

The Jurassic in North America

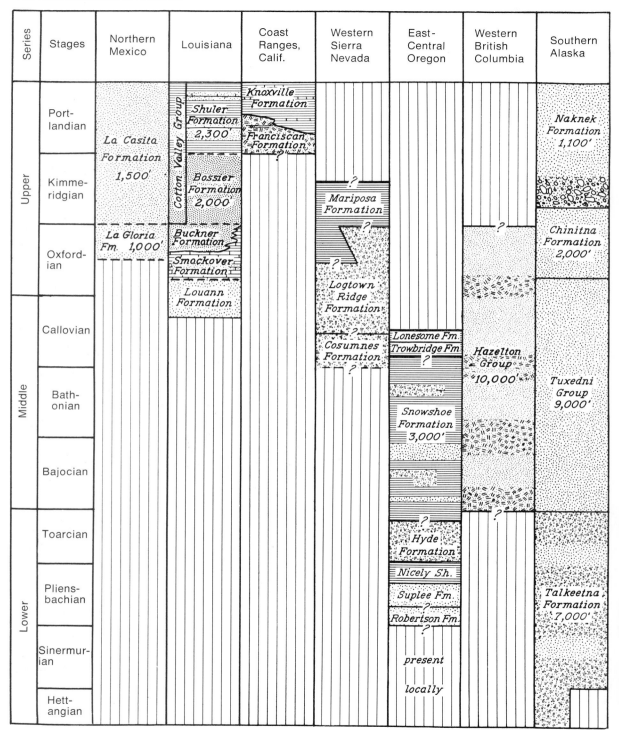

Series	Stages	Northern Mexico	Louisiana	Coast Ranges, Calif.	Western Sierra Nevada	East-Central Oregon	Western British Columbia	Southern Alaska
Upper	Port-landian	La Casita Formation 1,500'	Cotton Valley Group — Shuler Formation 2,300'	Knoxville Formation / Franciscan Formation				Naknek Formation 1,100'
Upper	Kimme-ridgian	La Casita Formation 1,500'	Bossier Formation 2,000		Mariposa Formation ?		?	
Upper	Oxford-ian	La Gloria Fm. 1,000'	Buckner Formation / Smackover Formation / Louann Formation		? Logtown Ridge Formation ?		? Hazelton Group 10,000'	Chinitna Formation 2,000'
Middle	Callovian				? Cosumnes Formation ?	Lonesome Fm. Trowbridge Fm. ?		Tuxedni Group 9,000'
Middle	Bath-onian					Snowshoe Formation 3,000'		
Middle	Bajocian							
Lower	Toarcian					? Hyde Formation	?	Talkeetna Formation 7,000'
Lower	Pliens-bachian					Nicely Sh. / Suplee Fm. ? / Robertson Fm. ?		
Lower	Sinermur-ian					present		
Lower	Hett-angian					locally		

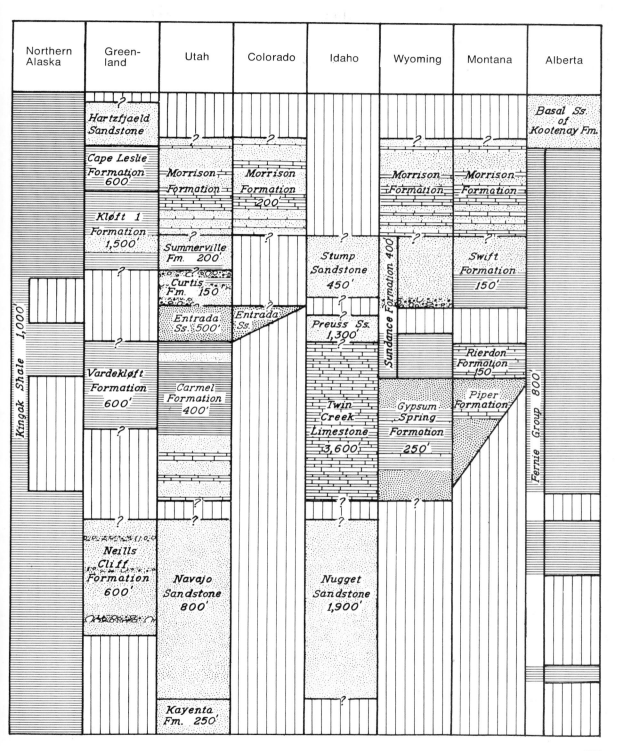

Northern Alaska	Greenland	Utah	Colorado	Idaho	Wyoming	Montana	Alberta
Kingak Shale 1,000'	Hartzfjaeld Sandstone						Basal Ss. of Kootenay Fm.
	Cape Leslie Formation 600	Morrison Formation	Morrison Formation 200		Morrison Formation	Morrison Formation	
	Kløft 1 Formation 1,500'	Summerville Fm. 200'		Stump Sandstone 450'	Sundance Formation 400	Swift Formation 150'	
		Curtis Fm. 150'					
		Entrada Ss. 500'	Entrada Ss.	Preuss Ss. 1,300'		Rierdon Formation 150	
	Vardekløft Formation 600'	Carmel Formation 400'		Twin Creek Limestone 3,600'	Gypsum Spring Formation 250'	Piper Formation	Fernie Group 800'
	Neills Cliff Formation 600'	Navajo Sandstone 800'		Nugget Sandstone 1,900'			
		Kayenta Fm. 250'					

663

The Jurassic outside North America

Series	Stages	England	South Germany	Central France	Kenya, Africa	Saudi Arabia	Moscow region, Russia	Himalayas, India
Upper	Port-landian	Purbeck beds / Portland beds		Portlandian			Volgian 20'	Lochambel beds / Chidamu beds
Upper	Kimmeridgian	Kimmeridge Clay	Hangende-Ls. 400 / Marl 100' / Ulmensis Ls. / limestone + dolomite 400'	Kimmeridgian	Chamgamwe Shale	Jubaila Limestone 300' ?	Lower-middle Kimmeridgian	Spiti Shales 500'
Upper	Oxford-ian	Corallian beds / Oxford Clay	Transversarius Marl	Oxfordian	Coroa-Mombasa Ls.+Sh. / Rabai Shale	Hanifa Formation 300' ?	Oxfordian 45'	Gerardi beds
Middle	Callovian	Kellaways beds / Cornbrash beds	Ornaten Sh. / Macro-cephalen Oolite 5'	Callovian	Miritini Shale / Kibiongoni beds	Tuwaiq Mountain Limestone 600'	Callovian	Sulcacutus beds 100'
Middle	Bath-onian	Great Oolite	Varians Marl 30' / Parkinsonia Oolite	Bathonian	Kambe Limestone	Dhruma Formation 1,200'	100'	
Middle	Bajocian	Inferior Oolite	Subfurkaten Sh. / Coronaten Sh. / Sowerbyi Ls. / Murchisonae Ss. / Opalinus Sh.	Bajocian	Posidonia Shale			
Lower	Toarcian	Upper Lias	Jurensis Marl 15' / Posidonien beds 24'	Toarcian 300'		Marrat Formation 300' ?		Upper Kioto Limestone (Tagling Stage) 1,500'
Lower	Pliens-bachian	Middle Lias	Amaltheen Clay 36' / Numismatis Marl 30'	Charmouthian 300'		Minjur Sandstone (age uncertain)		
Lower	Sinemur-ian	Lower Lias 400'	Turneri Clay 60' / Arieten Limestone 12'	Sinemurian 100'				
Lower	Hett-angian		Angulaten Ss. 18 / Psilonoten Sh. 6'	Hettangian 30'		?		

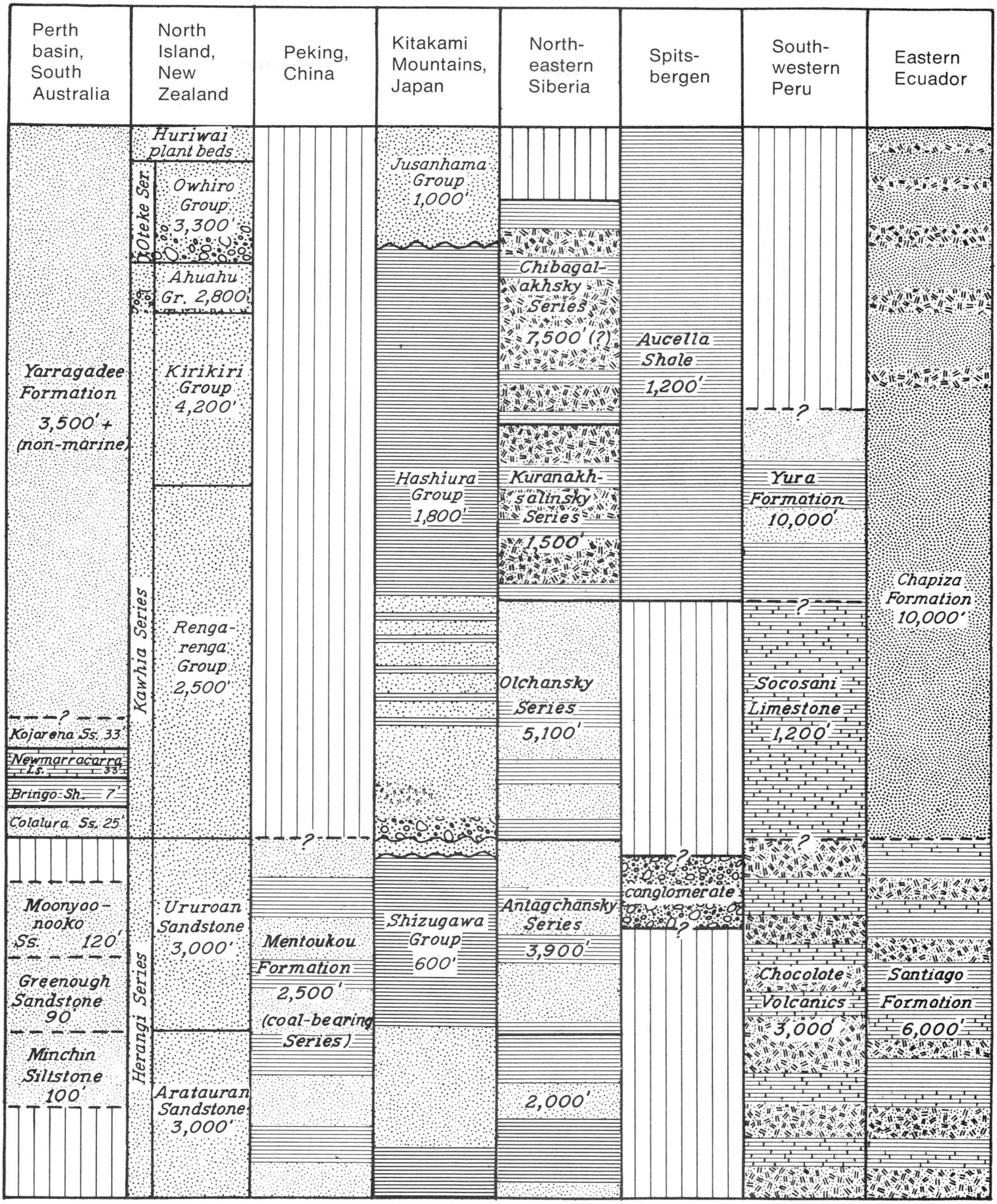

Perth basin, South Australia	North Island, New Zealand	Peking, China	Kitakami Mountains, Japan	North-eastern Siberia	Spits-bergen	South-western Peru	Eastern Ecuador

Huriwai plant beds

Oteke Ser.

Owhiro Group 3,300'

Jusanhama Group 1,000'

Ahuahu Gr. 2,800'

Chibagal-'akhsky "Series" 7,500' (?)

Aucella Shale 1,200'

Kirikiri Group 4,200'

Yarragadee Formation 3,500' + (non-marine)

Kawhia Series

Renga-renga Group 2,500'

Hashiura Group 1,800'

Kuranakh-'salinsky Series 1,500'

Yura Formation 10,000'

Chapiza Formation 10,000'

?

Socosani Limestone 1,200'

Olchansky Series 5,100'

?

Kojarena Ss. 33'
Newmarracarra Ls. 33'
Bringo Sh. 7'
Colalura Ss. 25'

?

?

Moonyoo-nooko Ss. 120'

Ururoan Sandstone 3,000'

Herangi Series

Mentoukou Formation 2,500' (coal-bearing Series)

Shizugawa Group 600'

Antagchansky Series 3,900'

?

conglomerate

Chocolote Volcanics 3,000'

Santiago Formation 6,000'

Greenough Sandstone 90'

Minchin Siltstone 100'

Aratauran Sandstone 3,000'

2,000'

The Cretaceous in North America

Series	Stages		New Jersey	Alabama	Texas	Sierra de Parras, Mexico	Western Colorado	Kansas	South-eastern Idaho
Paleo-cene	Danian					?	Animas Formation 700'		
Upper	Maas-trichtian		Monmouth Group 200'	Prairie Bluff Ch. / Ripley Fm.	Navarro Group 700'	Difunta Formation 12,000'	McDermott Fm. 300 / Kirkland Sh. / Fruitland Fm. / Picture Cliffs Ss / Lewis Shale / Mesaverde Gr.	Pierre Shale	
Upper	Campanian		Matawan Group 120'	Selma Chalk 700	Taylor Marl 1200				?
Upper	Santonian		Magothy Formation		Gulf Series / Austin Chalk 400'	Parras Shale 5,000'	Mancos Shale 3,000'	Niobrara Formation 200'	
Upper	Coniacian			Eutaw Formation		Caracol Fm. 3,000'			
Upper	Turonian			?	Eagle Ford Shale 300'	Indidura Formation 1,500'		Carlile Shale / Greenhorn Ls. 35 / Graneros Sh. 200	
Upper	Cenomanian		Ravitan Fm.	Tuscaloosa Formation 1,000'	Woodbine Fm.	Cuesta del Cura Ls. 200'	Dakota Sandstone 200'	Dakota Ss.	Wayan Formation 11,800'
Lower	Albian		Potomic Group		Comanche Series / Washita Gr. 350' / Fredericks-burg Gr. 400 / Trinity Gr. 1,500'	Aurora Ls. 300'		Kiowa Shale / Cheyenne Ss.	Bear River Fm.
Lower	Aptian					La Peña Formation 1,000'			Red Sh. / Draney Ls. / Bechler Congl. / Peterson Ls.
Lower	Neocomian	Barrem-ian				Cupido Ls. 1,000'			Gannett Group / Ephraim Con-glomerate 1,000'
Lower	Neocomian	Hauteriv-ian				?			
Lower	Neocomian	Valang-inian				Taraises Formation 1,000'			
Lower	Neocomian	Berrias-ian							

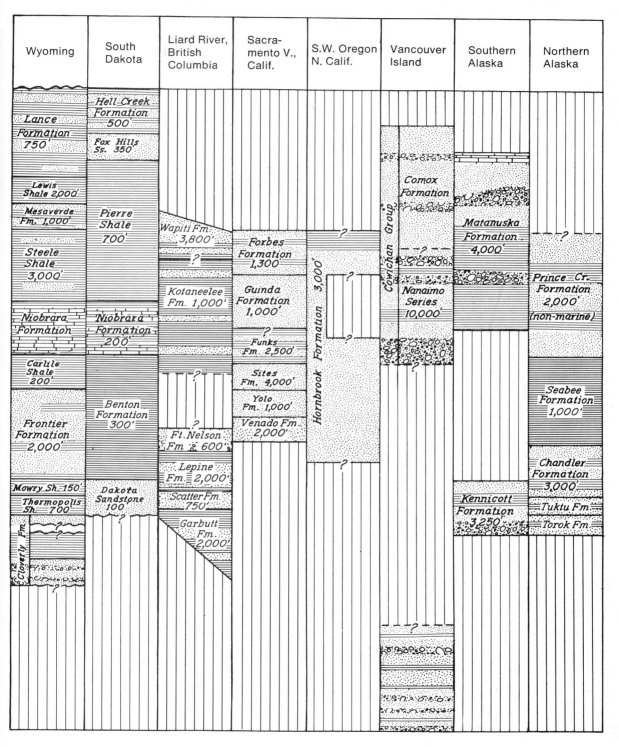

Wyoming	South Dakota	Liard River, British Columbia	Sacramento V., Calif.	S.W. Oregon N. Calif.	Vancouver Island	Southern Alaska	Northern Alaska
Lance Formation 750'	Hell Creek Formation 500'						
	Fox Hills Ss. 350'				Comox Formation		
Lewis Shale 2,000'						Matanuska Formation 4,000'	
Mesaverde Fm. 1,000'	Pierre Shale 700'	Wapiti Fm. 3,800'	Forbes Formation 1,300		Cowichan Group		?
Steele Shale 3,000'		?			Nanaimo Series 10,000'		Prince Cr. Formation 2,000' (non-marine)
		Kotaneelee Fm. 1,000'	Guinda Formation 1,000'	Hornbrook Formation 3,000			
Niobrara Formation	Niobrara Formation 200'		?				
			Funks Fm. 2,500'				
Carlile Shale 200'		?	Sites Fm. 4,000'				Seabee Formation 1,000'
	Benton Formation 300'	?	Yolo Fm. 1,000'				
Frontier Formation 2,000'		Ft. Nelson Fm. 600'	Venado Fm. 2,000'				Chandler Formation 3,000'
		Lepine Fm. 2,000'		?		Kennicott Formation 3,250'	
Mowry Sh. 150'	Dakota Sandstone 100	Scatter Fm. 750'					Tuktu Fm.
Thermopolis Sh. 700'							Torok Fm.
Cloverly Fm. ?	?	Garbutt Fm. 2,000'			?		

The Cretaceous outside North America

Series	Stages		England	Northern Europe	Swiss Alps	Saharan Atlas Mts., Algeria	Saudi Arabia	Moscow region, Russia	Himalayas, India
Paleocene	Danian			Faxe Limestone				?	Flysch Series
Upper	Maas-trichtian			Mucronaten Chalk			Aruma Formation 450'	? / white limestone with some phosphoritic sandstones 100'	Chikkim Series 250' ?
Upper	Campanian		Upper Chalk 1,000'	Quadraten Chalk	Wang beds 300' / Amden Shale 300'				
Upper	Santonian			Granulaten Chalk		? / 2,000'		marl in N., siliceous shale in S. 300'	
Upper	Coniacian			Emscher Marl	Seewen Shale 150'				
Upper	Turonian		Middle Chalk 150'	Pläner Formation	Seewen Limestone 300'	400'		limestone 75'	
Upper	Cenomanian		Lower Chalk 200'			300'	Wasi'a Formation	glauconite ss. with sandy ls. and phosphorites 100'	
Lower	Albian		Upper Greensand / Gault Clay 150'			300' / non-marine 1,000'	Biyadh Sandstone 800'	dark shale 12' / phosphoritic ss. 25	?
Lower	Aptian		Lower Greensand 400'	Hils Clay 1,000'		400'	Buwaib Formation 100'	dark shale and light sandstone 60'	Giumal Series 300'
Lower	Neocomian — Barremian		Wealden beds 1,000'		limestone 600'	non-marine 2,000'	? / Yamama Formation 150'	iron sandstone 40'	
Lower	Neocomian — Hauterivian				limestone + marls 600'	400'			
Lower	Neocomian — Valanginian			Wealden facies	limestone 600'	600'	? / Sulaiy Formation 600'	sandstone 10'	
Lower	Neocomian — Berriasian				Oehrli Limestone 600'			plant-bearing sandstone 40'	

Stratigraphic correlation chart

East-central Australia	New Zealand	Hokkaido, Japan	Mongolia	N. Siberia	Spitsbergen	Central Peru	Eastern Peru
Mt. Howie Ss. ? ? ?	Mata Series: Wangalcan Stage, Teurian Stage, Haumurian Stage, Peripauan Stage 1,000'; Raukumara Series: Teratan Stage 2,000', Mangoatanean Stage 700', Arowhanan Stage 1,000'; Ngaterian Stage 2,000'; Clarence Series: Moytuan Stage, Urutawan Stage, Korangan Stage; Taitai Series: Mekoiwian Stage	Hakabuchi Group 2,500'; Upper Yeso Group 3,000'; Middle Yeso Group 6,000'; Lower Yeso Groups 4,000'; Sorachi Group 3,000'	Gashato Formation 300'; Irendabasu Formation 150'; Djadochta Formation 300'; ? Ondaisair Formation 500'; Oshih Formation 500'	sandstone + coal; Baculites beds 300'; shale + sandstone; shale; Khetsky Fm. 300'; Ledyanaya Series 1,000'; shale + sandstone 1,000'; Labiatus beds; Dolgansky Series; Yakovlevsky Series; Malakhetsky Formation; sandstone; sandstone 1,500'	? ; marine facies; continental facies 400'; marine facies	Celendin Formation 300'; Jumasha Formation 1,800'; Pariatambo Fm. 350; Chulec Fm. 300'; Goyllarisquisga Formation 1,500'; ?	Cachiyacu Formation 100'; Vivian Formation 200'; Chonta Formation 400'; Huaya; Agua Caliente; Paco; Esperanza; Aguanuya; Cushabatay; Oriente Formation 6,000'
Winton Fm. 1,000'; Tambo Fm. 650'; Roma Fm. 500'; Blythesdale Group							

The Gondwana (Karroo, Santa Catharina) System

Systems in Europe	Antarctica	Falkland Islands	Brazil — Paraná	Brazil — Santa Catharina	Brazil — Rio Grande do Sul	South — Cape region
Lower Cretaceous			*São Bento Series:* Paraná Basalts 1,200'; Botucatu Ss. 450'	Paraná Basalts 2,000'; Botucatu Ss. 200'	Paraná Basalts 100'; Botucatu Ss. 100'	
Jurassic	Rhyolites and tuff; ? Shales ?; Conglomerate					*Stormberg Series:* Drakensberg Basalts 4,500; Cave Sandstone 1,000; Redbeds 1,600
Triassic	Beacon Sandstone 5,000'+	West Lafonian beds; Bay of Harbour beds 8,000'			Santa Maria 250'	Molteno beds 2,000'; *Beaufort Series:* Beaufort beds 12,000'
Permian		Choiseul Sound bed; Lafonian Sandstone 300'; Black Rock Slates; Lafonian Tillite; Bluff Cove beds	*Passa Dois Series:* Rio do Rasto 300'; Estrada Nova 450'; Irati 200'. *Tubarão Series:* Guatá 550'; Itararé 1,000'	Rio do Rasto 100'; Estrada Nova 300'; Irati 180'; Guatá 600'; Itararé 150'	Rio do Rasto 450'; Estrada Nova 350'; Irati 150'; Guatá 450'; Itararé 60'	*Ecca Series:* Ecca beds 10,000'; Upper shales 650'. *Dwyka Series:* Dwyka Tillite 2,000'; Lower shales 750'
Upper Carboniferous	?	Port Stanley beds				

Santa Catharina System spans the Brazil columns. *Karroo System* spans the Cape region columns.

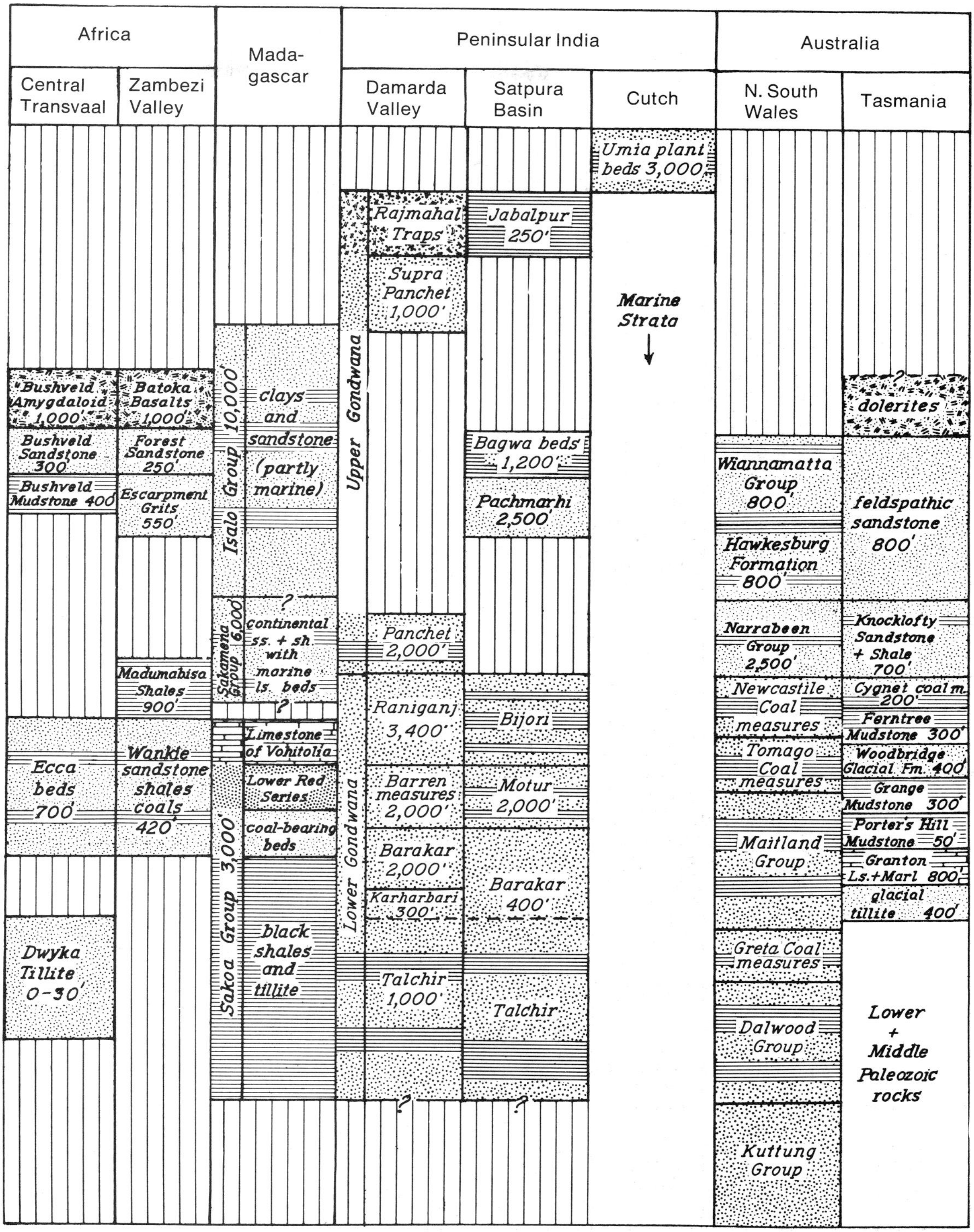

Africa		Mada-gascar	Peninsular India			Australia	
Central Transvaal	Zambezi Valley		Damarda Valley	Satpura Basin	Cutch	N. South Wales	Tasmania

Chart content (top to bottom):

Cutch: *Umia plant beds 3,000*

Damarda Valley / Satpura Basin: *Rajmahal Traps* — *Jabalpur 250'*

Supra Panchet 1,000'

Marine Strata ↓ (Cutch)

Africa — Central Transvaal: *Bushveld Amygdaloid 1,000'*; *Bushveld Sandstone 300*; *Bushveld Mudstone 400*; *Ecca beds 700*; *Dwyka Tillite 0–30'*

Zambezi Valley: *Batoka Basalts 1,000'*; *Forest Sandstone 250'*; *Escarpment Grits 550'*; *Madumabisa Shales 900'*; *Wankie sandstone shales coals 420'*

Madagascar: *Isalo Group 10,000'* — *clays and sandstone (partly marine)*; *Sakamena Group 6,000'* — ? *continental ss. + sh. with marine ls. beds* ?; *Limestone of Vohitolia*; *Lower Red Series*; *coal-bearing beds*; *Sakoa Group 3,000'* — *black shales and tillite*

Damarda Valley (Lower Gondwana): *Panchet 2,000'*; *Raniganj 3,400'*; *Barren measures 2,000'*; *Barakar 2,000'*; *Karharbari 300'*; *Talchir 1,000'*

Satpura Basin: *Bagwa beds 1,200'*; *Pachmarhi 2,500'*; *Bijori*; *Motur 2,000'*; *Barakar 400'*; *Talchir*

N. South Wales: *Wiannamatta Group 800*; *Hawkesburg Formation 800*; *Narrabeen Group 2,500'*; *Newcastle Coal measures*; *Tomago Coal measures*; *Maitland Group*; *Greta Coal measures*; *Dalwood Group*; *Kuttung Group*

Tasmania: *feldspathic sandstone 800'*; *dolerites* ?; *Knocklofty Sandstone + Shale 700'*; *Cygnet coal m. 200'*; *Ferntree Mudstone 300'*; *Woodbridge Glacial Fm. 400'*; *Grange Mudstone 300'*; *Porter's Hill Mudstone 50'*; *Granton Ls.+Marl 800'*; *glacial tillite 400'*; *Lower + Middle Paleozoic rocks*

(Upper Gondwana / Lower Gondwana labels span the India columns)

Series	Maryland	Western Florida	Mississippi	Texas	Central Calif.	N. of San Francisco, Calif.	Western Oregon
Pleistocene	sand + clay	sand	clay + sand	clay + sand	Tulare Fm.	Millerton Fm.	
Pliocene		Citronelle Formation 300	Citronelle Formation	Willis Sand 125 / Goliad Sand	San Joaquin Formation / Etchegoin Formation 3,600 / Jacalitos Formation 2,600	Sonoma Volc. / Petaluma Fm. / Tolay Volc. / Neroly Fm.	
Miocene	Yorktown Fm. 100' / St. Marys Fm. 140 / Choptank Fm. / Calvert Fm.	Choctaw-hatchee Formation 40' / Alum Bluff Group / Tampa Ls. 100	? / Pascagoula Clay 450 / Hattiesburg Clay 450' / Catahoula Ss.	Largato Clay / Oakville Sandstone 400 / Catahoula Ss.	Reef Ridge Shale 600 / McLure Shale 800 / Temblor Formation / ?	Monterey Shale 1,000'+	Astoria Formation 1,400 / Nye Shale 2,100
Oligocene		Suwannee Ls. / Byram Marl / Marianna Ls.	Chickasawhay Marl / Vicksburg Group 100' / Forest Hill Ss.	? / Frio Clay / ?	?	? / shaly sandstone / ?	Yaquina Formation 4,000 / Toledo Formation 2,800
Eocene	Nanjemoy Fm 125' / Aquia Fm 100'	Ocala Limestone 250'	Yazoo Clay / Moodys Marl 200' / Yegua Fm. / Lisbon Formation 400 / Tallahatta Fm. / Hatchetigbee Fm. / Bashi Fm. / Holly Springs Ss. / Salt Mt. Ls. / Nanafalia Fm. / Ackerman Fm. / Fern Springs Fm.	Jackson Formation / Yegua Fm. / Cook Mt. Fm. / Mt. Selman Fm. / Carrizo Sand / Wilcox Group 800'	Kreyenhagen Shale 900' / ? / Avenal Ss. / ?	Markley Formation 3,300' / Domengine Fm. / Capay Shale	Eocene sandstone + shale
Paleocene	Brightseat Fm. 8' ?		Naheola Fm. / Porters Creek Clay 250' / Clayton Formation 225	Wills Point Formation 280' / Kincaid Formation 100'		Martinez Formation 3,500'	Metchosin Volcanics 3,500'

Western Washington	Alaskan Peninsula	Montana	South-western Wyoming	Utah	S. Dakota and Nebraska	Kansas	North-western Texas
Quaduilt Fm. 2,300	plant-bearing beds	Flaxville Gravel 70			sand + clay	Meade Fm.	Tule Fm.
Montesano Formation 5,000	? Unga Conglomerate 600	Deep R. Formation			Ogallala Group 200	Ogallala Formation 200	Blanco
Astoria Formation 1,400							Hemphill 550
	?	Fort Logan Formation			Laverne Fm.	Laverne Fm.	Clarendon 400
Twin River Formation 2,000					Hemingford Gp. — Sheep Cr. Formation 100'		
Blakeley Fm. 8,000		Oreodon beds			Marsland Fm.		
Lincoln Formation					Arikaree Group 500		
	?	Pipestone Cr. Fm.			White R. Series — Brule Formation 600		
Cowlitz Formation 500'	volcanics + sedimentary rocks 5,000			Duchesne River Formation	Chadron Fm.		
				Uinta Formation 600'			
			Bridger Formation 2,500	Green River Formation 3,200			
?			Green R. Fm.				
	?		Knight Fm.	Wasatch Formation			
			Wasatch Group — Fowkes Fm.	Flagstaff Limestone 700'			
Metchosin Volcanics 4,000	?	Fort Union Series 4,000	Almy Fm. 2,100	North Horn Formation	Fort Union Series 4,000		
			Evanston Formation 1,500				

The Cenozoic Outside North America

Series	South-eastern England	Belgium	Paris Basin	Northern Switzerland	North Apennines, Italy	Sicily	Saudi Arabia
Pleistocene	Red Craig 30'	Amstelien				Agrigento 150'	
Pliocene	Coralline Crags 30'	Scaldisien 100'			sands + clays	Narbona 600'	?
Pliocene	Lenham beds 15	Deistien 60'		upper fresh-water molasse		1,800'	Hofuf Formation 300
						Trubi 300'	
						Gypsum series 500'	
						San Cataldo 600'	Dam Formation 300
Miocene	Boxstones	Anversien 60'	Touraine marls 50'	upper marine molasse	Marnoso-arenacea Formation	Barbara 3,000'	?
		Bolderien 30'				Valledolmo 3,000'	Hadrukh Formation 200
			Orléanais sands 50'	lower fresh-water molasse		Tavernola 1,800'	?
Oligocene			Beauce ls. Aquitanian Chattien		Macigno Formation	Alia 2,000'	
	Chattien 75'	Rupelien 100'	Stampian 100'	lower marine molasse			
	Tongrien 100'	Sannosian 50'		?			
	Asschesh 100'	Ludian marls 60'					
	Wemmel Sand 50'						
Eocene	Barton beds 200	Lédien 30'	Ledian Cresnes sands / Beauchamp sands 75'		Scisti Policromi Formation	Barracu 450'	Dammam Formation 75'
	Brackelsham beds 300	Bruxellien 60'	Lutetian Provins Ls. / Coarse Ls. 100'			?	Rus Formation 150
	Bagshot beds 100	Ypres Clay 300'	Ypresian Cuise sands 100'				
	London Clay 300						
Paleocene	Reading beds 100'	Landenian non-marine	Sparnacian Soissons Lignite 100' / Plastic Clay 100				Umm Er Radhuma Formation 600
	Thanet sands 45'	Landenian marine	Thanetian Bracheux sands 100'				
		Mons Limestone 150'	Montian Meudon Limestone 15'				

674

Pakistan		Northern Sumatra	Pacific Coast, S.W. Japan	Northern Siberia	Spitsbergen	Lake Maracaibo, Venezuela	Southern Peru
Baluchistan	Salt Range						

Baluchistan	Salt Range	Northern Sumatra	Pacific Coast, S.W. Japan	Northern Siberia	Spitsbergen	Lake Maracaibo, Venezuela	Southern Peru
Manchhar Series 10,000'	Siwalik Series 17,000'	Djulor Rajeuh Fm. 4,000'	Terrace dep.	Cherlaksky Formation 100'			Glacial dep.
		Seurula Formation 3,000'	Sogo Formation				Río Azangaro 300'
			Kakegawa Group				Sillapaca Volcanics 1,500'
		Keutapang Fm. 3,000'	Sagara Group	Pavlobarsky Formation 300'			
		Borderclay Fm. 1,500'					
Gaj Series 1,500'	Murree Series 8,000'	Pennulin Fm. 2,000'	Saigo Group	Ishimsky Formation 300'		La Villa Formation	Tacaza Volcanics 12,000'
		Black Claystone Fm. 2,000'	Kurami Group			Los Ranchos Fm.	
		Micaceous sandstone Fm. 1,200'	Megami Formation	Nekrasovsky Series		Lagunillas Formation	
Nari Series 6,000'		basal quartz breccia formation		Chegansky Series		La Rosa Formation	
Kirthar Series			Setogawa Group	Liulinborsky Series 600'	Upper Plant-bearing sandstone series 1,500'		Puno Group 20,000'
					Flaggy sandstone series 600'	Concepción Formation	
Laki Series 3,000'	Laki Series 600'				Upper black shale series 800'		
				Kliuchevsky Series 600'	Green sandstone series 600'		
Ranikot Series 2,000'	Ranikot Series 700'				Lower black shale series 300'	Guasare Formation	
					Lower light ss. series 400'		

675

INDEX TO CORRELATION CHARTS

INDEX

Page numbers referring to important concepts and technical terms are in boldface type